Alzheimer's Disease

Cellular and Molecular Aspects of Amyloid β

Subcellular Biochemistry

Volume 38

SUBCELLULAR BIOCHEMISTRY

SERIES EDITOR

J. ROBIN HARRIS, Institute of Zoology, University of Mainz, Mainz, Germany

ASSISTANT EDITOR

B. B. BISWAS, University of Calcutta, Calcutta, India

Recent Volumes in This Series

Volume 28 **Cholesterol: Its Functions and Metabolism in Biology and Medicine**
Edited by Robert Bittman

Volume 29 **Plant–Microbe Interactions**
Edited by B. B. Biswas and H. K. Das

Volume 30 **Fat-Soluble Vitamins**
Edited by Peter J. Quinn and Valerian E. Kagan

Volume 31 **Intermediate Filaments**
Edited by Harald Herrmann and J. Robin Harris

Volume 32 **α-Gal and Anti-Gal: α1,3-Galactosyltransferase, α-Gal Epitopes, and the Natural Anti-Gal Antibody**
Edited by Uri Galili and José Luis Avila

Volume 33 **Bacterial Invasion into Eukaryotic Cells**
Edited by Tobias A. Oelschlaeger and Jörg Hacker

Volume 34 **Fusion of Biological Membranes and Related Problems**
Edited by Herwig Hilderson and Stephen Fuller

Volume 35 **Enzyme-Catalyzed Electron and Radical Transfer**
Edited by Andreas Holzenburg and Nigel S. Scrutton

Volume 36 **Phospholipid Metabolism in Apoptosis**
Edited by Peter J. Quinn and Valerian E. Kagan

Volume 37 **Membrane Dynamics and Domains**
Edited by Peter J. Quinn

Volume 38 **Alzheimer's Disease: Cellular and Molecular Aspects of Amyloid β**
Edited by J. Robin Harris and Falk Fahrenholz

A Continuation Order Plan is available for this series. A continuation order will bring delivery of each new volume immediately upon publication. Volumes are billed only upon actual shipment. For further information please contact the publisher.

Alzheimer's Disease

Cellular and Molecular Aspects of Amyloid β

Subcellular Biochemistry Volume 38

Edited by

J. Robin Harris
Institute of Zoology
University of Mainz
Mainz, Germany

and

Falk Fahrenholz
Institute of Biochemistry
University of Mainz
Mainz, Germany

Springer

This series is a continuation of the journal *Sub-Cellular Biochemistry*,
Volumes 1 to 4 of which were published quarterly from 1972 to 1975

ISSN 0306-0225
ISBN 0-387-23225-7
eISBN 0-387-23226-5

Printed in the United States of America. (BS/DH)

9 8 7 6 5 4 3 2 1

springeronline.com

Contributors

Christian Behl, Institute for Physiological Chemistry & Pathobiochemistry, Faculty of Medicine, Johannes Gutenberg-University Mainz, Duesbergweg 6, D-55099 Mainz, Germany

Peter Borghgraef, Experimental Genetics Group, Department of Human Genetics, K.U.Leuven, B-3000 Leuven, Belgium

Ayelén Bulloj, Instituto de Química y Fisicoquímica Biológicas (IQUIFIB), CONICET, Cátedra de Química Biológica Patológica, Departamento de Química Biológica, Facultad de Farmacia y Bioquímica, Universidad de Buenos Aires, Junin 956, C113AAD, Buenos Aires, Argentina

Ashley I. Bush, Harvard Medical School and Massachusetts General Hospital, Boston MA 02114, USA

Miguel Calero, Centro Nacional de Microbiología, Instituto de Salud Carlos III, 28220 Majadahonda, Madrid, Spain

Mireia Coma, Unitat de Senyalització Cel lular, Departament de Ciències Experimentals i de la Salut, Universitat Pompeu Fabra (UPF), 08003-Barcelona, Spain

Donald B. Carter, CNS Therapeutics, Pfizer Corporation, Kalamazoo, MI, USA

Eduardo M. Castaño, Instituto de Química y Fisicoquímica Biológicas (IQUIFIB), CONICET, Cátedra de Química Biológica Patológica, Departamento de Química Biológica, Facultad de Farmacia y Bioquímica, Universidad de Buenos Aires, Junin 956, C113AAD, Buenos Aires, Argentina

Marcela Colombres, FONDA-Biomedical Center, P. Catholic University of Chile, Santiago, Chile

Math P. Cuajungco, Harvard Medical School and Massachusetts Eye and Ear Infirmary, Boston MA 02114, USA

Ilse Dewachter, Experimental Genetics Group, Department of Human Genetics, K.U.Leuven, B-3000 Leuven, Belgium

Lisbell Estrada, Protein Misfolding Disorders Laboratory, Department of Neurology, University of Texas Medical Branch, Galveston, TX, USA

Gunter P. Eckert, Department of Pharmacology, ZAFES, Biocenter University of Frankfurt, Germany

Christopher Exley, Birchall Centre for Inorganic Chemistry and Materials Science, Keele University, Staffordshire, ST5 5BG, UK

Falk Fahrenholz, Institute of Biochemistry, Johannes Gutenberg-University, Becherweg 30, Mainz D-55099, Germany

Blas Frangione, Department of Pathology, New York University School of Medicine, New York, NY 10016, USA

Christopher, J. Frederickson, NeuroBioTex, Inc. and University of Texas Medical Branch, Galveston TX 77550 USA

Jorge Ghiso, Department of Pathology, Room-TH432, New York University School of Medicine, 550 First Avenue, New York, NY 10016, USA

Charles C. Glabe, University of California, Irvine, Irvine, CA 92697, USA

J. Robin Harris, Institute of Zoology, University of Mainz, D-55099 Mainz, Germany

Tobias Hartmann, Center for Molecular Biology Heidelberg (ZMBH), University of Heidelberg, Im Neuenheimer Feld 282, D-69120 Heidelberg, Germany

Nibaldo C. Inestrosa, FONDA-Biomedical Center, P. Catholic University of Chile, Santiago 114-D, Chile

Vernon M. Ingram, Department of Biology, M.I.T, 31 Ames Street, Cambridge, MA 02139,USA

Hideyo Inouye, Department of Biology, Higgins Hall 510, Boston College, 140 Commonwealth Avenue, Chestnut Hill, MA 02467-3811, USA

Daniel A. Kirschner, Department of Biology, Boston College, Chestnut Hill, MA 02467-3811, USA

Elżbieta Kojro, Institute of Biochemistry, Johannes Gutenberg-University, Becherweg 30, Mainz D-55099, Germany

María Celeste Leal, Instituto de Química y Fisicoquímica Biológicas (IQUIFIB), CONICET, Cátedra de Química Biológica Patológica, Departamento de Química Biológica, Facultad de Farmacia y Bioquímica, Universidad de Buenos Aires, Junin 956, C113AAD, Buenos Aires, Argentina

Nathaniel G. N. Milton, Department of Clinical Neurosciences, Royal Free & University College Medical School, University College London, Rowland Hill Street, London NW3 2PF, UK

Laura Morelli, Instituto de Química y Fisicoquímica Biológicas (IQUIFIB), CONICET, Cátedra de Química Biológica Patológica, Departamento de Química Biológica, Facultad de Farmacia y Bioquímica, Universidad de Buenos Aires, Junin 956, C113AAD, Buenos Aires, Argentina

Werner E. Müller, Department of Pharmacology, ZAFES, Biocenter University of Frankfurt, Frankfurt, Germany

Francisco José Muñoz, Unitat de Senyalització Cel lular, Departament de Ciències Experimentals i de la Salut, Universitat Pompeu Fabra (UPF), Calle Dr Aiguader 80, 08003-Barcelona, Spain

Agueda Rostagno, Department of Pathology, New York University School of Medicine, New York, NY 10016, USA

Juan Paulo Sagal, FONDA-Biomedical Center, P. Catholic University of Chile, Santiago, Chile

Montserrat Solé, Unitat de Senyalització Cel lular, Departament de Ciències Experimentals i de la Salut, Universitat Pompeu Fabra (UPF), 08003-Barcelona, Spain

Claudio Soto, Protein Misfolding Disorders Laboratory, Department of Neurology, University of Texas Medical Branch, Galveston, TX, USA

Tom van Dooren, Experimental Genetics Group, Department of Human Genetics, K.U.Leuven, B-3000 Leuven, Belgium

Fred van Leuven, Experimental Genetics Group, Department of Human Genetics, K.U.Leuven, Campus Gasthuisberg, O & N6, B-3000 Leuven, Belgium

Robert Vassar, Department of Cell and Molecular Biology, The Feinberg School of Medicine, Northwestern University, 303 E. Chicago Avenue, Chicago, IL 60611, USA

W.G. Wood, Department of Pharmacology, University of Minnesota School of Medicine, Minneapolis, USA

Katsuhiko Yanagisawa, National Institute for Longevity Sciences, Gengo 36-3, Morioka, 474-8522 Obu, Japan

Preface

To understand Alzheimer's disease (AD) is one of the major thrusts of present-day clinical research, strongly supported by more fundamental cellular, biochemical, immunological and structural studies. It is these latter that receive attention within this book. This compilation of 20 chapters indicates the diversity of work currently in progress and summarizes the current state of knowledge.

Experienced authors who are scientifically active in their fields of study have been selected as contributors to this book, in an attempt to present a reasonably complete survey of the field. Inevitably, some exciting topics for one reason or another have not been included, for which we can only apologize.

Standardization of terminology is often a problem in science, not least in the Alzheimer field; editorial effort has been made to achieve standardization between the Chapters, but some minor yet acceptable personal / author variation is still present, *i.e.* β-amyloid/amyloid-β; Aβ42/Aβ1-42/$A\beta_{1\text{-}42}$!

The book commences with a broad survey of the contribution that the range of available microscopical techniques has made to the study of Alzheimer's amyloid plaques and amyloid fibrillogenesis. This chapter also serves as an Introduction to the book, since several of the topics introduced here are expanded upon in later chapters. Also, it is significant to the presence of this chapter that the initial discovery of brain plaques, by Alois Alzheimer, utilized light microscopy, a technique that continues to be extremely valuable in present-day AD research. Then follow 19 further chapters dealing with interesting areas of research that have a bearing upon

Alzheimer's disease. The authors present their own data within the context of a review of related work from others in their field of study.

Transgenic mouse models for AD are increasingly important and widely used, as is the understanding of the enzymology and biochemistry of amyloid β production from the amyloid precursor protein (APP) by the β- and γ-secretases, and the non-amyloidogenic APP cleavage products due to α-secreatases. Similarly, discussion of the enzymes responsible for the natural degradation of amyloid-β is included. Oxidative stress in AD and the possible protective role of vitamin E have emerged as highly significant topics. Likewise, the role of the metals, aluminium copper and zinc in AD is thoroughly addressed, along with the possible value of metal chelation. Current concepts as to the fundamental importance of amyloid-β oligomers *versus* Aβ fibrillar deposits (diffuse and senile plaques) in the pathogenesis of AD is presented, along with the likely role of tissue and cellular cholesterol in the promotion of Aβ fibrillogenesis, and Congo red binding to Aβ fibrils.

Several naturally occurring human proteins have the ability to bind to Aβ and are found in AD diffuse and senile plaques. Apolipoprotein E4, clusterin (ApoJ) and acetylcholinesterase are considered in depth. The direct effects of Aβ on neuronal membranes, in terms of neuronal membrane fluidity changes and calcium ion transport are dealt with, then the role of amyloid inhibitors and β-sheet breaking drugs is included. The likely important beneficial effects of the cholesterol-lowering drugs, the statins, in lowering brain Aβ and consequently the therapeutic prevention of AD by reducing Aβ deposition in plaques is discussed thoroughly. Finally, an exciting new concept is advanced, namely the possible significance of phosphorylated Aβ in Alzheimer's disease. From *in vitro* experiments Aβ phosphorylation appears to increase the proportion of Aβ oligomeric forms, an observation which correlates with the higher cytotoxicity of this species, compared to non-phosphorylated Aβ.

The editors hope that this book will be of interest and value to both medical and scientific research communities working on AD, and to others with a more general interest in the understanding of this devastating neurodegenerative disorder of the elderly.

Prof. J. Robin Harris
Prof. Falk Fahrenholz

University of Mainz
August, 2004

Contents

Alzheimer's Disease

Cellular and Molecular Aspects of Amyloid β

Subcellular Biochemistry
Volume 38

Chapter 1

The Contribution of Microscopy to the Study of Alzheimer's Disease, Amyloid Plaques and Aβ Fibrillogenesis

J. Robin Harris
Institute of Zoology, University of Mainz, D-55099 Mainz, Germany

Abstract: A broad survey is presented in this chapter, dealing with the impact that microscopy has made to the study of Alzheimer's disease, amyloid plaques and amyloid-β fibrillogenesis. This includes classical light microscopy and the modern immunolabelling and confocal microscopies, together with the contribution of transmission electron microscopy and atomic force microscopy. Whilst usefully standing alone, the individual microscopies often contribute most effectively when they are integrated with cellular, biophysical and molecular approaches.

Key words: Amyloid-β, Aβ, diffuse plaque, senile plaque, light microscopy, electron microscopy, atomic force microscopy, oligomer, fibril, fibrillogenesis.

1. INTRODUCTION

Since the initial post-mortem definition of cerebral amyloid plaques in the brains of ageing demented patients by Alois Alzheimer in the early 1900s from light microscopical (LM) observations, tremendous advances have occurred. Today much greater microscopical precision is now applied to brain intracellular neurofibrillary tangles and the extracellular diffuse and dense-cored/senile amyloid plaques; emphasis will be placed primarily on extracellular amyloid throughout this chapter.

Whole brain non-invasive 3D neuroimaging by Roentgen/X-ray computed tomography, magnetic resonance imaging and single photon/positron emission computed tomography has recently provided useful information enabling the early diagnosis and progression of Alzheimer's Disease (AD) to be assessed at the gross anatomical and histological level (Heckl, *et al.*, 2004; Klunk *et al.*, 2003; Masdeu, 2004; Mathis *et al.*, 2004; Petrella *et al.*, 2003). Wadghiri *et al.* (2003), using transgenic (Tg) mouse animal models for AD presented a method to detect Aβ plaques by whole brain magnetic resonance microimaging (microMRI), correlated with immunohistochemistry. This technique may well be of future significance for the detection of plaques in AD patients. Similarly, the use of radioiodinated thioflavine derivatives as tracers for imaging amyloid plaques by autoradiography holds considerable potential (Kung *et al.*, 2003). Such approaches cannot be considered further here. Rather, it is my intention to emphasise the contribution of light microscopical (LM) imaging, using conventional tissue staining and specific staining for amyloid deposits, and then lead on to contribution made by transmission electron microscopy (TEM) and atomic force microscopy (AFM) at the cellular and macromolecular level.

A distinction can be made between AD studied *in vivo,* with samples taken from man or an animal and then studied experimentally *ex vivo*, such as during cell culture or biochemical analysis, with experiments that are more truly *in vitro* because they use purified cloned proteins or chemically synthesised peptides. Whilst for most researchers the simple distinction between *in vivo* and *in vitro* is often sufficient, the recent use of the term *ex vivo* sometimes provides a useful element of precision for the description of some experimental systems.

Although most LM studies were initially performed on human autopsy samples, increasingly LM is being applied to the brains of animal AD models (*e.g.* gene knock-out and transgenic mice) and neuronal tissue in cell culture. The newer techniques of LM, such as fluorescence, confocal and multiphoton imaging are also increasingly being used and provide a meaningful continuity from the earlier LM data, now spanning almost a 100 years of AD research.

TEM has also made a useful contribution to studies on post-mortem brain samples from AD patients and Tg animal AD models, by providing supportive evidence for the existence and structure of extracellular diffuse and compact neuritic/senile plaques. With the increasing availability of synthetic amyloid-β (Aβ) peptides (*e.g.* 1-40, 1-42 and 1-43) and smaller fragments thereof, together with Aβ mutants and their synthetic equivalents, several studies relate to Aβ fibrillogenesis and the subsequent investigation of fibril stability, drug and protein interaction. In parallel with TEM studies on Aβ, a wealth of generally supportive structurally relevant information has come from studies on other amyloid systems, where pathological protein/peptide alteration involves a transformation of an initial α-helix-rich native soluble peptide into an unfolded crossed β-sheet-rich insoluble fibril-forming peptide. Usually, this related literature will only be commented upon briefly. The full impact of TEM immunolabeling for the study of amyloid plaque, *ex vivo* fibrils isolated from brain and *in vitro*-formed fibrils from synthetic Aβ peptides has yet to be seen; more extensive single and double immunolabeling studies can be expected in the future.

With the ready availability of commercially constructed equipment and the increasing expertise in its use, atomic force microscopy (AFM) and the related scanning tunnelling microscopy (STM) have been increasingly applied to the study of Aβ oligomers, fibrils and fibrillogenesis over the past ~10 years. The data produced does in general conform well with and extend from that produced by TEM, but the possibility of directly performing dynamic experiments on oligomer formation and fibril polymerisation, and drug-induced disruption of these events, makes this microscopical technique extremely valuable.

The contribution of TEM and AFM in this field cannot be considered in isolation from the range of available biophysical, biochemical and biological techniques. When appropriate, some discussion of these additional approaches will be included. Often researchers utilize and integrate a number of approaches, with TEM sometimes representing only a minor experimental component that fails to do full justice to the possible contribution of this technique. Because of the diversity of the content of this chapter, it can to some extent be considered as an introductory survey to much that will appear in the subsequent chapters.

2. LIGHT MICROSCOPY

2.1 Background

Light microscopy has been widely used for investigations into Alzheimer's disease since the earliest studies in this field. In this chapter, only aspects relating directly to the formation and properties of Aβ-containing diffuse and senile plaques will be emphasized, leaving many other significant LM applications unmentioned. LM studies have utilized human (postmortem) and animal brain samples as well as cultured cells (neurones, microglia). Recently, studies using Tg animal models for AD (mice and *Caenorhabditis elegans*) have made a strong impact, in many cases using LM alongside cellular and biochemical approaches.

Conventional LM with hematoxalin and eosin (H&E), silver and Congo red staining, as well as fluorescence microscopy with Congo red, thioflavine-S/T, an improved lipophilic thioflavine-T derivative (Kung et al., 2002) and dimethylamino-fluorenes (Lee *et al.*, 2003), fluorescein- or biotin-conjugated antibodies to Aβ and other components of the extracellular amyloid plaques and immunoperoxidase labeling, have all been successfully used. Advanced double-labeling immunofluorescence microscopy and confocal microscopy are making an increasing contribution. Classification of the progressive plaque formation in brain has been achieved by LM. Initially, immature/primitive plaques represent a loose accumulation of Aβ surrounded by abnormal neurones. Diffuse plaques are composed of an amorphous (non-fibrillar) or finely fibrillar accumulation of Aβ. Such diffuse plaques tend to stain only lightly with Congo red and thioflavine-S, but stain strongly with antibodies to Aβ. Mature/neuritic/senile plaques, which stain strongly with Congo red and thioflavine-S, tend to have a dense core of fibrillar amyloid surrounded by reactive astrocytes (macroglia) but no abnormal dystrophic neurones. Microglial (macrophage) cells are usually associated with all plaque types. Alzheimer senile plaques are found to be distributed throughout the cerebral cortex, cerebellum, amygdala, substantia innominata, hypothalamus, claustrum, tegmentum of the midbrain and rostral pons. In addition, senile plaques often occur adjacent to cerebral capillaries or larger vessels that have accumulated Aβ within their walls (cerebral amyloid angiopathy). These extracellular structural features can be detected by LM in the brains of non-demented persons, with or without head injury, but in a more pronounced manner in individuals suffering from AD, Down's syndrome (DS) and Creutzfeldt-Jacob disease (CJD) (Mochizuki *et al.*, 1996). A useful grading system for AD lesions has been advanced by Metsaars *et al.* (2003), within which the chronological sequence of neocortical lesions was graded 1 to 4, based on H&E and Congo red histology, and immunolabeling with anti-tau, -Aβ8-17 peptide, -APP, -ubiquitin and -cathepsin D. Immunohistochemistry was performed using biotinylated secondary antibodies and the streptavidin-biotin-peroxidase method, and the LM data was assessed by morphometric methods to minimize any subjective interpretation.

2.2 Application of light microscopy to cellular and extracellular studies on Aβ

Despite increasing claims that oligomeric forms of Aβ are responsible for nerotoxicity (*e.g.* Lambert *et al.*, 1998), an extensive literature exists, often including LM studies, showing that fibril formation may be important (Lorenzo and Yankner, 1994). Thus, in the present state of knowledge, it is best to keep a relatively open mind; both aspects will be duly mentioned within this chapter.

The prime location of Aβ oligomerization and fibril polymerization, *i.e.* cellular (neuronal/non-neuronal) or extracellular, has not been finally decided. Blood platelets are accepted as a major source of soluble plasmatic Aβ peptide, but it is not certain that the platelet-derived peptide is a principal source of brain Aβ plaques. Al-Khan (2002) discussed the concept that microglial cells may be responsible for the cellular processing of Aβ, via their lysosomal compartment. This was supported by EM studies of Weigel *et al.* (2000; 2001b) (*see below*). Accumulation of the amyloid precursor proten (APP) inside multivesicular bodies in cultured human brain pericytes, leptomeningal smooth muscle cells and vascular endothelial cells was shown

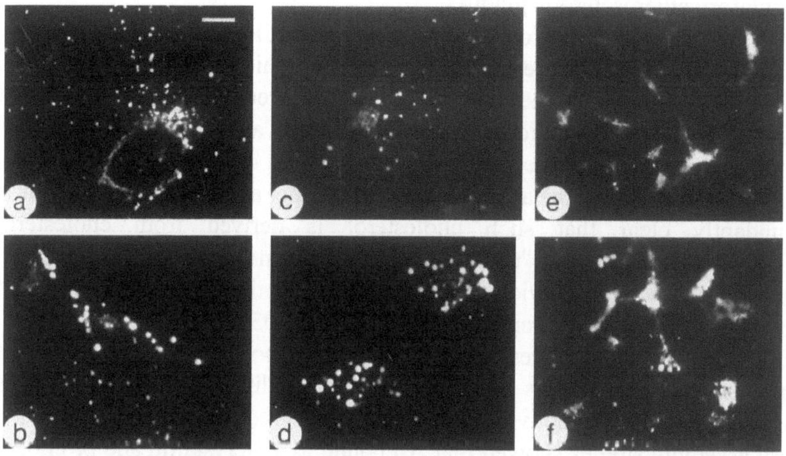

Figure 1. Immunofluorescence staining of untreated (A,C,E) and IFN-γ treated (B,D,F) cultures of human brain pericytes (A,B) human leptomeningeal arterial smooth muscle cells (C,D) and human microvascular endothelial cells (E,F) with anti-APP MAb P2-1. Large APP-positive intracellular organelles are present in all cell types after exposure to IFN-γ. Bar = 3 μm. Reproduced, with permission, from Verbeek *et al.*, Accumulation of the amyloid-β precursor protein in multivesicular body-like organelles. *Journal of Histochemistry & Cytochemistry* 50: 681-690, 2002.

using LM (Figure 1), including elegant double labeling immunofluorescence and EM by Verbeek *et al.* (2002). They suggested that multivesicle bodies function as an intermediate location in the intracellular trafficking of APP. That microglia could be involved in the clearance of exogenous Aβ1-42 was advanced by Takata *et al.* (2003), together with the fact that this cellular activity was facilitated by heat shock protein-90 (Hsp90). Furthermore, accumulation of Aβ1-42 within astocytes and the subsequent formation of astrocytic amyloid plaques in AD brains was shown using impressive LM immunohistochemistry by Nagele *et al.* (2003). Exclusive localization of Aβ in small astrocytic plaques to the subpial portion of the molecular layer of the cerebrocortex and co-localization with glial fibrillary acidic protein are features which distiguish these plaques from those derived from neurones.

Activation of microglial cells within amyloid plaques appears to be associated with an inflammatory response. Lue *et al.* (2001) cultured microglia from post-mortem brain tissue from AD and non-demented (ND) individuals. They showed significant differences between AD and ND individuals with respect to the secretion of M-CSF and complement C1q. Although microglia are phagocytic there is no evidence for the internalization of extracellular oligomeric or fibrillar Aβ or for ingestion and loading with cell fragments and cholesterol-rich membranes from dead neurones, as occurs in inflammatory macrophage/foam cells associated with circulatory atherosclerotic plaques.

Others implicate neuronal cells in the formation of amyloid plaques. Wang *et al.* (2002) suggested that the α7 nicotinic acetylcholine receptor could be involved in Aβ accumulation in Purkinje cells and the formation of cerebrellar diffuse amyloid plaques. Whether the associated neuronal lysis could also account for the presence of significant quantities of membrane-derived cholesterol in the amyloid plaques was not addressed. However, it is abundantly clear that such cholesterol is derived from cholesterol synthesized within brain tissue rather than from the blood plasma. An interesting although not widely appreciated fact is the link between AD and cataract formation in the lens. Goldstein *et al.* (2003) have found that in AD individuals Aβ accumulates as electron dense deposits in the cytoplasm of supranuclear cortical lens fibre cells of AD individuals. These deposits contain Aβ immunoreactivity and show Congo red birefringence. *In vitro* fibril assembly studies showed that Aβ bound to αB-crystallin and promoted aggregation of this lens protein.

Application of LM to the study of Aβ in cerebellar plaques was performed by Tamaoka *et al.* (1995). The detection of Aβ1-42 and Aβ1-43 as major components of diffuse-initial phase plaques implied the importance of these two Aβ species in AD pathogenesis. Immunohistochemical analysis of hippocampal sections from AD brains by Shimizu et al. (2002) combined with dot and Western blotting and *in vitro* Aβ peptide aggregation indicated that acceleration of spontaneous isomerization of the Asp23 might be significant for the formation of fibrillar amyloid in sporadic AD and that mutations or modifications of Glu22 and Asp23 could play a pathogenic role (Figure 2). The advanced LM technique of laser confocal microscopy was used by Dickson and Vickers (2001) to define morphologically the characteristic substructure of diffuse, fibrillar and dense-cored Alzheimer plaques. In this work, the plaque-associated dystrophic neurites labeled strongly for tau and neurofilament proteins.

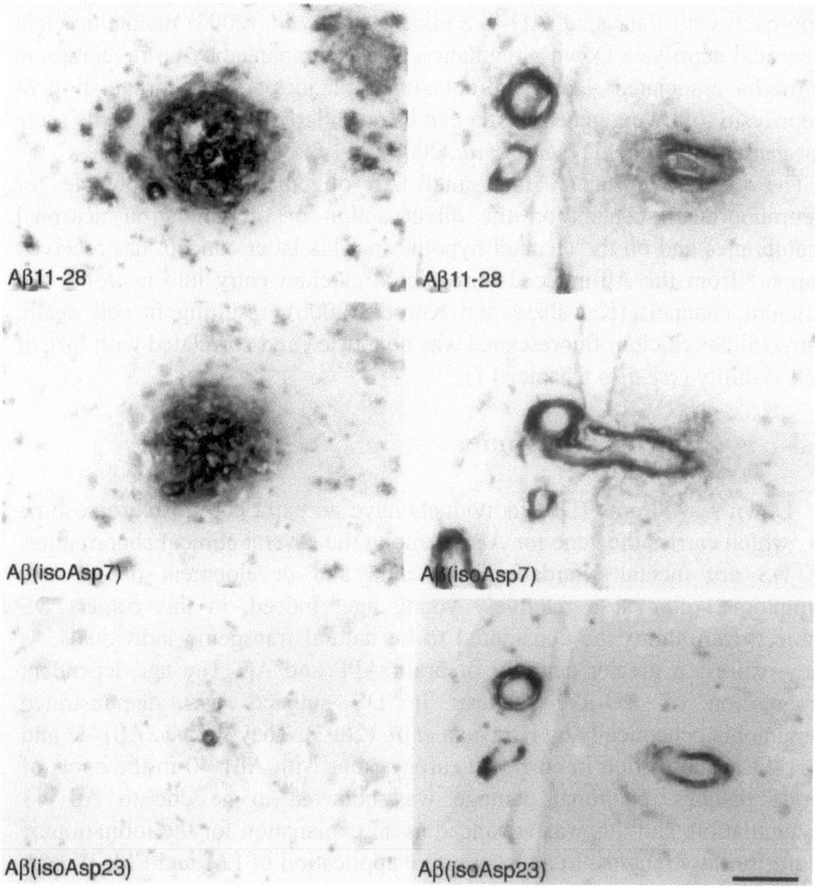

Figure 2. Immunohistochemical analysis of AD brain. Positively stained senile plaques and amyloid-bearing vessels in hippocampal sections. Bar = 50 μm. Reprinted by permission of Wiley-Liss, Inc., a subsidiary of John Wiley & Sons, Inc. from Shimizu *et al.*, Isoaspartate formation at position 23 of amyloid beta peptide enhanced fibril formation and deposited onto senile plaques and vascular amyloids in Alzheimer's disease. *J.Neurosci. Res.* 70: 451-461 © 2002.

From a detailed morphometric LM analysis of elderly non-demented and AD subjects Wegiel *et al.* (2001a) concluded that even at the end stage of AD the process of fibrillar plaque formation persists. Using quantitative immunohistochemical analysis activation of the Aβ-degrading endopeptidase, neprilysin (a zinc metalloproteinase), has been found to be reduced during aging of mouse hippocampus (Iwata *et al.*, 2002). This Down's regulation of neprilysin was considered to be relevant to AD pathology and to Aβ deposition during normal aging. Using a similar approach with transgenic AD-like mice, Apelt *et al.* (2003) maintained that neuronal neprilysin Down's-regulation was accompanied by up-regulation in astroglia associated with Aβ-containing plaques. The up-regulation of neprilysin following gene transfer can be considered to be one possible route for gene therapy in AD (Marr *et al.*, 2003).

The alternative theory, that small Aβ oligomers are responsible for neuronal death centers on the direct action of oligomers on neuronal membranes and on the channel hypothesis. This latter concept has received support from the Aβ-induced unregulated calcium entry into neurones via calcium channels (Kawahara and Kuroda, 2000), resulting in cell death. Intracellular calcium fluorescence was quantified and correlated with loss of cell viability (*see also* Chapter 17).

2.3 Down's syndrome

Down's syndrome (DS) individuals have an extra copy of chromosome 21, which carries the gene for APP. Among the several clinical abnormalites of DS are mental retardation, dementia and development of AD-like symptoms, often at a relatively young age. Indeed, in this respect DS subjects can almost be considered to be natural transgenic individuals, as they express a greater quantity of brain APP and Aβ. The age-dependent progression of AD-like features in DS subjects was demonstrated immunohistochemically by Hirayana *et al.* (2003). They showed Aβ1-42 and Aβ1-43 accumulation in cortical neurones, but with Aβ1-40 in the cores of senile plaques. Neuronal damage was believed to be due to Aβ1-43 accumulation, and this was advanced as an explanation for the formation of neurofibrillary tangles. In an impressive application of LM and EM, Wegiel *et al.* (2002) correlated the development of vascular fibrosis and calcification in the hippocampus of normal, DS and AD individuals. Although AD subjects showed marked hippocampal vascular fibrosis, this was significantly less in DS subjects. An age-dependent association of the first complement component C1q with thioflavine-S positive mature plaques in hippocampus and frontal entorhinal cortex of DS subjects was shown using immunohistochemistry by Head *et al.* (2001). Neurones were also C1q positive and were associated with activated microglia. This again implicates inflammation in the development of AD neuropathology.

2.4 Transgenic animals

Although early transgenic (Tg) mouse models for β-amyloidosis were not fully equivalent to the pathology of AD, Walker *et al.* (2002) considered that seeding of Aβ *in vivo* by injection of AD brain extracts into mouse brain might potentiate the pathological similarities. Indeed, Kuo *et al.* (2001) maintained that SDS-solubility of amyloid plaques from Tg mice was much greater than AD plaques. This was apparently accounted for by a reduced level of post-translational modification, such as N-terminal degradation, isomerization, racemization, pyroglutamyl formation, oxidation and covalent

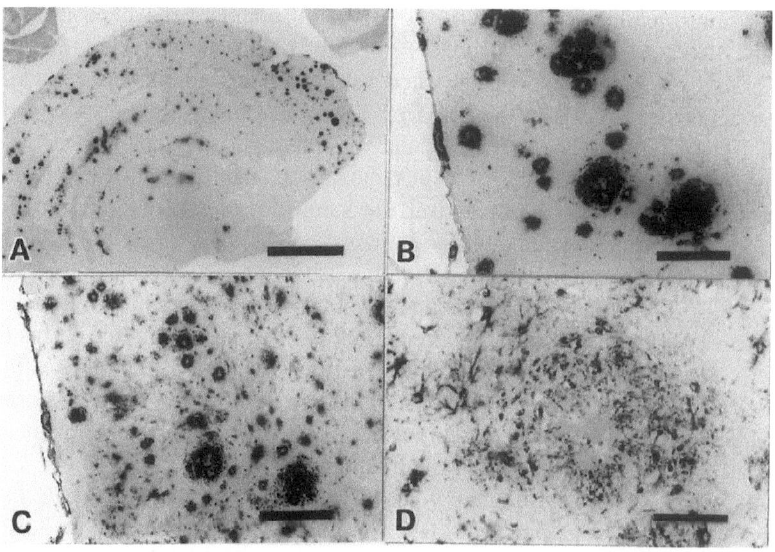

Figure 3. Amyloid deposits in the Tg2576 mouse brain. Senile plaques with large dense cores and amyloid angiopathy in 23-month-old transgenic mice visualized by Aβ1 (a), Aβ40 (b) and Aβ42 (c) antiboides. Note astrogliosis in close proximity to the dense-cored plaques in 20-month-old Tg2576 mice visualized using double staining Congo red and glial fibrillary acidic protein immunohistochemistry (d). Bars = 1 mm (a), 200 μm (b,c) and 50 μm (d). Reprinted, with permission, from Sasaki *et al.*, Amyloid cored plaques in Tg2576 transgenic mice are characterized by giant plaques, slightly activated microglia, and the lack of paired helical filament-typed, dystrophic neurites. *Virchows Arch.* 441: 358-367, © Springer-Verlag 2002.

linkage of Aβ dimers, required to fully stabilize the fibrillar Aβ within transgenic mouse plaques. Nevertheless, when brains from transgenic mice carrying the human APP (Swedish mutation) were compared with AD brain samples by LM immunohistochemistry and EM (Sasaki *et al.*, 2002), extremely large cored plaques were found in the older mice (18-29 months), which also contained slightly activated microglia (Figure 3). With a triple-transgenic AD mouse model (for APP, presenilin 1 and tau) Oddo *et al.*

(2003a) attempted to clarify the pathogenic role of intraneuronal Aβ in synaptic plasticity, within the context of both Aβ plaques and neurofibrillary tangles. In a further study, using the same triple transgenic model, (Oddo *et al.*, 2003b) concluded that amyloid deposition precedes tau tangle formation. An impressive combination of behavioural, and light and electron microscopical studies by Richardson *et al.* (2003), using the TAS10 Tg mouse line, have shown that definable changes precede amyloid deposition. Elegant LM immunohistochemistry of the hippocampus and entorhinal cortex using the PDAPP Tg mouse model, with 3D reconstruction of Aβ deposition, enabled Reilly *et al.* (2003) to differentiate the deposition of diffuse Aβ and compact plaques. A firm link between inflammation, assessed by the presence of activated astrocytes around plaques, and the early deposition of Aβ in the brain of TgCRND8 mice, was advanced by Dudal *et al.* (2004). The low molecular mass heparin, enoxaparin, has been found to reduce the size of Aβ deposits in the brains of Tg mice (Bergamaschini *et al.*, 2004) and at the same time markedly reduced the number of surrounding activated astrocytes.

Methoxy-X04 is a lipophilic derivative of Congo red, and has been used to label Aβ fibrils in AD plaques and in Tg mice. Systemic administration of methoxy-X04 enabled production of fluorescent images from the brains of living transgenic mice by multiphoton microscopy, thereby revealing individual plaques (Klunk et al., 2002). Furthermore, these authors proposed that methoxy-X04 might be suitable as an *in vivo* amyloid imaging agent in AD patients using photon emission tomography.

A doubly Tg mouse model, prone to both atherosclesosis and AD has been established by Li *et al.* (2003). Significantly, these mice show aortic atherosclesosis and cerebral β-amyloidosis, both conditions being enhanced by a lipid/cholesterol-rich atherogenic diet, along with impairment of learning capacity. Such considerations correlate with the views of de la Torre (2004), who believes that AD is more closely related to vascular disease than is commonly thought.

Transgenic *Caenorhabditis elegan* has been engineered to express human Aβ, produce amyloid deposits that stain with Congo red and thoflavin-S (Fay et al. (1998). In this animal model system for AD, $Leu^{17}Pro$ and $Meth^{35}Lys$ substitutions blocked the formation of thioflavine-S-reactive deposits, indicating an inability of Aβ to polymerize. The fluorescent, amyloid-specific dye X-34 (also a Congo red derivative) enabled *in vivo* staining of amyloid deposits to be monitored over a period of time (Link *et al.*, 2001). Transgenic *C. elegans* expressing a non-fibrillar Aβ peptide variant did not stain with X-34, but was positive with anti-Aβ antibodies. *Drosophila* expressing human Aβ1-40/42 have also been uses as a potential model for AD. Iijima *et al.* (2004), by immunostaining, showed the distribution of Aβ deposits in the fly brain, with late onset of neuronal loss the the Aβ1-42 brains.

Some further consideration of the increasing use of transgenic animals will be given elsewhere in this chapter (section 2.5 and 2.6, and see also Chapter 2). However, the fact that all mouse AD Tg models are by no means equal, as emphasized by Lehman *et al.* (2003), must not be lightly overlooked. Furthermore, Schwab et al. (2004) considered that Tg mice overexpressing Aβ are an incomplete model for AD.

2.5 Oxidative stress

The possible involvement of reactive oxygen species (ROS) in the deposition of fibrillar Aβ and the neurotoxicity of Aβ is a concept that has been discussed for several years (Behl, 1999). As with other Aβ-related topics, LM continues to make an important contribution in this area. McLellan *et al.* (2003) have used both the AD Tg mice model and *ex vivo* human AD brain tissue for investigations on ROS. By multiphoton microscopy for the detection of oxidised fluorogenic compounds they showed that only dense cored/senile plaques produced oxygen free radicals. This supported the earlier claims of Matsuoka *et al.* (2001) using more conventional LM for the detection of the natural oxidative stress markers 3-nitrotyrosine and 4-hydroxy-2-nonenal. However, it was maintained by Drake *et al.* (2003), using the Tg *C. elegans* Alzheimer model, that oxidative stress precedes fibril formation. Because of the ability of Aβ to bind metal ions (Cu/Zn/Fe/Al) it is likely that oxidative events in SP cores should not be considered in isolation from the chemical properties of these metals and the way they may mediate oxidative damage promoted by Aβ (Dong *et al.*, 2003). Although Aβ neurotoxicity, apparently not mediated by hydrogen-peroxide, was implied by the study of Saito *et al.* (2001), a dircct potentation of Aβ fibrillogenesis *in vitro* by hydrogen peroxide has recently been observed (*see below*, **3.3.3**). Immunocytochemical evidence for impairment of several endogenous enzymic antioxidant systems in brain by Aβ infusion has been demonstrated by Kim *et al.* (2003). That oestrogen and related compounds might be beneficial antioxidants is under intensive investigation (Behl and Mossmann, 2002), (*see also* Chapter 3).

2.6 Senile plaque-associated proteins

Several proteins have been found to associate with fibrillar Aβ within senile/dense cored plaques; they are often not present to the same extent within diffuse plaques. As already mentioned (see **2.2**, above) lens αB-crystallin is one of these proteins; others are acetylcholinesterase (AchE) the apolipoprotein epsilon 4 genotype ($ApoE_4$), Apolipoprotein J (ApoJ; clusterin), α1-antichymotrypsin, heparan and chondriotin sulphate proteoglycans, catalase, complement C1q, the two pentraxins serum amyoid P component (SAP) and C-reactive protein (CRP), the complement receptor CD11b, glial fibrillar acidic protein (GFAP), cystatin C, vascular endothelial growth factor (VEGF) and others. In most cases LM has contributed significantly to the assessment of presence and localization of these Aβ-associated proteins in senile plaques, but this was not the case for catalase, where Milton (1999) found the affinity of this anti-oxidant enzyme for Aβ on microplates.

The $ApoE_4$ genotype has been shown by LM to be linked to the formation of classic/senile AD plaques in preference to diffuse plaques (Oyewole *et al.*, 1999). Also, immunohistochemicaal localization of apolipoprotein E in cerebral blood vessels in association with amyloid deposits was demonstrated by Navarro *et al.* (2003). Thioflavine-S-reactive plaques in doubly transgenic mice expressing APP and human AchE (Rees *et al.*, 2003) were found to be formed more rapidly than in singly APP transgenic mice. AchE was shown histologically to be associated with the plaques and thus it was suggested that it might play a role in the pathogenesis of AD. Indeed, the C-terminal region of acetylcholinesterase has been implicated in amyloid fibril formation (Cottingham *et al.*, 2003). That cholinergic dysfunction is a feature of AD was emphasized by the study of Luth *et al.* (2003). These workers used LM and TEM of thin sectioned brain to demonstrate choline acetyl transferase immunoreactivity of nerve fibres in the vicinity of Aβ-containing plaques.

Boyett *et al.* (2003) found that when complement C1q was injected into the brains of doubly transgenic APP, PS1 mice, it resulted in an increase number of Congo red staining fibrillar deposits. This *in vivo* potentiation of amyloid fibrillogenesis agreed with earlier *in vitro* data showing that C1q promoted Aβ nucleation and aggregation (Webster *et al.*, 1995). Aβ1-42 induced activation of microglia by the plaque-associated proteins C1q and SAP was shown by Veerhuis *et al.* (2003); the pro-inflammatory cytokines IL-1α, IL-1β, IL-6, and TNFα were all increased. These authors suggested that microglial activation precedes the neurodegenerative changes associated with AD. A generalized role for the pentraxins (SAP and CRP) in inflammatory diseases, including AD, was advanced by McGeer *et al.* (2001), leading to the concept that the pentraxins could be targets for therapeutic intervention, as expanded upon by Pepys *et al.* (2002) within the context of peripheral amyloidosis rather than for amyloid-β plaques in the brain.

2.7 Cholesterol in amyloid plaques

Accumulation of cholesterol in dense-cored fibrillar plaques in AD and transgenic APP (SW) mice was shown initially by Mori *et al.* (2001), using fluorimetric staining for filipin and enyzmically using cholesterol oxidase. This work was extended by Burns *et al.* (2003), who showed co-localization of cholesterol and ApoE in hippocampus of transgenic PS1/APP mice (Figure 4). Data from both these groups showed that cholesterol was not associated with diffuse plaques, indicating Aβ fibril-specific cholesterol binding (*see also* section **3.3.5**).

It is known that the epsilon 4 isoform of ApoE is a predominant genetic risk factor in both familial and sporadic AD (reviewed by Poirier, 2003). The binding and transport of cholesterol by $ApoE_4$ to microglia and astrocytes may provide a cellular pathway for non-esterified cholesterol to be recovered from injested neuronal membranes, released and transported to amyloid plaques. It is likely that cholesterol is bound to fibrils in a molecular or micellar form rather than a microcrystalline form (as is known to occur in atherosclerotic plaques, in association with calcium phosphate). Also, conversion of brain cholesterol to 24S-hydroxycholesterol, due to the presence of a microglial oxidative cytochrome P-450 (termed CYP46), was advanced by Bogdanovic *et al.* (2001) as a significant factor in relation to the mechanism for cholesterol efflux from microglia.

Using LM immunohistochemistry and image analysis Pappola *et al.* (2003) assessed the amyloid load in human brain and correlated this with the presence of cholesterolemia. They were able to conclude that serum hypercholesterolemic could be an early risk factor in AD development, but from older subjects their data was not clear-cut (the data may have been biased by deaths due to cardiovascular disease). Conclusive evidence for the link between CSF cholesterol and brain amyloid has come from thc gcnctic study of Wollmer *et al.* (2003) who linked the incidence of polymorphism of the gene encoding acyl-coenzyme A: cholesterol acyltransferase with a low CSF cholesterol level and low brain amyloid load.

However, a note of caution comes from the work of Park *et a*l. (2003) who showed that in female transgenic mice, lovastatin actually enhanced Aβ production and senile plaque deposition. Nevertheless, the cholesterol-fed rabbit has been used as a model system for AD (*e.g.* Sparks, 1996). Furthermore, that trace quantities of copper are required for the formation of amyloid plaques in cholesterol-fed rabbits has been shown by Sparks and Schreurs (2003), using immmunohistochemistry to define senile plaques. The link between cholesterol and iron-dependent oxidative damage in both atherosclerosis and neurodegeneration has been reviewed by Ong and Halliwell (2004), with the conclusion that this might explain why

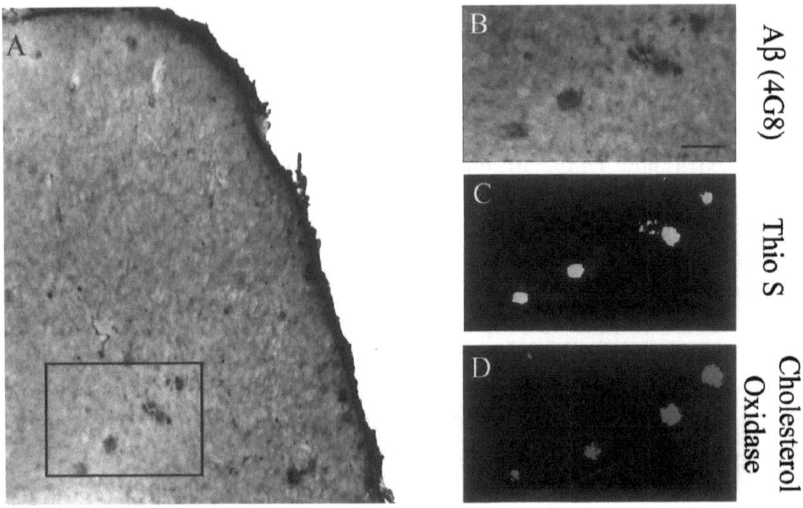

Figure 4. Co-localization of fibrillar amyloid with cholesterol oxidase in the cortex of a 12-month-old PS/APP mouse. (A) A low-power overview of amyloid plaques stained with 4G8. (B-D) Plaques from within the boxed areas double-labeled for Aβ (B) and thioflavine-S (C), and a consecutive section labeled for cholesterol oxidase (D). Scale bar = 100 um. Reprinted from Burns *et al.*, Co-localization of cholesterol, apolipoprotein E and fibrillar Aβ in amyloid plaques. *Molec. Brain Res.* **110:** 119-125, copyright (2003), with permission from Elsevier.

hypercholesterolemia is a risk factor in AD. In addition, differential expression of cholesterol hydroxylases has been advanced as a mechanism to explain the pathophysiology of AD (Brown *et al.*, 2004).

2.8 LM and *in vitro* Aβ fibrillogenesis

In their review on the structure of amyloid fibrils, Gorman and Chakrabartty (2001) show elegantly how LM usefully provides a basis from which the more molecular approaches used to study Aβ fibrillogenesis extend. Indeed, by direct observation the growth of amyloid fibrils can be monitored light microscopically by thioflavine-T fluorescence (Ban *et al.*, 2003). These workers used total internal reflection fluorescence microscopy (TIRFM) in a study of fibril dynamics for several different amyloid fibrils and were able to determine the exact length of growing Aβ fibrils. That slightly acidic conditions, such as pH 5.8, promote fibril formation from the Aβ peptide was shown with fluorescein-conjugated Aβ1-42 (Su and Chang, 2001), using the protocol established by Chang and Su (2000). This work correlates well with the suggestion from cellular studies that Aβ present in acidic (lysosomal) organelles in neurones or microglia may initially form neurotoxic fibrils more readily than Aβ in neutral subcellular compartments.

3. TRANSMISSION ELECTRON MICROSCOPY

3.1 Technical background

The various techniques available to prepare specimens for transmission electron microscopy (TEM) have enabled extensive studies to be performed on human post-mortem and animal brain samples, and also on cultured neurones, microglia, astrocytes and other cells used for AD studies. Quite separate from these tissue and cellular studies, TEM has contributed widely to knowledege of reconstituted Aβ and other amyloid fibre-forming systems *in vitro*. Brain and cell culture studies have generally utilized conventional thin sectioning of resin-embedded samples, but more recently the addition of pre- and post-embedding antibody labeling has increasingly made a useful contribution. This latter represents a higher resolution extension to immunofluorescence LM and often enables precise localization of colloidal gold or other label on intracellular or extracellular structures definable by TEM study.

Oligomerization and fibrillogenesis of the Aβ peptides (primarily Aβ1-40, 1-42, and fragments) has received considerable attention from electron microscopists, often using the negative staining technique and more recently by cryo-TEM of unstained frozen-hydrated specimens; *see* Harris (1997) and Harris and Scheffler (2002) for techniques. The more specialist approach of scanning transmission electron microscopy (STEM) has also made a contribution. This technique enables a direct measurement of the mass of macromolecules to be made and determination of the mass/unit length of fibrils (Antzutkin *et al.*, 2002)). Combination of TEM with one or more biophysical/biochemical techniques (X-ray fibre diffraction, electrospray ionization MS, CD, NMR and solid state NMR spectroscopy, fluorescence spectroscopy, dynamic light scattering, small angle neutron scattering, SDS-PAGE, analytical ultracentrifugation) often provides the strongest possible approach for the study of fibrillogenesis. However, when applied with care, experience and knowledge of the biological samples, TEM alone has great potential, as will be demonstrated below.

3.2 Thin sectioning TEM studies of brain

At the subcellular level, thin section TEM often supports and extends the data available from parallel LM studies. This is impressively shown in several of the publications from Jerzy Wegiel and his colleagues (Figure 5). The role of microglial cells in fibrillar plaque formation in the brains of man and transgenic APP_{SW} mice was shown by Wegiel *et al.*, 2000; 2001a. This interpretation was subsequently supported by the study of Sasaki *et al.* (2002) using the Tg2576 transgenic mouse. Further evidence that vascular inflammatory microglial infiltration in transgenic mice contributes to the formation of cored Aβ plaques also comes for the ultrastructural study of Wegiel *et al.* (2003), using computer-aided 3D reconstruction of the tissue to assist the image interpretation. The LM study on hippocampal vascular fibrosis and calcification on normal aging, AD and DS subjects (Wegiel *et al.*, 2002) was supplemented by TEM, which strengthens the link between brain vascular pathology and sclerosis in AD.

Ultrastructural and immunocytochemical studies on neurones have been usefully used to assess the synaptic location and quantification of APP and presenilins in rodent brains (Ribaut-Barassin *et al.*, 2003). Understanding the change in synaptic distribution of these proteins during development and aging may be of use for the understanding of synaptic loss that precedes the deposition of extracellular Aβ. Supplementary to this, Takahashi *et al.* (2004) showed by immunoelectron microscopy that oligomerized Aβ accumulates within the neuronal processes and synapses of Tg2576 transgenic mice.

Using the transgenic *C. elegans* Alzheimer model Link *et al.* (2001) extended their extensive LM observations by immunogold TEM labeling. They provided convincing evidence for the location of extracellular Aβ deposits, which at high magnifications clearly showed a fibrillar ultrastructure. Simlarly, Verbeek *et al.* (2002) using cultured human brain pericytes, also showed multivesicle-like bodies by thin section TEM, with accumulation of APP.

Furthermore, the intracellular accumulation of Aβ within multivesicle bodies in neurones of normal human, rat and mouse brain was shown by Takahashi *et al.* (2002). For Tg 2576 mice and human AD brain the Aβ accumulation in microvesicles was more pronounced and increased with aging. This intraneuronal deposition of Aβ initially occurred with normal synaptic morphology and prior to extracellular plaque formation, thereby suggestive of a crucial role in the development of AD. In their TAS10 transgenic mouse AD model Richardson *et al.* (2003) also showed dystrophic neurons containing multi-vesicle/multi-lamellar inclusions and activated microglia, alongside amyloid plaque.

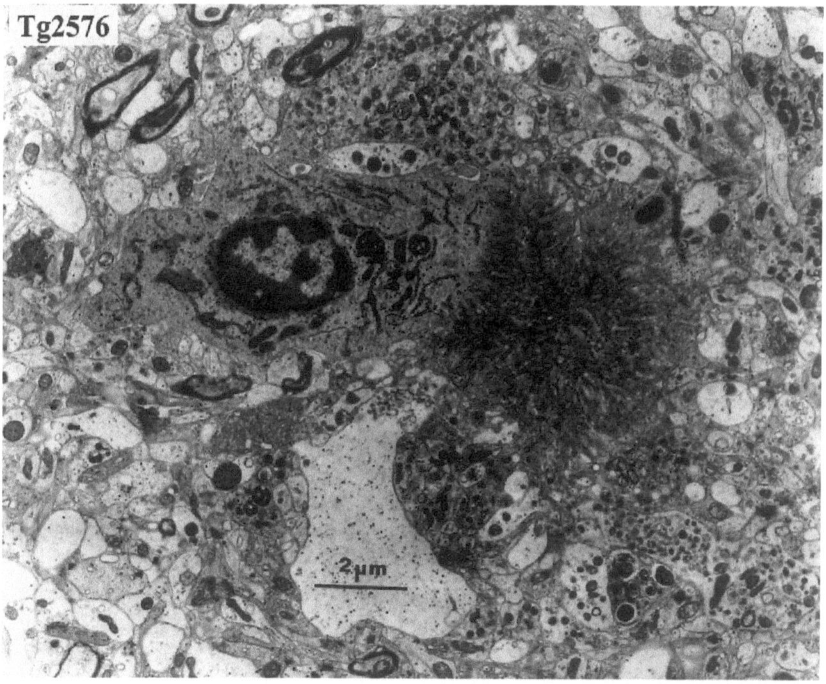

Figure 5. A small plaque in the thin sectioned brain of transgenic mouse (Tg2576). The micrograph show the body of a microglial cell, an immunolabeled amyloid star, dystrophic neurities and a few astrocytic processes. Previously unpublished micrograph, courtesy of Jerzy Wegiel.

Although there is evidence that Congo red and thioflavine-T (Th-T) inhibit cerebral Aβ fibrillogenesis, both these drugs fail to pass the blood-brain barrier (BBB). Use of polymeric nanoparticles as drug carriers, which have potential to pass the BBB has been investigated by Härtig *et al.* (2003). Fluorescent Th-T incorporated into latex nanoparticles was injected into the hippocampus of mice. After the photoconversion of Th-T it was detectable by LM and TEM, following injestion by microglia and subsequent aggregation. These authors speculated that such nanoparticles might be useful delivery system to target both intracellular and extracellular Aβ in the brain.

3D-Reconstruction from electron micrographs of amyloid deposits and microglial cells in the APP232 transgenic mouse model (Stalder *et al.*, 2001) conclusively showed that microglia do not phagocytose extracellular amyloid deposits and that there is no intracellular amyloid production in microglia. TEM immunogold-labelling of amyloid plaques and associated inflammation, in PS2APP transgenic mice, was impressively correlated with progressive age-dependent cognitive deficits (Richards *et al.*, 2003). Using confocal LM and TEM, Hu *et al.* (2003) showed preferential infiltration of dystrophic cholinergic presynaptic boutons into fibrillar amyloid aggregates, in doubly (APP + PSI) transgenic mice. The implication of this observation was claimed to relate to the degeneration of cholinergic nerve terminals in advance of those of other neurotransmitter systems.

3.3 Amyloid-β oligomerization and fibrillogenesis

Despite some considerable overlap within the literature, it is appropriate to present this topic in three sections, dealing in turn with Aβ oligomers, protofilaments and fibrils.

3.3.1 Aβ Oligomers

Possible variability of the commerically available synthetic Aβ peptides was commented upon by Soto *et al.* (1995), with respect to their ability to form oligomers, protofilaments and fibrils. This was expressed as a varying ability to transform from α-helix to β-sheet. Since that time, reasonable standardization appears to have been achieved, yet data produced by different groups performing *in vitro* fibrillogenesis is not always in accord or mutually supportive.

A careful comparative analysis of oligomerization of the Aβ40 and Aβ42 peptides was performed by Bitan *et al.* (2003). They showed by SDS-PAGE that cross-linked oligomers revealed distinct small molecular mass distributions, with Aβ42 forming somewhat higher mass oligomers than Aβ40. TEM analysis by negative staining showed the presence of ~5 to 20 nm spherical particles (termed paranuclei), which aggregated to form larger elongated oligomers and clusters. Essentially similar convincing TEM evidence was provided earlier by Nybo *et al.* (1999), using Aβ42. Immunodetection of neurotoxic oligomers by their epitope specificity, this epitope being absent on the Aβ peptide and fibrils, may prove to be diagnostically useful (El-Agnaf *et al.*, 2003; Kayed *et al.*, 2003). Also, it is likely that future immunoelectron microscopical studies using the anti-Aβ-oligomer antibody will have considerable potential.

An α-helix containing Aβ oligomer was defined by Kirkadze *et al.* (2001). By utilizing Aβ peptides of difference amino acid sequence they concluded that asparagine and histidine are important for the conversion to protofilaments. Support for this concept also comes from the study of Fezoui and Teplow (2002), in which they proposed that a partly unfolded Aβ intermediate, still containing α-helix, is involved in fibrillogenesis.

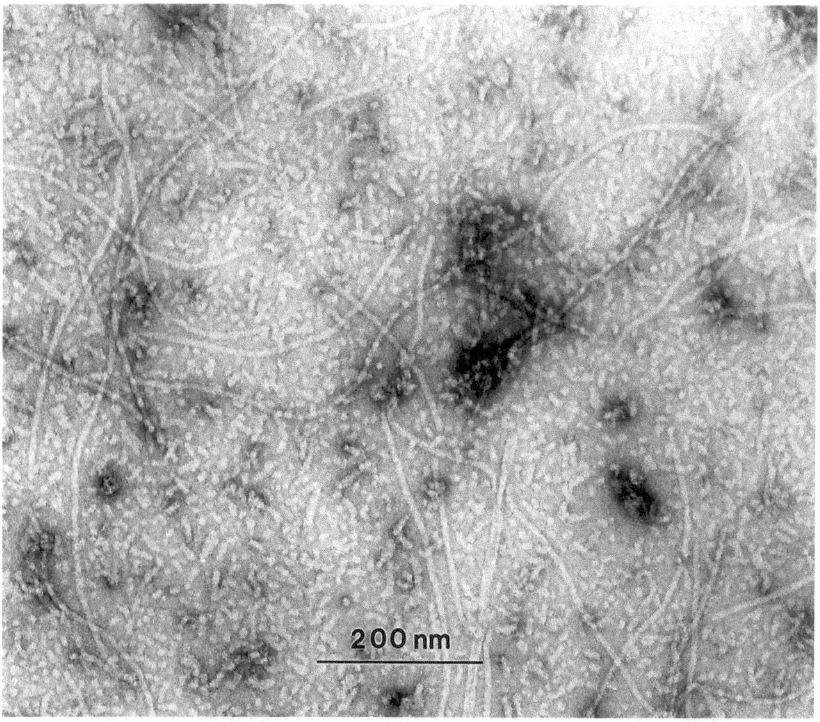

Figure 6. Oligomerization of human Aβ1-42 following incubation for 24h at 37°C in the presence of water alone. This negatively stained image shows that oligomers and flexuous protofibrils are present, but few mature fibrils (JRH, previously unpublished data).

Contrary to this, Huang *et al.* (2000) showed by CD spectroscopy that the very small Aβ oligomers contain an irregular secondary structure, i.e. no α-helix and no β-sheet, but that larger ~0.94 MDa oligomers do contain a β-sheet structure. This important observation may pin-point the Aβ oligomerization stage at which the α-helix to β-sheet conversion occurs, and thus enable the subsequent polymerization of a stable structural intermediate to proceed in order to generate the ~5 nm protofilaments and ~10 nm fibrils (*see below*). Although the TEM data discussed here does not provide any clear evidence that this intermediate (paranucleus) is a ring-like structure, others have produced averaged images of ring-like/annular oligomers or short protofilaments (Lashuel *et al.*, 2002a), which correlated with the concept that such oligomers could be responsible for penetration of the neuronal surface membrane. Our own data (Figure 6), using the human Aβ1-42 peptide shows that after 24 h incubation a range of short protofilament and smaller, possibly annular oligomers are present. From small angle neutron scattering data Yong *et al.* (2002) concluded that a stable micelle-like oligomeric Aβ intermediate has a spherocylindrical shape (radius ~2.4 nm and height ~11 nm), but it is not clear that this relates directly to the annular oligomers of Lashuel *et al.* (2002).

Further evidence that Aβ oligomers can form ion channels comes for experimental work using lipid bilayers, liposomes, neurones, oocytes and other cells (Kagen *et al.*, 2002; Lin *et al.*, 2001); *see also* Chapter 17. The possibility that cell membrane cholesterol/ganglioside/sphingomyelin-rich microdomains (rafts) could be the predominant sites of Aβ interaction was advanced by Kakio *et al.* (2003).

A parallel is shown by the scrapie prion protein, which oligomerizes to form stable ~10 nm annular structures, that have the ability to form 2D crystalline arrays (Willie *et al.*, 2002). These prion annuli correlated with the interpretation that they stack or polymerize longitudinally to form the amyloid-like prion rods. Interestingly, a rapidly-formed but transient nano-crystalline from of a 14-amino acid Aβ peptide has been described by Otzen and Oliveberg (2004). Using TEM these workers showed that the nanocrystalline form of this peptide leads to the formation of a tangled aggregate (hours) and amyloid fibres (days).

3.3.2 Aβ protofilaments

Definition of the β-sheet-containing protofilament that can be formed by Aβ and several other fibril-forming amyloidogenic peptides is by no means clear and may differ from the putative annular oligomer described above. Indeed, Lashuel *et al.* (2000) considered that the protofilament is ~2-3 nm in width, representing a single linear crossed β-sheet assembly of the peptide, without any indication of a globular or annular repeating structure. It is likely that the different amyloidogenic peptides may not behave in an identical manner when they undergo the folding change necessary to generate 5-6 nm single or 10-12 nm double helical filaments and in some cases large multi-filament ribbons.

In a series of publications, many including TEM studies, David Teplow and his colleagues have defined the Aβ protofilament intermediates involved in fibrillogenesis (Walsh *et al.*, 1997; 1999). Again a combination of biophysical techniques was used, including size-exclusion chromatography to isolate the Aβ polymers of increasing mass. Their sequential model for fibrillogenesis is essentially similar to that of Lashuel *et al.* (2000), although a globular 'nucleus' or monomer aggregate (equivalent to the 'paranucleus' of Bitan *et al.*, 2003) is included prior to protofilament assembly.

In a recent study Lashuel *et al.* (2003), using the Arctic Aβ1-40 variant (E22G) as well as wild-type Aβ1-40, from extensive negative stain TEM studies concluded that the pathogenic E22G variant more readily produced transformation of protofibrils to fibrils than did a mixture of E22G and wild-type.

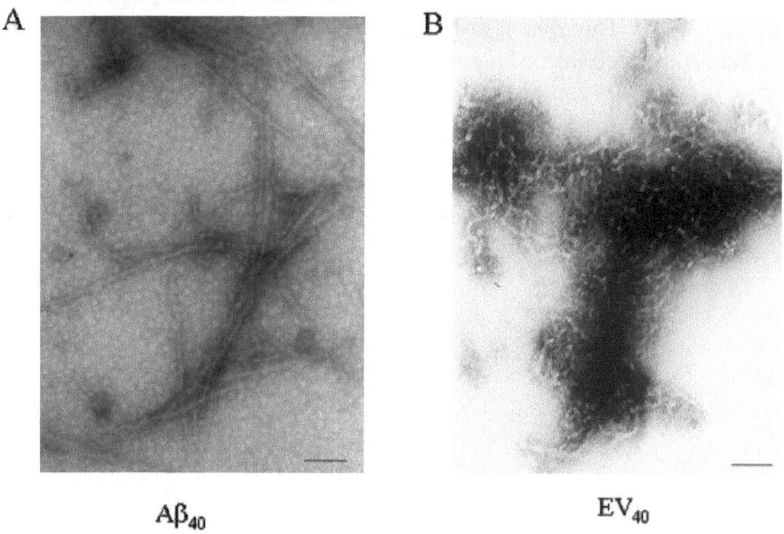

Figure 7. Negatively stained images of A1-40 (A) and EV40 (B) aggregates following incubation for 48h. Reprinted, with permission, from Qahwash *et al.*, Identification of a mutant amyloid peptide that predominantly forms neurotoxic protofibrillar aggregates. *J. Biol. Chem.* **278:** 23187-23195, © (2003).

Using a synthetic Aβ1-40, termed EV40, that had substitutions in the first two amino acids, Quhawash *et al.* (2003) showed by TEM that this variant could only form protofilaments, rather than fully-formed/mature Aβ fibrils (Figure 7). Remarkably, the EV40 peptide was even more toxic than Aβ1-40 towards neuroblastoma cells, and cortical and hippocampal neurones, suggesting the possible importance of this EV40 intermediate, beit protofilament or oligomer.

3.3.3 Aβ fibrils

In vitro conversion of clearly defined ~5 nm Aβ protofilaments to mature ~10 nm fibrils has not been intensively studied. It is acknowledged that some variability or fibril polymorphism exists for the different Aβ peptides, as for other fibril-forming peptides. In general, the mature Aβ fibril appears to be straight, with a smooth surface and a clear helical periodicity, generated from two protofilaments each of which itself contains two linear strands of helical crossed β-sheets. That this amyloid fibril then resembles a hollow tubule was advanced by Malinchik *et al.* (1998), from X-ray diffraction and TEM data (negative stain and thin section). A representative example of mature helical Aβ1-42 fibrils produced by incubation in the presence of 1 mM hydrogen peroxide is given in Figure 8. Aggregation of Aβ1-40 fibrils to form large ball-like aggregates was shown by Westlind-Danielsson and Arnerup (2001). This distinguished the polymorphic behaviour of Aβ1-40 from Aβ1-42, which formed parallel fibril bundles.

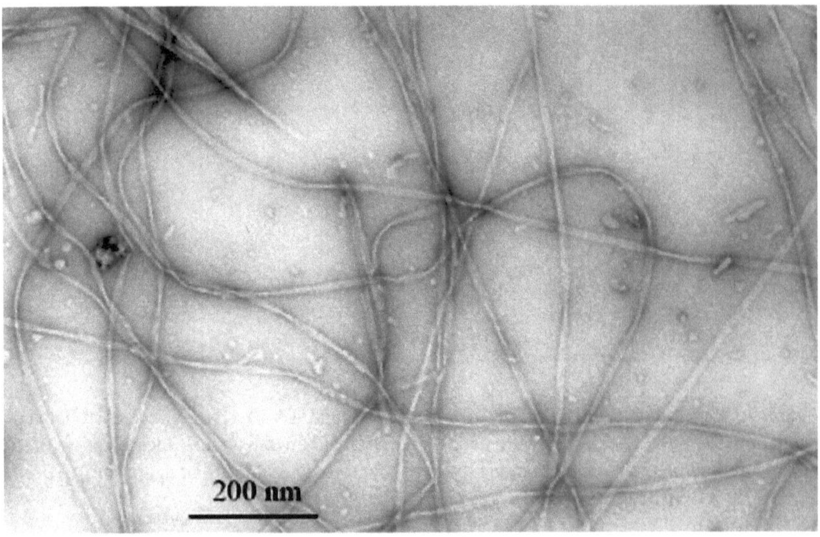

Figure 8. Amyloid-β1-42 fibrils formed after 24h 37°C incubation of the peptide (rat) in water in the presence of 1mM hydrogen peroxide (JRH, previously unpublished data).

The influence of pH on the fibrillogenesis of Aβ1-40 was assessed by Wood *et al.* (1996) using a range of biophysical techniques, including TEM. They concluded that at pH 5.8 Aβ1-40 tended to form turbid aggregates but at pH 7.4 transparent fibrils were formed. However, Abe *et al.* (2002), using the Aβ12-24 fragment found that 10-12 nm fibrils and 17-20 nm ribbons were formed at low pH. Others have attempted to define which amino acid residues are important for fibril formation (Fraser *et al.*, 1994; Bond *et al.*, 2003; Tjernberg *et al.*, 2002). The role of Aβ1-42 Gly33 and Meth35 in relation to neurotoxicity and free radical attack was advanced by Kanski *et al.* (2001). These workers also implicated the hydrophobic environment of Meth35 in the neurotoxicity and oxidative properties of Aβ1-42 (Kanski *et al.*, 2002). TEM and STEM were combined with solid state NMR to assess the fibrillogenesis of Aβ10-35, Aβ1-40 and Aβ1-42 by Antzutkin *et al.*

(2002). The value of mass/unit length measurement by STEM in relation to determination of the number of single filament crossed-β-sheets and protofilaments present in any one mature fibril was clearly demonstrated.

Using negative stain TEM as one of their approaches, Romano *et al.* (2003) assessed the structure of both synthetic and neuronally secreted Aβ fibrils. Both types of fibrils labeled with monocolonal antibodies against Aβ amino acid epitopes 1-17 and 17-24, as shown by immunonegative staining. Significantly, however, only the naturally secreted neuronal fibrils interfered with long-term memory loss in a crab learning model. Definition of the aa sequence of Aβ responsible for fibril formation has been investigated by Williams *et al.* (2004). By scanning proline mutagenasis they shwoed that the Aβ15-36 β-sheet-rich core region could create a hydrogen-bonded stack of peptides. Again, negative stain EM contributed to the understanding of the structure of the different proline mutant Aβ fibrils.

Further TEM and STEM work on short synthetic Aβ fragments of known length and sequence, as well as on the naturally occurring mutant Aβ peptides, is likely to continue to provide useful data.

3.3.4 Aβ fibril-associated proteins

From biochemical microplate-binding studies Hamazaki (1995) showed a calcium-dependent affinity of Aβ1-40 for serum amyloid P component (SAP). This has been confirmed using TEM by others, including the author, and appears to be of significance for the inflammatory response and the coating of Aβ fibrils with SAP may also prevent or reduce the natural proteolytic degradation of Aβ deposits. Nielsen *et al.* (2000) have shown by TEM of negatively stained specimens that SAP and heparan sulphate proteoglycan both bind to pre-formed fibrils. Binding was confirmed by immunostaining with 5 nm Protein A/G-gold particles after on-grid incubation with polyclonal anti-SAP. Our own data essentially agrees with that of Holm Nielsen *et al.* (2000), with the additional observation that mature Aβ fibrils bind SAP more efficiently than protofibrils (Figure 9). This may correlate with the varying binding of SAP in diffuse and dense cored amyloid plaques.

A link between SAP complement C1q and microglial activation was advanced by Veerhuis *et al.* (2003), using LM, TEM and ELISA assays. A marked increase in Aβ1-42 fibril formation was claimed in the presence of C1q and SAP. This implies that the inflammatory response may actually potentiate Aβ fibrillogenesis. That α1-antichymotrypsin regulates Aβ fibrillogenesis was shown by Eriksson *et al.* (1995) and the cysteine protease inhibitor cystatin C has also been localized in Aβ deposits (Vattemi *et al.*, 2003).

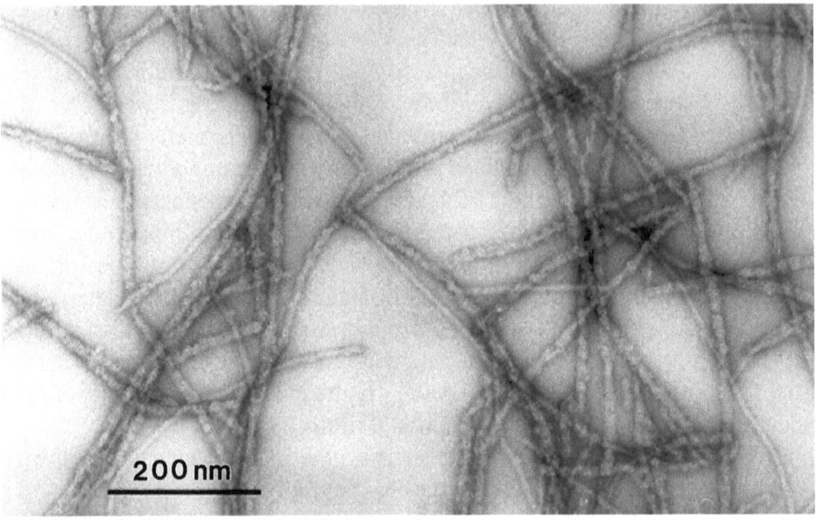

Figure 9. Amyloid-β1-42 fibrils (human aa sequence) formed in the presence of 0.5 mM H_2O_2 with bound SAP (+ 0.5 mM calcium and magnesium chloride). Note the coating of the mature paired helical fibrils by SAP, but the absence of binding to the single protofibrils (JRH, previously unpublished data).

The biochemical evidence for catalase binding to Aβ peptide and fibrils (*see also* section **2.6**) has recently been supported by TEM studies, as shown in Figure 10. Although the amino acid sequence of Aβ responsible for the affinity of catalase has been defined, it is not clear how this might influence fibrillogenesis, other than by inhibition of the anti-oxidant activity of this enzyme.

Acetylcholinesterase (AChE) has been shown to accelerate Aβ fibrillogenesis (Di Ferrari *et al.*, 2000; Inestrosa *et al.*, 1996). In both these publications TEM evidence was presented for binding of AchE to Aβ fibrils (Figure 11). Inhibition of AchE and AchE-induced Aβ fibrillogenesis by the drug 3-(4-{[benzyl(methyl)amino]methyl}-phenyl)-6,7-dimethoxy-2H-2-chromenone (AP2238) (Piazzi *et al.*, 2003), designed initially to bind to both the catalytic and peripheral sites of AchE, indicates the possible value of AP2238 for AD therapy.

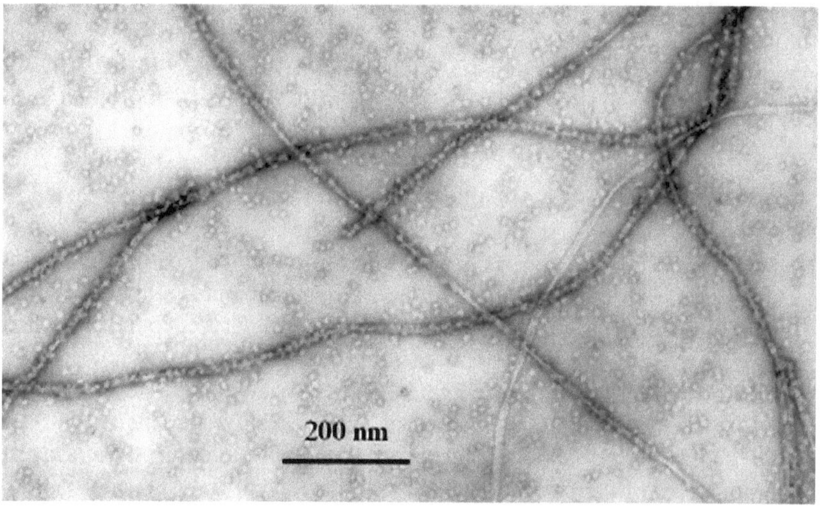

Figure 10. Amyloid-β1-42 fibrils (human aa sequence) formed in the presence of 0.5 mM H_2O_2, with bound human erythrocyte catalase (JRH, previously unpublished data).

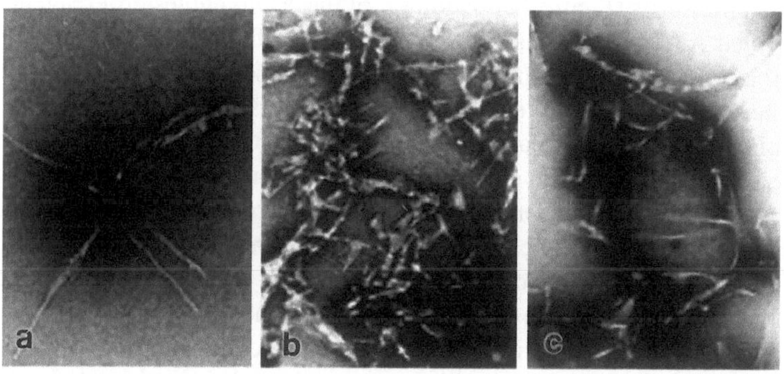

Figure 11. Amyloid –β fibrils (human aa sequence) with bound acetylcholinesterase. (a) Aβ1-40 fibrils alone. (b) Aβ fibrils assembled in the presence of *Torpedo* AChE. (c) Aβ fibrils assembled in the presence of the hydrophobic AChE fragment $H_{274-308}$. Reprinted with permission from Di Ferrari *et al.*, A structural motif of acetylcholinesterase that promotes amyloid β-peptide fibril formation. *Biochemistry* 40: 10447-10457. Copyright (2001) American Chemical Society.

The LM study of Goldstein *et al.* (2003) on αB-crystalin (see also section **2.6**) was supplemented by successful TEM double immunogold labeling of Aβ protofilament/fibril aggregates with bound αB-crystalin. Binding of Aβ by mitochondrial alcohol dehydrogenase (Lustbader *et al.*, 2004) has been linked to mitochondrial dysfunction and neuronal oxidative stress in AD.

3.3.5 Aβ fibrillogenesis and cholesterol

Despite the considerable LM and biochemical evidence for cholesterol localization in Aβ deposits in brain, the available supportive evidence from TEM is limited. This may be because of the difficulty encountered when attempting to stain for cholesterol in thin sectioned tissue and the lack of knowledge regarding the chemical state (molecular, micellar or microcrystalline) of any cholesterol that is bound to or associated with Aβ fibrils, and the esterification state of any such cholesterol. However, direct addition of cholesterol microcrystals or micellar "soluble cholesterol" (polyoxyethyl cholesterylsebacate) to Aβ1-42 peptide at the start of the fibrillogenesis incubation has been found to potentiate fibril formation (Harris, 2002), and provided evidence for cholesterol binding/association with mature fibrils (Figure 12). Addition of these two cholesterol reagents to pre-formed Aβ fibrils has also shows the binding of Aβ fibrils to cholesterol. Note that this affinity of Aβ for cholesterol is present in the absence of the cholesterol-binding lipoprotein $ApoE_4$, which has itself an affinity for Aβ.

McLaurin *et al.* (2000) assessed the action of several proteins and lipids on the secondary structural characteristics of Aβ and on Aβ fibrillogenesis. They provided strong supportive evidence that interaction of chondriotin and heparan sulphate proteoglycans enhances transformation to crossed β-sheet, with fibril formation. Among the phospholipids they investigated, both phosphatidylserine and phosphatidylinositol enhanced Aβ fibril formation, as did iron, copper and zinc ions; however, the potentiating effects of cholesterol and hydrogen peroxide was not included in this study. The involvement of heparan sulphate proteoglycans in AD has been reviewed by van Horssen *et al.* (2003), with emphasis upon LM and biochemical aspects.

3.3.6 Drug inhibition of Aβ fibrillogenesis

Numerous compounds have been claimed to retard or inhibit Aβ fibrillogenesis. Often studies rely upon reduction of Congo red or thioflavine-T (Th-t) fluorescence, as a measure of reduced oligomerization

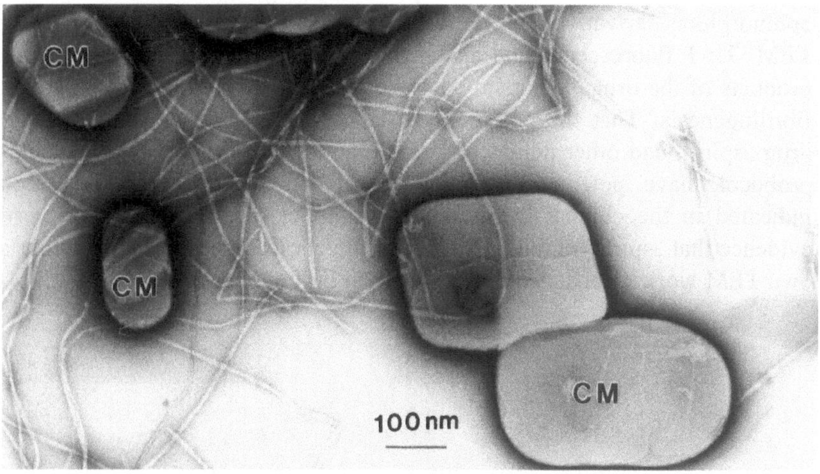

Figure 12. Amyloid-β1-42 fibrils (rat) assembled in the presence of cholesterol microcrystals (CM). Modified from Harris (2002).

and fibril formation; only work relating to or including TEM assessment will be mentioned here (for LM, see section **2.8**). Indeed, an inhibitory action of Congo red itself was found by Lorenzo and Yankner (1994) when this compound was added to the Aβ peptide at the start of *in vitro* fibrillogenesis. The hormone melatonin has been shown by Pappola *et al.* (1998) to inhibit Aβ fibrillogenesis, but the related compound 5-hydroxy-N-acetyl-tryptane had no effect, neither did the powerful non-specific anti-oxidant N-t-butyl-α-phenylnitrone.

In Switzerland, the EM groups of Jacques Dubochet and Ueli Aebi have independently both been involved in significant investigations on inhibition of Aβ fibril assembly, in collaboration with pharmaceutical companies. Goldberg *et al.* (2000) provided both TEM and STEM data that enabled characterization of the polymorphic Aβ1-40 fibril assemblies to be assessed. From their extensive data a scheme advanced for the development of fibril polymorphism was advanced, which is essentially similar to that of Lashuel *et al.* (2000). The inhibitory interaction of several morpholinotriazines was considered by Bohrmann *et al.* (2000), from their TEM and biophysical data, to occur at the level of small Aβ1-42 structural intermediates.

Remarkably, Kiuchi *et al.* (2002a) maintained that type IV collagen inhibits Aβ1-40 fibrillogenesis, and Ueda *et al.* (2002) showed that high concentrations of sucrose (1.5 M) retarded the fibrillogenesis of Aβ1-42. Significantly, in the presence of sucrose, a compact amorphous Aβ1-42 aggregate remained which did not produce Th-T fluorescence (*cf* the effect of aspirin, below). The effect on Aβ1-40 fibrillogenesis of a new class of apomorphine derivatives was investigated by Lashuel *et al.* (2002), using TEM, Th-T fluorescence and analytical ultracentrifugation. Auto-oxidation products of the drugs apparently acted directly on the Aβ peptide to inhibit fibrillogenesis. That the commonly used analgesic and anti-inflammatory drug aspirin, and other non-sterol anti-inflammatory drugs (NSAIs) such as probucol, have potential as inhibitors of Aβ fibrillogenesis has been indicated in the clinical literature for a number of years. Direct *in vitro* evidence that aspirin inhibits Aβ1-42 fibrillogenesis comes from the author's own TEM work (Harris, 2002), as shown in Figure 13.

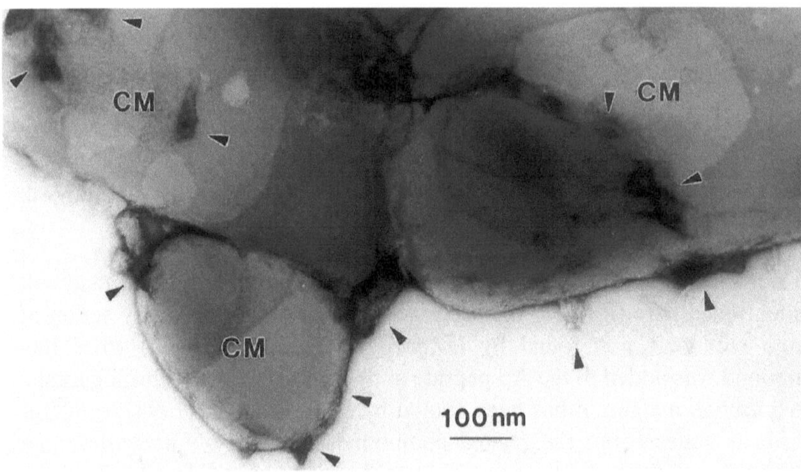

Figure 13. The inhibition of cholesterol-potentiated Aβ1-42 (rat) fibrillogenesis by 1 mM aspirin. Note the absence of fibrils and the presence of small Aβ clusters (arrowheads) binding to the cholesterol microcrystals (CM). Modified from Harris (2002).

Because of the therapeutic potential, in addition to basic understanding of the inhibition of Aβ fibrillogenesis, considerable interest has also be directed towards the possibility that drugs and other compounds may produce disassembly of pre-formed Aβ fibrils, detectable by TEM and correlated with reduction of Th-T fluorescence. In our own studies, only a slow and possibly insignificant effect was detected with aspirin and other compounds require further careful investigation using pre-formed mature Aβ fibrils prepared in the presence of cholesterol or low concentrations of hydrogen peroxide.

Basement membrane components (collagen IV, entactin and liminin) have been shown to be reactive by producing disassembly of pre-formed Aβ1-42 (Kiuchi *et al.* 2002b). This correlates with the work of Kiuchi *et al.*

(2002a), mentioned above, where they inhibited fibrillogenesis using collagen IV alone. Similarly, the anti-oxidant nordihydroguaiaretic acid can apparently both inhibit Aβ fibrillogenesis and produce breakDown's of pre-formed fibrils (Naiki *et al.* 1998; Ono *et al.*, 2002a). The same group also showed that nicotine prevents Aβ fibrillogenesis and induced fibril breakDown's (Ono *et al.*, 2002b), due specifically to the N-methylpyrrolidine moiety. If this is the case in man, it would seem possible that nicotine could readily be administered to Alzheimer patients orally or transdermally, but addiction is likely and other side effects would need to be carefully monitored. An alternative explanation for the action of nicotine has been advanced by Dickerson and Janda (2003). These workers linked Aβ glycation as the likely mechanism for the neuroprotection by nicotine. That both the D-(+) and L-(-) enantiomers of nicotine interfere with Aβ fibrillogeneis and cytotoxicity was shown by Moore *et al.* (2004). A more disrupted type of Aβ aggregate, shown by negative staining, was formed in the presence of nicotine, rather than well defined fibrils. The reduction of insoluble Aβ deposits in the brain of APPsw transgenic mice by relatively short-term (10 day) administration of nicotine (Hellstrom-Lindhal *et al.*, 2004) implies a direct action of this drug on amyloid fibres and an influence on their physiological clearance.

4. ATOMIC FORCE MICROSCOPY

4.1 Tcchnical background

When the scanning probe microscopies (SPMs), atomic force microscopy (AFM) and scanning tunneling microscopy (STM), were first applied to biological samples the images produced were often difficult to interpret and fraught with artefacts. Today, this situation has largely been overcome. Confidence can be placed in the validity of the images, especially due to correlative studies where both TEM and SPM have been used, as is often the case within the Alzheimer and other fields of study. It is also apparent that the quality of SPM images has improved significantly throughout the past ~10 years, and new modes of imaging have extended the possibilities for the high resolution study of biological structures. For technical detail the reader is referred to the recent review Yang *et al.* (2003). As with the TEM studies on Alzheimer fibrils, SPM studies are often linked to LM and the various biophysical approaches, in particular light scattering, the spectroscopies, analytical ultracentrifugation and SDS-PAGE.

The first SPM contributions to Aβ fibrillogenesis appeared in the mid-1990s. Since then there has been a marked expansion of the available data, particularly from AFM. Although there is considerable overlap and agreement between the data from AFM and TEM, only the former has so far provided useful information on Aβ oligomerization dynamics and the formation of protofilaments; both have provided data on the oligomerization and the structure and polymorphism of mature Aβ fibrils.

Instead of subdividing the specific contributions from AFM and STM in the manner of section **3** above, it is more appropriate to now summarize the available SPM data as a whole. An element of chronological order will be included, with indication of the larger groups that have made major contributions and the smaller groups whose contributions may be no less significant.

4.2 AFM: Aβ oligomerization and fibrillogenesis

One of the first assessments of progressive Aβ1-40 filament assembly came from Shivji *et al.* (1995), using STM with platinum-carbon-coated samples. The AFM was used by Oda *et al.* (1995) to show inhibition of Aβ1-42 fibril formation by clusterin (ApoJ) (*see also* Chapter 14). Smaller, more slowly sedimenting Aβ complexes were produced in the presence of clusterin, which induced greater oxidative stress and neurotoxicity, compared to untreated Aβ fibrils. Elongation of Aβ1-40 protofilaments by addition of monomers (and oligomers) and the lateral association of protofilaments was monitored by Nichols *et al.* (2002) using light scattering and AFM.

In a series of articles Peter Lansbury and colleagues have presented AFM data on comparative fibrillogenesis, with emphasis on the formation and structure of Aβ1-40 protofilaments (Harper *et al.*, 1997a,b: Harper *et al.*, 1999; Koo *et al.*, 1999). More recently, JoAnne McLaurin and colleagues have published extensive AFM data demonstrating the structure of soluble oligomers of Aβ1-40 (Huang *et al.*, 2000) and the interaction of Aβ1-28, Aβ1-40 and Aβ1-42 with lipid bilayers (Yip and McLaurin, 2001, Yip *et al.*, 2002). Their review on modulating factors and Aβ fibrillogenesis (McLaurin *et al.*, 2000) remains a valuable contibution to the subject that presents both TEM and AFM data. The incorporation of both Aβ1-40 and Aβ1-42 into mixed pre-fibrillar aggregates was assessed by fluorescence, CD spectroscopy and AFM (Frost *et al.*, 2003). This approach is important in view of the varying deposition of Aβ1-40 and Aβ1-42 in diffuse and senile plaques.

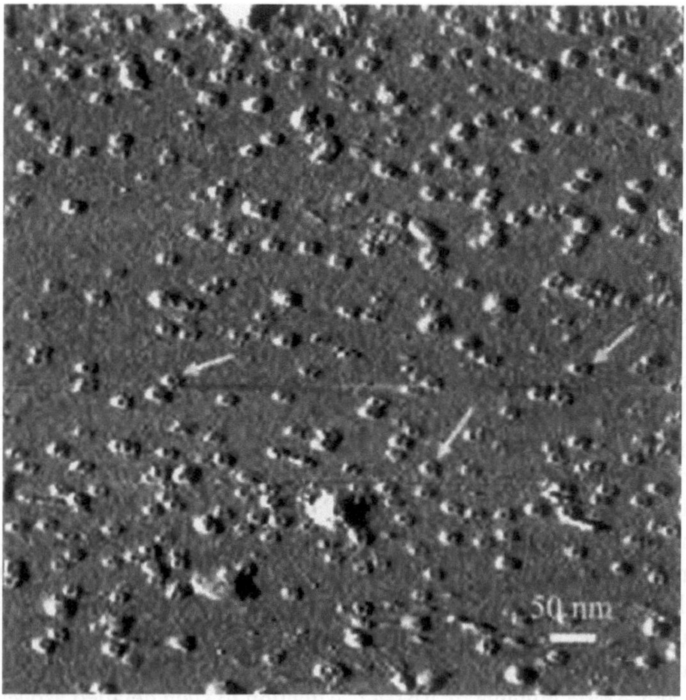

Figure 14. An AFM image Amyloid–β1-42 reconstituted on a planar bilayer membrane. Individual donut-shaped structures protruding from the membrane surface represent individual channels (arrows). Reprinted, with permission, from Lin *et al.*, Amyloid β protein forms ion channels: implications for Alzheimer's disease pathophysiology. *FASEB J.* 15: 2433-2444. Copyright (2001).

By spreading Aβ1-42 for AFM on the surface of mica (hydrophilic) and graphite (hydrophobic) Kowalewsk and Holtzman (1999) showed structural differences in Aβ assembly. On graphite, sheet-like parallel linear aggregates were formed, whereas on mica Aβ globules and short globular protofibril chains were produced. These workers suggested that their observations could have relevance for Aβ interaction with the hydrophobic membrane bilayer and lipoproteins. More recently, the same group (Legleiter *et al.*, 2004), assessed the effect of two sequence-specific anti-Aβ monocolonal antibodies on Aβ fibrillogenesis by AFM. An antibody against the N-terminal aa residues 1-5 slowed fibril formation, whereas an antibody against aa residues 13-28 prevented fibril formation. Liu *et al.* (2004a) using ThT fluorescence and AFM showed that single chaing variabale antibody fragment (scFv) to aas 17-28 of Aβ inhibited both Aβ aggregation and neurotoxicity. Aβ aggregation on hydrophobic graphite was also studied by high resolution AFM and STM by Wang *et al.* (2003), who suggested that the Aβ fibrils had a characteristic right-handed helical twist with an underlying domain texture due to association of protofilaments and monomers.

Lin *et al.* (2001) who showed by AFM that globular oligomers of Aβ1-42 formed in solution at short incubation times, but when the peptide was incorporated into planar lipid bilayers 8-12 nm diameter ring-like ion channels were formed (Figure 14). These workers also correlated their data with cellular studies using cultured neurones and LM. Changes in neurone morphology and viability were assessed following the addition of Aβ1-42, to emphasize the likely significance of the ion channels with respect to production of neuronal death.

AFM has considerable potential for dynamic "real-time" aqueous imaging of Aβ oligomerization and fibril growth, as demonstrated by Goldsbury *et al.* (2001) and Parbhu *et al.* (2002). By TEM, samples can readily be made at increasing time intervals from the start of Aβ incubation, but by AFM the growth of the same individual fibril can be monitored over a period of time (Figure 15).

An impressive combination of AFM with SDS-PAGE, determination of oligomer mass by ultracentrifugation and neuronal viability testing has recently been published by Dahlgren *et al.* (2002) and Stine *et al.* (2003). Particular care was taken by these workers to standardize their oligomerization and fibrillogenesis protocols, with Aβ1-40, Aβ1-42, and the Dutch and Arctic mutants of Aβ1-42. Attention to such detailed aspects is of undoubted importance for the accurate structural definition of the small and larger Aβ oligomeric intermediates and the correlation of oligomer structure and fibrillogenesis with neurotoxicity.

Fibril-free stable neurotoxic globular Aβ1-42 oligomers, produced at 4°C by Chromy *et al.* (2003), were analysed in detail by AFM and non-denaturing gel electropheresis. Furthermore, these workers showed that neuroprotection by *Ginkgo biloba* extracts correlated with inhibition of Aβ oligomer formation. By co-incubation of Aβ1-40 with the fragments Aβ25-35 and Aβ17-40 Liu *et al.* (2004b) concluded from AFM studies that the residues 17-20 and 30-35 play a critical role in the formation of oligomers and protofibrils.

5. CONCLUDING COMMENTS

The diversity of the microscopical data presented above indicates the many exciting aspects of current research into amyloid-β peptide oligomerization and polymerization to form fibrils and Aβ involvement in the pathogenesis of Alzheimer's disease. Nevertheless, some may readily dismiss the contribution of microscopy; this chapter attempts to rectify this

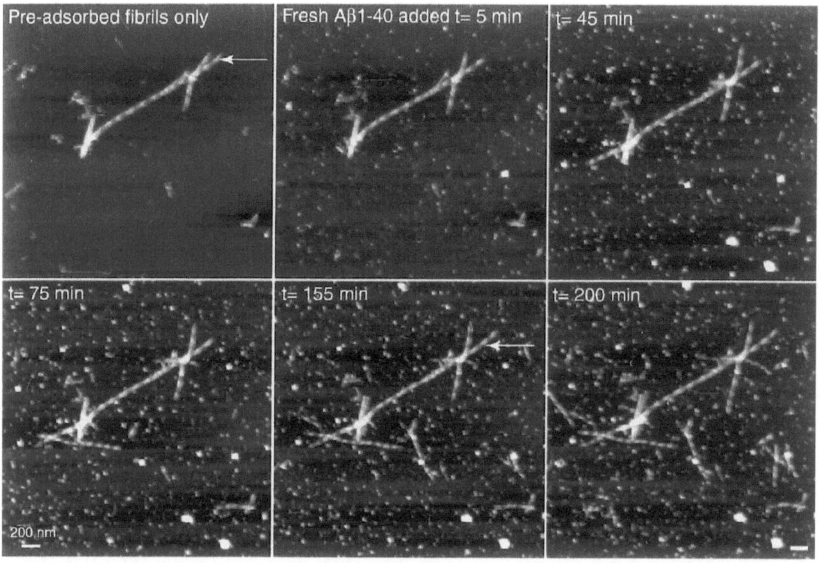

Figure 15. Seeded growth of Aβ1-40 fibrils on a mica surface, imaged by AFM. Reprinted from Goldsbury *et al.*, Visualizing the growth of Alzheimer's Aβ amyloid-like fibrils. *TRENDS in Molec. Med.* 7: 582. Copyright (2001), with permission from Elsevier.

misguided opinion. LM, TEM and AFM all continue to make a major contribution to the Alzheimer field. This is best exemplified by quoting the most recent publications where microscopy makes a major contribution (*e.g.*, Anderson *et al.*, 2004; Li *et al.*, 2003; Miu and Andreescu, 2003; Ribaut-Barassin *et al.*, 2003; Richardson *et al.*, 2003; Schwab *et al.*, 2004). Indeed, the combination of several microscopical techniques, alongside NMR and electron paramagnetic resonance, was reviewed by Antzutkin (2004), for the study of Aβ fibrillogenesis and metal binding by amyloid fibrils.

Although the modern microscopies can be stand-alone techniques, their strength is best shown when the data they produce is integrated with that from cellular, biochemical and biophysical approaches. This important aspect is not neglected above. There is little doubt that the combination of immunology with microscopy, as demonstrated by the use of monospecific polyclonal and monoclonal antibodies as labels for subcellular structures and for molecular-level labeling of Aβ oligomers, protofilaments and fibrils *in vitro*, and the other amyloid plaque-associated proteins, has even more to contribute.

REFERENCES

Abe, H., Kawasaki, K., and Nakanishi, H., 2002, pH-Dependent aggregate forms and conformation of Alzheimer amyloid β-peptide (12-24). *J. Biochem.* **132:** 863-874.

Ali-Khan, Z., 2002, Searching for an *in vivo* site for nascent amyloid fibril formation. *J. Alzh. Dis.* **4:** 105-114.

Anderson, D.H., Tlaga, K.C., Rivest, A.J., Barron, E., Hageman, G.S., and Johnson, L.V., 2004, Characterization of β amyloid assemblies in drusen: the deposts associated with aging and age-related macular degeneration. *Exp. Eye Res.* **78:** 243-256.

Antzutkin, O.N., Leapman, R.D., Balback, J.J., and Tycko, R., 2002, Supramolecular structural constraints on Alzheimer's β-amyloid fibrils from electron microscopy and solid-state nuclear magnetic resonance. *Biochemistry* **41:** 15436-15450.

Antzutkin, O.N., 2004, Amyloidosis of Alzheimer's Abeta peptides: solid-state nuclear magnetic resonance, electron paramagnetic resonance, transmission electron microscopy, scanning transmission electron microscopy and atomic force microscopy studies. *Mag. Reson. Chem.* **42:** 231-246.

Apelt, J., Ach, K., and Schliebs, R., 2003, Aging-related Down's-regulation of neprilysin, a putative β-amylolid-degrading enzyme, in transgenic Tg2576 Alzheimer-like mouse brain is accompanied by an astroglial upregulation in the vicinity of β-amyloid plaques. *Neurosci. Lett.* **339:** 183-186.

Ban, T., Hamada, D.., Hasegawa, K., Naiki, H., and Goto, Y., 2003, Direct observation of amyloid fibril growth monitored by thioflavine T fluorescence. *J. Biol. Chem.* **278:** 16462-16465.

Bergamaschini, L., Rossi, E., Storini, C., Pizzimenti, S., Distaso, M., Perego, C., De Luigi, A., Vergani, C., Grazia De Simoni, M., 2004, Peripheral treatment with enoxaparin, a low molecular weight heparin, reduces plaques and beta-amyloid accumulation in a mouse model of Alzheimer's disease. *J. Neurosci.* **24:** 4181-4186.

Behl, C., 1999, Alzheimer's disease and oxidative stress: Implications for novel therapeutic approaches. *Prog. Neurobiol.* **57:** 301-323.

Behl, C., and Moosmann, B., 2002, Antioxidant protetion in Alzheimer's disease as preventive and therapeutic approach. *Free Rad. Biol. Med.* **33:** 182-191.

Bitan, G., Kirkitadze, M.D., Lomakin, A., Vollers, S.S., Benedek, G.B., and Teplow, D.B., 2003, Amylid β-protein (Aβ) assembly: Aβ40 and Aβ42 oligomerize through distinct pathways. *Proc. Natl. Actd. Sci. USA* **100:** 330-335.

Bogdanovic, N., Bretillon, L., Lund, E.G., Diczfalusy, U., Lannfelt, L., Winblad, B., Fussell, D.W., and Björkhem, I., 2001, On the turnover of brain cholesterol in patients with Alzheimer's disease. Abnormal induction of the cholesterol-catabolic enzyme CYP46 in glial cells. *Neurosci. Lett.* **314:** 45-48.

Bohrmann, B., Adrian, M., Dubochet, J., Kuner, P., Müller, F., Huber, W., Nordstedt, C., and Döblei, H., 2000, Self-assembly of β-amyloid 42 is retarded by small molecular ligands at the stage of structural intermediates. *J. Struct. Biol.* **130:** 232-246.

Bond, J.P., Deverin, S.P., Inouye, H., El-Agnaf, O.M.A., Teeter, M.M., and Kirshner, D.A., 2003, Assemblies of Alzheimer's peptides Aβ25-35 and Aβ31-35: reverse-turn conformation and side-chain interactions revealed by X-ray diffraction. *J. Struct. Biol.* **141:** 156-170.

Boyett, K.W., DiCarlo, G., Jantzen, P.T., Jackson, J., O'Leary, C., Wilcock, D., Morgan, D., and Gordon,M.N., 2003, Increased fibrillar β-amyloid in response to human C1q injections into hippocampus and cortex of APP+PS1 transgenic mice. *Neurochem. Res.* **28:** 83-93.

Brown, J., Thiesler, C., Silberman,S., Magnuson, D., Gottardi-Littell, N., Lee, J.M., Yager, D., Cowley, J., Sambamurti, K., Rahman, M.M., Reiss, A.B., Eckman, C.B., and Wolozin,

B., 2004, Differential expression oof cholesterol hydroxylases in alzheimer's disease. *J. Biol. Chem.* In press.

Burns, M.P., Noble, W.J., Olm, V., Gaynor, K., Casey, E., LaFrancois, J., Wang, L., and Duff, K., 2003, Co-localization of cholesterol, apolipoprotein E and fibrillar Aβ in amyloid plaques. *Molec. Brain Res.* **110:** 119-125.

Chang, P.-T., and Su, Y., 2000, Protocol for quantitative and qualitative analysis of the in vitro aggregation of synthetic β-amyloid: A method applicable to the identification of substances that may have therapeutic efficacy for Alzheimer's disease. *Brain Res. Protocols* **6:** 6-12.

Chromy, B.A., Nowak, R.J., Lambert, M.P., Viola, K.L., Chang, L., Velasco, P., Jones, B.W., Fernandez, S.J., Lacor, P.N., Horowitz, P., Finch, C.E., Krafft, G. A., and Klein, W. L., 2003. Self-assembly of $A\beta_{1-42}$ into globular neurotoxins. *Biochemistry* **42:** 12749-12760.

Cottingham,M.G., Voskuil,J.L., and Vaux,D.J., 2003, The intact human acetylcholiesterase C-terminal oligomerization domain is alpha-helical in situ and in isolation, but a shorter fragment forms beta-sheet-rich amyloid fibrils and protofibrillar oligomers. *Biochemistry* **42:** 10863-10873.

Dahlgren, K.N., Mandelli, A.M., Stine, W.B., Baker, L.K., Kraffit, G.A., and LaDu, M.J., 2002, Oligomeric and fibrillar species of Amyloid-β peptides differentially affect neuronal viability. *J. Biol.Chem.* **277:** 32046-32053.

De Ferrari, G.V., Canales, M.A., Shin, I., Weiner, L.M., Silman, I., and Inestrosa, N.C., 2001, A structural motif of acetylcholinesterase that promotes amyloid β-peptide fibril formation. *Biochemistry* **40**: 10447-10457.

De la Torre, J.C., 2004, Is Alzheimer's disease a neurodegenerative or a vascular disorder? Data, dogma, and dialectics. *The Lancet* ***Neurology*** **3:** 184-190.

Dickerson, T.J., and Janda, K.D., 2003, Glycation of the amyloid B-protein by a nicotine metabolite: A fortuitous chemical dynamic between smoking and Alzheimer's disease.*Proc. Natl. Acad. Sci.USA* **100:** 8182-8187.

Dickson, T.C., and Vickers, J.C., 2001, The morphological phenotype of β-amyloid plaques and associated neuritic changes in Alzheimer's disease. *Neuroscience* **105:** 99-107.

Dong, J., Atwood, C.S., Anderson, V.E., Siedlak, S.L., Smith, M.A., Perry, G., and Carey, P.R., Metal binding and oxidation of amyloid-β within isolated senile plaque cores: Raman microscopic evidence. *Biochemistry* **42:** 2768-2773.

Drake, J., Link, C.D., and Butterfield, D. A., 2003, Oxidative stress precedes fibrillar deposition of Alzheimer's disease amyloid β-peptide (1-42) in a transgenic *Caenorhabditis elegans* model. *Neurobiol. Aging* **24:** 415-420.

Dudal, S., Krzywkowski, P., Paquette, J., Morissette, C., Lacombe, D., Tremblay, P., and Gervais, F., 2004, Inflammation occurs early during the Aβ deposition process in TgCRND8 mice. *Neurobiol. Aging*, In press.

El-Agnaf, M.A., Walsh, D.M., and Allsop, D., 2003, Soluble oligomers for the diagnosis of neurodegenerative diseases. *The Lancet Neurology* **2:** 641-462

Eriksson, E., Janciauskiene, S., and Lannfelt, L, 1995, α_1-Antichymotrypsin regulates Alzheimer β-amyloid peptide fibril formation. *Proc. Natl. Acad.Sci. USA* **92:** 2313-2317.

Fay, D.S., Fluet, A., Johnson, C.J., Link, C.D., 1998, In vivo aggregation of β-amyloid peptide variants. *J. Neurochem.* **71:** 1616-1625.

Fezoui, Y., and Teplow, D.B., 2002, Kinetics of Amyloid β-protein assembly. *J. Biol. Chem.* **277:** 36984-36954.

Fraser, P.E., McLachlan, D.R., Surewicz, W.K., Mizzen, C.A., Snow, A.D., Nguyen, J.T., and Kirschner, D.A., 1994, Conformation and fibrillogenesis of Alzheimer Aβ peptides with selected substitution of charged residues. *J. Mol. Biol.* **244**: 64-73.

Frost, D., Gorman, P.M., Yip, C.M., and Chakrabartty, A., 2003, Co-incorporation of Aβ40 and Aβ42 to form mixed pre-fibrillar aggregates. *Eur. J. Biochem.* **270:** 654-663.

Goldsbury, C., Aebi, U., and Frey, P., 2001, Visualizing the growth of Alzheimer's Aβ amyloid-like fibrils. *TRENDS in Molec. Med.* **7:** 582.

Goldsbury, C.S., Wirtz, S., Müller, S.A., Sunderji, S., Wicki, P., Aebi, U., and Frey, P., 2000, Studies on the *in vitro* assembly of Aβ 1-40: Implications for the search for Aβ fibril formation inhibitors. *J. Struct. Biol.* **130:** 217-231.

Goldstein, L.E., Muffat, J. A., Cherny, R. A., Moir, R.D., Ericsson, M.H., Huang, X., Marvros, C., Coccia, J.A., Faget, K.Y., Fitch, K.A., Masters, C.L., Tanzi, R.E., Chylack, L.E., and Bush, A. I., 2003, Cytosolic β-amyloid deposition and supranuclear cataracts in lenses from people with Alzheimer's disease. *Lancet* **361:** 1258-1265.

Gorman, P.M., Chakrabartty, A., 2002, Alzheimer β-amyloid peptides: Structures of amyloid fibrils and alternate aggregation products. *Biopolymers (Peptide Sci.)* **60:** 381-394.

Hamazaki, H., 1995, Ca^{2+}-dependent binding of human serum amyloid P component to Alzheimer β-amyloid peptide. *J. Biol. Chem.* **270:** 10392-10394.

Harper, J.D., Lieber, C.M., and Lansbury, P.T., 1997a, Atomic force microscopic imaging of seeded fibril formation and fibril branching by the Alzheimer's disease amyloid-beta peptide. *Chem. Biol.* **4:** 951-959.

Harper, J.D., Wong, S.S., Lieber, C.M., and Lansbury, P.T., 1997b, Observation of metastable A beta amyloid protofibrils by atomic force microscopy. *Chem. Biol.* **4**: 119-125.

Harper, J.D., Wong, S.S., Lieber, C.M., and Lansbury, P.T., 1999, Assembly of Aβ amyloid protofibrils: An in vitro model for a possible early event in Alzheimer's disease. *Biochemistry* **38:** 8972-8980.

Harris, J.R., 1997, Negative staining and cryoelectron microscopy. RMS Microscopy Handbook No. 35, Bios Scienctific Publishers, Oxford.

Harris, J.R., and Scheffler, D., 2002, Routine preparation of air-dried negatively stained and unstained specimens on holey carbon support films: a review of applications. *Micron* **33:** 461-480.

Harris, J.R., 2002, In vitro fibrillogenesis of the amyloid $\beta_{1\text{-}42}$ peptide: cholesterol potentiation and aspirin inhibition. *Micron* **33:** 609-626.

Härtig, W., Paulke, B.-R., Varga, C., Seeger, J., Harkany, T., and Kacza, J., 2003, Electron microscopic analysis of nanoparticles delivering thioflavine-T after intrahippocampal injection in mouse: implications for targeting β-amyloid in Alzheimer's disease. *Neurosci. Lett.* **338:** 174-176.

Head, E., Azizeh, B.Y., Lott, I.T., Tenner,A. J., Cotman, C.W., and Cribbs, D. H., 2001, Complement association with neurons and β-amyloid deposition in the brains of aged individuals with Down's syndrome. *Neurobiol. Disease* **8:** 252-265.

Heckl, S., Pipkorn, R., Nagele, T., Vogel, U., Kuker, W., Voight, K., 2004, Molecular imaging: Bridging the gap between neuroradiology and neurohistology. *Histol. Histopathol.* **19:** 651-668.

Hellstrom-Lindahl, E., Court, J., Keverne, J., Svedberg, M., Lee, M., Marutle, A., Thomas, A., Perry, E., Bednar, I., and Nordberg, A., 2004, Nicotine reduces Abeta in the brain and cerebral vessels of APPsw mice. *Eur. J. Neurosci.* **19:** 2703-2710.

Hirayama, A., Horikoshi, Y., Maeda, M., Ito, M., and Takashima,. S., 2003, Characteristic developmental expression of amyloid β40, 42, and 43 in patients with Down's syndrome. *Brain Devel.* **25:** 180-185.

Holm Nielsein, E., Nybo, M., Junker, K., Toftedal Hansen, P., Rasmussen, I.M., and Svehag, S.-E., 2002, Localization of human serum amyloid P component and heparan sulfate proteoglycan in *in vitro*-formed Aβ fibrils. *Scand. J. Immunol.* **52:** 110-112.

Hu, L., Wong, T. P., Côté, S. L., Bekk, K.F. S., and Cuello, A.C., 2003, The impact of Aβ-plaues on cortical cgholinergic and non-cholinergic presynaptic boutons in alzheimer's disease-like transgenic mice. *Neruoscience* **121:** 421-432.

Huang, T.H.J., Yang, D.-S., Plaskos, N.P., Go, S., Yip, C.M., Fraser, P.E., and Chakrabartty, A., 2000, Structural studies of soluble oligomers of the Alzheimer β-amyloid peptide. *J.Mol. Biol.* **297:** 73-87.

Iijima, K., Liu, H.-P., Chiang, A.-S., Hearn, S.A., Konsolaki, M., and Zhong, Y., 2004, Dissecting the pathological effects of human Aβ40 and Aβ42 in *Drosophila*: A potential model for Alzheimer's disease. *Proc. Natl. Acad. Sci. USA* **101:** 6623-6628..

Inestrosa, N.C., Alvarez, A., Pérez, C.A., Moreno, R.D., Vicente,M., Linker, C., Casanueva, O.I., Soto, C., and Garrido, J., 1996, Acetylcholinesterase accelerates assembly of amyloid-β-peptides in Alzheimer's fibrils: Possible role of the peripheral site of the enzyme. *Neuron* **16:** 881-891.

Iwata, N., Takaki, Y., Fukami, S., Tsubuki, S., and Saido, T.C., 2002, Region-specific reduction of Aβ-degrading endopeptidase, neprilysin, in mouse hippocampus upon aging. *J. Neurosc. Res.* **70:** 493-500.

Kagan, B.L., Hirakura, Y., Azimov, R., Azimova, R., and Lin, M.-C., 2002, The channel hypothesis of Alzhemier's disease: current status. *Peptides* **23***:* 1311-1315.

Kakio, A., Nishimoto, S., Kozutsumi, Y., and Matsuzaki, K., 2003, Formation of a membrane-active form of β-protein in raft-like model membranes. *Biochem. Biophys. Res. Commun.* **303:** 514-518.

Kanski, J., Aksenova, M., and Butterfield, D. A., 2002, The hydrophobic environment of Met35 of Alzheimer's Aβ(1-42) is important for the neurotoxic and oxidative properties of the peptide. *Neurotoxicity Res.* **4:** 219-223.

Kanski, J., Varadarajan, S., Aksenova, M., and Butterfield, D.A., 2001. Role of glycine-33 in Alzheimer's amyloid β-peptide 1-42-associated oxidative stress and neurotoxicity. *Biochim. Biophys. Acta* **1586:** 190-198.

Kayend, R., Head, E., Thompson J.L., McIntire, T.M., Milton, S.C., Cotman, C.W., and Glabe, C.G., 2003, Common structure of soluble amyloid oligomers implies common mechanism of pathogenesis. *Science* **300:** 486-489.

Kawahara, M., and Kuroda, Y., 2000, Molecular mechanism of neurodegeneration induced by Alzheimer's β-amyloid proten: Channel formation and disruption of calcium homeostasis. *Brain Res. Bull.* **53:** 389-397.

Kim, H.-C., Yamada, K., Nitta, A., Olariu, A., Tran, M.H., Mizuno, M., Nakajima, A., *et al.*, 2003, Immunocytochemical evidence that amyloid β (1-42) impairs endogenous antioxidant systems *in vivo*. *Neuroscience* **119:** 399-419.

Kirkitadze, M.D., Condron, M.M., and Teplow, D.B., 2001. Identification and characterization of key kinetic intermediates in amyloid β-protein fibrillogenesis. *J. Mol. Biol.* **312:** 1103-1119.

Kiuchi, Y., Isobe, Y., Fukushima, K., 2002a, Type IV collagen prevents amyloid β-protein fibriul formation. *Life Sciences* **70:** 1555-1564.

Kiuchi, Y., Isobe, Y., Fuklushima, K., and Kimura, M., 2002b, Disassembly of amyloid β-protein fibril by basement membrane components. *Life Sciences* **70:** 2421-2431.

Klunk, W.E., Bacskai, B.J., Mathis, C.A., Kajdasz, S.T., McLellan, M.E., Frosch, M.P., Debnath, M.L., Holt, D.P., Wang, Y., and Hyman, B.T., 2002, Imaging Aβ plaques in living transgenic mice with multiphoton microscopy and methoxy-X04, a systemically administered Congo red derivative. *J. Neuropathol. Exper. Neurol.* **61:** 797-805.

Klunk, W.E., Engler, H., Nordberg, A., Bacskai, B.J., Wang, Y., Price, J.C., Bergstrom, M., Hyman, B.T., Langstrom, B., and Mathis, C.A., 2003, Imaging the pathology of

Alzheimer's disease: amyloid-imaging with positron emission tomography. *Neuroimaging Clin. N. Am.* **13:** 781-789.

Koo, E.H., Lansbury, P.T., and Kelly, J.W., 1999, Amyloid diseases: Abnormal protein aggregation in neurodegeneration. *Proc.Natl. Acad.Sci. USA* **96:** 9989-9990.

Kowalewski, T., and Holtzman, D.M., 1999, *In situ* atomic force microscopy study of Alzheimer's β-amyloid peptide on different substrates: New insights into mechaism of β-sheet formation. *Proc. Natl. Acad.Sci. USA* **96:** 3688-3693.

Kuo, Y.-M., Kokjohn, T.A., Beach, T.G., Sue, L.I., Brune, D., Lopez, J.C., Kalback, W.M., Abramowski,D., Sturchler-Pierrat, C., Staufenbiel, M., and Roher, A.E., 2001, Comparative analysis of Amyloid-β chemical structure and amyloid plaque morphology of transgenic mouse and Alzheimer's disease brains. *J. Biol.Chem.* **276:** 12991-12998.

Kung, M.-P., Hou, C., Zhuang, Z.-P., Zhang, B., Skovronsky, D., Trojanowski, J.Q., Lee, V. M.-Y., and Kung, H.F., 2002, IMPY: an improved thoflavin-T derivative for in vivo labeling of β-amyloid plaques, *Brain Res.* **956:** 202-210.

Kung, H.F., Kung, M.-P., Zhuang, Z.P., Hou, C., Lee, C.-W., Plössl, K., Zhuang, B., Skovronsky, D.M., Lee, V.M.-Y., and Trojanowski, J.Q., 2003, Iodinated tracers for imaging amyloid plaques in the brain. *Molec. Imag. Biol.* **5:** 418-426.

Lambert, M.P., Barlow, A.K., Chromy, B.A., Edwards, C., Freed, R., Liosatos, M., Morgan, T. E., Rozovisky, I., Trommer, B., Viola, K.L., Wals, P., Zhang, C., Finch, C.E., Frafft, G. A., and Klein, W. L., 1998, Diffusible, nonfibrillar ligands derived from $A\beta_{1-42}$ are potent central nervous system neurotoxins. *Proc. Natl. Acad.Sci. USA* **95:** 6448-6453.

Lashuel H.A., Hartley, D., Petre, B.M., Walz, T., and Lansbury, P.T., 2002a, Amyloid pores from pathogenic mutations. *Nature* **418:** 291.

Lashuel, H.A., Hartley, D.M., Balakhaneh, D., Aggarwal, A., Teichberg, S., and Callaways, D.J.E., 2002b, New class of inhibitors of amyloid-β fibril formation. *J. Biol. Chem.* **277:** 42881-42890.

Lashuel, H.A., Hartley, D. M., Petre, B. M., Wall,. J. S., Simon, M. N., Walz, T., and Lansbury, P. T., 2003, Mixtures of wild-type and pathogenic (E22G) form of Aβ40 *in vitro* accumulate protofibrils, including amyloid pores. *J. Mol. Biol.* **332:** 795-808.

Lashuel , H.A., LaBrenz, S.R., Woo, L., Serpell, L.C., and Kelly, J.W., 2000, Protofilaments, filaments, ribbons, and fibrils from peptidomimetic self-assembly: Implications for amyloid fibril formation and mateials science. *J. Am. Chem.Soc.* **122:** 5262-5277.

Lee, C.-W., Kung, M.-P., Hou, C., and Kung, H.F., 2003, Dimethylamino-fluorenes: ligands for detecting β-amyloid plaques in the brain. *Nuc. Med. Biol.* **30:** 573-580.

Legleiter, J., Czilli, D.L., Gitter, B., MeMattos, R.B., Holtzman, D.M., and Kowalewski, T., 2004, Effect of different anti-Aβ antibodies on Aβ fibrillogenesis as assessed by atomic force microscopy. *J. Mol. Biol.* **335:** 997-1006.

Lehman, E. J., Kulnane, L. S., and Lamb, B. T., 2003, Alterations in beyt-amyloid production and deposition in brain regions of two transgenic models. *Neurobiol. Aging* **24:** 645-653.

Li., L., Cao, D., Garber, D.W., Kim, H., and Fukuchi, K., 2003, Association of aortic atherosclerosis with cerebral β-amyloidosis and learning deficits in a mouse model of Alzheimer's disease. *Amer. J. Pathol.* **163:** 2155-2164.

Lin, H., Bhatia, R., and Lal, R., 2001, Amyloid β protein forms ion channels: implications for Alzheimer's disease pathophysiology. *FASEB J.* **15:** 2433-2444.

Link, C.D., Johnson, C.J., Fonte, V., Paupard, M.-C., Hall, D. H., Styren, S., Mathis, C. A., and Klunk, W.E., 2001, Visualization of fibrillar amyloid deposits in living, transgenic *Caenorhabditis elegans* animals using the sensitive amyloid dye, X-34. *Neurobiol Aging* **22:** 217-226.

Liu, R., Yuan, B., Emadi, S., Zameer, A., Schulz, P., McAllister, C., Lyubchenko, Y., Gound, G., and Sierks, M.R., 2004, Single chain variable fragments against B-amyloid (AB) can inhibit AB aggregation and prevent AB-induced neurotoxicity. *Biochemistry* **42:** 6959-6967.

Liu, R., McAllister, C., Lyubchencko, Y., and Sierks, M.R., 2004b, Residues 17-20 and 30-35 of beta-amyloid play critical roles in aggregation. *J. Neurosci. Res.* **75:** 162-171.

Lorenzo, A., and Yankner, B.A., 1994, β-Amyloid neurotoxicity requires fibril formation and is inhibited by Congo red. *Proc. Natl. Acad. Sci. USA* **91:** 12234-12247.

Lue, L.-F., Walker, D.G., and Rogers, J., 2001, Modeling microglial activation in Alzheimer's disease with human postmortem microglial cultures. *Neurobiol. Aging* **22:** 945-956.

Lustbader, J.W., Maurizio, C., Lin, C., Xu, H.W., Takuma, K., Wang, N., Caspersen, C., *et al.*, 2004, ABAD directlylinks Aβ to mitochondrial toxicity in Alzheimer's disease. *Science* **304:** 448-452.

Luth, H.J, Apelt, J., Ihunwo, A.O., Arendt, T., and Schliebs, R., 2003, Degeneration of beta-amyloid-associated cholinergic structures in transgenic APP(SW) mice. *Brain Res.* **977:** 16-22.

Malinchik, S.B., Inouye, H., Szumowski, K.E., and Kirschner, D.A., 1998, Structural analysis of Alzheimer's β(1-40) amyloid: Protofilament assembly of tubular fibrils. *Biophys. J.* **74:** 537-545.

Marr, R.A., Rockstein, E., Mukherjee, A., Kindy,M.S., Hersh, L.B., Gage, F.H., Verma, I.M., and Masliah, E., 2003, Neprilysin gene transfer reduced human amyloid pathology in transgenic mice. *J. Neurosci.* **23:** 1992-1996.

Matsuoka, Y., Picciano, M., La Francois, J., and Duff, K., 2001, Fibrillar β-amyloid evokes oxidative damage in a transgenic mouse model of Alzheimer's disease. *Neuroscience* **104:** 609-613.

Masdeu, J., 2004, Neuroimaging in Alzheimer's disease: an overview. *Rev. Neurol.***38:** 1156-1165.

Mathis, C.A., Wang, Y., and Klunk, W.E., 2004, Imaging beta-amyloid plaques and neurofibrillary tangles in the aging human brain. *Curr. Pharm. Des.* **10:** 1469-1492.

McGeer, E.G., Yasojima, K., Schwab, C., and McGeer, P.L., The pentraxins: possible role in Alzheimer's disease and other innate inflammatory diseases. *Neurobiol. Aging* **22:** 843-848.

McLaurin, J., Yang, D.-S., Yip, C.M., and Fraser, P.E., 2000, Review: Modulating factors in amyloid-β fibril formation. *J.Struct. Biol.* **130:** 259-270.

McLellan, M.E., Kajdasz, S.T., Hyman, B.T., and Bacskai, B.J., 2003, In vivo imaging of reactive oxygen species specifically associated with thioflavinee S-positive amyloid plaques by multiphoton microscopy. *J. Neurosci.* **23:** 2212-2217.

Metsaars, W.P., Hauw, J.-J., van Welsem M.E., and Duyckaerts, C. , 2003, A grading system on Alzheimer disease lesions in neocortical areas. *Neurobiol. Ageing* **24:** 563-572.

Milton, N.G.N., 1999, Amyloid-β binds catalase with high affinity and inhibits hydrogen peroxide breakDown's. *Biochem J.* **344:** 293-296.

Miu, A.C., and Andrreescu, C.A., A behavioural and histological study of the effects of long-term exposure of adult rats to aluminium. *Intern. J. Neuroscience* **113:** 1197-1211.

Mochizuki, A., Peterson, J.W., Mufsuon, E. J., and Trapp, B.D., 1996, Amyloid load and neural elements in Alzhemier's disease and nondemented individuals with high amyloid plaque density. *Exper. Neurol.* **142:** 89-102.

Moore, S.A., Huckerby, T.N., Gibson, G.L., Fullwood, N.J., Turnbull, S., Tabner, B.J., El-Agnaf, O.M.A., and Allsop, D., 2004, Both the D-(+) and L-(-)enatiomers of nicotine inhibit Aβ aggregation and cytotoxicity. *Biochemistry* **43:** 819-826.

Mori, T., Paris, D., Rojiani A.M., Sparks, D.L., Delledonne, A., Crawford, F., *et al.*, Cholesterol accumulates in senile plaques of Alzheimer disease patients and in transgenic APP(SW) mice. *J. Neuropathol. Exp. Neurol.* **60:** 778-785.

Naiki, H., Hasegawa, K., Yamaguchi, I., Nakamura, H., Gejyo, F., and Nakakuki, K., 1998, Apolipoprotein E and antioxidants have different mechanisms of inhibiting Alzheimer's β-amyloid fibril formation *in vitro*. *Biochemistry* **37:** 17882-17889.

Navarro, A., del Valle, E., Astudillo, A., González del Rey, C., and Tolivia, J., 2003, Immunohistochemical study of the distribution of apolipoproteins E and D in human cerebral β amyloid deposits. *Exp. Neurol.* **184:** 697-704.

Negele, R.G., D'Andrea, M.R., Lee, H., Venkataraman, V., and Wang, H.-Y., 2003, Astrocytes accumulate Aβ42 and give rise to astrocytic amyloid plaques in Alzheimer's disease brains. *Brain Res.* **97:** 197-209.

Nichols, M.R., Moss, M.A., Reed, D.K., Lin, W.-L., Mukhopadhyay, R., Hoh, J.H., and Rosenberry, T.L., 2002, Growth of β-amyloid(1-40) protofibrils by monomer elongation and lateral association. Characterization of distinct products by light scattering and atomic force microscopy. *Biochemistry* **41:** 6115-6127.

Nybo, M., Svehag, S.-E., Nielsen, E.H., 1999, An ultrastructural study of amyloid intermediates in $A\beta_{1-42}$ fibrillogenesis. *Scand. J. Immunol.***49:** 219-223.

Qahwash, I., Weiland, K.L., Lu, Y., Sarver, R.W., Kletzien, R.F., NS Yan, R., 2003, Identification of a mutant amyloid peptide that predominantly forms neurotoxic protofibrillar aggregates. *J. Biol. Chem.* **278:** 23187-23195.

Oda, T., Wals, P., Osterburg, H.H., Johnson, S.A., Pasinetti, G.M., Morgan, T.E., Rozovsky, I., Stine, W.B., Snyder, S.W., Holzman, T.F., Krafft, G.A., and Finch, C.E., 1995, Clusterin (apoJ) alters the aggregation of amyloid β-peptide ($A\beta_{1-42}$) and forms slowly sedimenting Aβ complexes that cause oxidative stress. *Exp. Neurol.* **136:** 22-31.

Oddo, S., Caccamo, A., Shepherd, J.D., Murphy, M.P., Golde, T.E., Kayed, R., Metherate, R., Mattson, M.P., Akbari, Y., and LaFerla, F.M., 2003a, Triple-transgenic model of Alzheimer's disease with plaques and tangles. Intracellular A-beta and synaptic dysfunction. *Neuron* **39:** 409-421.

Oddo, S, Caccamo, A., Kitazawa, M., Tseng, B.P., and La Ferla, F. M., 2003b, Amyloid deposition precedes tangle formation in a triple transgenic model of Alzhemier's disease. *Neurobiol. Ageing* **24:** 1063-1070.

Ong, W.-Y., and Halliwell, B., 2004, Iron, Atherosclerosis and Neurodegeneration. *Ann. N.Y. Acad. Scil.* **1012:** 51-64.

Ono, K., Hasegawa, K., Yoshiike, Y., Takashima, A., Yamada, M., and Naiki, H., 2002a, Nordihydroguaiaretic acid potently breaks Down's pre-formed Alzheimer's β-amyloid fibrils *in vitro*. *J. Neurochem.* **81:** 434-440.

Ono, K., Hasegawa, K., Yamada, M., and Naiki, H., 2002b, Nicotine breaks Down's preformed Alzheimer's β-amyloid fibrils in vitro. *Biol. Psych.* **52:**880-886.

Otzen, D.E., and Oliveberg, M., 2004, Transient formation of nano-crystalline structures during fibrillation of an Aβ-like peptide. *Protein Science* **13:** 1417-1421.

Oyewole, D., Mann, D.M.A., and Royston, M.C., 1999, Apolipoprotein E epsilon 4 genotype and the morphology of amyloid plaques in Alzheimer's disease. *Alzheimers Rep.***2:** 267-270.

Pappolla, M., Bonzer, P., Soto, C., Shao, H., Robakis, N.K., Zagorski, M., Fragione, B., and Ghiso, J., 1998, Inhibition of Alzheimer β-fibrillogenesis by melatonin. *J. Biol.Chem.* **278:** 7185-7188

Pappolla, M.A., Bryant-Thomas, T.K., Herbert, D., Pacheco, J., Fabra Garcia,M., Manjon,M., Girones, X., Hentry, T.L., Matsubara, E., Zambon, D., Wolozir, B., Sano,M., Cruz-Sanchez, F.F., Thal, L.J., Petanceska, S.S., and Refolo, L.M., 2003, Mild hypercholesterolemia is an early risk factor for the development of Alzheimer amyloid pathology. *Nerurology* **61:** 199-205.

Park, I.-H., Hwang, E. M., Hong, H.S., Boo, J. H., Oh, S. S., Lee, J., Jung, M. W., Bang, O. Y., Kim, S. U., and Mook-Jung, I., 2003, Lovastatin enhances Aβ production and senile plaque deposition in female Tg2576 mice. *Neurobiol. Aging* **24:** 637-643.

Pepys, M.B., Herbert, J., Hutchinson, W.L., Tennent, G.A., Lachmann, H.J., *et al.*, 2002, Targeted pharmacological depletion of serum amyloid P component for treatment of human amyloidosis. *Nature* **417:** 254-259.

Petrella, J.R., Coleman, R.E., Doraiswamy, P.M., 2003, Neuroimaging and early diagnosis of Alzheimer disease: A look to the future. *Radiology* **266:** 315-336.

Piazzi, L., Rampa, A., Bisi, A., Gobbi, S., Belluti, F., Cavalli, A., Bartolini, M., Andrisano, V., Valenti, P., and Recanatini, M., 2003, 3-(4-{[benzyl(methyl)amino]methyl}-phenyl)-6,7-dimethoxy-2H-2-chromenmone (AP2238) inhibits both acetylcholinesterase and acetylcholinesterase-induced B-amyloid aggregation: A dual function lead for Alzheimer's disease therapy. *J.Med.Chem.* **46:** 2279-2282.

Poirier, J., 2003, Apolipoprotein E and cholesterol metabolism in the pathogenesis and treatment of Alzheimer's disease. *TRENDS Molec. Med.* **9:** 94-101.

Parbhu, A., Lin, H., Thimm, J., and Lal, R., 2002, Imaging real-time aggregation of amyloid beta protein (1-42) by atomic force microscopy. *Peptides* **23: 1265-1270.**

Rees, T., Hammond, P.I., Soreq, H., Younkin, S., and Brimijoin, S., 2003, Acetylcholinesterase promotes beta-amyloid plaques in cerebral cortex. *Neurobiol. Aging* **24:** 777-787.

Reilly, J. F., Games, D., Rydel, R. E., Freedman, S., Schenk, D., Young, W. G., Morrison, J. H., and Bloom, F.E., 2003, Amyloid deposition in the hippocampus and entorhinal cortex: Quantitative analysis of a transgenic mouse model. *Proc. Natl. Acad. Sci. USA* **100:** 4837-4842.

Ribaut-Barassin, C., Dupont, J.L., Haeberle, A.M., Bombared,G., Huber, G., Moussaoui, S., Mariani, J., and Bailly, Y., 2003, Alzheimer's disease proteins in cerebellar and hippocampal synapses during postnatal development and aging of the rat. *Neuroscience* **120:** 405-423.

Richards, J.G., Higgins, G.A., Ouagazzal, A.-M., Ozmen, L., Kew, J.N.C., Bohrmann, B., Malherbe, P., Loetscher, H., Czech, C., Huber, G., Bluethmann, H., Jacobsen, H., and Kemp, J.A., 2003, PS2APP transgenic mice, coepressing hPS2mut and hAPPswe, show age-related cognitive deficits associated with discrete brain amyloid deposition and inflammation. *J. Neuarosci.* **23:** 8989-9003.

Richardson, J.C., Kendal, C.E., Anderson, R., Priest, F., Gower, E., Soden, P., Gray, R., Topps, S., Howlett, D.R., Lavender, D., Clarke, N.J., Barnes, J.C., Haworth, R., Stewart, M.G., and Rupniak, H.T.R., 2003. Ultrastructural and behavioural changes precede amyloid deposition in a transgenic model of Alzheimer's disease. *Neuroscience* **122:** 213-228.

Romano, A., Serafino, A., Krasnowska, E., Ciotti, M.T., Calissano, P., Ruberti, F., and Calli, C., 2003, Neuronal fibrillogenesis: amyloid fibrils from primary neuronal cultures impair long-term membory in the crab *Chasmagnathus.Behav. Brain Res.* **147:** 73-82.

Sasaki, A., Shoji, M., Harigaya, Y., Kawarabayashi, T., Ikeda, M., Naito, M., Matsubara, E., Abe, K., and Nakazato, Y., 2002, Amyloid cored plaques in Tg2576 transgenic mice are characterized by giant plaques, slightly activated microglia, and the lack of paired helical filament-typed, dystrophic neurites. *Virchows Arch.* **441:** 358-367.

Saito, T., Kijima, H., Kiuchi, Y., Isobe, Y., and Fukushima, K., 2001. β-amyloid induced caspase-dependent early neurotoxic change in PC12 cells: correlation with H_2O_2 neurotoxicity. *Neurosc. Lett.* **305:** 61-64.

Schwab., C., Hosokawa, M., and McGeer, P.L., 2004, Transgenic mice overexpressing amyloid beta protein are an incomplete model of Alzheimer disease. *Expetl. Neurol.* **188:** 52-64.

Shimizu, T., Fukada, H., Muryama, S., Izumiyama, N., and Shirasawa, T., 2002, Isoaspartate formation at position 23 of amyloid beta peptide enhanced fibril formation and deposited onto senile plaques and vascular amyloids in Alzheimer's disease. *J.Neurosci. Res.* **70:** 451-461.

Sivjhi, A.P., Brown, F., Davies, M.C., Jennings, K.H., Roberts, C.J., Tendler, S.J.B., Wilkinson, M.J., Williams, P.M., 1995, Scanning tunnelling microscopy studies of β-amyloid fibril structure and assembly. *FEBS Lett.* **371:** 25-28.

Soto, C., Castaño, E.M., Kumar, R.A., Beavis, R.C., and Fragione, B., 1995. Fibrillogenesis of synthetic amyloid-β peptides is dependent on their initial secondary structure, *Neurosci. Lett.* **200:** 105-108.

Sparks, D. L., 1996, Intraneuronal β-amyloid immunoreactivity in the CNS. *Neurobiol. Aging* **17:** 291-299.

Sparks, D. L., and Schreurs, B.G., 2003. Trance amounts of copper in water induce β-amyloid plaques and learning deficits in a rabbit model of Alzheimer's disease. *Proc. Natl. Acad. Sci. USA.* **100:** 11065-11069.

Stalder, M., Deller, T., Staufenbiel, M., and Jucker, M., 2001, 3D-Reconstruction of microglia and amyloid in APP23 transgenic mice: no evidence of intracellular amuyloid. *Neurobiol. Age.* **22:** 427-434.

Stine, W.B.,Dahlgren, K.N., Krafft, G.A., and LaDu, M.J., 2003, In vitro characterization of conditions for amyloid-B peptide oligomerization and fibrillogenesis. *J.Biol. Chem.* **278:** 11612-11622.

Su, Y., and Chang, P.-T., 2001, Acidic pH promotes the formation of toxic fibrils from β-amyloid peptide. *Brain Res.* **893:** 287-291.

Takahashi, R.H., Milner, T.A., Li, F., Nam, E.E., Edgar, M.A., Yamaguchi, H., Beal, M.F., Xu, H., Greengard, P., and Gouras, G.K., 2002, Intraneuronal Alzheimer Aβ42 accumulates in multivesicular bodies and is associated with synaptic pathology. *Am. J. Pathol.* **161:** 1869-1879.

Takahashi, R.H., Almeida, C.G., Kearney, P.F., Yu, F., Lin, M.T., Milner, T.A., and Gouras, G.K., 2004. Oligomerization of Alzheimer's beta-amyloid within processes and synapses of cultured neurones and brain.

Takata, K., Kitamura, Y., Tsuchiya, D., Kawasaki, T., Tanigushi, T., and Shimohama, S., 2003, Heat shock protein-90-induced microglial clearance of exogenous amyloid-$\beta_{1\text{-}42}$ in rat hippocampus in vivo. *Neurosci. Lett.* **344:** 87-90.

Tamaoka, A., Sawamura, N., OsK,., Auzuki, N., Mizusawa, H., Ahoji, S., and Mori, H., 1995, Amylid β protein 1-42/43 (Aβ 1-42/43) in cerebellar diffuse plaques: enzyme-linked immunosorbent assay and immunocytochemical study. *Brain Res.* **679:** 151-156.

Tjernberg, L.O., T Tjernberg, A.T., Bark, N., Ahi, Y., Ruzsicska, B.P., Bu, Z., Thyberg, J., and Callaways, D.J.E., 2002, Assembling amyloid fibrils from designed structures containing a significant amyloid β-peptide fragment. *Biochem. J.* **366:** 343-351.

Ueda, T., Nagata, M., Monji, A., Yoshida, I., Tashiro, N., and Imoto, T., 2002, Effect of sucrose on formation of the β-amyloid fibrils and D-aspartic acids in Aβ1-42. *Biol. Pharm. Bull.* **25:** 375-378.

Van Horssen, J., Wesseling, P., van den Heuvel, L.P.W.J., de Waal, R.M.W., and Verbeek, M.M., 2003, Heparan sulphate proteoglycans in Alzheimer's disease and amyloid-related disorders. *The Lancet Neurology* **2:** 482-492.

Vattemi, G., King Engel, W., McFerrin, J., and Askanas, V., 2003, Cystatin C colocalizes with amyloid-beta and coimmunoprecipitates with amyloid-beta precursor protein in sporadic inclusion-body myositis muscles. *J. Neurochem.* **85:** 1539-1546.

Verbeek, M. M., Otte-Höller, I., Fransen, J.A.M., and de Waal, R.M.W., 2002, Accumulation of the amyloid-β precursor protein in multivesicular body-like organelles. *J. Histochem. Cytochem.* **50:** 681-690.

Veerhuis, R., van Breemen, M.J., Hoozemans, J.J.M., Morbin, M., Ouladhadj, J., Tagliavini, F., and Eikelenboom, P., 2003, Amyloid β plaque-associated proteins C1q and SAP enhance the $A\beta_{1-42}$ peptide-induced cytokine secretion by adult human microglia in vitro. *Acta Neuropathol.* **105:** 135-144.

Wadghiri, Y.Z., Sigurdsson, E.M., Sadowski, M., Elliott, J.I., Li, Y., Scholtzova, H., Tang, C.Y., Aguinaldo, G., Pappolla, M., Duff,K., Wisniewski, T., and Turnbull, D.H., 2003, Detection of Alzheimer's amyloid in transgenic mice using magnetic resonance microimaging. *Magn. Reson. Med.* **50:** 293-302.

Walker, L.C., Bian, F., Callahan, M.J., Lipinski, W.J., Durham, R.A., and LeVine, H., 2002, Modelling Alzheimer's disease and other proteopathies in vivo: Is seeding the Key? *Amino acids* **23:** 87-93.

Walsh, D.M., Hartley, D.M., Kusumoto, Y., Fezoui, Y., Condron, M.M., Lomakin, A., Benedek, G.B., Selkoe, D.J., and Teplow, D.B., 1999, Amyloid β-protein fibrillogenesis: Structure and biological activity of protofibrillar intermediates. *J. Biol. Chem.* **274:** 25945-25952.

Walsh, D.M., Lomakin, A., Benedek, G.B., Condron, M.M., and , D.B., 1997. Amyloid β-protein fibrillogenesis: detection of a protofibrillar intermediate. *J. Biol. Chem.* **272:** 22364-22372.

Wang, H.-Y., D'Andrea, M.R., and Nagele, R.G., 2002, Cerebellar diffuse amyloid plaques are derived from dendritic Aβ42 accumulations in Purkinje cells. *Neurobiol. Ageing* **23:** 213-223.

Wang, Z., Zhou, C., Wang, C., Wan, L., Fang, X., and Bai, C., 2003, AFM and STM study of β-amyloid aggregation on graphite. *Ultramicroscopy* **97:** 73-79.

Webster, S., Glabe, C., and Rogers, J., 1995, Multivalent binding of complement component C1q to the amyloid β-peptide (Aβ) promotes the nucleation phase of Aβ aggregation. *Biochem.Biophys.Res.Commun.* **217:** 869-875.

Wegiel, J., Bobinski, M., Tarnawski, M., Dziewiatkowski, J., Papovitch, E., Bobinski, M., Lach, B., Reisberg, B., Miller, D. C., de Santi, S., and de Leon, M.J., 2001a, Shift from fibrillar to nonfibrillar Aβ deposits in the neocortex of subjects with Alzheimer disease. *J. Alzheimer's Dis.* **3:** 49-57.

Wegiel, J., Imaki, H., Wang, K.-C., Wegiel, J., Wronska, A., Osuchowski, M., Rubenstein, R., 2003, Origin and turnover of microglial cells in fibrillar plaques of APPsw transgenic mice. *Acta Neuropathol.* **105:** 393-402.

Wegiel, J., Wang, K.-C., and Tarnawski, M., 2000, Microglial cells are the driving force in fibrillar plaque formation, whereas astrocytes are a leading factor in plaque degradation. *Acta Neuropathol.* **100:** 356-364.

Wegiel, J., Wang, K.-C., Imaki, H., Rubenstein, R., Wronsk, A., Osuchowski, M., Lipinski, W.J., Walker, W.J., and LeVine, H., 2001b, The role of microglial cells and astrocytes in fibrillar plaque evolution in transgenic APP_{SW} mice. *Neurobiol. Aging* **22:** 49-61.

Wegiel, J., Kuchna, I., Wisniewski, T., de Leon, M.J., Reisberg, B., Pirttila, T., Kivimaki, T., and Lehtimaki, T., 2002, Vascular fibrosis and calcification in the hippocampus in aging, Alzheimer disease, and Down's syndrome. *Acta Neuropathol.* **103:** 333-343.

Westlind-Danielsson, A., and Arnerup, G., 2001, Spontaneous in vitro formation of supramolecular β-amyloid structures, "βamy balls", by β-amyloid 1-40 peptide. *Biochemistry* **40:** 14736-14743.

Wille, H., Michelitsch, M.D., Guénebaut, V., Supattapone, S., Serban, A., Cohen, F.E., Agard, D.A., and Prusiner, S.B., 2002, Structural studies of the scrapie prion protein by electron crystallography. *Proc. Natl. Acad.Sci. USA* **99:** 3563-3568.

Williams, A.D., Portelius, E., Kheterpal, I., Guo, J.-T., Cook, K.D., Xu, Y., and Wetzel, R., 2004, Mapping Aβ amyloid fibril secondary structure using sanning proline mutagenesis. *J. Mol. Biol.* **335:** 833-842.

Wollmer, M.A., Streffer, J.R., Tsolaki, M., Grimaldi, L.M., Lutjohann, D., Thal,D., Von Bergman, K., Nitsch, R.M., Hock, C., and Papassitiropoulos, A., (2003), Genetic association of acyl-coenzyme A: cholesterol acyltransferase with cerebrospinal fluid cholesterol levels, brain amyloid load, and the risk for Alzheimer's disease. *Mol. Psychiatry* **8:** 635-638.

Wood, S.J., Maleeff, B., Hart, T., and Wetzel, R., 1996, Physical, morphological and functional differences between pH 5.8 and 7.4 aggregates of the Alzheimer's amyloid peptide Aβ. *J. Mol. Biol.* **256:** 870-877.

Yang, Y., Wang, H., and Erie, D.A., 2003, Quantitative characterization of biomolecular assemblies and interactions using atomic force microscopy. *Methods* **29:** 175-187.

Yip, C.M. and McLaurin, J., 2001, Amyloid-β peptide assembly: A critical step in fibrillogenesis and membrane disruption. *Biophys. J.* **80:** 1358-1371.

Yip, C.M., Darable, A.A., and McLaurin, J., 2002, Aβ42-peptide assembly on lipid bilayers. *J.Mol. Biol.* **318:** 97-107.

Yong, W., Lomakin, A., Kirkitadze, M.D., Teplow, D.B., Chen, S.-H., and Benedek, G.B., 2002, Structure determination of micelle-like intermediates in amyloid β-protein fibril assembly by using small angle neutron scattering. *Proc. Natl. Acad.Sci. USA* **99:** 150-154.

Chapter 2

Transgenic Mouse Models for APP Processing and Alzheimer's Disease: Early and Late Defects

Tom van Dooren, Ilse Dewachter, Peter Borghgraef and Fred van Leuven
Experimental Genetics Group, Department of Human Genetics, K.U.Leuven, B-3000 Leuven, Belgium

Abstract: Transgenic mice with neuronal expression of human AD-mutant APP[V717I] in their brain recapitulate robustly the amyloid pathology as seen in Alzheimer's disease (AD) patients. The AD related pathological phenotype consisting of amyloid plaques and vascular amyloid pathology, develop progressively and relative late in ageing APP transgenic mice, between 10 and 15 months of age. In contrast to the late - and clinically irrelevant - amyloid plaque-pathology, the early cognitive defects and behavioural features are clinically more interesting. This review discusses the generation and in depth phenotypic characterization of both aspects of the APP[V717I] transgenic mice. Attention is focussed on the relation of biochemical data of the different APP fragments and amyloid peptides to the formation of the typical early defects and the late parenchymal and vascular amyloid depositions. The APP[V717I] transgenic mice are a perfect model to characterize and investigate early biochemical and cognitive aspects and a potential resource to define pathological interactions of different factors known to be involved in AD. Finally, any therapeutic intervention can be directly tested and explored in these transgenic mice as excellent pre-clinical models

Key words: Transgenic mice, Alzheimer's disease (AD), amyloid-plaque-pathology.

1. INTRODUCTION: GENETICS AND PATHOLOGICAL FEATURES OF ALZHEIMER'S DISEASE

Alzheimer patients are characterized by mostly a-typical clinical features during life and by very typical pathological lesions in their brain, observed post-mortem: extra-cellular amyloid deposited as parenchymal plaques and vascular angiopathy, in addition to intra-cellular neurofibrillary tangles. Because of the ill-defined and largely unknown relations between the clinical and pathological problems on the one hand, and between both pathological lesions on the other hand, this chapter will first concentrate on the brain pathology and its robust recapitulation in transgenic mouse brain. Those late defects will then be related and traced back to early defects, mainly referring to behavioral and cognitive aspects that are obvious even at the "pre-amyloid" stage of the disease in APP transgenic mice.

Inherited familial forms of AD are very rare (<1% of all cases) and mainly caused by mutant Presenilin genes. Most AD cases are sporadic of unknown ethiology (for reviews see St. George-Hyslop, 2000; Selkoe, 2001), but all AD cases comprise by definition the pathological combination of amyloid plaques and intra-neuronal tau-aggregates (Braak and Braak, 1991; Delacourte and Buée, 2000). Independent of their genetic make-up or epi-genetic history, all familial and sporadic AD patients present with the same pathological lesions in their brain at autopsy, *i.e.* extracellular amyloid senile plaques and intracellular neurofibrillary tangles, both concentrated in the hippocampus and the cerebral cortex. This is the pathological definition of this neurodegenerative disorder, brought "to light" almost a century ago by silver impregnation of brain sections (Alzheimer, 1906). Practically all AD-patients have also amyloid deposits in the wall of small blood vessels in the brain, although comparable cerebral amyloid angiopathy (CAA) is also evident in about a third of all elderly people. In this chapter we will concentrate on the amyloid pathology in brain parenchym and vasculature and on its successful recapitulation in the brain of APP[V717I] transgenic mice - and we will explore how the amyloid hypothesis is tested and extended towards the early defects. It must be remembered that knowledge of these early defects, not only in clinical and cognitive terms, but especially in functional and molecular detail, is essential for the development of early detection and treatment of this devastating diseases, *i.e.* before irreversible brain damage has occurred.

1.1 Amyloid plaques in brain

The types of amyloid deposits that are evident in the parenchym of the human brain are named as diffuse and as neuritic or senile plaques. The latter are very typical and diagnostic for AD. They consist predominantly as deposits with a central core of amyloid that is up to 100 μm across and are surrounded by abnormal neuronal processes originating from neighboring cells. These swollen and dystrophic neurites are distended and contain a variety of degenerating cellular organelles, mitochondria and lysosomes, besides being immuno-reactive for many synaptic and cytoskeletal markers. Interspersed among the neurites are processes and occasionally also cell-bodies of activated microglia and of activated astrocytes. This is taken as evidence for an ongoing inflammatory reaction in and surrounding the plaques that is attracting more and more attention in terms of its negative pathological involvement and as potential therapeutic target.

1.2 Cerebral amyloid angiopathy (CAA)

The abluminal deposition of amyloid peptides in cerebral vessel walls is inherent and a diagnostic element of the AD pathology. CAA occurs sporadically and is observed in a third of all elderly people over 60. The vascular amyloid is deposited most commonly in meningeal and cortical arteries and arterioles, and less frequently in veins and capillaries. Vascular amyloid disturbs the smooth muscle cells and the basal lamina of the vessel walls, causing degeneration and weakening of the vessel walls and provoking small and large aneurysms. These lesions have been estimated to be responsible for 15% of all hemorrhagic strokes in elderly people. The CAA as an essential pathological feature in AD patients might constitute a direct link to the closely related and clinically difficult to distinguish entity of vascular dementia, the ethiology of which is unknown but involves defects in the vascular wall. In daily clinical practice, the difficult differential diagnosis of AD and vascular dementia has led to the assumption that these are two extremes of a similar pathology, with a mixed type of dementia in between that is hard or even impossible to define precisely.

1.3 Neurofibrillary tangles

Neurofibrillary tangles (NFT) consist mainly of protein tau and typify not only AD but many neurodegenerative diseases, *e.g.* fronto-temporal dementia (FTD), corticobasal degeneration, Pick's disease, ... collectively termed tau-opathies (Delacourte and Buée, 2000). Primary tau-opathies are caused by diverse mutations in the tau gene and named fronto-temporal dementia with Parkinsonism linked to chromosome 17 (FTDP-17) (Heutink, 2000; Delacourte and Buée, 2000; Ingram and Spillantini, 2002). AD is not a primary, but by far the most prevalent tau-opathy, however it is not caused by mutations in the tau gene.

Aggregated protein tau is invariably hyper-phosphorylated supporting the assumption that hyper-phosphorylation is inherent to the mechanism of tangle formation. In addition, hyper-phosphorylated protein tau readily forms paired helical filaments *in vitro* while de-phosphorylation decomposes tau-filaments into soluble tau-monomers. Conversely, it cannot be excluded that in diseased human brain, neuronal protein tau forms aggregates that subsequently become hyper-phosphorylated and thereby stabilized, causing or contributing to their neurotoxicity by disturbing axonal transport (for review see Terwel *et al.*, 2002).

2. MAJOR QUESTIONS IN AD

Three phases are recognized in the clinical condition of AD patients: (i) mild cognitive impairment (MCI) with minor short-term (recent) memory problems, progressing into (ii) obvious and then massive deficits in learning and memory with important repercussions in personality, emotional state and character (intermediate clinical stages), rapidly worsening into (iii) deep dementia with associated "brain atrophy" (late clinical phase).

The first and major clinical feature or symptom that is observed but not typical for AD patients is the inability to learn and remember new facts. In the absence of objective biochemical or molecular methods to diagnose AD early and reliably, the clinical diagnosis is uncertain and heavily dependent on the actual stage, evolving from "probable" to "possible" to practically "definite" AD only in the late clinical stages. It must be remembered that definite diagnosis can only be obtained post-mortem, by pathological examination of brain sections after silver staining (*cf.* A. Alzheimer, 1906).

It is evident that objective and early diagnosis is absolutely needed, not only to solve the social and clinico-medical problems, but even more so to allow the pharmacological and/or other developments of effective therapy. The many problems encountered in this respect are beyond the scope of this chapter, but readers should be aware of their utmost importance.

The major problem in AD remains to decipher what is cause, correlation and consequence of the disease, despite - or perhaps due to - the multiplicity of molecular and biochemical defects and brain lesions that have been and are being reported. Unanswered questions remain whether dementia and neurodegeneration in patients are equal to loss of neurons or only of synapses, or whether only a functional disturbance of synaptic transmission is at stake. Other important issues in the overall pathogenic process – which emerge in many hypotheses – are the apoptosis of neurons and loss of synapses and the malfunction of the glutamate neurotransmitter system, resulting in excitotoxicity.

Here, we will concentrate on transgenic mice that recapitulate robustly the amyloid related neuro-pathology as seen in AD patients. This is a major contribution to understand the biochemical and physiological aspects of amyloid precursor protein (APP), of α- and β-secretase and of presenilin in the complex context of the brain. It is generally believed that mishaps in processing of APP, in which presenilin is intimately involved (De Strooper *et al.*, 1998) will eventually lead to amyloid peptide production as the central problem in AD (Hardy, 1997; Selkoe, 2001).

Many aspects of the pathogenesis and the role of amyloid plaques and of cerebral amyloid angiopathy in AD remain for a large part unclear. The fact that senile plaques and vascular amyloid are diagnostic in all AD patients and that they consist mainly of amyloid peptides, has been taken to suggest that they are essential and central in the overall process of "amyloidogenesis" in AD. All other and non-specific clinical and pathological features of AD are thought to be secondary to or caused by amyloid peptides, *i.e.* synaptic loss, gliosis, neuronal loss, brain-shrinkage, and even neurofibrillary tangle formation. Nevertheless, the in-depth characterization of APP transgenic mice, and of multiple transgenic derived strains, have yielded strong evidence that the amyloid pathology does not *per se* involve physical deposition, but that the concept of the amyloid cascade must incorporate the early stages where the amyloid peptides are present as dimers, oligomers and small aggregates. How these cause the early defects in synaptic transmission or synapse loss, probably underlying MCI and early stage AD, constitutes the evident challenge for this research-field.

3. A ROBUST MODEL FOR AMYLOID PATHOLOGY: APP[V717I] TRANSGENIC MICE

We generated and have characterized in depth the transgenic mouse strain named APP-London or APP[V717I] that shows all the robust and reproducible features of the late amyloid pathology as observed in the brain of AD patients: starting with diffuse amyloid plaques that evolve with ageing into numerous thioflavine-S-positive senile plaques, followed by extensive cerebral amyloid angiopathy. This transgenic mouse strain was obtained by overexpression of the London-mutant of human amyloid precursor protein (APP) *i.e.* APP[V717I], carrying the mutation just Down'sstream of the γ-secretase cleavage site. This human gene was incorporated into a construct based on the neuron-specific elements of the murine thy-1 gene promoter (Moechars *et al.*, 1996, 1999). In the brain of heterozygous APP[V717I] transgenic mice, the level of over-expression of human APP is about 2 to 3 times higher than that of endogenous murine APP. Significant for the operational aspects of this model, the expression of human APP is restricted exclusively to the neurons of the central nervous system, as demonstrated by *in situ* hybridization and by immuno-histochemistry (Moechars *et al.*, 1999; Van Dorpe *et al.*, 2000) and the expression is moreover appreciable only in the second or third week after birth, thereby eliminating any interference with the development of the brain.

3.1 Occurrence and progression of diffuse and senile amyloid plaques

At about one year of age, the APP[V717I] transgenic mice begin to develop pathological features that are strikingly similar to those observed in the brain of AD patients. From that point onwards, the APP[V717I] transgenic mice develop diffuse and neuritic plaques with a morphology that is characteristic for and practically identical to those observed post-mortem at autopsy in human brain. The senile or neuritic plaques contain a core of amyloid that stains intensely with thioflavine-S (Figure 1) and with Congo red, surrounded by numerous dystrophic neurites. Ultrastructurally, the core consist of bundled amyloid fibers that are 8 to 10 nm in diameter that are immuno-stained with practically all antibodies directed against amyloid peptides of either 40 or 42 amino acids, *i.e.* Aβ40 and Aβ42. Diffuse plaques in APP[V717I] transgenic mice are detected immuno-histochemically with a pan-Aβ antibody (Figure 2) as well as with antibodies directed against Aβ40 and Aβ42. Similar to AD patients, these diffuse plaques are quantitatively more important - by about a factor of 10 - than the senile, thioflavine-S-positive plaques in ageing APP[V717I] transgenic mice. Amyloid plaques develop first in the subiculum and spread to other regions of the hippocampus and the cerebral cortex. With ageing they continue to increase in numbers, up to about 20-24 months of age of the transgenic mice, and remain preponderantly in the hippocampus and neocortex, but are also evident in thalamus and other brain regions, but are never observed in the cerebellum (Moechars *et al.*, 1999; Van Dorpe *et al.*, 2000; Dewachter *et al.*, 2002). These qualitative and quantitative features of the amyloid plaques, *i.e.* type, evolution and regional distribution approach very closely what is known of the amyloid pathology in AD patients, as far as it can be progressively traced by post-mortem analysis of occasional cases, dying from accidental or unrelated causes.

3.2 Cerebral amyloid angiopathy

Besides and in addition to the amyloid plaques in the brain parenchym, the older APP[V717I] transgenic mice progressively develop very abundant, thioflavine-S-positive amyloid deposits in their cerebral blood vessels. The phenomenon of cerebral amyloid angiopathy (CAA) is at first evident in the leptomeningeal pial blood-vessels, and further develops in cortical, thalamic, and hippocampal vessels (Figure 3). The accumulation of amyloid in blood-vessels is observed some few months later than the amyloid depositions in the brain parenchym, *i.e.* in APP[V717I] transgenic mice that are older than 15 months of age. This indicates that they form later than the amyloid plaques and, similar to AD patients, show variation in individual APP[V717I] transgenic mice, both in distribution and extent.

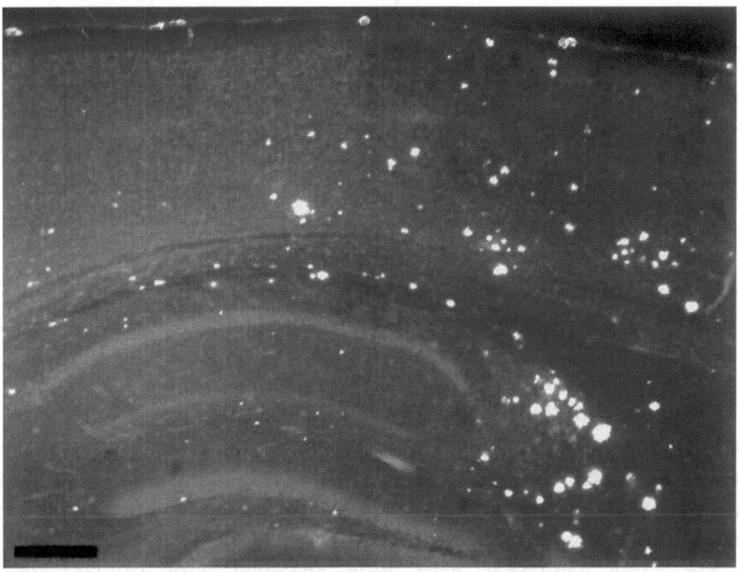

Figure 1. Thioflavine-S staining of typical neuritic plaques in the subiculum, hippocampus and the neocortex of a 22 months old APP[V717I] mouse. Bar=300 μm.

A minor difference is that neither diffuse nor senile plaques, nor vascular amyloid deposits were ever observed in the cerebellum of APP[V717I] transgenic mice. This is due to the low expression of the human transgene in cerebellum relative to forebrain. This is an intrinsic characteristic of the mouse thy1 gene promoter construct, but since in AD patients cerebellar pathology is also not well developed, this aspect is in fact a bonus of the APP[V717I] transgenic model.

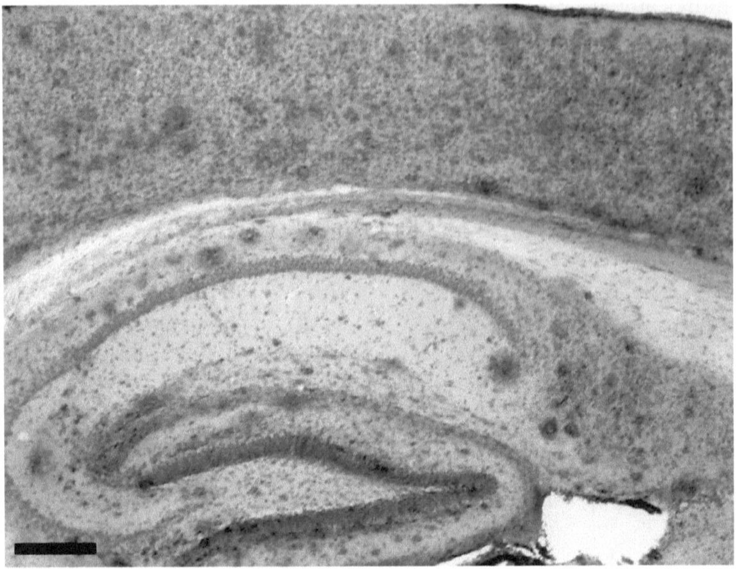

Figure 2. Immunohistochemical detection of diffuse and senile plaques with pan-AB in a 22 months old APP[V717I] transgenic mouse. Bar=300 μm

The staining with thioflavine-S reveals the most severely affected arteries to exhibit amyloid-based fluorescence in patterns of concentric rings, while less affected vessels show focal abluminal accumulations similar to human CAA. Arteries are more affected than veins that show only small focal accumulations. Capillaries are rarely affected in the APP[V717I] mouse brain but some capillaries have been observed with amyloid depots penetrating into the neuropil, resembling dyshoric amyloid in AD patients. Very similar to human CAA, the large arteries at the base of the transgenic mouse brain constituting the circle of Willis, nor the extra cranial blood vessels were affected (Van Dorpe *et al.*, 2000). Vessels with important amyloid deposition show loss of smooth muscle cells, while rarely the internal elastic lamina damaged. The vessel wall damage leads to dilation and aneurysms, again as in human patients, but hemorrhage was not evident in brain of APP[V717I] transgenic mice. Moreover, the cerebral blood flow was measured by laser-Doppler flowmetry and proved unaffected even in very old APP[V717I] transgenic mice.

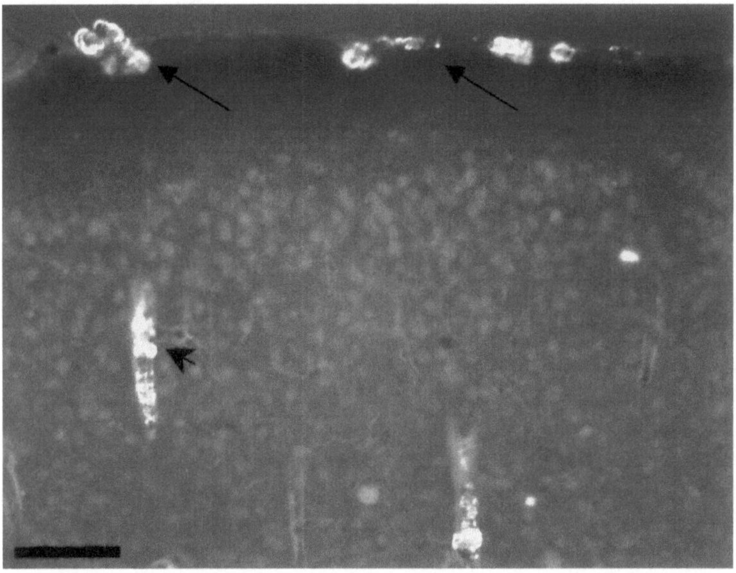

Figure 3. Cerebral amyloid angiopathy (CAA) in a 22 months old APP[V717I] transgenic mouse. At this age, thioflavine-S-positive amyloid deposits are abundantly present in the leptomeningeal pial (arrows) and cortical (arrowhead) blood vessel. Bar=100 μm.

3.3 Ageing and amyloid pathology, a paradox of pathology versus biochemistry

In APP[V717I] transgenic mice, both vascular and plaque amyloid load progressively increased with age. Quantitatively, amyloid plaquc pathology begins to develop in the 10-12 months age window, while vascular pathology forms not before the age of 15 months. Both types of amyloid pathology progress further and continue to increase quantitatively during the 2nd year of life of these mice. At all age-points analyzed, the thioflavine-S and immuno-reactive amyloid plaque load correlated closely with the number of vessels containing vascular amyloid (Van Dorpe *et al.*, 2000). In APP[V717I] transgenic mice 2 years of age, more than 90% of the leptomeningeal arterioles over the cerebrum were abundantly loaded with amyloid deposits. At that age, the subiculum of these old mice is literally inundated with diffuse and senile plaques, similar to some cortical regions.

In humans, ageing remains the most important, but least understood parameter or risk factor for dementia and for AD. The APP[V717I] transgenic mice were therefore compared at ages of 3 months, 6-9 months and 15 months, to measure APP and its different products of proteolytic processing, *i.e.* the membrane bound precursor (APPm), the soluble and plaque-associated amyloid peptides Aβ40 and Aβ42, the secreted ectodomain of APP processed by α- or β-secretase (APPsα, APPsβ) and the C-terminal transmembrane and cytoplasmatic domain resulting from β-secretase cleavage, referred to as "β-C-stubs" (Dewachter *et al.*, 2000). The comparative analysis revealed that ageing did not appreciably affect the normalized levels of either the α-secretase cleaved ectodomain nor the residual β-secretase cleaved C-terminal stubs of APP. Between the age of 3 to 15 months, the most pronounced effect concerned the dramatic and expected increase in insoluble amyloid peptides, Aβ40 and Aβ42, as well as an increased Aβ42/40 ratio.

Ageing of the APP[V717I] transgenic mice caused progressively more aggregation of the amyloid peptides in the physical form of amyloid plaques, due to increased levels of soluble amyloid peptides, especially of the least soluble Aβ42. This phenomenon did however, never occur before the age of 10-12 months in the APP[V717I] transgenic mice, as observed in unrelated APP transgenic mouse models in which Aβ also increased with age (Hsiao *et al.*, 1996; Johnson-Wood *et al.*, 1997). Measurements of the mRNA coding for endogenous murine APP and for human transgene APP[V717I] increased with age in the brain of transgenic mice (Moechars *et al.*, 1999). Membrane-bound APP protein increased in close parallel with mRNA levels, with levels reaching 122% and 169% respectively at 6-9 months and 15 months (Dewachter *et al.*, 2000).

The combined data demonstrate that in brain of the APP[V717I] transgenic mice, ageing *per se* did not markedly affect the normal processing of APP as mediated by α- and β-secretase. Therefore, the marked increase in both soluble and plaque-associated amyloid peptides, essentially situated at the age of 12 months, was not a direct consequence of a disturbed or tilted balance in the two competing proteolytic events that govern APP metabolism, *i.e.* the non-amyloidogenic pathway by ADAM10 as α-secretase (Postina *et al.*, 2004; *see also* Chapter 6) and the amyloidogenic pathway mediated by the combined actions of β- and γ-secretases (reviewed by Dewachter and Van Leuven, 2002). Whereas amyloid deposition is unequivocally further increasing with age, the production of amyloid peptides is not increasing! This apparent paradox between pathology and biochemistry points to another mechanism, *i.e.* that degradation and/or clearance of the amyloid peptides must be impaired in the APP[V717I] transgenic mice - and by extrapolation, also in the brain of AD patients (see Chapter 7). This conclusion implies, among other factors, an operational mechanism of liquid drainage from the brain into the CSF, functionally similar to the lymphatic system in peripheral organs (Weller and Nicoll,

2003) which thereby directly connects the amyloid depositions in parenchym and in the peri-vascular spaces, *i.e.* the amyloid plaques and the CAA pathology, as discussed in the following section.

3.4 Pathogenesis of cerebral amyloid angiopathy

A long-standing debate concerns the origin of the vascular amyloid depositions. In human and mouse brain, as well as in other tissues, endogenous APP is ubiquitously expressed by many cell types, including these of the vasculature. Some proposed that Aβ in the vessel walls would derive from vascular smooth muscle cells or pericytes since pial vessels are most often affected, implying local production as potential source of Aβ. This hypothesis failed, however, to explain the specific neuroanatomical pattern of cerebral amyloid angiopathy and its exclusive localization in intracranial vessels.

We demonstrated by Western blotting that blood plasma of aged APP[V717I] transgenic mice did not contain detectable levels of amyloid peptides (*i.e.* lower than picogram-levels) as opposed to CSF that contains Aβ in the range of 10-20 ng/ml (Van Dorpe et al, 2000). This finding, combined with the strong evidence for the exclusive neuronal expression of the APP transgene in this mouse strain (Moechars *et al.*, 1999) and the abluminal deposition of vascular amyloid (Van Dorpe *et al.*, 2000), unequivocally demonstrated that amyloid peptides originating exclusively from neurons within the brain are the direct cause of amyloid plaques in the brain parenchym as well as of the CAA.

CAA appears in vessels that do not express the transgene, but evidently requires and implies a means of transport of amyloid peptides to the deposition sites. This problem is less evident or obvious in the formation of senile plaques. Amyloid peptides are present in the CSF of normal and AD individuals and drainage pathways must be involved along the perivascular space surrounding intracortical and leptomeningeal arteries. These channels eventually connect with nasal lymphatics draining to the cervical lymph nodes. The suggestion that significant amounts of Aβ drain along this pathway in humans (Weller *et al.*, 1998) is fully confirmed by the objective analysis of all data in the APP[V717I] transgenic mice (Van Dorpe *et al.*, 2000). Recently, in AD patients the hypothesis of the neuronal origin of Aβ and the drainage of Aβ via the interstitial fluid from the central nervous system to capillary walls is confirmed in AD patients (Roher *et al.*, 2003; Attems *et al.*, 2004).

3.5 Aβ42 is the primary cause of amyloid deposition in parenchym as well as in blood vessels

In brain of AD patients, the ratio of Aβ42 to Aβ40 is higher in plaques than in the vascular amyloid deposits, which is recapitulated in the APP[V717I] transgenic mice that also have much higher Aβ42/Aβ40 ratios in the amyloid plaques than in the vascular amyloid (Van Dorpe *et al.*, 2000). The purely neuronal origin of the amyloid peptides then allows us to define their path and their fate.

After secretion by neuronal cells (by as yet unknown mechanisms) the level of the amyloid peptides in the intercellular and interstitional spaces increases to reach levels at which the least soluble Aβ42 peptides begin to aggregate into oligomers (*see also* Chapter 1). These form the nuclei for further aggregation in which also Aβ40 peptides participate, as well as less abundant Aβ38 and eventual other species. These aggregates develop into diffuse and progress later into senile plaques by accruing more amyloid and other insoluble proteins that are either trapped or otherwise become lodged into these plaques, that can reach diameters of up to 1 mm.

At the same time, excess peptides are being drained by the perivascular route into the CSF, eventually taking also some of the smaller aggregates to flow along and become trapped in the perivascular spaces. These form the nuclei of the CAA, whereby evidently the more soluble Aβ40 peptides diffuse faster and further than the least soluble Aβ42 peptides. A gradient of Aβ42 thereby must form *de facto*, with highest levels in the neurons and decreasing towards the CSF - explaining also why in AD patients the Aβ42 concentration in CSF actually decreases with the progression of the pathology (Blennow *et al.*, 2003).

Hence, similar to senile plaque formation, the Aβ42 is first to become deposited in the vessel walls (nucleation) while the more soluble Aβ40 is subsequently entrapped (growth). This is further supported by the earlier and increased formation of senile plaques and of cerebral amyloid angiopathy in double transgenic APP[V717I] x PS1[A246E] mice, obtained by crossing the respective single transgenic mice. In these double transgenic mice, the extra incorporation of the human mutant PS1 transgene caused a selective increase in production of Aβ42 over Aβ40 (Dewachter *et al.*, 2000).

More recent evidence in BACE x APP[V717I] double transgenic mice add further weight to this hypothesis, demonstrating that N-terminal truncated forms of the Aβ peptides that are even less soluble than Aβ42, actually increase amyloid deposition in brain parenchym while reducing CAA in the vessel walls (Willem *et al.*, 2004). The fact that N-terminal truncated Aβx-40/42 aggregate faster and more easily than the full-length counterparts (Liu *et al.*, 2002) further supports the hypothesis as well as the negative correlation between plaque and CAA pathology in human AD brain (Tian *et al.*, 2003).

Although APP[V717I] transgenic mice express APP at higher levels than in human brain, the striking similarity to human cerebral amyloid angiopathy is more than a strong suggestion that similar mechanisms cause the vascular amyloid deposition in patients and transgenic mice, alike.

4. EARLY COGNITIVE AND BEHAVIOURAL DEFECTS OF APP[V717I] TRANSGENIC MICE

Besides the specified and characteristic pathological defects caused by insoluble Aβ40 and Aβ42, the APP[V717I] transgenic mice additionally display early phenotypic behavioural and cognitive impairments (Moechars *et al.*, 1999). With the revival of the hypothesis of "synaptic deficit" in AD (Selkoe, 2003) these early defects were and will be more and more interesting as typical characteristics of a mildly impaired cognitive phenotype and as read-out to define effects of genetic or epigenetic factors and treatment (Dewachter *et al.*, 2002; Dewachter and Van Leuven, 2002).

As opposed to plaque formation that develops after 10-12 months of age, disturbed behaviour and cognitive defects occur at an age as early as 3-6 months. Profound disturbances in behaviour are evident as hyperactivity, anxiety, aggression and neophobia. Some of these obvious and typical characteristics were observed in APP[V717I] transgenic mice as young as 8 weeks and came progressively more evident with age. Cognition, spatial learning and memory as estimated by the Morris water maze paradigm, by novel object recognition and by contextual d-fear conditioning are significantly impaired compared to age and sex-matched non-transgenic mice with the same genetic background (Moechars *et al.*, 1998, 1999; Dewachter *et al.*, 2000; 2002). In addition, a progressive decay of long term potentiation (LTP), an accepted parameter or even model for synaptic plasticity, was evident long before any plaques were formed in the brain parenchym. Intriguingly, the early behavioural and cognitive defects were also observed to a large extend in transgenic mice that overexpressed the human APP695 wild-type isoform (Moechars *et al.*, 1999), providing strong evidence for a direct role of APP metabolites and against amyloid plaques that never formed in these mice. The APP[V717I] transgenic mice studied over the last years and as presented here, are therefore perfect tools to study the early cognitive defects in relation to the early and late biochemical characteristics of the induced AD-like amyloid pathology.

5. MODULATING AMYLOID PEPTIDE PRODUCTION: THERAPEUTIC VALUE?

Presenilin-1 (PS1) is essential in the generation of amyloid peptides from APP (De Strooper *et al.*, 1998) either by exerting the proteolytic activity itself or by acting as an essential component of the complex that constitutes γ-secretase. The crucial and causal role of PS1 in the development of amyloid pathology has also made it a primary therapeutic target, despite the fact that inactivation of the PS1 gene resulted in embryonic death due to pleomorphic effects, *i.e.* disturbed somitogenesis, cranial hemorrhages, impaired neurogenesis, cerebral cavitations and malformation of the brain (Wong *et al.*, 1997; Shen *et al.*, 1997; De Strooper *et al.*, 1998).

To allow us to study the function of PS1 in adult brain, we have generated conditional PS1-knockout mice, with a post-natal, neuron-specific deletion of PS1 not impairing their health, fertility nor normal life-span up to the age of 2 year (Dewachter *et al.*, 2002). Subsequently, these PS1(n-/-) mice were crossed with the APP[V717I] transgenic mice described above to define the effect on the amyloid pathology. In the APP[V717I] x PS1(n-/-) double transgenic mice, the neuronal absence of PS1 effectively prevented the formation of all amyloid pathology. In contrast to the parental APP[V717I] transgenic mice, no thioflavine-S-reactive amyloid plaques nor immuno-reactive diffuse amyloid deposits were detected in the brain of double transgenic mice. The fact that the neuron-specific inactivation of PS1 *in vivo* was positive and had no major adverse effects on these animals proved that adult neurons can manage without PS1, at least in mice and possibly because PS2 took over some functions. Moreover, the prevention of both types of amyloid pathology, *i.e.* amyloid plaques and CAA, again demonstrated their common origin (Dewachter *et al.*, 2002).

The inactivation of PS1 effectively inhibited the formation of amyloid peptides but caused the accumulation of their obligate precursor, *i.e.* the C99 or β-CTF or β-C-terminal fragments of APP. Although it is hardly likely that these have a physiological function on their own, it has been postulated, based on cellular studies, that these CTF can be neurotoxic (for review see Dewachter and Van Leuven, 2002). That could explain our unexpected observations that the inhibition of amyloid peptide formation did actually not improve, but aggravated the cognitive defects even of young APP[V717I] x PS1(n-/-) double transgenic mice, relative to age matched APP[V717I] single transgenic mice (Dewachter *et al.*, 2002). That outcome confirmed the negative expectations from short-term experiments with γ-secretase inhibitors that these were not effective, and moreover very toxic (Wong *et al.*, 2004).

In another vein, a recent collaborative study aimed to define the role of the desintegrin and metalloproteinase ADAM10 as the major α-secretase in the non-amyloidogenic pathway, by cleaving membrane bound APP in the middle of the amyloid region to prevent formation of amyloid peptides. In double transgenic mouse models, we provided the first evidence that ADAM10 effectively acted as an α-secretase in vivo (Postina *et al.*, 2004; see also Chapter 6)). Neuronal co-expression of wild type ADAM10 or of a catalytically inactive dominant-negative mutant together with APP[V717I in double transgenic mice, respectively stimulated and inhibited the non-amyloidogenic processing of APP. Beneficial results of enhanced α-secretase cleavage of APP were the increased production of secreted neurotrophic APPsα, concomitant with reduced formation of amyloid peptides and an almost complete prevention of their deposition into amyloid plaques at old age. Moreover, the cognitive defects of the parental APP[V747I] transgenic mice were improved by co-expression of ADAM10 (Postina *et al.*, 2004). Inversely, co-expression of the dominant-negative mutant of ADAM10 led to increased amyloid plaques in the brain of double transgenic mice.

6. CONCLUDING REMARKS AND PROSPECTS FOR AD MODELS: A CASE FOR MULTIPLE TRANSGENIC MICE?

The senile plaques and cerebral amyloid angiopathy in the APP[V717I] mouse exhibit a very striking similarity to those observed in familial and sporadic AD patients at old age. The morphological pattern, the ultrastructural aspects and the biochemical composition of senile plaques and vascular amyloid deposits in humans are reproduced in an almost identical manner in the APP[V717I] transgenic mice (Moechars *et al.*, 1999; Van Dorpe *et al.*, 2000; Dewachter *et al.*, 2002). In humans, AD is characterized by a protracted clinical course covering up to 20 years while the lifespan of a laboratory mouse does not exceed more than 2 years. It is therefore surprising that the entire pathological history of the disease can be compressed into about 18 months, but presenting opportunities and advantages for further investigations into the pathogenesis of this devastating neurodegenerative disease. In particular, we anticipate to be able to identify new markers, targets and procedures for the early and objective diagnosis, allowing us to test novel therapeutic drugs and strategies in these pre-clinical models. They will help to accelerate drug discovery, leading to the recognition of therapeutic agents that are effective in postponing the onset or slowing the progression of neurodegenerative disease.

Whereas amyloid pathology in all its aspects in the brain of AD patients has now been approximated or even "copied" almost exactly by overexpression of mutant forms of APP, with or even without PS1, a similar strategy failed to yield neurofibrillary tangles in brain of transgenic mice - thus far. Generating an animal model for Alzheimer's disease that contains or develops all the neuropathological features was and is an enormus challenge. Following up the current amyloid success, models are needed to include the evident progression with age as in patients, with matching topology and regional distribution and including quantitative characteristics that parallel the human disease.

The successful generation of mice with amyloid plaques and angiopathy that actually seem to miss only the neurofibrillary tangles to show the complete AD-pathology, is evidently very encouraging. Moreover, the combination with other transgenic mouse strains, *i.e.* PS1 mutant, PS1(n-/-), ADAM10, BACE, APP-/-, ApoE4, have provided us with invaluable data, impossible to otherwise obtain. This evidently makes a strong case for more multiple transgenic mice.

To further explore the relation between the amyloid and tau-pathologies, we have generated APP[V717I] x PS1 x tau-4R triple transgenic mice. Surprisingly, the resulting offspring suffered very high premature mortality, surviving between 6 to maximally 20 weeks; this prevented extensive analysis of the phenotype of adult and aged mice. We confirmed that these mice are extremely sensitive to external environmental stress as reported by others, although the exact cause of death was not evident (Pedersen *et al.*, 1999). Subsequently we have modeled FTD by generating tau-P301L transgenic mice (Terwel *et al.*, unpublished results) and combined these with the APP[V717I] into double transgenic mice. Although these experiments are ongoing, these mice do not appear to develop a really synergistic pathology (Muyllaert *et al.*, unpublished results), as claimed to be observed in similar models (Lewis *et al.*, 2001).

Despite the lack of neurofibrillary tangles in any of our APP single or double transgenic mice, the mice allowed us to study and explore APP-processing *in vivo*, in particular at a young age. The double transgenic APP[V717I] x BACE, APP[V747I] x ADAM10 and APP[V717I] x PS1[A246E] mice are excellent tools to further investigate the origin and repercussions of the amyloid peptides, both at young and old age.

The mechanism(s) by which neurons become non-viable and are eventually "executed" in AD is still in need of detailed understanding at the molecular level. The reported and discussed findings do emphasize the role and importance of clearance pathways of amyloid peptides (Vekrellis et al, 2000; Li et al, 1999) and of the involvement of ApoE-lipoproteins and its receptors in brain (Tesseur *et al.*, 2000a,b; May and Herz, 2003). It is clear that our continued effort to model all aspects of the pathology of AD will eventually have to answer these and more fundamental questions that are bound to surface. We believe that a continued effort in transgenesis and in combined mouse models will further prove of great value to understand the interactions of human genes, or rather proteins *in vivo,* that can only be studied in multiple transgenic mice.

REFERENCES

Attems J., Jellinger K.A., 2004, Only cerebral capillary amyloid angiopathy correlates with Alzheimer pathology – a pilot study. *Acta Neuropathol. (Berl).* **107:** 83-90.

Blennow K., Hampel H.,2003, CSF markers for incipient Alzheimer's disease. *Lancet Neurol.* **2:** 605-613.

Braak H., Braak E., 1991, Neuropathological stageing of Alzheimer-related changes. *Acta Neuropathol. (Berl)* **82** : 239-259.

De Strooper B., Saftig P., Craessaerts K., Vanderstichele H., Guhde G., Annaert W., Von Figura K., Van Leuven F., 1998, Deficiency of presenilin-1 inhibits the normal cleavage of amyloid precursor protein. *Nature* **391:** 387-390.

Delacourte A., Buee L., 2000, Tau pathology: a marker of neurodegenerative disorders. *Curr. Opin. Neurol.* **13:** 371-376.

Dewachter I., Van Dorpe J., Smeijers L., Gilis M., Kuiperi C., Laenen I., Caluwaerts N., Moechars D., Checler F., Vanderstichele H. and Van Leuven F., 2000, Aging increased amyloid peptide and caused amyloid plaques in brain of old APP/V717I transgenic mice by a different mechanism than mutant presenilin1. *J. Neurosc. 20:* 6452-6458.

Dewachter I., Reverse D., Caluwaerts N., Ris L., Kuiperi C., Van den Haute C., Spittaels K., Umans L., Serneels L., Thiry E., Moechars D., Mercken M., Godaux E., Van Leuven F., 2002, Neuronal deficiency of presenilin 1 inhibits amyloid plaque formation and corrects hippocampal long-term potentiation but not a cognitive defect of amyloid precursor protein [V717I] transgenic mice. *J. Neurosci..* **22:** 3445-3453.

Dewachter I., Van Leuven F., 2002, Secretases as targets for the treatment of Alzheimer's disease : the prospects. *Lancet Neurol..* **1:** 409-416.

Hardy J., 1997, The Alzheimer family of diseases : many etiologies, one pathogenesis? *Proc. Natl. Acad. Sci. USA* **94:** 2095-2097.

May P. and Herz J., 2003, LDL Receptor-related proteins in Neurodevelopment. *Traffic,* **203:** 291-301.

Heutink P., 2000, Untangling tau-related dementia. *Hum. Mol. Genet..* **9:** 979-986.

Hsiao K., Chapman P., Nilsen S., Eckman C., Hariqaya Y., Younkin S., Yang F., Cole G., 1996, Correlative memory deficits, Abeta elevation, and amyloid plaques in transgenic mice. *Science* **274:** 99-102.

Ingram E.M., Spillantini M.G., 2002, Tau gene mutations : dissecting the pathogenesis of FTDP-17. *Trends Mol. Med.* **8:** 555-562.

Jin L.W., Hua D.H., Shie F.S., Maezawa I., Sopher B., Martin G.M., 2002, Novel tricyclic pyrone compounds prevent intracellular APP C99-induced cell death. *J. Mol. Neurosci..* **19:** 57-61.

Johnson-Wood K., Lee M., Motter R., Hu K., Gordon G., Barbour R., Khan K., Gordon M. Tan H., Games D., Lieberburg I., Schenk D., Seubert P., McConlogue L., 1997, Amyloid precursor protein processing and Abeta42 deposition in a transgenic mouse model of Alzheimer disease. *Proc. Nat.l Acad. Sci. USA* **94:** 1550-1555.

Lewis J., Dickson D.W., Lin W.L., Chisholm L., Corral A., Jones G., Yen S.H., Sahara N., Skipper L., Yager D., Eckman C., Hardy J., Hutton M., McGowan E., 2001, Enhanced neurofibrillary degeneration in transgenic mice expressing mutant tau and APP. *Science* **293:** 1487-1491.

Li Q.X., Maynard C., Cappai R., McLean C.A., Cherny R.A., Lynch T., Culvenor J.G., Trevaskis J., Tanner J.E., Bailey K.A., Czech C., Bush A.I., Beyreuther K., Masters C.L., 1999, Intracellular accumulation of detergent-soluble amyloidogenic A beta fragment of Alzheimer's disease precursor protein in the hippocampus of aged transgenic mice. *J. Neurochem..* **72:** 2479-2787.

Liu K., Doms R.W. and Lee V.M.Y., 2002, Glu11 site cleavage and N-terminally truncated Aβ production upon BACE overexpression. *Biochemistry* **41:** 3128-3136.

Moechars D., Lorent K., De Strooper B., Dewachter I. and Van Leuven F., 1996, Expression in brain of amyloid precursor protein mutated in the α-secretase site causes disturbed behavior, neuronal degeneration and premature death in transgenic mice. *EMBO J.* **15:** 1265-1274.

Moechars D., Gilis M., Kuiperi C. Laenen I., Van Leuven F., 1998, Aggressive behaviour in transgenic mice expressing APP is alleviated by serotonergic drugs. *NeuroReport* **9:** 3561-3564.

Moechars D., Dewachter I., Lorent K., Reversé D., Baekelandt V., Naidu A., Tesseur I., Spittaels K., Van Den Haute C., Checler F., Godaux E., Cordell B. and Van Leuven F., 1999a, Early phenotypic changes in transgenic mice that overexpress different mutants of Amyloid Precursor Protein in brain. *J. Biol. Chem.* **274:** 6483-6492.

Moechars D., Lorent K., and Van Leuven F., 1999b, Premature death in transgenic mice that overexpress mutant Amyloid precursor protein is preceded by severe neurodegeneration and apoptosis. *Neuroscience* **91**: 819-830, 1999.

Pedersen W.A., Culmsee C., Ziegler D., Herman J.P., and Mattson M.P., 1999, Aberrant stress response associated with severe hypoglycemia in a transgenic mouse model of Alzheimer's disease. *J. Mol. Neurosci.* **13:** 159-165.

Postina R., Schroeder A., Dewachter I., Bohl J., Schmitt U., Kojro E., Prinzen C., Endres K., Hiemke C., Blessing M., Flamez P., Dequenne A., Godaux E., Van Leuven F., Fahrenholz F., 2004, A disintegrin-metalloproteinase prevents amyloid plaque formation and hippocampal defects in an Alzheimer's disease mouse model. *J. Clin. Invest., in press.*

Roher A.E., Kuo Y.M., Esh C., Knebel C., Weiss N., Kalback W., Luehrs Childress J.L., Beach T.G., Weller R.O., Kokjohn T.A., 2003, Cortical and leptomeningeal cerebrovascular amyloid and white matter pathology in Alzheimer's disease. *Mol. Med..* **9:** 112-22.

Selkoe D.J., 2001, Alzheimer's disease: genes, proteins, and therapy. *Physiol. Rev.* **81:** 741-766.

Selkoe D.J., 2002, Alzheimer's disease is a synaptic failure. *Science.* **298:** 789 –791.

St. George-Hyslop P.H., 2000, Genetic factors in the genesis of Alzheimer's Disease. *Ann. N. Y. Acad. Sci.* **924:** 1-7.

Terwel D., Dewachter L., Van Leuven F., 2002, Axonal transport, tau protein, and neurodegeneration in Alzheimer's disease. *Neuromolecular Med.* **2:** 151-165.

Tesseur I., Van Dorpe J., Bruynseels K., Bronfman F., Sciot R., Van Lommel A., Van Leuven F., 2000, Prominent axonopathy and disruption of axonal transport in transgenic mice expressing human apolipoprotein E4 in neurons of brain and spinal cord. *Am. J. Pathol.* **157:** 1495-1510.

Tesseur I., Van Dorpe J., Spittaels K., Van den Haute C., Moechars D., Van Leuven F., 2000, Expression of human apolipoprotein E4 in neurons causes hyperphosophorylation of protein tau in the brains of transgenic mice. *Am. J. Pathol.* **156:** 951-964.

Tian J., Shi J., Bailey K., Mann D.M., 2003, Negative association between amyloid plaques and cerebral amyloid angiopathy in Alzheimer's disease. *Neurosci. Lett.* **352:** 137-140.

Van Dorpe J., Smeijers L., Dewachter I., Nuyens D., Spittaels K., Van den Haute C., Mercken M., Moechars D., Laenen I., Kuiperi C., Bruynseels K., Tesseur I., Loos R., Vanderstichele H., Checler F., and Van Leuven F., 2000, Prominent cerebral amyloid angiopathy in transgenic mice overexpressing the London mutant of human APP in neurons. *Am. J. Pathol.* **157:** 1283-98.

Vekrellis K., Ye Z., Qiu W.Q., Walsh D., Hartley D., Chesneau V., Rosner M.R., Selkoe D.J., 2000, Neurons regulate extracellular levels of amyloid beta-protein via proteolysis by insulin-degrading enzyme. *J. Neurosci.* **20:** 1657-1665.

Weller R.O., Massey A., Newman T.A., Hutchings M., Kuo Y.M. and Roher A.E., 1998, Amyloid β accumulates in putative interstitial fluid drainage pathways in Alzheimer's Disease. *Am. J. Pathol.* **153:** 725-733.

Weller R.O. and Nicoll J.A., 2003, Cerebral amyloid angiopathy: pathogenesis and effects on the ageing and Alzheimer brain. *Neurol. Res.* **25:** 611-616.

Willem M., Dewachter I., Smyth N.,Van Dooren T.,Borghgraef P., Haass C., Van Leuven F., 2004, BACE-1 increases amyloid deposition in brain parenchym but reduces cerebrovascular amyloid angiopathy in BACE x APP[V717I] double transgenic mice. Submitted.

Wong P.C., Zheng H., Chen H., Becher M.W., Sirinathsinghji D.J., Trumbauer M.E., Chen H.Y., Price D.L., Van der Ploeg L.H., Sisodia S.S., 1997, Presenilin 1 is required for Notch1 and DII1 expression in the paraxial mesoderm. *Nature* **387:** 288-292.

Wong G.T., Manfra D., Poulet F.M., Zhang Q., Josien H., Bara T., Engstrom L., Pinzon-Ortiz M., Fine J.S., Lee H.J., Zhang L., Higgins G.A., Parker E.M., 2004, Chronic treatment with the gamma-secretase inhibitor LY-411,575 inhibits beta-amyloid peptide production and alters lymphopoiesis and intestinal cell differentiation. J. Biol. Chem. **279:**12876-12882.

Chapter 3

Oxidative stress in Alzheimer's Disease: Implications for Prevention and Therapy

Christian Behl

Institute for Physiological Chemistry & Pathobiochemistry, Faculty of Medicine, Johannes Gutenberg University, Mainz, Duesbergweg 6, D-55099 Mainz, Germany

Abstract: Oxidative stress is a marker of neurodegeneration and has been recently shown to be also involved in the early stages of the pathogenesis of various neurodegenerative disorders. In general, all biomolecules of the cell can be oxidized and thereby damaged. Consequently, the concept of neuroprotection by antioxidants has been developed. In many cases the direct scavanging of free radicals have been used as a strategy to prevent oxidative stress damage and a variety of physiological and synthetic antioxidant molecules have been identified and synthesized including the female sex homone estrogen. In Alzheimer's Disease amyloid-β protein on its way to brain deposition can also induce oxidative changes rendering nerve cells more vulnerable to additional insults. In addition, inflammatory mediators are attracted by amyloid deposits that can further speed up the generation of an oxidative micro-environment. Based on recent clinical data the use of a combination of various antioxidants might indeed be effective in preventing Alzheimer's Disease. Nevertheless, the exact molecular mechanisms and the real impact of oxidative stress on the development and progression of Alzheimer's Disease as well as of other neurodegenerative disorders still needs to be further investigated.

Key words: Oxidative stress, free radicals, lipid peroxidation, protein oxidation, DNA oxidation, amyloid-β protein, inflammation, vitamin E, vitamin C, anti-inflammatory drugs, estrogen, blood-brain-barrier.

1. INTRODUCTION

A variety of human neurodegenerative disorders and symptoms have been associated with damage to essential biochemical structures of the cells of the nervous systems. While the specific sources of these oxidations and their main targets clearly vary from one pathology to another, there are some general patterns of experimental observation which lead to the conclusion that free radical damage to nervous system cell types may constitute a final common pathway of the pathogenesis of many neurodegenerative disorders. These diseases include Amyotrophic lateral sclerosis, Demyelinating disease, Down's syndrome, HIV neuropathy, Huntington's disease, Parkinson's disease, Prion diseases, Stroke, and most of all also Alzheimer's disease (AD), the most common form of dementia in the elderly.

Oxidative damage to proteins, lipids and DNA is caused by highly reactive oxygen and nitrogen species (ROS, RNS). There is a constant production of ROS and RNS in the cells and under homeostatic conditions accumulating ROS and RNS are efficiently detoxified by potent antioxidant defence systems that include biochemical structures such as vitamins (vitamin E and vitamin C) and antioxidant enzymes (*e.g.* glutathione peroxidase, catalase).

Induced by exogenous signals, such as neurotoxins or UV-light or by intracellular changes that are based on genetic conditions such as mutations this balance of free radical production and detoxification is significantly disturbed leading to the accumulation of ROS and RNS. This overflow of reactive species leads to rapid oxidation of cellular structures leaving remaining molecules non-functional (for review see Halliwell and Gutteridge 2003).

For AD many factors have been discovered that either directly or indirectly induce disturbance of the oxidative homeostasis, including the amyloid-β protein (Aβ), that is one major hallmark protein of Alzheimer's neuropathology. The significance of these findings will be discussed below but it has to be mentioned here that changes in protein cleavage and biochemistry, such as in AD, have also been discussed for other human neurodegenerative diseases, so that various neurodegenerative disorders can now be considered as protein aggregation disorders (*e.g.* AD, Huntington's disease, Parkinson's disease) suggesting that aggregation of protein and misfolding of protein leads to changes in protein degradation and also subsequent changes in oxidative conditions. While oxidative stress has long been believed to be a downstream event of pathological conditions in neurodegeneration recent findings strongly indicate that oxidation can occur, very frequently upstream in the disease process and experimental evidence is mounting that it can indeed play a direct causative role.

2. OXIDATIVE CHANGES IN ALZHEIMER'S DISEASE PATHOGENESIS

The manifestations of free radical damage observed in AD have recently been reviewed in detail (Markesberry and Carney 1999; Joseph *et al.*, 2001). In general, it can be summarized that specific oxidative changes can be observed in pathological neurodegenerative disorders. These include:

- Levels of oxidatively altered metabolites generally rise in case of neurodegenerative disease.
- Afflicted cells in neurodegenerative disorders show specific characteristics of an oxidative stress response.
- Disturbances of energy metabolism, especially with respect to mitochondria, can be observed, which are accompanied by an increased production of free radicals.
- Animal models reproducing specific neuropathological hallmarks of neurodegenerative disease concomitantly show signs of oxidative stress.
- Certain neurodegenerative disorders can present as sporadic or familiar forms. In some of these familiar forms, mutations of antioxidant enzymes or mutations in proteins which indirectly serve to maintain oxygen or nitrogen radical homeostasis have been found.

The fact that many neurodegenerative processes in the brain are characterised by oxidation and oxidative destruction suggest that the brain is particularly vulnerable to oxidative damage. Indeed, it can be argued that the brain is indeed more prone to oxidation because it has high energy requirements and a high physiological oxygen consumption rate. Moreover the brain is enriched in non-saturated fatty acid side chains in its diverse cellular membranes, it contains high levels of transition metals which can act as catalysts for the formation of ROS and it has a relative low level of antioxidants, compared to other tissues in the body (for review see Floyd, 1999).

Since AD is an age-associated disorder the involvement of oxidative stress in the pathogenesis of this disorder is further supported by the free radical hypothesis of aging which states that the age-related accumulation of reactive oxygen species results in damage to major cell components

(Harman, 1992; Ames *et al.*, 1993). Quantitative markers of oxidative stress in AD brain have been described and include:

- Lipid peroxidation (*e.g.* lipid hydroperoxids, 4-hydroxynonenal, isoprostanes)
- Protein oxidation (*e.g.* protein carbonyls, nitrotyrosine)
- DNA oxidation
- Glyco-oxidation (*e.g.* advanced glycation end products).

During lipid peroxidation the non-saturated carbohydrate fatty acid side chains of membrane phospholipids are peroxidized during the attack by ROS. Indeed markers of brain lipid peroxidation have been the most studied biochemical finding of oxidative stress in AD. In quantitative post mortem studies lipid peroxidation has been quantified by measuring malonedialdeyde levels. In addition, lipid hydroperoxide and isoprostane levels have been measured in AD tissue. Protein oxidation results in the formation of protein carbonyls or nitrotyrosine. Again quantitative post mortem studies can be preformed using standardised oxidation assays and have been presented in various studies (for review see Pratico and Delanty, 2000).

The more recent findings that markers of oxidative stress precede the appearance of other, preferable more disease-specific pathological hallmarks, has been of great value in proving a causative role of oxidative stress in AD. In this respect, increased 8-hydroxyguanosine and nitrotyrosine levels have been documented to precede the formation of amyloid β plaques in Down's syndrome, which is frequently discussed as a human model of AD disease caused by the overproduction of Aβ (Nunomura *et al.*, 2000). In an experimental transgenic mouse model of AD overexpressing amyloid precursor protein, an increased level of isoprostane, a well known marker of lipid peroxidation, has been reported to occur before the appearance of immuno-stainable amyloid plaques (Pratico *et al.*, 2001). But what is the experimental evidence that oxidation directly or indirectly caused by Aβ is indeed involved in the early pathogenesis of AD?

AD is possibly the best characterized neurological disorder with respect to the involvement of oxidative stress and free radical changes (for review see Markesbery and Carney, 2001; Harman, 1996; Christen, 2000). As for many years it has been found that oxidation end products of lipids proteins and DNA do indeed occur in the pathological tissue of AD and that there are increased levels of a variety of oxidation products when Alzheimer tissue is compared with age-matched controls, more recently focus has been put on the kinetic analysis of the occurrence of oxidative products during the AD process. Most interestingly, recent kinetic analysis of the time point of occurrence of oxidatively altered metabolites in Down's syndrome, AD and some animal models of both disorders, have suggested that oxidative stress might constitute a rather early event in the cause of these diseases. These changes may even trigger some of the neurodegenerative cascades in these disorders (Nunomura *et al.*, 2000; Pratico *et al.*, 2001, Nunomura *et al.*, 2001). Various oxidative mechanisms to initiate neuronal degeneration have been clearly described, for instance the oxidative cross-linking of some amyloid-β peptides to form a first proteolysis-resistant aggregation weed (Galeazzi *et al.*, 1999; Koppaka and Axelsen, 2000, Perutz and Windle, 2001). So, biochemistry can explain the fact that oxidative changes can cause downstream neurodegenerative events.

A large number of animal models for AD have been developed in order to study the pathogenesis of the disease. Generally they used the strategy of transgenic overexpression of APP, the precursor protein of amyloid-β protein. Very frequently mutant forms of APP, derived from familiar cases of AD, are used for the generation of transgenic animals. Frequently, combination with a mutation in another gene that is associated with familiar forms of AD, such as the presenilin gene, is also used (Games *et al.*, 1995; Sturchler-Pierrat and Abramowski, 1997; Holcomb *et al.*, 1998). Indeed, these animal (usually mouse) models are of great value to study the process of amyloid formation and deposition in the brain tissue (see also Chapter 2). Unfortunately, these animal models do not fully represent the pathogenetic process of AD, since additional factors are obviously necessary to develop a full AD neurodegeneration. In all of these animal models the nerve cell death that does occur is only very limited, if occurring at all. Nevertheless, as mentioned before they provide biochemical tools to study amyloid processing under *in vivo* conditions. These transgenic mice mirror single neuroanatomical or physiological characteristics of the disease, such as Aβ deposition, changes in synaptic formation, presence of an inflammatory reaction, hyperphosphorylation of tau, some behavioural deficits and a general age-dependence of these individual pathological manifestations to a varying degree. Unfortunately, they have usually not been able to reproduce the combined major features of AD, *i.e.* Aβ and tau pathology, concomitantly (for review see Dewachter *et al.*, 2000). Apart form these animal models, a major model used to test and identify potentially beneficial AD drugs is the direct and indirect toxicity of Aβ in cell and tissue culture. Since first described by Bruce Yankner (Yankner *et al.*, 1989) much effort has been put into the identification of the possible toxicity mechanisms of the Aβ peptides. Of course, it was very intriguing that the Aβ protein that is deposited in high amounts in the brain also induces toxicity at the cellular level. Based on tist, we have been able to show that under certain cell culture conditions nerve cells undergo an oxidative degeneration induced by Aβ and this could be prevented by free radical scavengers such as vitamin E (α-tocopherol) that directly prevent lipid peroxidation (Behl *et al.,* 1992; 1994). Many groups confirmed the finding of a direct oxidative toxicity but the concept of Aβ toxicity *in vitro* developed some weakness, such as the dependences of the toxic effect on relatively soft influences (peptide sources, peptide conformation and the exact application procedure). Subsequently, the toxicity model was further modified and it was described that perhaps it is not only a direct toxic effect of the Aβ on neurons in cell culture but rather it is the attraction of inflammatory mediators that cause secondary oxidation, frequently seen during inflammation in tissue (for review see McGeer and McGeer, 1996). The strongest evidence that argues against a direct detrimental toxic effects of aggregated Aβ deposits on the nerve cells is given by the APP transgenic animals, since there is no significant neurodegeneration surrounding the plaques formed in these animals. Still, the importance of Aβ misprocessing by certain secretases for AD pathogenesis is strongly supported by much experimental data (for review see Selkoe, 1999), most of all by the fact that familiar cases of AD cause an overproduction of Aβ in an autosomal dominant manner. More recently the concept of amyloid toxicity has been further developed and it has been found that it is not the fully aggregated Aβ that can be found in senile amyloid plaques in AD brain tissue but rather early forms of amyloid aggregation such as Aβ oligomers and protofibrils. Indeed, it has been found that only such early Aβ oligomers and protofibrils on their way to be deposited as amyloid plaque in the brain, can disturb synaptic function, as documented by the adverse changes in synaptic transmission during long-term potentiation (Walsh *et al.*, 2002). Despite some problems in the *in vitro* amyloid toxicity model it is still used as an AD relevant paradigm of neurodegeneration, and is useful in order to develop novel antioxidant compounds. But how are antioxidants defined, how do they function and what is the experimental and clinical evidence for a role of antioxidant compounds in the prevention and perhaps in the therapeutic treatment of AD?

3. ANTIOXIDANTS AND ALZHEIMER'S DISEASE

3.1 Biochemical classification of antioxidant compounds

When discussing *antioxidants* the term *free radicals scavengers* is frequently used synonymously. In fact the potentially toxic cellular overload with free radicals and other electrophils can essentially be alleviated by three mechanisms that can be clearly distinguished (for review see Sies, 1993). Those strategies are:

- The inhibition of free radical formation.
- The direct scavenging of free radicals by antioxidant compounds.
- The strengthening self-defence of the cell and its damage repair capacity.

While each of these strategies may be equally promising most data with respect to neuroprotection and prevention against neurodegenerative diseases in vitro and *in vivo,* including clinical studies, the focus is mostly put on compounds acting according to the second strategy, substances that directly act at the biochemical level with reactive free radical species.

Compounds that chemically interfere with formed free radicals are also called *direct antioxidants* and include a variety of chemically different compounds such as tocopherols and other monophenols, flavonoids and other polypheonls, arylamines and indoles, ebselen and other selenium containing compounds, carotin, lyocopen, retinol and other polyenes, and finally manganese containing mimetics of catalase/superoxide dismutase.

The action of direct antioxidants does not rely on endogenous cellular macromolecules to exert their primary effect, since they react with a damaging free radical molecule itself. The major classes of antioxidants in this group are chain breaking antioxidants such as phenols, which can be divided into monopheolic compounds (tocopherol, estrogen, serotonin) which are mainly endogenous and polyphenolic compounds, which are usually exogenous such as flavonoids, stilbenes and hydrochinons. Low molecular weight enzyme mimetics constitute a slightly different group of antioxidant compounds since their mechanism of reaction is more dependent on endogenous reducing systems, compared to the chain breaking antioxidants. Enzyme mimetics react with small diffusible free radicals at relatively high velocities and are very effective, but they need to be cyclically re-reduced in order to exhibit true catalytic efficacy. Today, enzyme mimetics for a few but important antioxidant enzymes are known, most notably ebselen and other selenium compounds and a variety of manganese containing complexes. Ebselen mimics glutathione peroxidase and scavenges peroxinitrite, while manganese complexes mimic superoxide dismutase and to some extent catalase (for review see Moosmann and Behl, 2002; Klotz and Sies, 2003).

3.2 Experimental and clinical evidence of antioxidant neuroprotection

A powerful neuroprotective role as an antioxidant in an *in vitro* model of oxidative neurodegeneration, caused by Aβ, has been initially assigned to vitamin E (α-tocopherol) (Behl and Davis, 1992). These *in vitro* studies were followed by experimental and clinical studies and therefore tocopherol has been the prime model antioxidant for AD. In a multi-centre, double blind, placebo controlled study on 341 patients with moderate to severe AD, 200 IU of α-tocopherol administered for two years lead to a slight but significant delay in reaching the primary outcome, which were generalized markers of neurological decline (Sano *et al.*, 1997). This initial result fuelled a number of other vitamin E intervention studies in AD. It is also worth mentioning another multi-centre clinical trial by the AD cooperative study on patients with mild cognitive impairment, presumably containing individuals with AD at a very early stage of the disease. The outcome of this particular study will be very interesting but it is not yet finalised. The relevant aspects of tocopherol treatment in AD have been reviewed in detail recently (Grundman, 2000).

Acetylcarnitine (3g/per day) as an antioxidant showed no significant effects in an one year clinical trial in 229 early onset AD patients which confirmed the outcome of two earlier trials of this compound (Thal *et al.*, 1996, 2000; Spangnolia *et al.*, 1991). Following these earlier trials the minor positive effects induced by acetylcarnitine could not subsequently be verified (Spangnolia *et al.*, 1991).

Idebenon, another metabolic antioxidant, which might beneficially affect mitochondrial function, has been employed in a relatively large clinical study. Idebenon which is structurally similar to ubichinon has been reported to slow down AD progression in dose-dependent fashion (Gutzmann and Hadler, 1998). The exact mechanism as to how this protective effect is reached is still not clear and needs to be investigated further in greater detail.

While there are a number of other antioxidants under consideration in clinical trials with AD patients a few other approaches should be mentioned here which also include aspects of antioxidant neuroprotection. These approaches include estrogen replacement therapy (ERT) (Paganini-Hill and Henderson, 1996; Tang *et al.*, 1996; Waring *et al.*,1999; Carlson *et al.*, 2001) and the chronic intake of anti-inflammatory drugs (in t' Veld *et al.*, 2001; Breitner and Zandi, 2001; Zandi and Breitner, 2001; Nourhashemi *et al.*, 2000). Estrogen is a well known neuroprotective compound and part of its protective activity is mediated *via* its function as a monophenolic direct antioxidant (for review see Behl, 2002). Based on the fact that estrogen is a multifunctional hormone and develops many estrogen receptor-dependent activities throughout the body, including the brain, it is not a prime antioxidant. Nevertheless, antioxidant activity produced by estrogen needs to be considered whenever possible oxidant damage occurs in tissue. Since it is known that estrogen can enter the brain quite readily to develop its activities in various target areas, it could function as a blueprint structure for the development of other estrogen-derived antioxidants which do not develop hormonal side-effects, as investigated by Moosmann and Behl (1999).

Anti-inflammatory drugs may interfere with immunological components in the development of AD (Zandi and Breitner, 2001; McGeer and Rogers, 1992; Akiyama *et al.*, 2000). Non-steroidal anti-inflammatory drugs (NSAIDs) inhibit cyclo-oxygenase enzymes which reduce the production of oxidants by the enzymes themselves as well by arachidonic acid attracted and activated microglia (Babior, 2000). In addition they may prevent excitotoxicity and oxidative glutamate toxicity both of which involve cyclo-oxygenase activities. The statistical significance of the negative association between AD and the long-term intake of NSAIDs is strongly supported by a recent prospective cohort study conducted in the Netherlands (the Rotterdam study) (in t' Veld *et al.*, 2001). The Rotterdam study reported a relative risk of AD of 0.2 for long-term (over 24 months) users of NSAIDs as compared to none-users. The effects on NSAIDs in the Rotterdam study appeared to be strongly dependent on extend period of drug use (in t' Veld *et al.*, 2001). Further clinical intervention studies have so far yielded only negative results, which do not necessarily contradict the results of epidemiological studies (Zandi and Breitner, 2001). But there is another interesting link between NSAIDs and AD pathogenesis, when focusing on APP processing. Weggen *et al.* (2001) have shown that NSAIDs directly interfere with the secretase process preventing the formation of Aβ aggregates. Therefore the neuroprotective effect of NSAIDs in AD could be mediated by their antioxidant activity by blocking cyclo-oxygenases as well as by their effect on APP processing. Whether both processes are linked is under investigation.

Additionally, other compounds with direct or indirect antioxidant activities have been used in experimental and clinical AD prevention and therapy including Ginkgo biloba extracts (*e.g.* Oken *et al.*, 1998), the pineal hormone melatonin (*e.g.* Brusco and Marquez, 2000) and the Cu/Zn chelator clioquinol. The latter used in APP transgenic mice at a concentration of 30 mg/kg/day resulted in an amelioration of general health parameters and a reduction of amyloid deposition in the brain (Gouras and Beal, 2001). Given the overwhelming evidence that metals cause oxidative stress in an Aβ bound form, such a treatment might prevent neuronal cell loss by preventing oxidant formation (Smith *et al.,* 1997; Huang *et al.,* 1999). One single-blind study with desferrioamine, a transition metal chelator administered intramuscularly, on 48 patients AD patients has reported essentially positive effects of this class of compounds on AD progression (Crapper McLachlan *et al.,* 1991).

In summary, despite the rather ambivalent general picture to-date, we expect that direct antioxidants will nevertheless become part of a future standard therapy for certain neurodegenerative disorders, including AD. Based on the biochemical evidence the highest potential may lie in blood-brain-barrier permeable chain-braking antioxidants of the tocopherol type and in enzyme mimetics of the glutathione peroxidase or SOD/catalase type.

Very likely, the best outcome of neuroprotection is when such compounds are used in a preventive fashion. Just recently a large clinical trial (Cache study) has shown that combined administration of vitamin E (α-tocopherol) and vitamin C (ascorbate) is associated with reduced prevalence and incidence of AD. The authors concluded that antioxidant supplements merit further studies as agents for the primary prevention of AD (Zandi *et al.*, 2004), strongly supporting the effort to identify and design novel neuroprotective antioxidant compounds for neurological disease prevention.

ACKNOWLEDGEMENTS

The work of the author is supported in part by the Deutsche Forschungsgemeinschaft, the European Union and the Peter-Beate-Heller Foundation.

REFERENCES

Akiyama, H., Barger, S., Barnum, S., Bradt, B., Bauer, J., Cole, G.M., Cooper, N.R., Eikelenboom, P., Emmerling, M., Fiebich, B.L., Finch, C.E., Frautschy, S., Griffin, W.S., Hampel, H., Hull, M., Landreth, G., Lue, L., Mrak, R., Mackenzie, I.R., McGeer, P.L., O'Banion, M.K., Pachter, J., Pasinetti, G., Plata-Salaman, C., Rogers, J., Rydel, R., Shen, Y., Streit, W., Strohmeyer, R., Tooyoma, I., Van Muiswinkel, F.L., Veerhuis, R., Walker, D., Webster, S., Wegrzyniak, B., Wenk, G., and Wyss-Coray, T., 2000, Inflammation and Alzheimer's disease. *Neurobiol. Aging.* **21:** 383-421.

Ames, B.N, Shigenda, M.K, and Hagen, T.M., 1993, Oxidants, antioxidants, and the degenerative diseases of aging. *Proc Natl Acad Sci USA* **90:** 7915-7922.

Babior, B.M., 2000, Phagocytes and oxidative stress. *Am. J. Med.* **109:** 33-44.

Behl, C., 2002, Oestrogen as a neuroprotective hormone. *Nat. Rev. Neurosci.* **3:** 433-442.

Behl, C., Davis, J., Cole, G.M., and Schubert, D., 1992, Vitamin E protects nerve cells from amyloid beta protein toxicity. *Biochem. Biophys. Res. Commun.* **186:** 944-950.

Behl, C., Davis, J.B., Lesley R., and Schubert, D., . 1994, Hydrogen peroxide mediates amyloid beta protein toxicity. *Cell* **77:** 817-827.

Breitner, J.C., and Zandi, P.P., 2001, Do nonsteroidal antiinflammatory drugs reduce the risk of Alzheimer's disease? *N. Engl. J. Med.* **345:** 1567-1568.

Brusco, L.I., Marquez, M., and Cardinali, D.P., 2000, Melatonin treatment stabilizes chronobiologic and cognitive symptoms in Alzheimer's disease. *Neuroendocrinol. Lett.* **21:** 39-42.

Carlson, M.C., Zandi, P.P., Plassman, B.L., Tschanz, J.T., Welsh-Bohmer, K.A., Steffens, D.C., Bastian, L.A., Mehta, K.M., and Breitner, J.C., 2001, Cache County Study Group. Hormone replacement therapy and reduced cognitive decline in older women: the Cache County Study. *Neurology* **57:** 2210-2216.

Christen, Y., 2000, Oxidative stress and Alzheimer disease. *Am. J. Clin. Nutr.* **71:** 621S-629S.

Crapper McLachlan, D.R., Dalton, A.J., Kruck, T.P., Bell, M.Y., Smith, W.L., Kalow, W., Andrews, D.F., 1991, Intramuscular desferrioxamine in patients with Alzheimer's disease. *Lancet* **337:** 1304-1308.

Dewachter, I., van Dorpe, J., Spittaels, K., Tesseur, I., Van Den Haute, C., Moechars, D., Van Leuven, F., 2000, Modeling Alzheimer's disease in transgenic mice: effect of age and of presenilin1 on amyloid biochemistry and pathology in APP/London mice. *Exp. Gerontol.* **35:** 831-841.

Floyd, R.A., 1999, Antioxidants, oxidative stress, and degenerative neurological disorders. *Proc. Soc. Exper. Biol. Me.d* **222:** 236-245.

Galeazzi, L., Ronchi, P., Franceschi, C., and Giunta, S., 1999, In vitro peroxidase oxidation induces stable dimers of beta-amyloid (1-42) through dityrosine bridge formation. *Amyloid* **6:** 7-13.

Games, D., Adams, D., Alessandrini, R., Barbour, R., Berthelette, P., Blackwell, C., Carr, T., Clemens, J., Donaldson, T., Gillespie, F,, *et al.*, 1995, Alzheimer-type neuropathology in transgenic mice overexpressing V717F beta-amyloid precursor protein. *Nature* **373:** 523-527.

Gouras, G.K., and Beal, M.F., 2001, Metal chelator decreases Alzheimer beta-amyloid plaques. *Neuron* **30:** 641-642.

Grundman, M.. 2000, Vitamin E and Alzheimer disease: the basis for additional clinical trials. Am. J. Clin. Nutr. **71:** 630S-636S.

Gutzmann, H., and Hadler, D., 1998, Sustained efficacy and safety of idebenone in the treatment of Alzheimer's disease: update on a 2-year double-blind multicentre study. *J. Neural. Transm.* Suppl. **54:** 301-310.

Halliwell, B., and Gutteridge, J.M.C., 2003, Free Radicals in Biology and Medicine. Oxford University Press.

Harman D., 1992, Free radical theory of aging. *Mutan Res.* **275:** 257-266.

Harman D., 1996, A hypothesis on pathogenesis of Alzheimer's disease. *Ann. NY Acad. Sci.* **786:** 152-168.

Holcomb, L., Gordon, M.N., McGowan, E., Yu, X., Benkovic, S., Jantzen, P., Wright, K., Saad, I., Mueller, R., Morgan, D., Sanders, S., Zehr, C., O'Campo, K., Hardy, J., Prada, C.M., Eckman, C., Younkin, S., Hsiao, K., and Duff, K., 1998, Accelerated Alzheimer-type phenotype in transgenic mice carrying both mutant amyloid precursor protein and presenilin 1 transgenes. *Nature Med.* **1:** 97-100.

Huang, X., Cuajungco, M.P., Atwood, C.S., Hartshorn, M.A., Tyndall, J.D., Hanson, G.R., Stokes, K.C., Leopold, M., Multhaup, G., Goldstein, L.E., Scarpa, R.C., Saunders, A.J., Lim, J., Moir, R.D., Glabe, C., Bowden, E.F., Masters, C.L., Fairlie, D.P., Tanzi, R.E., and Bush, A.I., 1999, Cu(II) potentiation of alzheimer abeta neurotoxicity. Correlation with cell-free hydrogen peroxide production and metal reduction. *J. Biol. Chem.* **274:** 37111-37116.

in t' Veld, B..A., Ruitenberg, A., Hofman, A., Launer, L.J., van Duijn, C.M., Stijnen, T., Breteler, M.M., and Stricker, B.H. 2001, Nonsteroidal antiinflammatory drugs and the risk of Alzheimer's disease. *N. Engl. J. Med.* **345:** 1515-1521.

Joseph, J, Skuitt-Hale, B, Denisova, N.A, Martin, A, Perry, G, and Smith, M.A, 2001, Copernicus revisited: amyloid β in Alzheimer's disease. *Neurobiol.. Aging* **22:** 131-146.

Klotz, L.O., Sies, H., 2003, Defenses against peroxynitrite: selenocompounds and flavonoids. *Toxicol. Lett.* **140-141:** 125-132.

Koppaka, V., and Axelsen, P.H., 2000, Accelerated accumulation of amyloid beta proteins on oxidatively damaged lipid membranes. Biochemistry 39: 10011-10016.

Markesberry, W.R., and Carney, J.M., 1999, Oxidative alterations in Alzheimer's disease. *Brain Pathol.* **9:** 133-146.

McGeer, P.L., and, McGeer, E.G., 1996, Anti-inflammatory drugs in the fight against Alzheimer's disease. *Ann. N. Y. Acad. Sci.* **777:** 213-220.

McGeer, P.L., and Rogers, J., 1992, Anti-inflammatory agents as a therapeutic approach to Alzheimer's disease. *Neurology* **42:** 447-449.

Moosmann, B., and Behl, C., 1999, The antioxidant neuroprotective effects of estrogens and phenolic compounds are independent from their estrogenic properties. *Proc. Natl. Acad. Sci. U S A* **96:** 8867-8872.

Moosmann, B., and Behl, C., 2002, Antioxidants as treatment for neurodegenerative disorders. *Expert Opin. Investig. Drugs.* **11:** 1407-1435.

Nunomura, A., Perry, G., Pappolla, M.A., Friedland, R.P., Hirai, K., Chiba, S., and Smith M.A., 2000, Neuronal oxidative stress precedes amyloid-beta deposition in Down syndrome. *J. Neuropathol. Ex.p Neurol.* **59:** 1011-1017.

Nunomura, A., Perry, G., Aliev, G., *et al.*, 2001, Cytosolic peroxidative damage is an early event in Alzheimer's disease. *J. Neuropathol. Exp. Neurol.* **60:** 759-767.

Nourhashemi, F., Gillette-Guyonnet, S., Andrieu, S., Ghisolfi, A., Ousset, P.J., Grandjean, H., Grand, A., Pous, J., Vellas, B., and Albarede, J.L., 2000, Alzheimer disease: protective factors. *Am. J. Clin. Nutr.* **71:** 643S-649S.

Oken, B.S., Storzbach, D.M., and Kaye, J.A., 1998, The efficacy of Ginkgo biloba on cognitive function in Alzheimer disease. *Arch. Neurol.* **55:** 1409-1415.

Paganini-Hill, A., and Henderson, V.W., 1996, Estrogen replacement therapy and risk of Alzheimer disease. *Arch Intern Med.* **156:** 2213-2217.

Perutz, M.F., and Windle, A.H., 2001, Cause of neural death in neurodegenerative diseases attributable to expansion of glutamine repeats. Nature **412:** 143-144.

Pratico, D., and Delanty, N., 2000, Oxidative injury in diseases of the central nervous system: focus on Alzheimer's disease. *Am. J. Med.* **109:** 577-585.

Pratico, D, Uryu, K, Leight, S., Trojanoswki, J.Q., and Lee, V.M., 2001, Increased lipid peroxidation precedes amyloid plaque formation in an animal model of Alzheimer amyloidosis. *J. Neurosci.* **21:** 4183-4187.

Sano, M., Ernesto, C., Thomas, R.G., Klauber, M.R., Schafer, K., Grundman, M., Woodbury, P., Growdon, J., Cotman, C.W., Pfeiffer, E., Schneider, L.S., and Thal, L.J., 1997, A controlled trial of selegiline, alpha-tocopherol, or both as treatment for Alzheimer's disease. The Alzheimer's Disease Cooperative Study. *N. Engl. J. Med.* **336:** 1216-1222.

Selkoe, D.J., 1999, Translating cell biology into therapeutic advances in Alzheimer's disease. *Nature* **399** (Suppl) A23-31.

Sies, H., 1993, Strategies of antioxidant defense. *Eur. J. Biochem.* **215:** 213-219.

Smith, M.A., Harris, P.L., Sayre, L.M., and Perry, G., 1997, Iron accumulation in Alzheimer disease is a source of redox-generated free radicals. *Proc. Nat.l Acad. Sci. U S A.* 1997 **94:** 9866-9868.

Spagnoli, A., Lucca, U., Menasce, G., Bandera, L., Cizza, G., Forloni, G., Tettamanti, M., Frattura, L., Tiraboschi, P., Comelli, M., *et al.*, 1991, Long-term acetyl-L-carnitine treatment in Alzheimer's disease. *Neurology* **41:** 1726-1732.

Sturchler-Pierrat, C., Abramowski, D., Duke, M., Wiederhold, K.H., Mistl, C., Rothacher, S., Ledermann, B., Burki, K., Frey, P., Paganetti, P.A., Waridel, C., Calhoun, M.E., Jucker, M., Probst, A., Staufenbiel, M., and Sommer, B., 1997, Two amyloid precursor protein transgenic mouse models with Alzheimer disease-like pathology. *Proc. Natl. Acad. Sci. USA* **94:** 13287-13292.

Tang, M.X., Jacobs, D., Stern, Y., Marder, K., Schofield, P., Gurland, B., Andrews, H., and Mayeux, R., 1996, Effect of oestrogen during menopause on risk and age at onset of Alzheimer's disease. *Lancet* **348:** 429-432.

Thal, L.J., Calvani, M., Amato, A., and Carta, A., 2000, A 1-year controlled trial of acetyl-l-carnitine in early-onset AD. *Neurology* **55:** 805-810.

Thal, L.J., Carta, A., Clarke, W.R., Ferris, S.H., Friedland, R.P., Petersen, R.C., Pettegrew, J.W., Pfeiffer, E., Raskind, M.A., Sano, M., Tuszynski, M.H., and Woolson, R.F., 1996, A 1-year multicenter placebo-controlled study of acetyl-L-carnitine in patients with Alzheimer's disease. *Neurology* **47:** 705-711.

Walsh, D.M., Klyubin, I., Fadeeva, J.V., Cullen, W.K., Anwyl, R., Wolfe, M.S., Rowan, M.J. ,and Selkoe, D.J., 2002, Naturally secreted oligomers of amyloid beta protein potently inhibit hippocampal long-term potentiation in vivo. *Nature* **416:** 535-539.

Waring, S.C., Rocca, W.A., Petersen, R.C., O'Brien, P.C., Tangalos, E.G., and Kokmen, E., 1999, Postmenopausal estrogen replacement therapy and risk of AD: a population-based study. *Neurology* **52:** 965-970.

Weggen, S., Eriksen, J.L., Das, P., Sagi, S.A., Wang, R., Pietrzik, C.U., Findlay, K.A., Smith, T.E., Murphy, M.P., Bulter, T., Kang, D.E., Marquez-Sterling, N., Golde, T.E., and Koo, E.H., 2001, A subset of NSAIDs lower amyloidogenic Abeta42 independently of cyclooxygenase activity. *Nature* **414:** 212-216.

Yankner, B.A., Dawes, L.R., Fisher, S., Villa-Komaroff, L., Oster-Granite, M.L., and Neve, R.L., 1989, Neurotoxicity of a fragment of the amyloid precursor associated with Alzheimer's disease. *Science* **245:** 417-420.

Zandi, P.P., Anthony, J.C., Khachaturian, A.S., Stone, S.V., Gustafson, D., Tschanz, J.T., Norton, M.C., Welsh-Bohmer, K.A., and Breitner, J.C., 2004, Cache County Study Group. Reduced risk of Alzheimer disease in users of antioxidant vitamin supplements: the Cache County Study. *Arch. Neurol.* **61:** 82-88.

Zandi, P.P., and Breitner, J.C., 2001, Do NSAIDs prevent Alzheimer's disease? And, if so, why? The epidemiological evidence. *Neurobiol Aging* **22:** 811-817.

Chapter 4

β-Secretase, APP and Aβ in Alzheimer's Disease

Robert Vassar
Department of Cell and Molecular Biology, The Feinberg School of Medicine, Northwestern University, Chicago Avenue, Chicago, Il, USA

Abstract: Amyloid plaques, hallmark neuropathological lesions in Alzheimer's disease (AD) brain, are composed of the β-amyloid peptide (Aβ). A large body of evidence suggests Aβ is central to the pathophysiology of AD and is likely to start this intractable neurodegenerative disorder. Mutations in three genes (amyloid precursor protein/APP, presenilin1, presenilin2) cause early on-set familial AD by increasing synthesis of the toxic 42 amino acid species of Aβ (Aβ42). Fibrillar Aβ in amyloid plaques appears to cause neurodegeneration, although recent studies suggest soluble Aβ oligomers may also be neurotoxic. Regardless, given the strong correlation between Aβ and AD, therapeutic strategies to lower cerebral Aβ levels should prove beneficial for the treatment of AD. Aβ is derived from APP via cleavage by two proteases, β- and γ-secretase. β-secretase, recently identified as the novel aspartic protease BACE1, initiates the formation of Aβ. Consequently, BACE1 in principle is an excellent therapeutic target for strategies to reduce the production of Aβ in AD. However, the discovery of the homologue BACE2 raised the question of whether it too may be a β-secretase. To settle this issue, our group and others have used gene targeting to generate BACE1 deficient (knockout) mice. These BACE1 knockout mice have been instrumental in validating BACE1 as the authentic β-secretase *in vivo*. Here, I review the roles of BACE1, APP, and Aβ in AD and discuss the implications of therapeutic approaches that target BACE1 for the treatment of AD.

Key words: β-secretase, amyloid precursor protein (APP), β- amyloid (Aβ), BACE1

1. INTRODUCTION

AD is a neurodegenerative disorder characterized by progressive dementia that inevitably leads to incapacitation and death. Synaptic loss and neuronal death occurs in AD brain regions critical for cognitive function, including cerebral cortex, entorhinal cortex, and hippocampus (reviewed in (Terry *et al.*, 1999)). The inexorable loss of neurons and synapses over the course of AD is responsible for the dementia that slowly robs AD patients of their memories, personalities, and eventually their lives. Clearly, understanding the pathophysiological mechanisms underling neurodegeneration in AD is essential for the rational design of therapies aimed at slowing or stopping disease progression.

Two characteristic neuropathological lesions define AD: 1. amyloid plaques, extracellular deposits primarily composed of the 4 kDa, 40-42 amino acid Aβ peptide (Glenner and Wong, 1984), a product of APP proteolysis, and 2. neurofibrillary tangles, intracellular aggregates of the microtubule associated protein tau (Lee *et al.*, 1991). The relationships between amyloid plaques, neurofibrillary tangles, and the pathogenic mechanisms causing AD are controversial. However, cumulative evidence from a large number of studies indicates that Aβ is critically involved at an early stage in AD pathogenesis.

A strong genetic correlation exists between early-onset familial forms of AD (FAD) and the 42 amino acid species of the Aβ peptide (Aβ42) (reviewed in Hutton *et al.*, 1998; Sisodia *et al.*, 1999; Younkin, 1998). Autosomal dominant mutations in the genes for APP, presenilinl (PS1), and presenilin2 (PS2) all increase production of Aβ42 and cause FAD with nearly 100% penetrance. Moreover, Down's syndrome (DS) patients, who all develop early-onset AD, have an extra copy of the APP gene on chromosome 21, thus causing Aβ42 overproduction. In FAD, the Aβ42 increase occurs years before AD symptoms arise, suggesting that Aβ42 is likely to initiate AD pathophysiology. The exact structural form of Aβ42 conferring pathogenicity in AD is unclear, since both fibrillar Aβ42 found in amyloid plaques and soluble Aβ42 oligomers appear to cause neurodegeneration *in vitro* and *in vivo* (see Section 9). Although the majority of AD cases appear to be sporadic, the strong association of Aβ42 with FAD argues in favor of a critical role for Aβ42 in the etiology of sporadic AD as well. Consequently, elucidating the molecular pathways responsible for the generation of Aβ, particularly Aβ42, is essential for developing rational therapeutic approaches that lower cerebral Aβ levels in AD.

The Aβ peptide is generated by the endoproteolysis of the large type I membrane protein APP (reviewed in (Selkoe, 2001; Vassar and Citron, 2000). APP is expressed ubiquitously, and Aβ is a normal product of APP metabolism in all cells studied to date. A protease called β-secretase first cleaves APP to form the N-terminus of Aβ at the Asp+1 residue of the Aβ sequence. Two β-secretase cleavage products are produced: a secreted ectodomain of APP named APPsβ and the C99 fragment, the membrane bound C-terminal 99 amino acids of APP. Following β-secretase cleavage, C99 is cut by a second protease called γ-secretase, which cleaves to generate the C-terminus of Aβ. Thus, the mature Aβ peptide is formed and is subsequently secreted from the cell. γ-secretase cleavage is not precise and produces a spectrum of Aβ peptides varying in length by several amino acids at the C-terminus, the majority of which end at Aβ amino acid 40. A third protease, α-secretase, cleaves APP in the middle of the Aβ domain (at Leu+17) and precludes the formation of Aβ (*see also* Chapter 5). α-Secretase cleavage produces two products: the secreted APPsα ectodomain, and the membrane bound C-terminal fragment C83, which in turn is cleaved by γ-secretase to form the non-amyloidogenic 3 kDa fragment, p3.

Interestingly, the mutations in APP that cause FAD are all located near the secretase cleavage sites and they directly affect the efficiency or position of secretase cleavage. For example, the Swedish mutation (discovered in a Swedish family), is the amino acid substitution LysMet→AsnLeu at the P2-P1 positions immediately N-terminal to the β-secretase cleavage site in APP (Mullan *et al.*, 1992a). This mutation makes APP a more efficient substrate for β-secretase and dramatically increases the rate of cleavage at the β-secretase site, thus leading to increased production of total Aβ. Several other FAD mutations have been identified near the γ-secretase site that shift the balance of γ-secretase cleavage toward greater production of Aβ42 (reviewed in Hutton *et al.*, 1998). In addition, some FAD mutations occur near the α-secretase site and appear to reduce the efficiency of α-secretase cleavage, thus providing more APP for β-secretase processing.

Until recently, the secretases were described only as APP-cleaving activities found in cells and tissues, but now molecular identities have been proposed for all. Two different proteases appear to be responsible for the α-secretase activity: TACE (TNF-α converting enzyme)(Buxbaum *et al.*, 1998) and ADAM-10 (a disintegrin and metalloprotrotease domain protein) (Lammich *et al.*, 1999). The presenilin proteins, PS1 and PS2 (Wolfe *et al.*, 1999a; Wolfe *et al.*, 1999b), appear to be components of the γ-secretase complex, along with nicastrin (Yu *et al.*, 2000), Aph1 and Pen2 (Francis *et al.*, 2002; Takasugi *et al.*, 2003). Finally, the β-secretase has been identified as the novel transmembrane aspartic protease BACE1 (for β-site APP Cleaving Enzyme 1) (Sinha *et al.*, 1999; Vassar *et al.*, 1999), also known as Asp2 (for novel aspartic protease 2) (Hussain *et al.*, 1999; Yan *et al.*, 1999) and memapsin2 (for membrane aspartic protease/pepsin 2) (Lin *et al.*, 2000).

2. THE FUNCTIONAL PROPERTIES OF β-SECRETASE

Following the discovery that Aβ was a product of APP endoproteolysis (Kang *et al.,* 1987; Tanzi *et al.*, 1987), a large number of studies were undertaken to define the properties of β-secretase activity in cells and tissues. This information was essential for the validation of BACE1, since the characteristics of a strong β-secretase candidate must match one-to-one with the previously established functional properties of β-secretase activity. Therefore, I summarize the major properties of β-secretase below.

β-secretase activity is present in the majority of cells and tissues of the body (Haass *et al.*, 1992), although maximal activity is found in neural tissue and neuronal cell lines (Seubert *et al.*, 1993). Interestingly, astrocytes exhibit less β-secretase activity than neurons (Zhao *et al.*, 1996). Therefore, it was predicted that the β-secretase enzyme would be widely expressed in many tissues and cell lines, but should be at higher levels in neurons of the brain.

β-secretase activity in cells efficiently cleaves membrane-bound APP substrates only: APP constructs lacking the transmembrane domain are not cleaved in transfected cells (Citron *et al.*, 1995). This implies that β-secretase is likely to be a membrane-bound protease or, alternatively, is tightly associated with a membrane protein.

β-secretase has maximal activity at acidic pH, since agents that disrupt intracellular pH inhibit β-secretase activity (Haass *et al.*, 1995a; Haass *et al.*, 1993; Knops *et al.*, 1995). Moreover, β-secretase activity is highest in the acidic subcellular compartments of the secretory pathway, including the Golgi apparatus and endosomes (Haass *et al.*, 1995b; Koo and Squazzo, 1994). These data suggest that the active site of β-secretase is located within the lumen of acidic intracellular compartments.

Site-directed mutagenesis of the amino acids surrounding the cleavage site in APP has defined the sequence preference of the β-secretase (Citron *et al.*, 1995). Substitutions of larger hydrophobic amino acids (such as Leu found in the Swedish FAD mutation) for the Met residue at P1 improve the efficiency of β-secretase cleavage. Conversely, substitution of the smaller hydrophobic amino acid Val at the same position inhibits cleavage. Many other substitutions at this site and at surrounding positions decrease cleavage, and indicate that the β-secretase is highly sequence-specific.

Radiosequencing demonstrates that Aβ isolated from amyloid plaques, as well as that produced in cell lines, predominantly begins at the Asp+1 residue of Aβ (Roher *et al.*, 1993), although minor Aβ species begin at Val-3, Ile-6, and Glu+11 (Haass *et al.*, 1992). Inhibitor studies suggest that the Val-3 and Ile-6 species are generated by a protease that is different than β-secretase (Citron *et al.*, 1996). However, the Glu+11 species is produced in parallel with Asp+1 Aβ (Gouras *et al.*, 1998), suggesting that β-secretase is responsible for cleaving at both positions. Interestingly, the Glu+11 species is the predominant form of Aβ made in rat primary neuron cultures (Gouras *et al.*, 1998). Finally, β-secretase is insensitive to pepstatin, an inhibitor of many (but not all) aspartic proteases.

3. THE CLONING AND CHARACTERIZATION OF BACE1

Three different methodologies were employed to identify BACE1. We used an expression cloning strategy designed to identify genes that increase Aβ production in cells (Vassar *et al.*, 1999). Following transfection with pools of ~100 random cDNAs from an HEK 293 cell cDNA expression library, Aβ levels were measured by enzyme linked immunosorbent assay (ELISA), positive pools were broken down to single clones by sib selection, and the full-length BACE1 cDNA was isolated and sequenced. Sinha and colleagues (Sinha *et al.*, 1999) employed an affinity purification approach to isolate the β-secretase from human brain. Their affinity ligand was a peptide-based transition-state analog that strongly inhibits and binds tightly to the enzyme. Using an *in vitro* assay to follow β-secretase cleavage, the group then performed several enrichment steps before the final affinity purification step with the substrate analog inhibitor. N-terminal sequencing revealed a single amino acid sequence, which was used to isolate the full-length cDNA by conventional cloning methods. The other groups employed various genomic strategies to identify novel aspartic proteases that were evaluated for β-secretase properties (Hussain *et al.*, 1999; Lin *et al.*, 2000; Yan *et al.*, 1999). For example, based on the sequence characteristics of the β-secretase cleavage site, Yan and colleagues (Yan *et al.*, 1999) reasoned that the β-secretase might be an aspartic protease and therefore searched the C. elegans genome database to identify novel aspartic proteases. Using the C. elegans protease sequences, they searched vertebrate expressed sequence tag (EST) databases and found four novel human aspartic proteases. The group then screened the candidates for β-secretase function using antisenese technology, and determined that only one protease had bona fide β-secretase activity. Most importantly, all the groups identified exactly the same protein and concur it is β-secretase, even though the groups used very different methodological approaches. This fact provides strong support for the hypothesis that BACE1/Asp2/memapsin2 (henceforth referred to as BACE1) is the authentic β-secretase.

BACE1 exhibits all the known functional properties and characteristics of the β-secretase. First, BACE1 is clearly a protease: the 501 amino acid sequence exhibits the hallmark features of eukaryotic aspartic proteases of the pepsin family. Two aspartic protease active site motifs of the sequence DTGS (residues 93-96) and DSGT (residues 289-292) are present in BACE1, and mutation of either aspartic acid residue inactivates the enzyme (Bennett *et al.*, 2000b; Hussain *et al.*, 1999). BACE1 has an N-terminal signal sequence (residues 1-21) and a pro-peptide domain (residues 22-45) that are removed post-translationally, so the mature enzyme begins at residue Glu46. Importantly, BACE1 has a single predicted transmembrane domain near its C-terminus (residues 455-480). Thus, BACE1 is a TypeI membrane protein with a lumenal active site, features predicted for β-secretase. The position of the BACE1 active site within the lumen of intracellular compartments provides the correct topological orientation for cleavage of APP at the β-secretase site. Similar to other aspartic proteases, BACE1 has several N-linked glycosylation sites and six lumenal cysteine residues that form three intramolecular disulfide bonds.

The expression pattern of BACE1 is consistent with that of β-secretase (Marcinkiewicz and Seidah, 2000; Vassar *et al.*, 1999; Yan *et al.*, 1999). The levels of BACE1 mRNA by Northern analysis are highest in pancreas and brain, and are significantly lower in most other tissues. Moreover, by *in situ* hybridization analysis, BACE1 is highly expressed in neurons but little, if any, is present in glial cells of the brain, as expected for β-secretase. The protein is abundant in both normal human and AD brain (Marcinkiewicz and Seidah, 2000; Vassar *et al.*, 1999). The high pancreatic mRNA expression was initially confusing, given the low levels of β-secretase activity in this tissue (Sinha *et al.*, 1999). However, recent reports indicate that a significant proportion of BACE1 mRNA in the pancreas consists of a splice variant missing the majority of exon 3 (Bodendorf *et al.*, 2001). This splice variant encodes a BACE1 isoform lacking β-secretase activity, thus providing an explanation for the low β-secretase activity found in the pancreas. The functional relevance of this pancreas-specific splice variant is unclear.

When transfected into stable APP-overexpressing cell lines, BACE1 induces a dramatic increase in β-secretase activity (Hussain *et al.*, 1999; Lin *et al.*, 2000; Sinha *et al.*, 1999; Vassar *et al.*, 1999; Yan *et al.*, 1999). The immediate products of β-secretase cleavage, APPsβ and C99, are increased several fold over levels found in untransfected cells. Aβ production is also elevated by BACE1 transfection in cells overexpressing wild-type APP (APPwt), but (surprisingly) not in cells overexpressing APP with the Swedish mutation (APPsw), even though C99 levels are dramatically increased. Possibly, the very high C99 levels produced by endogenous BACE1 in APPsw cells may saturate γ-secretase and prevent processing of the additional C99 made after BACE1 transfection. Interestingly, APPsα levels are reduced upon BACE1 transfection of APPwt and APPsw cells, suggesting that α- and β-secretases compete for APP substrate in cells. Contrary to the effects of BACE1 transfection in cells, treatment of APP-overexpressing cells with BACE1 antisense oligonucleotides decreases BACE1 mRNA and inhibits β-secretase activity (Vassar *et al.*, 1999; Yan *et al.*, 1999). BACE1 antisense inhibition reduces production of APPsβ, C99, and Aβ in cells; conversely, APPsα generation is elevated.

BACE1 cleaves APP only at the known β-secretase sites of Asp+1 and Glu+11 of Aβ, as determined by radiosequencing of Aβ and APP C-terminal fragments from APP and BACE1 co-expressing cells (Vassar *et al.*, 1999). Moreover, purified recombinant BACE1 directly cleaves APP substrates at these same sites *in vitro*, demonstrating that the BACE1 molecule intrinsically exhibits protease activity (Vassar *et al.*, 1999; Yan *et al.*, 1999). The sequence specificity of purified BACE1 is the same as β-secretase: it cleaves Swedish mutant APP substrate much more efficiently than wild-type, and does not cleave a P1 Met→Val mutant substrate that is resistant to β-secretase cleavage when expressed in cells. Like β-secretase activity, BACE1 has optimal activity at ~pH 4.5, is resistant to inhibition by pepstatin, and is localized within acidic subcellular compartments of the secretory pathway, primarily the Golgi apparatus and endosomes. Taken as a whole, the properties of BACE1 correlate extremely well with the previously established functional characteristics of β-secretase in cells and tissues.

4. THE HOMOLOGUE BACE2

Soon after the discovery of BACE1, searches of EST databases with the BACE1 sequence identified a homologous novel aspartic protease, BACE2 (also termed Asp1, mempapsin1, and DRAP) (Acquati *et al.*, 2000; Bennett *et al.*, 2000a; Lin *et al.*, 2000; Saunders *et al.*, 1999; Solans *et al.*, 2000; Yan *et al.*, 1999). BACE1 and BACE2 have ~64% amino acid similarity, and both have two aspartic protease active site motifs, six conserved lumenal cysteine residues, a C-terminal transmembrane domain, N-linked glycosylation sites, and other similar structural characteristics. Although BACE1 and BACE2 are most closely related to the pepsin family, they possess features that clearly set them apart from other aspartic proteases. First, BACE1 and BACE2 share only ~40-44% amino acid similarity to individual pepsin family members, while similarity within the pepsin family is significantly higher (~52-69%). Moreover, the positions of only two of the six lumenal cysteine residues of BACE1 and BACE2 are conserved with those of the pepsins. Thus, the disulfide bond structure of BACE1 and BACE2 is very different than that of the pepsin family, and may influence enzyme properties such as stability, activity, or substrate specificity (Haniu *et al.*, 2000). Most importantly, BACE1 and BACE2 are the only aspartic proteases identified that have a C-terminal extension with a predicted transmembrane domain. Membrane attachment may facilitate intracellular localization or may serve to increase the local concentration of enzyme in the lipid bilayer for the processing of membrane-bound substrates. Taken together, the characteristics of BACE1 and BACE2 define a novel family of transmembrane aspartic proteases distinct from the pepsins and the evolutionarily more ancient retroviral aspartic proteases that include the human immunodeficiency virus (HIV) protease (Bennett *et al.*, 2000a).

The high degree of similarity between BACE1 and BACE2 initially suggested that BACE2 might also function as a β-secretase. The BACE1 gene is localized on chromosome 11q23.3 (Saunders *et al.*, 1999). So far, no mutations in the BACE1 gene have been identified that strongly associate with AD (Murphy *et al.*, 2001), although a weak association between a polymorphism in BACE1 exon5 and AD in individuals carrying an ApoE4 allele has been recently reported (Nowotny *et al.*, 2001). Intriguingly, the BACE2 gene is located in the obligate DS region on chromosome 21 (Saunders *et al.*, 1999). Thus, a third copy of the BACE2 gene (and the APP gene) is present in DS and suggests a potential role for BACE2 in the early-onset AD of DS patients. Indeed, cell transfection studies demonstrate that BACE2 cuts APP at the β-secretase site (Farzan *et al.*, 2000; Hussain *et al.*, 2000; Yan *et al.*, 2001). However, BACE2 cleaves with higher efficiency at two other positions within the Aβ domain near the α-secretase cleavage site, Phe+19 and Phe+20. Interestingly, the Flemish FAD mutation of APP (Ala→Gly at +21 of Aβ) is adjacent to the Phe+20 cleavage site and causes an increase in Aβ production that is mediated by BACE2 but not BACE1 in transfected cells (Farzan *et al.*, 2000). Thus, BACE2 may play a role in the pathogenesis of Flemish FAD. However, BACE2 acts like an alternative α-secretase on wild-type APP, so that APP processing by BACE2 typically reduces Aβ production in cells.

BACE2 mRNA is expressed in most tissues at moderate to low levels, but is nearly undetectable in whole brain by Northern analysis (Bennett *et al.*, 2000a; Marcinkiewicz and Seidah, 2000). At the cellular level, *in situ* hybridization studies reveal an intriguing distribution of BACE2 mRNA in the brain: the message is very low or undetectable in most brain regions, but appears high in neurons of a small number of discrete nuclei including ventromedial hypothalamus, mammilary body, and isolated nuclei of the brain stem (Bennett *et al.*, 2000a). The neuroanatomical pattern of BACE2 mRNA expression does not reveal a clear physiological function for BACE2, but implies a role in the processing of substrates important for these specific neuronal populations. The overall expression pattern of BACE2 mRNA in the brain contrasts markedly with that of BACE1 mRNA, which is highly expressed in neurons of most brain regions (Vassar *et al.*, 1999). Consequently, the high levels of β-secretase activity found in the brain are inconsistent with the very low levels of cerebral BACE2 mRNA. This observation together with the α-secretase-like cleavage activity of BACE2 argues against a major role for BACE2 in Aβ generation.

5. THE GENERATION AND CHARACTERIZATION OF BACE1 KNOCKOUT MICE

Soon after the discovery of BACE1, a flood of data strongly suggested BACE1 was the authentic β-secretase. However, it was still formally possible that BACE2 or an as yet unidentified enzyme might have β-secretase activity and thus contribute to Aβ generation *in vivo*. To address this issue, it was necessary to disrupt the BACE1 gene (BACE1 knockout) in the mouse in order to unequivocally prove that BACE1 was the bona fide β-secretase in the brain. BACE1 knockout mice were also required to determine whether BACE1 had a vital function *in vivo*, or if it was dispensable. These questions are of critical importance to investigators interested in the therapeutic development of BACE1 inhibitors.

Given the importance of the knockouts, several groups undertook efforts to generate BACE1 deficient mice by gene targeting. Four knockout strategies were used to inactivate the BACE1 gene: 1. deletion of exon1, thus removing the ATG start codon (Cai *et al.*, 2001); 2. insertion of a β-galactosidase reporter gene immediately downstream of the ATG start codon (Roberds *et al.*, 2001); 3. deletion of exon2, thus removing the N-terminal protease active site motif (Luo *et al.*, 2001); 4. deletion of exon4 through exon8, causing the removal of the C-terminal half of the protease domain (Roberds *et al.*, 2001).

All of the knockout strategies produced viable, fertile BACE1 deficient ($BACE1^{-/-}$) mice at the expected Mendelian frequency (Cai *et al.*, 2001; Luo *et al.*, 2001; Roberds *et al.*, 2001). Lack of expression of the BACE1 gene in the knockouts did not appear to adversely affect embryonic development, nor did it significantly affect the morphology, physiology, biochemistry, and behavior of post-natal or adult mice. Detailed analysis of the phenotype of adult $BACE1^{-/-}$ mice revealed no discernable abnormalities in tissues (morphology, weights, histology), hematology, or blood and urine chemistries, as compared to wild-type mice (Luo *et al.*, 2001; Roberds *et al.*, 2001). Since brain expresses high levels of BACE1, brain histology from the knockout mice was closely examined and no microscopic differences with wild-type brain was observed, irrespective of brain region (Luo *et al.*, 2001; Roberds *et al.*, 2001).

To more closely investigate brain function in the knockout mice, Roberds and colleagues (Roberds *et al.*, 2001) investigated gross behavioral and neuromuscular parameters of BACE1$^{-/-}$ mice and found that no demonstrable differences existed, as compared to wild-type mice. For example, BACE1$^{-/-}$ mice exhibit normal locomotor activity, gait, and exploratory behavior, and are neither hyperactive nor sedated. In addition, the knockouts have normal grip strength, righting reflex, geotaxis, eye-blink reflex, and reactions to tactile stimuli. Overall, it is clear from these studies that the absence of BACE1 is well tolerated *in vivo* and does not appear to cause untoward effects in the embryonic, post-natal, or adult mouse.

Importantly, β-secretase activity is abolished in brains and cultured neurons of BACE1$^{-/-}$ mice. Since endogenous Aβ is difficult to detect in the mouse, we mated BACE1$^{-/-}$ mice with Swedish APP-overexpressing transgenic mice (Tg2576) (Hsiao *et al.*, 1996), which produce robust levels of Aβ in the brain and develop cerebral amyloid deposits with age. BACE1$^{-/-}$•Tg2576 bigenic mice were generated, and brain extracts were analyzed for APP soluble ectodomains (APPsα, APPsβ), C-terminal fragments (C83, C99) and Aβ species (total Aβ, Aβ40, Aβ42) (Luo *et al.*, 2001). BACE1$^{-/-}$•Tg2576 mice lacked all forms of Aβ in the brain, as well as APPsβ and C99, as compared to BACE1$^{+/-}$•Tg2576 or BACE1$^{+/+}$•Tg2576 mice. In other words, all products of APP processing by β-secretase, including Aβ, were abolished in BACE1 knockout brain. This result unequivocally proves that BACE1 is the major, if not only, β-secretase responsible for Aβ generation in the brain.

The work of Cai (Cai *et al.*, 2001) and Roberds (Roberds *et al.*, 2001) also demonstrated that BACE1 is the principle β-secretase. Cai and colleagues infected cultures of BACE1$^{-/-}$ embryonic neurons with APP-expressing adenovirus and determined that Aβ and C99 are abolished in these cells by using mass spectrometry and gel electrophoresis analysis, respectively. In BACE1$^{-/-}$ cultures, the absence of Aβ species starting at Phe+19 or Phe+20 (the major sites of BACE2 cleavage) indicated that BACE2 is not significantly involved in APP cleavage in neurons (Cai *et al.*, 2001). In addition, Roberds and colleagues found no measurable Aβ by ELISA or β-secretase activity using an *in vitro* assay in extracts of whole brain or cultured neurons from BACE1$^{-/-}$ mice. Peptide-based statine inhibitors of BACE1 showed essentially the same IC50s for both purified human BACE1 and brain extracts from wild-type mice, demonstrating that mouse and human BACE1 have similar enzymatic properties, as expected (Roberds *et al.*, 2001).

Interestingly, the α-secretase cleavage products APPs α, C83, and p3 were dramatically elevated in BACE1$^{-/-}$•Tg2576 brain, demonstrating a competition between α- and β-secretases for cleavage of APP *in vivo* (Cai *et *al.*, 2001; Luo *et al.*, 2001). These results were similar to those obtained with BACE1 antisense inhibition experiments that show an elevation of α-secretase cleavage in BACE1 antisense treated cultured cells (Vassar *et al.*, 1999; Yan *et al.*, 1999).

By co-infecting BACE1$^{-/-}$ neurons with APP and BACE1 adenoviruses, Cai (Cai *et al.*, 2001) determined that β-secretase cleavage at Glu+11 is species-specific. Mass spectrometry analysis revealed that both the Asp+1 and Glu+11 cleavages of APP caused by BACE1 are abolished in BACE1$^{-/-}$ primary neuronal cultures. By co-expressing different combinations of human or mouse BACE1 with human or mouse APP, Cai and colleagues found that the Asp+1 site is always cleaved, regardless of whether the substrate or enzyme comes from man or mouse. In other words, the Asp+1 site shows no BACE1 species selectivity. However, the Glu+11 site is cleaved only when BACE1 enzyme and APP substrate are from the same species. These observations make sense in light of the fact that the sequence surrounding the Asp+1 site is absolutely conserved between mouse and human, while the Glu+11 sequence is divergent (Cai *et al.*, 2001). Moreover, these results suggest that APP and BACE1 have co-evolved within a species to preserve cleavage at Glu+11, although the functional significance of Glu+11 cleavage is unknown. Since Glu+11 is a major site of β-secretase cleavage in neurons, and Glu+11 Aβ appears to be more fibrillogenic and neurotoxic than Asp+1 Aβ *in vitro* (Pike *et al.*, 1995), Aβ species starting at Glu+11 may play a significant, but currently under-appreciated, role in AD pathophysiology.

Recently, we have shown BACE1$^{-/-}$•Tg2576 mice not only lack cerebral Aβ, but also fail to develop amyloid plaques with age (Luo *et al.*, 2003). Tg2576 mice begin to deposit amyloid in the brain at ~9-12 months of age. Conversely, BACE1$^{-/-}$•Tg2576 bigenic mice show no evidence of amyloid deposits even at 13 months of age. Wong and colleagues have also performed experiments similar to our own and have obtained the same result (International Alzheimer's Conference, 2002; Abstract#560). Taken together, these results demonstrate BACE1 is required for amyloid formation. Since BACE1$^{-/-}$•Tg2576 mice have elevated α-secretase cleavage, as judged by increased APPsα and C83, their p3 levels are also likely to be increased. Although it remains to be confirmed that BACE1$^{-/-}$•Tg2576 mice have increased p3 levels, the apparent lack of amyloid deposits in these mice implies that p3 is in fact non-amyloidogenic, as expected. Wong and colleagues are currently investigating whether partial inhibition of BACE1, as occurs in heterozygous BACE1$^{+/-}$ mice, significantly delays the on-set of amyloid formation in BACE1$^{+/-}$•APP mice. If so, it implies that only partial (~50%) therapeutic inhibition of BACE1 may be sufficient to delay AD pathogenesis in human patients.

As a whole, the results of the BACE1 knockout experiments conclusively demonstrate that BACE1 is the major, if not only, β-secretase *in vivo*. Therefore, BACE2 and other proteases can be excluded from serious consideration as β-secretase candidates (with the possible exception of a role for BACE2 in Flemish FAD; (Farzan *et al.*, 2000)). The normal appearing phenotype of BACE1$^{-/-}$ mice indicates that BACE1 is dispensable for normal development and physiological functions *in vivo*. In contrast, the ablation of the PS1 gene, which is required for γ-secretase function, causes embryonic lethality. This suggests the possibility that, unlike γ-secretase inhibitors, BACE1 inhibitors may not be associated with mechanism-based toxicity in human beings. Finally, the lack of Aβ generation in the brains of BACE1 deficient mice indicates that therapeutic inhibition of BACE1 should reduce Aβ levels and amyloid development, an outcome widely believed to be beneficial for the treatment of AD.

6. BACE1 OVEREXPRESSING TRANSGENIC MICE

Recently, transgenic mice that overexpress BACE1 in the brain have been generated (Bodendorf *et al.*, 2002). The rationale for these experiments was 1.) to determine if elevated BACE1 expression would increase the steady-state levels of Aβ in the brain, and 2.) to test whether increased steady-state levels of Aβ caused by BACE1 overexpression could accelerate amyloid deposition. The transgenic mice of Bodendorf and colleagues overexpress human BACE1 driven by the neuron-specific mouse Thy1 promoter. Both single BACE1 transgenics and mice bigenic for both BACE1 and APP transgenes show a significant increase in the steady state levels of all APP cleavage products made by β-secretase (APPsβ, C99, C89, Aβ40 and Aβ42), but increases are relatively modest (~1.5-2 fold) probably due to high levels of clearance in the rodent brain. Levels of the APP metabolites C99, APPsβ, and Aβ all show a similar increase in BACE1 transgenic brain, indicating a direct relationship between amyloidogenic processing of APP and an increase in Aβ. This suggests that β-secretase cleavage of APP by BACE1, rather than γ-secretase cleavage, is the rate-limiting step in the production of Aβ *in vivo*.

In cultured cells, BACE1 not only cleaves full-length APP, but also cuts C99 at Glu+11 to produce C89 (Farzan *et al.*, 2000). To explore whether this occurs *in vivo*, Bodendorf (Bodendorf *et al.*, 2002) compared the steady-state levels of C89 and C99 in transgenic mice expressing matched levels of either wild-type APP (APPwt) or APP with the Swedish mutation (APPsw). As expected, since APPsw is a better substrate for BACE1 than APPwt, the APPsw mice expressed about 4 fold more C99 than the APPwt mice.

Interestingly, the levels of C89 were the same in both APPsw and APPwt mouse brains. Since C89 levels were not increased in APPsw mice, the authors concluded that, in contrast to the results observed in cell culture, *in vivo* C89 is not made by BACE1 cleavage of C99, but rather C89 is generated directly from full-length APP. Finally, the study by Bodendorf and colleagues found no evidence for the species-specificity of the Glu+11 cleavage, again in contrast to cell culture experiments (Cai *et al.*, 2001). Clearly, the resolution of the discrepancies surrounding Glu+11 generation and species-specificity must await further investigation.

Thus far, it has not yet been reported whether BACE1 overexpression accelerates amyloid deposition in APP transgenic mice, analogous to the effects of mutant PS1 transgenes. These experiments will provide insight into the potential role of BACE1 overexpression in humans as a pathophysiological initiator or accelerant during AD. In the future, it will be useful to introduce a human BACE1 transgene into BACE1 knockout mice in order to test the specificity and efficacy of BACE1 inhibitors *in vivo*, thus preventing any confounding effects from endogenous mouse BACE1. In addition, testing BACE1 inhibitors in $BACE1^{-/-}$ mice will also be informative for investigating possible non-specific side effects that are not directly due to inhibition of BACE1.

7. BACE1 X-RAY STRUCTURE

Structural information about the interaction of substrate with the active site of BACE1 would greatly facilitate the rational design of small molecule BACE1 inhibitors. Toward this end, Sauder et al. (Sauder *et al.*, 2000) used molecular modeling to simulate the BACE1 active site bound with wild-type or mutant APP substrates. Since the basic structure of most aspartic protease active sites is well conserved, the X-ray structure of pepsin was used to model BACE1. X-ray structural information of a peptide inhibitor bound to rhizopuspepsin was also incorporated to model the interaction with APP. The molecular modeling identified several residues in BACE1 that potentially contribute to substrate specificity. In particular, Arg296 forms a salt-bridge with the P1' Asp+1 residue of the β-secretase cleavage site, thus explaining the unusual preference of BACE1 among aspartic proteases for substrates that are negatively charged at this position. In addition, several hydrophobic residues in BACE1 form a pocket for the hydrophobic P1 residue. The model also showed that the Swedish FAD mutation, LysMet→AsnLeu at P2-P1, interacts more favorably with Agr296 and the hydrophobic pocket of BACE1 than does wild-type substrate, providing an explanation for the enhanced cleavage of this mutation. Conversely, the substitution of Met→Val at P1 blocks the catalytic Asp93 residue, explaining the lack of cleavage of this mutation by BACE1.

Shortly after the molecular modeling study, the X-ray structure of the BACE1 protease domain co-crystalized with a transition-state inhibitor was determined to 1.9 Å resolution (Hong *et al.*, 2000). As expected, the BACE1 catalytic domain is similar in structure to pepsin and other aspartic proteases, despite the relatively low sequence similarity. Interestingly, the BACE1 active site is more open and less hydrophobic than that of other aspartic proteases. Four hydrogen bonds from the catalytic aspartic acid residues (Asp93 and Asp289) and ten additional hydrogen bonds from various residues in the active site are made with the inhibitor, most of which are conserved in other aspartic proteases. The X-ray structure indicates that Arg296 and the hydrophobic pocket of the active site play an important role in substrate binding, confirming the results of the molecular modeling study. In addition, the bound inhibitor has an unusual kinked conformation from P2' to P4'. The BACE1 X-ray structure suggests that small molecules targeting Arg296 and the hydrophobic pocket residues should inhibit β-secretase cleavage. Moreover, mimicking the unique P2'-P4' conformation of the bound inhibitor may increase the selectivity of inhibitors for BACE1 over BACE2 and the other aspartic proteases.

8. BACE1 INHIBITOR DEVELOPMENT

The challenges facing BACE1 inhibitor development are significant, but not insurmountable. Two transition-state analog inhibitors of BACE1 modeled on the β-secretase cleavage site of the Swedish mutation have been reported with relatively low IC50 concentrations. The first inhibitor, P10-P4'StatVal, contains Asn at P2 (like the Swedish mutation), a statine group at P1, and Val at P1' and has an IC50 of ~30nM (Sinha *et al.*, 1999). The second, called OM99-2, has an IC50 of ~1.6nM and is P4-P4' with AsnLeu at P2-P1, Ala at P1', and a hydroxyethylene isostere between P1 and P1' (Ghosh *et al.*, 2000). The former inhibitor was used as the affinity ligand for the purification of BACE1 protein from human brain, and the later inhibitor was co-crystallized with BACE1 for the X-ray structure determination. Enzyme inhibitors with therapeutic potential are preferably smaller than 700 daltons and have low nanomolar IC50 concentrations or better, so these large peptide-based inhibitors are not viable drug candidates. However, they are starting points for rational drug design efforts and are useful reagents for studying the enzymatic properties of BACE1. For example, one group has recently used P10-P4'StatVal to study the kinetics of BACE1 inhibition (Marcinkeviciene *et al.*, 2001). They concluded that inhibition involves a two-step process: a fast initial association between inhibitor and enzyme, followed by a slower "tightening up" of the complex due to conformational changes in the flap region and displacement of the catalytic water in the active site of BACE1. Such studies enhance our mechanistic understanding of BACE1 inhibition and provide the foundation for further advances in BACE1 inhibitor development.

Several other factors must be taken into consideration for the development of a viable BACE1 drug candidate: 1. in addition to small size, high potency, and low toxicity, the candidate should exhibit favorable pharmacokinetic properties; 2. the inhibitor must possess sufficient lipophilicity to a) efficiently cross the blood-brain barrier and achieve high concentrations in the brain, and b) to traverse two lipid bilayers and reach BACE1 localized in the TGN/endosomal lumen (although the periodic cycling of BACE1 to the cell surface may facilitate enzyme-inhibitor binding); 4. since it is not yet known whether BACE2 is dispensable *in vivo*, it may be necessary to design BACE1-selective drugs that exhibit minimal cross-inhibition of BACE2 and other aspartic proteases. Regarding this latter point, the high degree of homology between BACE1 and BACE2 suggests that the active sites of the two proteases are quite similar, thus making the design of BACE1-selective drugs potentially difficult.

9. SOLUBLE Aβ OLIGOMERS AS PATHOGENIC AGENTS IN AD

The amyloid plaque is a deposit in the parenchyma of the brain composed mainly of insoluble fibrils of the Aβ peptide (reviewed in Selkoe, 2001). The more fibrillogenic Aβ42 is the predominant species of Aβ in amyloid plaques, and is increased in production in autosomal dominant forms of FAD (see Section 1). Given these associations, and the fact that Aβ fibrils appear neurotoxic *in vitro*, it was postulated that amyloid plaques are pathogenic and cause neurodegeneration in AD, as stated in the original amyloid cascade hypothesis (reviewed in Hardy and Selkoe, 2002). However, direct evidence that amyloid plaques are pathogenic *in vivo* has been difficult to obtain. Transgenic mice that overexpress APP with FAD mutations have elevated cerebral Aβ levels and develop amyloid deposits with age (reviewed in Hsiao Ashe, 2001). APP transgenics have many of the neuropathological characteristics of AD, including neurodegenerative changes, and they exhibit memory deficits. On the other hand, the mice do not develop neurofibrillary tangles, nor do they have significant neuronal loss, both important features of AD. Moreover, plaque number poorly correlates with severity of dementia in AD (Terry *et al.*, 1999). Although the total amount of Aβ in the brain (Aβ load) directly correlates with dementia (Cummings and Cotman, 1995; Parvathy *et al.*, 2001), the fact that plaque number does not correlate, together with the incomplete recapitulation of AD pathology in APP transgenic mice, calls into question the role of amyloid plaques in AD pathogenesis.

Recently, research has begun to focus on soluble, non-fibrillar oligomeric forms of Aβ as potential toxic agents in AD. The first indication that soluble forms of Aβ may play a role in AD came from Roher and colleagues who determined that the amount of water-soluble Aβ in AD brain tissue was significantly greater than in control tissue (Kuo *et al.*, 1996). Interestingly, evidence suggests amyloid deposits in APP transgenic mice are dynamic and can increase in size or be cleared over time (Bacskai *et al.*, 2001). These results imply that Aβ can make the transition from soluble (oligomer) to insoluble (plaque) form in a reversible fashion. Given the high concentrations of soluble Aβ in AD brain, and that it may diffuse large distances through the parenchyma, soluble Aβ may have great neurotoxic potential.

Next, studies investigated whether soluble Aβ was toxic to neurons, causing neuronal dysfunction and neurodegenerative changes (reviewed in (Klein *et al.*, 2001)). Aβ derived diffusible ligards (ADDLs) (Lambert *et al.*, 1998) and protofibrils (Hartley *et al.*, 1999) are non-fibrillar Aβ assemblies that are soluble in physiological solutions. Although the precise structures of ADDLs and protofibrils are still under investigation, gel electrophoresis and atomic force microscopy indicates they are small oligomers probably composed of less than ~10 Aβ molecules in ADDLs and perhaps a few dozen Aβ molecules in protofibrils (*see also* Chapter 1). Preparations of these Aβ assemblies are potent neurotoxins and kill neurons at nanomolar concentrations *in vitro* (Hartley *et al.*, 1999; Lambert *et al.*, 1998). Interestingly, ADDLs and protofibrils disrupt neuronal physiology including long-term potentiation, a cellular correlate of memory and learning, in hippocampal slices. Thus, soluble Aβ oligomers appear toxic to neurons and interfere with physiological mechanisms of memory that are likely to be important in AD.

Evidence that soluble Aβ oligomers may be toxic to neurons *in vivo* and could impair memory has begun to emerge recently. APP transgenic mice develop memory deficits prior to the formation of amyloid deposits, implying that increased cerebral levels of soluble Aβ disrupt memory function (Hsiao Ashe, 2001). Moreover, studies with these APP transgenic mice have demonstrated that anti-Aβ antibodies, either generated endogenously by Aβ vaccination (Schenk *et al.*, 1999) or administered exogenously by peripheral injection (Bard *et al.*, 2000), reduce amyloid deposition and Aβ levels in the brain. Aβ-immunized APP transgenic mice also show improved performance in memory tests as compared to non-immunized transgenics (Janus *et al.*, 2000; Morgan *et al.*, 2000). Passive immunization of APP transgenics with anti-Aβ monoclonal antibodies demonstrated rapid memory improvement (Dodart *et al.*, 2002; Kotilinek *et al.*, 2002), in one study occurring as early as 24 hours after a single peripheral injection of antibody (Dodart *et al.*, 2002). Since amyloid burden was unaffected by passive immunization in these studies, the authors concluded that some form of soluble Aβ (rather than the amyloid deposit) was likely to cause memory deficits in the APP transgenic mice and in AD. Although the underlying mechanisms of memory rescue in these cases are unknown, presumably antibody-mediated sequestration and clearance of soluble Aβ in the brain is responsible.

Finally, we have recently conducted studies to determine whether BACE1 deficiency, and the consequent ablation of Aβ, is sufficient to rescue memory deficits in Tg2576 APP transgenic mice (Ohno *et al.*, 2004). Since Tg2576 mice develop memory impairments at an early age before the onset of amyloid deposition (Hsiao Ashe, 2001), we chose to study young pre-deposit mice to analyze the contribution of soluble Aβ to memory dysfunction. Our work demonstrated that memory deficits and cholinergic dysfunction in the hippocampus did not develop in $BACE1^{-/-}$•Tg2576 bigenic mice that lacked Aβ, while florid deficits were apparent in Aβ-overproducing Tg2576 monogenics. Because the Aβ in Tg2576 at the time of testing was non-fibrillar and soluble, we concluded that soluble Aβ assemblies rather than amyloid plaques are responsible for at least some aspects of AD-related memory deficits. Moreover, our work is further validation of BACE1 as a prime therapeutic target for AD.

To put these studies into perspective, it was recently discovered that soluble Aβ potentially has a normal function as a negative regulator of excitatory synaptic transmission (Kamenetz *et al.*, 2003). Upon electrical stimulation of glutaminergic pathways in hippocampal slices, presynaptic terminals release Aβ, which then inhibits postsynaptic neurons from further depolarization. Such a mechanism may be important for "putting on the brakes" on excitatory transmission in order to protect against over-stimulation under certain conditions. Most importantly, these results convincingly suggest a normal function for soluble Aβ and indicate that cerebral Aβ levels may need to be maintained within a narrow margin, or else memory function may become disrupted.

Taken together, the evidence suggests that soluble Aβ oligomers are an important pathogenic agent in AD. However, significant work remains before we can confidently conclude that soluble Aβ is at the root of AD etiology. In fact, amyloid plaques are also likely to contribute to pathogenicity in AD, since clear signs of inflammation and neurodegeneration such as dystrophic neurites are increased in the immediate vicinity of amyloid deposits in both AD and APP transgenic mice (Selkoe, 2001). In the end, it is likely that both soluble Aβ oligomers and amyloid plaques are involved in the pathophysiology and progression of AD.

10. CONCLUSIONS

A large body of evidence has demonstrated that BACE1 is the authentic β-secretase. Yet, these are still early days in the study of BACE1 (and BACE2) and its potential as a drug target in AD. The lack of Aβ production and amyloid deposition in BACE1 deficient mice clearly validates BACE1 as the β-secretase and indicates that BACE1 inhibitors should reduce cerebral Aβ levels, but questions remain. What percentage of BACE1 inhibition is required to significantly delay amyloid deposition? In addition, does the other β-secretase cleavage product, Glu+11 Aβ, have any role in amyloid deposition? These questions have important implications for both AD pathogenesis and therapeutic development and thus require further investigation.

Despite progress in characterizing the molecular and cellular properties of BACE1 and BACE2, little is known about the substrates (other than APP) and the biological functions of these two proteases. Recent results suggest that a Golgi-resident sialyltransferase, ST6Gal I (Kitazume *et al.*, 2001), and P-selectin glycoprotein ligand-1 (Lichtenthaler *et al.*, 2003) are BACE1 substrates, however the functional significance of these data is unknown and further work in this area is required. Although the apparently normal phenotype of BACE1 knockout mice is uninformative regarding BACE1 function *in vivo*, it is possible that subtle effects of BACE1 deficiency may be revealed with challenge. Phenotypes uncovered by different challenges or stresses could provide a clue to the biological function of BACE1 and would guide predictions regarding potential side effects of BACE1 inhibitors in humans under certain conditions.

Regarding BACE2 function, the intriguing pattern of BACE2 expression in the brain suggests an important role for BACE2 in specific neuronal subpopulations. The generation of BACE2 knockout mice will be instrumental for addressing this question. BACE1 and BACE2 are highly homologous and have expression patterns that partially overlap, implying that these enzymes may be functionally redundant in some tissues. The analysis of $BACE1^{-/-}$•$BACE2^{-/-}$ compound knockout mice could provide insight into this issue. Clearly, identification of other substrates for BACE1 and BACE2 will vastly increase our understanding of the functional role of this novel family of transmembrane aspartic proteases.

To date, no mutations in the BACE1 gene have been identified that definitively associate with AD. Still, it is possible that mutations in the BACE1 gene could increase the risk of AD by elevating either BACE1 gene expression or enzyme activity. Such BACE1 mutations would be expected to increase the production of Aβ, thus potentially contributing to AD pathogenesis. Increasing the level of Aβ by 50% is enough to cause early onset AD in DS, and it is possible that even much smaller Aβ increases may have profound effects over time to cause some forms of late-onset AD. In this light, it is interesting to note that BACE1 levels appear to be increased by ~2.7 fold or more in the brains of at least some late-onset AD cases, as compared to age-matched controls, and that C99 levels were ~2 fold higher (Holsinger *et al.*, 2002). A recently identified polymorphism in exon5 of the BACE1 gene has been found to weakly associate with AD (Nowotny *et al.*, 2001) and one may speculate that such mutations may increase BACE1 and Aβ levels, although currently there is no evidence to support this hypothesis. In any case, further analysis of BACE1 protein levels and BACE1 gene sequences from AD patients is warranted and may eventually reveal mutations that increase the risk of AD.

Finally, as the key enzyme that initiates Aβ formation *in vivo*, BACE1 is a prime drug target for inhibiting the production of Aβ. Regardless of whether soluble or fibrillar Aβ is the pathogenic agent in AD, therapeutic inhibition of BACE1 is expected to lower cerebral Aβ levels, which should prove beneficial for AD. Although the creation of pharmaceutically viable BACE1 inhibitors will be challenging, it is likely that BACE1 drugs will be developed with time. Drugs that inhibit other therapeutically important aspartic proteases such as renin and the HIV protease have been successfully developed, and these drugs provide paradigms for the rational design of BACE1 inhibitors for the treatment of AD.

ACKNOWLEDGEMENTS

This chapter would not have been possible without the dedicated work of the many scientists involved in Alzheimer's and β-secretase research. Regrettably, due to the brief nature of this review, it was impossible to include all the studies of β-secretase, APP, and Aβ, and I apologize to those who's research was not included.

REFERENCES

Acquati, F., Accarino, M., Nucci, C., Fumagalli, P., Jovine, L., Ottolenghi, S., and Taramelli, R., 2000, The gene encoding DRAP (BACE2), a glycosylated transmembrane protein of the aspartic protease family, maps to the Down'scritical region. *FEBS Lett.* **468:** 59-64.

Bacskai, B.J., Kajdasz, S.T., Christie, R.H., Carter, C., Games, D., Seubert, P., Schenk, D., and Hyman, B.T., 2001, Imaging of amyloid-beta deposits in brains of living mice permits direct observation of clearance of plaques with immunotherapy. *Nat Med.*, **7:** 369-372.

Bard, F., Cannon, C., Barbour, R., Burke, R.L., Games, D., Grajeda, H., Guido, T., Hu, K., Huang, J., Johnson-Wood, K., Khan, K., Kholodenko, D., Lee, M., Lieberburg, I., Motter, R., Nguyen, M., Soriano, F., Vasquez, N., Weiss, K., Welch, B., Seubert, P., Schenk, D., and Yednock, T., 2000, Peripherally administered antibodies against amyloid beta-peptide enter the central nervous system and reduce pathology in a mouse model of Alzheimer disease. *Nat. Med.* **6:** 916-919.

Bennett, B.D., Babu-Khan, S., Loeloff, R., Louis, J.-C., Curran, E., Citron, M., and Vassar, R., 2000a, Expression analysis of BACE2 in brain and peripheral tissues. *J. Biol. Chem.* **275:** 20647-20651.

Bennett, B.D., Denis, P., Haniu, M., Teplow, D.B., Kahn, S., Louis, J.-C., Citron, M., and Vassar, R., 2000b, A furin-like convertase mediates propeptide cleavage of BACE, the Alzheimer's β-secretase. *J. Biol. Chem.* **275:** 37712-37717.

Bodendorf, U., Danner, S., Fischer, F., Stefani, M., Sturchler-Pierrat, C., Wiederhold, K.H., Staufenbiel, M., and Paganetti, P., 2002, Expression of human β-secretase in the mouse brain increases the steady-state level of β-amyloid. *J. Neurochem.*, **80:** 799-806.

Bodendorf, U., Fischer, F., Bodian, D., Multhaup, G., and Paganetti, P., 2001, A splice variant of β-secretase deficient in the amyloidogenic processing of the amyloid precursor protein. *J. Biol. Chem.* **276:** 12019-12023.

Buxbaum, J.D., Liu, K.N., Luo, Y., Slack, J.L., Stocking, K.L., Peschon, J.J., Johnson, R.S., Castner, B.J., Cerretti, D.P., and Black, R.A., 1998, Evidence that tumor necrosis factor alpha converting enzyme is involved in regulated alpha-secretase cleavage of the Alzheimer amyloid protein precursor. *J. Biol. Chem.* **273:** 27765-27767.

Cai, H., Wang, Y., McCarthy, D., Wen, H., Borchelt, D.R., Price, D.L., and Wong, P.C., 2001, BACE1 is the major β-secretase for generation of Aβ peptides by neurons. *Nature Neurosci.* **4:** 233-234..

Citron, M., Diehl, T.S., Capell, A., Haass, C., Teplow, D.B., and Selkoe, D.J., 1996, Inhibition of amyloid ß-protein production in neural cells by the serine protease inhibitor AEBSF. *Neuron* **17:** 171-179.

Citron, M., Teplow, D.B. and Selkoe, D.J., 1995, Generation of amyloid β-protein from its precursor is sequence specific. *Neuron* **14:** 661-670.

Cummings, B.J. and Cotman, C.W., 1995, Image analysis of beta-amyloid load in Alzheimer's disease and relation to dementia severity. *Lancet* **346:** 1524-1528.

Dodart, J.C., Bales, K.R., Gannon, K.S., Greene, S.J., DeMattos, R.B., Mathis, C., DeLong, C.A., Wu, S., Wu, X., Holtzman, D.M., and Paul, S.M., 2002, Immunization reverses memory deficits without reducing brain Aβ burden in Alzheimer's disease model. *Nat. Neurosci.* **5:** 452-457.

Farzan, M., Schnitzler, C.E., Vasilieva, N., Leung, D., and Choe, H., 2000, BACE2, a β-secretase homolog, cleaves at the β site and within the amyloid-β region of the amyloid-β precursor protein. *Proc. Natl. Acad. Sci. USA* **97:** 9712-9717.

Francis, R., McGrath, G., Zhang, J., Ruddy, D.A., Sym, M., Apfeld, J., Nicoll, M., Maxwell, M., Hai, B., Ellis, M.C., Parks, A.L., Xu, W., Li, J., Gurney, M., Myers, R.L., Himes, C.S., Hiebsch, R., Ruble, C., Nye, J.S., and Curtis, D., 2002, aph-1 and pen-2 are required for

Notch pathway signaling, gamma-secretase cleavage of betaAPP, and presenilin protein accumulation. *Dev. Cell* **3:** 85-97.

Ghosh, A.K., Shin, D., Downs, D., Koelsch, G., Lin, X., Ermolieff, J., and Tang, J., 2000, Design of potent inhibitors for human brain memapsin 2 (β-secretase). *J. Am. Chem. Soc.* **122:** 3522-3523.

Glenner, G.G., and Wong, C.W., 1984, Alzheimer's disease: initial report of the purification and characterization of a novel cerebrovascular amyloid protein. *Biochem. Biophys. Res. Commun.* **120:** 885-890.

Gouras, G.K., Xu, H., Jovanovic, J.N., Buxbaum, J.D., Wang, R., Greengard, P., Relkin, N.R., and S., G., 1998, Generation and regulation of β-amyloid peptide variants by neurons. *J. Neurochem.* **71:** 1920-1925.

Haass, C., Capell, A., Citron, M., Teplow, D.B., and Selkoe, D.J., 1995a, The vacuolar H+ ATPase inhibitor bafilomycin A1 differentially affects proteolytic processing of mutant and wild-type β-amyloid precursor protein. *J. Biol. Chem.* **270:** 6186-6192.

Haass, C., Hung, A.Y., Schlossmacher, M.G., Teplow, D.B., and Selkoe, D.J., 1993, β-amyloid peptide and a 3-kDa fragment are derived by distinct cellular mechanisms. *J. Biol. Chem.* **268:** 3021-3024.

Haass, C., Lemere, C.A., Capell, A., Citron, M., Seubert, P., Schenk, D., Lannfelt, L., and Selkoe, D.J., 1995b, The Swedish mutation causes early-onset Alzheimer's disease by β-secretase cleavage within the secretory pathway. *Nature Med.* **1:** 1291-1296.

Haass, C., Schlossmacher, M.G., Hung, A.Y., Vigo-Pelfrey, C., Mellon, A., Ostaszewski, B.L., Lieberburg, I., Koo, E.H., Schenk, D., Teplow, D.B., and Selkoe, D.J., 1992, Amyloid β-peptide is produced by cultured cells during normal metabolism. *Nature* **359:** 322-325.

Haniu, M., Denis, P., Young, Y., Mendiaz, E.A., Fuller, J., Hui, J.O., Bennett, B.D., Kahn, S., Ross, S., Burgess, T., Katta, V., Rogers, G., Vassar, R, and Citron, M., 2000, Characterization of Alzheimer's β-secretase protein BACE. *J. Biol. Chem.* **275:** 21099-21106.

Hardy, J., and Selkoe, D.J., 2002, The amyloid hypothesis of Alzheimer's disease: progress and problems on the road to therapeutics. *Science* **297:** 353-356.

Hartley, D.M., Walsh, D.M., Ye, C.P., Diehl, T., Vasquez, S., Vassilev, P.M., Teplow, D.B., and Selkoe, D.J., 1999, Protofibrillar intermediates of amyloid beta-protein induce acute electrophysiological changes and progressive neurotoxicity in cortical neurons. *J. Neurosci.* **19:** 8876-8884.

Holsinger, R.M.D., McLean, C.A., Beyreuther, K., Masters, C.L., and Evin, G., 2002, Increased expression of the amyloid precursor β-secretase in Alzheimer's disease. *Ann. Neurol.* **51:** 783-786.

Hong, L., Koelsch, G., Lin, X., Wu, S., Terzyan, S., Ghosh, A.K., Zhang, X.C., and Tang, J., 2000, Structure of the protease domain of memapsin 2 (β-secretase) complexed with inhibitor. *Science* **290:** 150-153.

Hsiao Ashe, K., 2001, Learning and Memory in Transgenic Mice Modeling Alzheimer's Disease. *Learn. Mem.* **8:** 301-308.

Hsiao, K., Chapman, P., Nilsen, S., Eckman, C., Harigaya, Y., Younkin, S., Yang, F., and Cole, G., 1996, Correlative memory deficits, Aβ elevation, and amyloid plaques in transgenic mice. *Science* **274:** 99-103.

Hussain, I., Powell, D., Howlett, D.R., Tew, D.G., Meek, T.D., Chapman, C., Gloger, I.S., Murphy, K.E., Southan, C.D., Ryan, D.M., Smith, T.S., Simmons, D.L., Walsh, F.S., Dingwall, C., and Christie, G., 1999, Identification of a novel aspartic protease (Asp 2) as β-Secretase. *Mol. Cell. Neurosci.* **14:** 419-427.

Hussain, I., Powell, D.J., Howlett, D.R., Chapman, G.A., Gilmour, L., Murdock, P.R., Tew, D.G., Meek, T.D., Chapman, C., Schneider, K., Ratcliffe, S.J., Tattersall, D., Testa, T.T., Southan, C., Ryan, D.M., Simmons, D.L., Walsh, F.S., Dingwall, C., and Christie, G., 2000, ASP1 (BACE2) cleaves the amyloid precursor protein at the β-secretase site. *Mol. Cell. Neurosci.* **16:** 609-619.

Hutton, M., Perez-Tur, J., and Hardy, J., 1998, Genetics of Alzheimer's disease. *Essays Biochem.* 33: 117-131.

Janus, C., Pearson, J., McLaurin, J., Mathews, P.M., Jiang, Y., Schmidt, S.D., Chishti, M.A., Horne, P., Heslin, D., French, J., Mount, H.T., Nixon, R.A., Mercken, M., Bergeron, C., Fraser, P.E., St George-Hyslop, P., and Westaway, D., 2000, Aβ peptide immunization reduces behavioural impairment and plaques in a model of Alzheimer's disease. *Nature* **408:** 979-982.

Kamenetz, F., Tomita, T., Hsieh, H., Seabrook, G., Borchelt, D., Iwatsubo, T., Sisodia, S. and Malinow, R., 2003, APP processing and synaptic function. *Neuron* **37:** 925-937.,

Kang, J., Lemaire, H.-G., Unterbeck, A., Salbaum, J.M., Masters, C.L., Grzeschik, K.-H., Multhaup, G., Beyreuther, K., and Muller-Hill, B., 1987, The precursor of Alzheimer's disease amyloid A4 protein resembles a cell-surface receptor. *Nature* **325:** 733-736.

Kitazume, S., Tachida, Y., Oka, R., Shirotani, K., Saido, T.C., and Hashimoto, Y., 2001, Alzheimer's β-secretase, β-site amyloid precursor protein-cleaving enzyme, is responsible for cleavage secretion of a Golgi-resident sialyltransferase. *Proc. Natl. Acad. Sci. USA* **98:** 13554-13559.

Klein, W.L., Krafft, G.A., and Finch, C.E., 2001, Targeting small Abeta oligomers: the solution to an Alzheimer's disease conundrum? *Trends Neurosci.* **24:** 219-224.

Knops, J., Suomensaari, S., Lee, M., McConlogue, L., Seubert, P., and Sinha, S., 1995, Cell-type and amyloid precursor protein-type specific inhibition of Aβ release by bafilomycin A1, a selective inhibitor of vacuolar ATPases. *J. Biol. Chem.* **270:** 2419-2422.

Koo, E.H., and Squazzo, S., 1994, Evidence that production and release of amyloid β-protein involves the endocytic pathway. *J. Biol. Chem.* **269:** 17386-17389.

Kotilinek, L.A., Bacskai, B., Westerman, M., Kawarabayashi, T., Younkin, L., Hyman, B.T., Younkin, S., and Ashe, K.H., 2002, Reversible memory loss in a mouse transgenic model of Alzheimer's disease. *J. Neurosci.* **22:** 6331-6335.

Kuo, Y.M., Emmerling, M.R., Vigo-Pelfrey, C., Kasunic, T.C., Kirkpatrick, J.B., Murdoch, G.H., Ball, M.J., and Roher, A.E., 1996, Water-soluble Aβ (N-40, N-42) oligomers in normal and Alzheimer disease brains. *J. Biol. Chem.* **271:** 4077-4081.

Lambert, M.P., Barlow, A.K., Chromy, B.A., Edwards, C., Freed, R., Liosatos, M., Morgan, T.E., Rozovsky, I., Trommer, B., Viola, K.L., Wals, P., Zhang, C., Finch, C.E., Krafft, G.A., and Klein, W.L., 1998, Diffusible, nonfibrillar ligands derived from Abeta1-42 are potent central nervous system neurotoxins. *Proc. Natl. Acad. Sci. U S A* **95:** 6448-6453.

Lammich, S., Kojro, E., Postina, R., Gilbert, S., Pfeiffer, R., Jasionowski, M., Haass, C., and Fahrenholz, F., 1999, Constitutive and regulated alpha-secretase cleavage of Alzheimer's amyloid precursor protein by a disintegrin metalloprotease. *Proc. Natl. Acad. Sci. U S A* **96:** 3922-3927.

Lee, V.M.-Y., Balin, B.J., Otvos, L. and Trojanowski, J.Q., 1991, A major subunit of paired helical filaments and derivatized forms of normal tau. *Science* **251:** 675-678.

Lichtenthaler, S.F., Dominguez, D.I., Westmeyer, G.G., Reiss, K., Haass, C., Saftig, P., De Strooper, B. and Seed, B., 2003, The cell adhesion protein P-selectin glycoprotein ligand-1 is a substrate for the aspartyl protease BACE1. *J. Biol. Chem.* **278:** 48713-4871.

Lin, X., Koelsch, G., Wu, S., Downs, D., Dashti, A., and Tang, J., 2000, Human aspartic protease memapsin 2 cleaves the β-secretase site of β-amyloid precursor protein. *Proc. Natl. Acad. Sci. USA* **97:** 1456-1460.

Luo, Y., Bolon, B., Damore, M.A., Fitzpatrick, D., Liu, H., Zhang, J., Yan, Q., Vassar, R., and Citron, M., 2003, BACE1 (β-secretase) knockout mice do not acquire compensatory gene expression changes or develop neural lesions over time. *Neurobiol. Dis.* **4:** 81-88.

Luo, Y., Bolon, B., Kahn, S., Bennett, B.D., Babu-Khan, S., Denis, P., Fan, W., Kha, H., Zhang, J., Gong, Y., Martin, L., Louis, J.-C., Yan, Q., Richards, W.G., Citron, M., and Vassar, R., 2001, Mice deficient in BACE1, the Alzheimer's β-secretase, have normal phenotype and abolished β-amyloid generation. *Nature Neurosci.* **4:** 231-232.

Marcinkeviciene, J., Luo, Y., Graciani, N.R., Combs, A.P., and Copeland, R.A., 2001, Mechanism of inhibition of β-site amyloid precursor protein-cleaving enzyme (BACE) by a statine-based peptide. *J. Biol. Chem.* **276:** 23790-23794.

Marcinkiewicz, M., and Seidah, N.G., 2000, Coordinated expression of β-amyloid precursor protein and the putative β-secretase BACE and α-secretase ADAM10 in mouse and human brain. *J. Neurochem.* **75:** 2133-2143.

Morgan, D., Diamond, D.M., Gottschall, P.E., Ugen, K.E., Dickey, C., Hardy, J., Duff, K., Jantzen, P., DiCarlo, G., Wilcock, D., Connor, K., Hatcher, J., Hope, C., Gordon, M., and Arendash, G.W., 2000, Aβ peptide vaccination prevents memory loss in an animal model of Alzheimer's disease. *Nature* **408:** 982-985.

Mullan, M., Crawford, F., Houlden, H., Axelman, K., Lilius, L., Winblad, B.,and Lannfelt, L., 1992a, A pathogenic mutation for probable Alzheimer's disease in the APP gene at the N-terminus of β-amyloid. *Nat. Genet.* **1:** 345-347.

Murphy, T., Yip, A., Brayne, C., Easton, D., Evans, J.G., Xuereb, J., Cairns, N., Esiri, M.M., and Rubinsztein, D.C., 2001, The BACE gene: genomic structure and candidate gene study in late-onset Alzheimer's disease. *Neuroreport* **12:** 631-634.

Nowotny, P., Kwon, J.M., Chakraverty, S., Nowotny, V., Morris, J.C., and Goate, A.M., 2001, Association studies using novel polymorphisms in BACE1 and BACE2. *Neuroreport* **12:** 1799-1802.

Ohno, M., Sametsky, E.A., Younkin, L.H., Oakley, H., Younkin, S.G., Citron, M., Vassar, R., and Disterhoft, J.F., 2004, BACE1 Deficiency Rescues Memory Deficits and Cholinergic Dysfunction in a Mouse Model of Alzheimer's Disease. *Neuron* **41:** 27-33.

Parvathy, S., Davies, P., Haroutunian, V., Purohit, D.P., Davis, K.L., Mohs, R.C., Park, H., Moran, T.M., Chan, J.Y., and Buxbaum, J.D., 2001, Correlation between Abetax-40-, Abetax-42-, and Abetax-43-containing amyloid plaques and cognitive decline. *Arch. Neurol.* **58:** 2025-2032.

Pike, C.J., Overman, M.J., and Cotman, C.W., 1995, Amino-terminal deletions enhance aggregation of β-amyloid peptides in vitro. *J. Biol. Chem.* **270:** 23895-23898.

Roberds, S.L., Anderson, J., Basi, G., Bienkowski, M.J., Branstetter, D.G., Chen, K.S., Freedman, S.B., Frigon, N.L., Games, D., Hu, K., Johnson-Wood, K., Kappenman, K.E., Kawabe, T.T., Kola, I., Kuehn, R., Lee, M., Liu, W., Motter, R., Nichols, N.F., Power, M., Robertson, D.W., Schenk, D., Schoor, M., Shopp, G.M., Shuck, M.E., Sinha, S., Svensson, K.A., Tatsuno, G., Tintrup, H., Wijsman, J., Wright, S., and McConlogue, L., 2001, BACE knockout mice are healthy despite lacking the primary β-secretase activity in brain: implications for Alzheimer's disease therapeutics. *Hum. Mol. Genet.* **10:** 1317-1324.

Roher, A.E., Lowenson, J.D., Clarke, S., Wolkow, C., Wang, R., Cotter, R.J., Reardon, I., Zurcher-Neely, H.A., Heinrikson, R.L., Ball, M.J., and Greenberg, B.D., 1993, Structural alterations in the peptide backbone of β-amyloid core protein may account for its deposition and stability in Alzheimer's disease. *J. Biol. Chem.* **268:** 3072-3083.

Sauder, J.M., Arthur, J.W., and Dunbrack, R.L., Jr., 2000, Modeling of substrate specificity of the Alzheimer's disease amyloid precursor protein β-secretase. *J. Mol. Biol.* **300:** 241-248

Saunders, A.J., Kim, T.-W., Tanzi, R.E., Fan, W., Bennett, B.D., Babu-Khan, S., Luo, Y., Louis, J.-C., McCaleb, M., Citron, M., Vassar, R., and Richards, W.G., 1999, *BACE* maps to chromosome 11 and a BACE homolog, *BACE2*, reside in the obligate Down'ssyndrome region of chromosome 21. *Science* **286:** 1255a.

Schenk, D., Barbour, R., Dunn, W., Gordon, G., Grajeda, H., Guido, T., Hu, K., Huang, J., Johnson-Wood, K., Khan, K., Kholodenko, D., Lee, M., Liao, Z., Lieberburg, I., Motter, R., Mutter, L., Soriano, F., Shopp, G., Vasquez, N., Vandevert, C., Walker, S., Wogulis, M., Yednock, T., Games, D., and Seubert, P., 1999, Immunization with amyloid-beta attenuates Alzheimer-disease-like pathology in the PDAPP mouse. *Nature* **400:** 173-177.

Selkoe, D.J., 2001, Alzheimer's disease: genes, proteins, and therapy. *Physiol. Rev.* **81:** 741-766.

Seubert, P., Oltersdorf, T., Lee, M.G., Barbour, R., Blomqist, C., Davis, D.L., Bryant, K., Fritz, L.C., Galasko, D., Thal, L.J., Lieberburg, I., and Schenk, D.B., 1993, Secretion of β-amyloid precursor protein cleaved at the amino-terminus of the β-amyloid peptide. *Nature* **361:** 260-263.

Sinha, S., Anderson, J.P., Barbour, R., Basi, G.S., Caccavello, R., Davis, D., Doan, M., Dovey, H.F., Frigon, N., Hong, J., Jacobson-Croak, K., Jewett, N., Keim, P., Knops, J., Lieberburg, I., Power, M., Tan, H., Tatsuno, G., Tung, J., Schenk, D., Seubert, P., Suomensaari, S.M., Wang, S., Walker, D., Zhao, J., McConlogue, L., and John, V., 1999, Purification and cloning of amyloid precursor protein β-secretase from human brain. *Nature* **402:** 537-540.

Sisodia, S.S., Kim, S.H., and Thinakaran, G., 1999, Function and dysfunction of the presenilins. *Am. J. Hum. Genet.* **65:** 7-12.

Solans, A., Estivill, X., and de La Luna, S., 2000, A new aspartyl protease on 21q22.3, BACE2, is highly similar to Alzheimer's amyloid precursor protein β-secretase. *Cytogenet. Cell Genet.* **89:** 177-184.

Takasugi, N., Tomita, T., Hayashi, I., Tsuruoka, M., Niimura, M., Takahashi, Y., Thinakaran, G., and Iwatsubo, T. (2003) The role of presenilin cofactors in the gamma-secretase complex. *Nature* **422:** 438-441.

Tanzi, R.E., Gusella, J.F., Watkins, P.C., Bruns, G.A.B., St. George-Hyslop, P.H., Van Keuren, M.L., Patterson, D., Pagan, S., Kurnit, D.M., and Neve, R.L., 1987, Amyloid β-protein gene: cDNA, mRNA distribution, and genetic linkage near the Alzheimer locus. *Science* **235:** 880-884.

Terry, R.D., Masliah, E., and Hansen, L.A., 1999, The neuropathology of Alzheimer disease and the structural basis of its cognitive alterations. In *Alzheimer Disease* (R.D. Terry, R. Katzman, K.L. Bick, and S.S. Sisodia, eds.), Lippincott Williams and Wilkins, Philadelphia, pp. 187-206.

Vassar, R., Bennett, B.D., Babu-Khan, S., Kahn, S., Mendiaz, E.A., Denis, P., Teplow, D.B., Ross, S., Amarante, P., Loeloff, R., Luo, Y., Fisher, S., Fuller, J., Edenson, S., Lile, J., Jarosinski, M.A., Biere, A.L., Curran, E., Burgess, T., Louis, J.-C., Collins, F., Treanor, J., Rogers, G., and Citron, M., 1999, β-Secretase cleavage of Alzheimer's amyloid precursor protein by the transmembrane aspartic protease BACE. *Science* **286:** 735-741.

Vassar, R., and Citron, M., 2000, Aβ-generating enzymes: recent advances in β- and γ-secretase research. *Neuron* **27:** 419-422.

Wolfe, M.S., Angeles, J.D.L., Miller, D.D., Xia, W., and Selkoe, D.J., 1999a, Are presenilins intramembrane-cleaving proteases? Implications for the molecular mechanism of Alzheimer's disease. *Biochemistry* **38:** 11223-11230.

Wolfe, M.S., Xia, W., Ostaszewski, B.L., Diehl, T.S., Kimberly, W.T., and Selkoe, D.J., 1999b, Two transmembrane aspartates in presenilin-1 required for presenilin endoproteolysis and gamma-secretase activity. *Nature* **398:** 513-517.

Yan, R., Bienkowski, M.J., Shuck, M.E., Miao, H., Tory, M.C., Pauley, A.M., Brashler, J.R., Stratman, N.C., Mathews, W.R., Buhl, A.E., Carter, D.B., Tomasselli, A.G., Parodi, L.A.,

Heinrikson, R.L., and Gurney, M.E., 1999, Membrane-anchored aspartyl protease with Alzheimer's disease β-secretase activity. *Nature* **402:** 533-537.

Yan, R., Munzner, J.B., Shuck, M.E., and Bienkowski, M.J., 2001, BACE2 functions as an alternative α-secretase in cells. *J. Biol. Chem.* **276:** 34019-34027.

Younkin, S.G., 1998, The role of Abeta 42 in Alzheimer's disease. *J. Physiol. (Paris)* **92:** 289-292.

Yu, G., Nishimura, M., Arawaka, S., Levitan, D., Zhang, L., Tandon, A., Song, Y.Q., Rogaeva, E., Chen, F., Kawarai, T., Supala, A., Levesque, L., Yu, H., Yang, D.S., Holmes, E., Milman, P., Liang, Y., Zhang, D.M., Xu, D.H., Sato, C., Rogaev, E., Smith, M., Janus, C., Zhang, Y., Aebersold, R., Farrer, L.S., Sorbi, S., Bruni, A., Fraser, P., and St George-Hyslop, P., 2000, Nicastrin modulates presenilin-mediated notch/glp-1 signal transduction and βAPP processing. *Nature* **407:** 48-54.

Zhao, J., Paganini, L., Mucke, L., Gordon, M., Refolo, L., Carman, M., Sinha, S., Oltersdorf, T., Lieberburg, I., and McConlogue, L., 1996, β-Secretase processing of the β-amyloid precursor protein in transgenic mice is efficient in neurons but inefficient in astrocytes. *J. Biol. Chem.* **271:** 31407-31411.

Chapter 5

The Non-Amyloidogenic Pathway: Structure and Function of α-Secretases

Elżbieta Kojro and Falk Fahrenholz
Institute of Biochemistry, Johannes Gutenberg University, Mainz, Germany

Abstract: The amyloid cascade hypothesis is the most accepted explanation for the pathogenesis of Alzheimer's disease (AD). APP is the precursor of the amyloid β peptide (Aβ), the principal proteinaceous component of amyloid plaques in brains of Alzheimer's disease patients. Proteolytic cleavage of APP by the α-secretase within the Aβ sequence precludes formation of amyloidogenic peptides and leads to a release of soluble APPsα which has neuroprotective properties. In several studies, a decreased amount of APPsα in the cerebrospinal fluid of AD patients has been observed. Three members of the ADAM family (a disintegrin and metalloproteinase) ADAM-10, ADAM-17 (TACE) and ADAM-9 have been proposed as α-secretases. We review the evidence for each of these enzymes acting as a physiologically relevant α-secretase. In particular, we focus on ADAM-10, which recently was shown in a transgenic mouse model for AD, to act as an α-secretase *in vivo*. We also discuss the pharmacological up-regulation of α-secretases as a possible therapeutic treatment for AD.

Key words: α-secretase, non-amyloidogenic pathway, Alzheimer's disease, ADAM-9, ADAM-10, ADAM-17, cholesterol, G-protein-coupled receptors, acetyl choline esterase inhibitors.

1. INTRODUCTION

The accumulation of the β-amyloid peptide (Aβ) in the brain is a central event leading to the development of Alzheimer's disease (AD). Aβ is a 40/42-residue fragment of the brain transmembrane protein β-amyloid precursor protein (APP), released by two proteases known as β- and γ-secretase. The Aβ-peptide sequence is located at the junction between the integral membrane domain and the extracellular domain of APP (Kang *et al.*, 1987). In the alternative non-amyloidogenic pathway APP is cleaved within the Aβ domain by α-secretase between amino acids 16(Lys) and 17(Leu) of the Aβ region (Haass and Selkoe, 1993; Selkoe, 1996). The action of the α-secretase not only precludes the formation of Aβ peptides but also induces release of the large APP ectodomain from the cell surface. Soluble N-terminal APP fragments of 105-125 kDa (APPsα) are released constitutively into vesicle lumens and from the cell surface; similar species are identified in human plasma and in the cerebrospinal fluid (Weidemann *et al.* 1989). Secreted APPsα appears to have beneficial effects, evoking coordinated responses in neuronal and some peripheral target cells. APPsα exerts proliferate effects in a variety of cell types as well as neurotrophic effects (Mucke *et al.*, 1996; Mattson *et al.*, 1993). The C-terminal 591-612 region of APPsα contains a heparin-binding domain that is lacking in APPsβ and that appears to play a key role in the neurotrophic and calcium regulation effects (Furukawa *et al.*, 1996). APPsα has potent memory-enhancing effects and blocks learning deficits induced by scopolamine in mice (Meziane *et al.*, 1998). It is also interesting to note that a reduction of APPsα is evident in the cerebrospinal fluid of AD patients (Lannfelt *et al.*, 1995; Sennvik *et al.*, 2000).

Stimulation of the α-secretase may have beneficial effects for two reasons: i) by preventing the formation of the neurotoxic Aβ; ii) by increasing the amount of the neuroprotective APPsα. Therefore, pharmacological up-regulation of the α-secretase could be a possible therapeutic treatment for AD.

In this article we review: 1) enzymes of ADAM family as α-secretase candidates, and 2) the pharmacological up-regulation of α-secretases as a possible therapeutic treatment for AD.

2. BIOCHEMICAL AND PHYSIOLOGICAL PROPERTIES OF α-SECRETASES

Processing of APP by α-secretases was the first proteolytic pathway of APP to be characterized in detail. The amyloid precursor protein belongs to a family of type I membrane-spanning glycoproteins and is constitutively expressed in many types of mammalian cells. The major proteolytic pathway of APP is the constitutive secretory pathway that involves cleavage by a putative α-secretase within the Aβ sequence at the cell surface (Sisodia, 1992; Haass *et al.*, 1992; Ikezu *et al.*, 1998), and in the trans-Golgi network (Kuentzel *et al.*, 1993; De Strooper *et al.*, 1993; Sambamurti *et al.*, 1992;

Tomita *et al.*, 1998). Cleavage of APP by the α-secretase generates the soluble N-terminal APP fragment (APPsα) of 105-125 kDa and a C-terminal fragment of 10 kDa, which is further cleaved by the γ-secretase to yield the p3 fragment. In most investigations p3 was not found in amyloid cores of classical plaques. Therefore, this pathway is often called the "non-amyloidogenic pathway" (Figure 1). Studies in various cell types confirmed that the major α-secretase cleavage site is between Lys-16 and Leu-17 in the Aβ domain, but multiple minor cleavages around this site have been observed (Zhong *et al.*, 1994; Simons *et al.*, 1996).

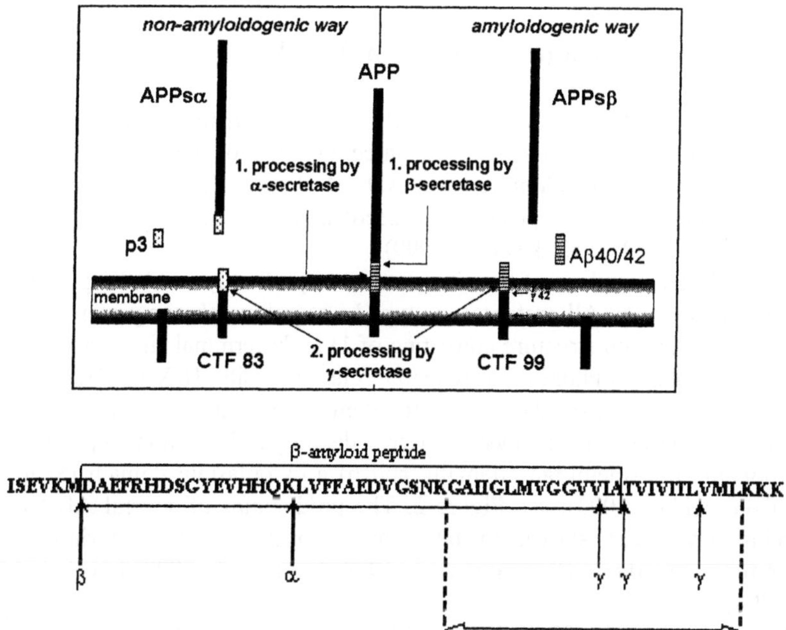

Figure 1. Processing of the amyloid precusor protein (APP)

Several studies have shown that the principal determinants of APP cleavage by α-secretase appear to be an α-helical conformation and the distance (12-13 residues) of the hydrolyzed bond from the membrane, and that the specificity of cleavage is independent of the primary sequence of the precursor (Sisodia, 1992; Sahasrabudhe *et al.*, 1992). To characterize the enzyme class of α-secretase several inhibitors were examined for their ability to inhibit the production of APPsα. Studies with the range of class-specific proteinase inhibitors demonstrated that only the zinc chelating agent 1,10-phenantrolin resulted in a significant reduction of APPsα secretion (Roberts *et al.*, 1994). Other studies have shown that hydroxamic acid-based compounds, such as TAPI, batimastat and marimastat, inhibited the release of APPsα in the medium of a variety of neural and peripheral cell lines (Arribas *et al.*, 1996; Parvathy *et al.*, 1998; Parkin *et al.*, 2002; Racchi *et al.*, 1999). The results of these studies classified the α-secretase to be a metalloprotease.

3. DISINTEGRIN METALLOPROTEASES (ADAMS) AS α-SECRETASES

3.1 Structure and functions of ADAMs

Proteinases of the ADAM family are the main candidates as physiologically relevant α-secretases. Three members of the ADAM family (a disintegrin and metalloproteinase) ADAM-10, ADAM-17 (TACE) and ADAM-9 have been proposed as α-secretases (Lammich *et al.*, 1999; Buxbaum *et al.*, 1998; Koike *et al.*, 1999).

ADAMs are members of the zinc protease superfamily and belong to the metzincin subgroup. All proteases are type I integral membrane proteins and have a multi-domain structure consisting of 1) an N-terminal signal peptide; 2) a prodomain containing a potential cleavage site [RX(R/K)R)] for prohormone convertase; 3) a catalytic domain typical of the metzincin metalloproteinases; 4) a cysteine-rich, disintegrin-like domain; 5) a hydrophobic transmembrane domain, and 6) a cytoplasmic domain (Seals and Courtneidge, 2003), see Figure 2. The prodomains of several ADAM proteins have been shown to be cleaved constitutively by furin-type proprotein convertases (Roghani *et al.*, 1999; Lum *et al.*, 1998; Loechel *et al.*, 1998).

Currently, there are 29 distinct ADAM family members, 17 are predicted to be catalytically active based of the presence of a conserved zinc binding sequence (HEXGHXXGXXHD; X represents any amino acid) in the protease domain. For most ADAM proteinases their physiological substrates are unknown.

The ADAM family has been implicated in the control of cytokine and growth factor shedding, membrane fusion and cell migration as well as in processes such as muscle development and fertilization. ADAM-mediated shedding of diverse membrane proteins occurs both constitutively and in response to a variety of stimuli, including peptide growth factors, changes in intracellular calcium concentration and protein kinase C (PKC) activation. Given this diversity of signals and substrates for sheddase action, it is possible that shedding is controlled by multiple regulatory mechanisms (Pandiella and Massague, 1991).

We will now focus on the three ADAMs known to be functional proteases involved in the cleavage of a variety of membrane proteins, releasing ("shedding") their extracellular domains from cells: ADAM-10, ADAM-17, and ADAM-9.

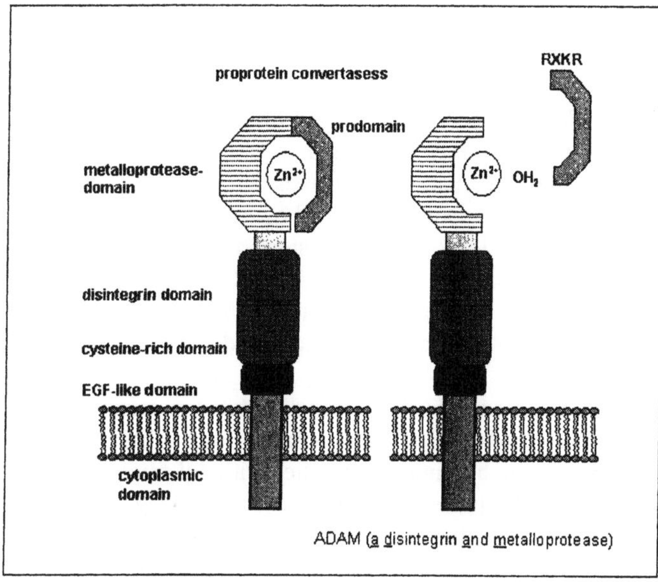

Figure 2. Domain structure of ADAM-10

3.2 ADAM-10

The enzyme ADAM-10 was purified in 1989 from bovine brain based on its ability to cleave myelin basic protein (Chantry *et al.*, 1989). ADAM-10 is closely related to Kuzbanian (Kuz), its *Drosophila* homologue which is involved in neurogenesis and myoblast differentiation (Fambrough *et al*, 1996; Rooke *et al.*, 1996,). High levels of ADAM-10 proteins are found in the brain and in other tissues such as heart, muscle, lung and kidney (Chantry and Glynn, 1990; Howard et al, 1996). ADAM-10 mRNA is also found in bone and cartilage (Dallas *et al.*, 1999). Putative ADAM-10 substrates are: precursor TNF-α (Rosendahl *et al.*, 1997; Lunn *et al.*, 1997), Notch receptor (Pan and Rubin, 1997) and Notch ligand Delta (Qi *et al.*, 1999; Six *et al.*, 2003), type IV collagen (Millichip *et al.*, 1998), ephrin-A2 (Hattori *et al.*, 2000), L1 adhesion molecule (Gutwein *et al.*, 2003) and chemokine fractalkine (CX3CL1) (Hundhausen *et al.*, 2003).

ADAM-10 is synthesized as an inactive zymogene and is processed to its mature form by cleavage of the prodomain by proprotein convertases, such as PC7 and furin (Anders *et al.*, 2001). We have purified membrane-bound metalloprotease ADAM-10 from bovine kidney. Characterizing the enzyme we noticed a striking similarity between the inhibition of ADAM-10 by various inhibitors, and the results reported for inhibition of APP cleavage by a putative α-secretase (Lammich *et al.*, 1999). We therefore investigated whether ADAM-10 protease has α-secretase activity and is involved in basal and stimulated ectodomain shedding of APP. Our studies showed that over-expression of ADAM-10 in HEK cells led to a several fold increase of APPsα and of the p10 fragment. The increase in shedding activity was stronger in cells with a higher expression level of ADAM-10. This enhanced α-secretase activity, caused by overexpression of ADAM-10, could be further increased by stimulation of protein kinase C with phorbol esters, a characteristic feature of the proposed α-secretase. The enhanced α-secretase activity was inhibited by hydroxamic acid-based zinc metalloprotease inhibitor BB-3103.

The endogenous basal and phorbol ester-stimulated α-secretase activity in HEK cells was inhibited by a dominant-negative form of ADAM-10 with a point mutation in the zinc-binding site. Studies with Aβ fragments showed that purified ADAM-10 protease from bovine kidney cleaved the octadecapeptide $A\beta_{11\text{-}28}$ spanning the α-secretase cleavage site at the Lys16-Leu17 bond. We also showed that the extent of peptide cleavage by ADAM-10 is conformation-dependent. The octadecapeptide $A\beta_{11\text{-}28}$ with α-helical conformation was a better substrate for ADAM-10 than a similar yet unstructured peptide $A\beta_{11\text{-}28}$ with the substitution Ala → Gly at position 21 corresponding to the Flemish mutation.

The colocalization of APP, BACE and ADAM-10 in mouse and human brains support the role of ADAM-10 and BACE as authentic α- and β-secretases (Marcinkiewicz and Seidah, 2000).

For investigating the role of ADAM-10 *in vivo*, ADAM-10-deficient mice were generated. They die at day 9.5 of embryogenesis with multiple defects in the developing central nervous system, somites and cardiovascular system. *In situ* hybridization revealed a reduced expression of the Notch target gene *hes-5* in the neural tube, and an increased expression of the Notch ligand *dll-1,* supporting an important role for ADAM-10 in Notch signaling in vertebrates. The early lethality precluded the establishment of primary neuronal cultures, and therefore APPsα generation was analyzed in embryonic fibroblasts. Fibroblasts from the knockout mice were found to have a preserved α-secretase activity in most cases, although there was a significant variation between clones, including several ones with increased α-secretase activity (Hartmann *et al.*, 2002). The early lethality of knock-out animals prevented a reliable analysis of the ADAM-10 function *in vivo*, especially in neuronal cells.

Recently we provided strong evidence for ADAM-10 to act as an efficient α-secretase of APP *in vivo*. We showed that even a moderate neuronal overexpression of ADAM-10 in mice transgenic for human APP [V717I] increased the production of secreted neurotrophic APPsα, reduced the formation of Aβ peptides and prevented their deposition in plaques. Functionally, impaired long-term potentiation and cognitive deficits were alleviated. Expression of the catalytically inactive ADAM-10 mutant led to an enhancement of the number and size of amyloid plaques in the brains of double-transgenic mice (Postina *et al.*, 2004).

Generation of mice overexpressing ADAM-10 provides strong evidence for ADAM-10 to act as an efficient α-secretase in vivo and to be a promising therapeutic target in AD. However, the existence of several α-secretases which might interact to cleave APP at the α-secretase cleavage site, cannot be excluded.

Recently it was shown by using expression silencing technique (RNA interference, RNAi) that three ADAMs, namely 9, 10 and 17, to a similar extent are involved in the α-secretase cleavage of APP in a human glioblastoma cells (Asai *et al.*, 2003).

3.3 ADAM-17

ADAM-17, often referred to as TACE, has been identified as the protease responsible for shedding of the transmembrane form of the tumor necrosis factor-α (TNF-α) at its physiological processing site (Black *et al.*, 1997; Moss *et al.*, 1997). TACE mRNA is found in most tissues, and expression of the protein is largely constitutive (Black *et al.*, 1997). The mature form of TACE is generated similarly to other proteolytically activc ADAMs from the proform by proteolytic removal of the prodomain by furin-type proprotein convertase (Schlöndorff *et al.*, 2000; Srour *et al.*, 2003; Endres *et al.*, 2003). The catalytic domain of TACE has been co-crystallized with a hydroxamic acid inhibitor and the structure solved at 2.0 Å resolution. The X-ray structure shows that the zinc environment and the placement of the major structural elements closely resemble those of snake venom metalloproteinases. Of all identified ADAMs only ADAM-10 has significant sequence homology with TACE; 91% of the ADAM-10 catalytic domain residues are identical to those of TACE (Maskos *et al.*, 1998).

Subsequent studies using mice and cell lines lacking TACE activity implicated TACE in the shedding of TGF-α, L-selectin, p75 TNF-receptor (Peschon *et al.*, 1998). TACE is also involved in the processing of Notch signaling (Brou *et al.*, 2000), the chemokine fractalkine (Garton et al, 2001), and the HER4/erbB4 receptor, a member of the epidermal growth factor receptor family (Rio *et al.*, 2000). TACE appears to be a sheddase with a wide range of substrates in adult organisms and additional substrates required for development.

The role for TACE as an α-secretase stems from the analysis of APP processing in primary embryonic fibroblasts derived from $TACE^{\Delta Zn/\Delta Zn}$ mice. The majority of $TACE^{\Delta Zn/\Delta Zn}$ mice died between embryonic day 17.5 (e17.5) and the first day after birth. All newborn $TACE^{\Delta Zn/\Delta Zn}$ mice could be identified by the presence of stunted vibrissae and open eyelids. Histological examination of $TACE^{\Delta Zn/\Delta Zn}$ fetuses on e17.5 revealed multiple additional defects, primarily in epithelial cell maturation and organization.

In primary embryonic fibroblasts derived from $TACE^{\Delta Zn/\Delta Zn}$ mice the constitutive release of APPsα was unaffected, but the regulated phorbol ester-stimulated α-secretase activity was strongly inhibited. The authors conclude that there are two classes of α-secretases, one involved in basal secretion, and another (probably TACE) involved in the regulated secretion. It was also shown that recombinant TACE could cleave a synthetic 10-mer peptide spanning the α-secretase cleavage site at the Lys16-Leu17 bond in the Aβ domain (Buxbaum *et al.*, 1998).

In contrast, it has been reported that TACE is involved in the constitutive α-secretase cleavage of APP in non-neuronal cells. In HEK cells transiently transfected with TACE, constitutive APPsα release was increased several fold, but the muscarinic- receptor activated APPsα release was not altered in TACE transfectants (Slack *et al.*, 2001).

In situ hybridization performed in mice during prenatal and postnatal development and in adulthood revealed an only partial overlap of the expression of TACE and APP (Marcinkiewicz and Seidah, 2000). According to an immunohistochemical analysis of human brain, TACE was present in neurons and co-localized with Aβ deposits in AD brains. These observations support a neuronal role for TACE and suggest a mechanism for its involvement in AD pathogenesis (Skovronsky *et al.*, 2001).

Until now, evidence is still lacking for a role of TACE in α-secretase processing of APP in neuronal cells. Generation of transgenic mice overexpressing TACE specifically in neurons will be necessary to answer this question.

3.4 ADAM-9

ADAM-9, also known as MDC 9 or meltrin-γ, is expressed in a variety of tissues and in hematological malignancies (Wu at al., 1997). ADAM-9 has been implicated in the ectodomain shedding of the heparin-binding EGF-like growth factor (HB-EGF) from epithelial cells (Izumi *et al.*, 1998), and it plays a role also in regulating the motility of cells by binding to the integrin $\alpha_6\beta_1$ (Nath *et al.*, 2000).

An α-secretase activity of ADAM-9 has been shown by co-expression of mouse ADAM-9 with APP in COS cells. Overexpression of both proteins led to an increase of constitutive and phorbol ester-stimulated APPsα secretion (Koike *et al.*, 1999). It has also been reported that the secreted form of human ADAM-9 (hADAM-9s), which lacks the carboxy-terminus, has α-secretase activity: when hADAM-9s was co-expressed in COS cells with APP and treated with phorbol ester an increased secretion of APPsα was observed (Hotoda *et al.*, 2002). However, recombinant ADAM-9 did not cleave a 12-mer peptide at the Lys16-Leu17 bond, but at the His14-Gln15 bond in the Aβ domain (Roghani *et al.* 1999). Alternative cleavage sites in APP have been identified in primary cultures of rat hippocampal neurons, one of which corresponds to the His14-Gln-15 bond in Aβ (Simons *et al.*, 1996). To learn more about ADAM-9 function, mice lacking functional ADAM-9 were generated. These animals had no evident abnormalities during development or adult life. There were no differences in the production of APPsα, p_3 fragment and Aβ in cultured hippocampal neurons from ADAM-9 knockout or wild-type mice, arguing against an essential major role of ADAM-9 as an α-secretase in mice (Weskamp *et al.*, 2002).

3.5 α-Secretases and Alzheimer's Disease

Several studies examined α-secretase activity in AD patients. A decreased amount of APPsα in the cerebrospinal fluid (CSF) of familial and sporadic AD patients was observed (Lannfelt *et al.*, 1995; Sennvik *et al.*, 2000). α-Secretase activity was also reduced in temporal cortex homogenates from AD patients (Tyler *et al.*, 2002). The studies of Colciaghi *et al.* suggest that alteration of the α-secretase activity might occur both in the neuronal compartment and in peripheral cells. ADAM-10 protein levels expressed in platelets obtained from AD patients were significantly reduced, and a decreased APPsα amount was observed both in thrombin-activated platelets and CSF in the same pool of AD patients (Colciaghi *et al.*, 2002).

The mechanism by which APPsα in CSF is decreased is not yet fully understood. One can speculate that the observed reduction of soluble APPsα might result from alterations in either the expression level or the activity of α- and β-secretases.

4. THE α-SECRETASE AS A TARGET FOR THE TREATMENT OF ALZHEIMER'S DISEASE

The cleavage of APP by α-secretase precludes the formation of Aβ, and released APPsα stimulates neurite outgrowth and has potent memory-enhancing effects (Furukawa *et al.*, 1996; Meziane *et al.*, 1998). This suggests that processing of APP through this pathway not only does not lead to AD pathology, but also may be neuroprotective. Therefore, pharmacological α-secretase stimulation may be a therapeutic interventional strategy in Alzheimer's disease.

4.1 Activation of G-protein-coupled receptors and downstream signaling pathways

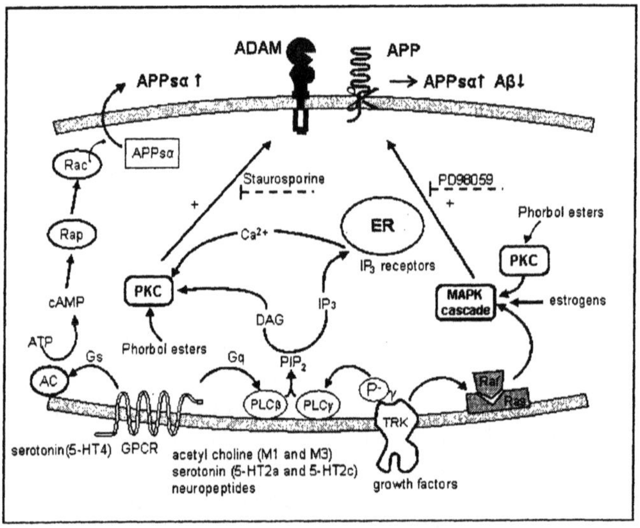

Figure 3. Stimulation of the non-amyloidogenic pathway

One approach for upregulation for α-secretase activity is the activation of G-protein-coupled receptors for neurotransmitters and neuropeptides in brain areas affected by Alzheimer's disease. In many cases in which an effect on α-secretase activity has been demonstrated, these receptors have been shown to be coupled to the activation of the phospholipase C and protein kinase C system.

The first receptors which have been shown to be connected to the stimulation of the α-secretase activity were muscarinic acetylcholine receptors

(Nitsch *et al.*, 1992). Stimulation of M1 and M3-receptor subtypes with carbachol increased the basal release of secreted APP within minutes. The receptor-activated APP release was blocked by staurosporin suggesting that protein kinase C mediates neurotransmitter receptor-controlled APP processing. These first experiments were performed in human embryonic kidney cells. The results were confirmed and extended to several animal experiments. Treatment of normal or aged rats and rats with severe basal forebrain cholinergic deficits with a muscarinic receptor agonist resulted in increased levels of secreted APP in the cerebrospinal fluid and in changes of cognitive functions (Lin *et al.*, 1999). In several animal models mimicking different aspects of AD, M1 muscarinic agonists restored cognitive impairments and, in selected cases, induced a decrease in brain Aβ levels (Fisher *et al.*, 2003). In a study with AD-patients, the selective muscarinic M1-agonist AF102B significantly decreased levels of total Aβ in the cerebrospinal fluid (Nitsch *et al.*, 2000). Until now, however, intensive clinical tests showing clear beneficial effects of muscarinic agonists are still lacking.

Several other studies demonstrated that APP processing and regulation of the α-secretase activity is under control of several major neurotransmitters. Metabotropic glutamate receptors stimulate APPsα secretion from peripheral cells via the receptor subtype 1α (Nitsch *et al.*, 1997), and from neural cells *via* subtypes 1α/β and 5α (Jolly-Tornetta *et al.*, 1998). Furthermore, it has been shown that stimulation of serotonin 5-HT2a and 5-HT2c leads to secretion of the amyloid precursor protein ectodomain (Nitsch *et al.*, 1996). In most studies, protein kinase C was involved in signal transduction, although the participation of other effector systems, like phospholipase A or the mitogen-activated protein kinase pathway, was evident.

The serotonin 5-HT(4) receptor plays a role in memory and cognition. Activation of this receptor has been shown to stimulate secretion of APPsα from peripheral and neural cells. In contrast to the receptors mentioned above, its activation leads to the production of cyclic AMP. This phenomenon involves the small GTPase Rac of the Rho (ρ)-family, which is regulated by cAMP in a PKA-independent manner (Maillet *et al.*, 2003; Robert *et al.*, 2001). The peptide hormone bradykinin induced an increase in the release of soluble APP from human fibroblasts, the effect appeared to be dependent on interaction of the ligand with the B2-type of bradykinin receptors, yet was independent of activation of protein kinase C (Racchi *et al.*, 1998).

It is now generally accepted that activation of protein kinase C by phorbol ester increases secretion of APPsα. In cultured cells, the α-secretase activity of ADAM-10 (Lammich *et al.*, 1999) and ADAM-17 (Buxbaum *et al.*, 1998) was stimulated by treatment with phorbol esters. Recently, new amide-bearing benzolactam-based protein kinase C modulators were developed that induce enhanced secretion of the APP metabolite APPsα. Since these compounds show a marked loss of tumor-promoting activity, they are therefore valuable candidate compounds in the search for Alzheimer's therapeutics capable of modulating amyloid processing (Kozikowski *et al.*, 2003). It has, however, been shown that constitutive overactivation of protein kinase C in guinea pig brain increases APPsα without decreasing Aβ generation. Apparently in some systems there is dissociation between the generation of APPsα and Aβ, the reasons for which are not yet understood (Rossner *et al.*, 2000).

4.2 Cholinesterase inhibitors

Inhibition of acetylcholine esterase increases the amount of neurotransmitters at the cholinergic synapses and consequently the cholinergic transmission. AChE-inhibitors represent the first generation of anti-AD agents. However, they only slow down but do not halt or reverse the progression of the disease.

There is evidence that they act *via* an additional mechanism, the activation of the non-amyloidogenic pathway. In neuronal cell cultures an elevation of released APP and PKC-levels was found after treatment with AChE inhibitors (Pakaski *et al.*, 2001). Other reports suggest an additional involvement of the MAP-kinase and tyrosine kinase-dependent pathway in the enhancement of the APPsα release by the AChE inhibitor (Youdim *et al.*, 2003). A short treatment of SH-SY5Y neuroblastoma cells with ganstigmine, a novel geneserine-derived AChE inhibitor, promoted an increase in the APPsα secretion (Mazzucchelli *et al.*, 2003). As a reasonable hypothesis, the α-secretase pathway may be stimulated by ganstigmine through indirect stimulation of muscarinic receptors.

4.3 Steroid hormones

A number of *in vitro* studies have shown that treatment of cells with physiological concentrations of estrogen (17 β-estradiol) is associated with accumulation in the conditioned medium of the amino-terminal cleavage product of APP, indicative of non-amyloidogenic processing (Jaffe *et al.*, 1994; Manthey *et al.*, 2001; Xu *et al.*, 1998). Xu *et al.* have reported that estrogen reduces the generation of Aβ, possibly by promoting the trafficking of APP out of the endoplasmic reticulum and Golgi, where β-secretase cleaves APP. In another report (Manthey *et al.*, 2001) it has been shown that a rapid secretion of APPsα might be induced via the mitogen-activated protein kinase pathway. They provide evidence that this pathway is independent of estrogen receptors. Testosterone also increases the production of APPsα and decreases Aβ generation from N2a cells and rat primary cerebrocortical neurons (Gouras *et al.*, 2000). Interestingly, according to a recent study, testosterone concentrations were lower in men who developed Alzheimer's disease and this difference occurred several years before AD diagnosis (Moffat *et al.*, 2004). Future research may determine whether higher endogenous free testosterone levels offer protection against Alzheimer's disease in older men.

4.4 Other pharmacological agents

Hormones and growth factors acting via tyrosine kinase receptors like epidermal growth factors (EGF) and insulin stimulate the secretion of APPsα from various cell lines. The results demonstrate that in A431 cells activation of the EGF receptor increases α-secretase activity by a mechanism that partly depends on PKC activity (Slack *et al.*, 1997). Insulin regulated APPsα production is mediated mainly via the phosphatidyl inositol 3-kinase pathway and may act at the level of vesicular trafficking (Solano *et al.*, 2000).

It has been proposed that treatment with non-steroidal anti-inflammatory drugs (NSAIDs) reduces the risk for AD. Various NSAIDs such as the cyclooxygenase inhibitors, nimesulide, ibuprofen and indomethacine, stimulate the generation of APPsα into the conditioned media of SH-SY5Y neuroblastoma and PC12 cells. This effect was mediated by protein kinase C and mitogen-activated protein (MAP) kinase (Avramovich *et al.*, 2002).

It has been reported that adding copper to Chinese-hamster ovary (CHO) greatly reduced the levels of Aβ and caused an increase in the secretion of the APP ectodomain; thus copper promoted the non-amyloidogenic pathway of APP (Borchardt *et al.*, 1999). Recent *in vivo*-experiments in APP-transgenic mice demonstrated that copper treatment lowered Aβ production and reduced amyloid plaques (Bayer *et al.*, 2003). Further studies with dietary copper in AD patients should reveal whether it has a beneficial effect.

4.5 Cholesterol-lowering drugs

Biochemical, epidemiological and genetic aspects demonstrate a link between cholesterol levels, Aβ production and Alzheimer's disease. The ε4 allele of apolipoprotein E, which is associated with higher plasma cholesterol levels, has been shown to be a risk factor for the disease (Corder at al., 1993), and elevated levels of serum low-density lipoprotein and total cholesterol positive correlate with $A\beta_{1\text{-}42}$ levels in the AD brain (Kuo *et al.*,

1998). Epidemiological studies indicate that the prevalence of AD is reduced among people taking a class of cholesterol-lowering medicines, termed statins, such as lovastatin and simvastatin (Wolozin *et al.*, 2000; Jick *et al.*, 2000). Both studies suggest that patients taking statins have an approx. 70% lower risk of developing AD. Statins are 3-hydroxy-3-methylglutaryl coenzyme A (HMG-CoA) reductase inhibitors that competitively inhibit HMG-CoA reductase, the enzyme catalyzing conversion of HMG-CoA to L-mevalonate, a key intermediate in cholesterol synthesis. Increasing evidence suggests that statins can inhibit Aβ production.

In our studies, we have shown that modulation of cholesterol concentration can influence the activity of the α-secretase; in particular we identified the α-secretase ADAM-10 as a major target of the cholesterol effects. Treatment of various peripheral and neural cell lines with either the cholesterol extracting agent methyl-β-cyclodextrin (MβCD) or lovastatin resulted in a drastic increase of secreted APPsα. In the human astroglioma cells overexpressing APP, the increase of α-secretase activity resulted in a decreased secretion of Aβ peptides. Several mechanisms were elucidated as being the basis of enhanced α-secretase activity: increased membrane fluidity and impaired internalization of APP were responsible for the effect observed with MβCD; treatment with lovastatin resulted in higher expression of the α-secretase ADAM-10. Our results demonstrate that cholesterol reduction promotes the non-amyloidogenic α-secretase pathway, increased formation of neuroprotective APPsα and reduction of Aβ production (Kojro *et al.*, 2001).

In accordance with this finding, it has been demonstrated that lovastatin and simvastatin reduce intracellular and extracellular levels of $A\beta_{40}$ and $A\beta_{42}$ peptides in primary cultures of hippocampal neurons and mixed cortical neurons. Reduction of Aβ levels was also observed *in vivo.* Guinea pigs treated with high concentration of simvastatin show up to 50% decrease in Aβ production within 3 weeks (Fassbender *et al.*, 2001). In addition, high dietary cholesterol accelerated AD-related pathologies in a transgenic mouse model, including Aβ deposition (Refolo *et al.*, 2000). Preliminary studies performed in humans analyzed the effects of statins on Aβ levels. Persons with elevated low-density lipoprotein cholesterol were treated with different doses of lovastatin for three months. In the serum of these patients a significant reduction of Aβ concentration up to 40% has been observed (Friedhoff *et al.*, 2001). In other studies, treatment with simvastatin in normocholesterolemic patients with Alzheimer's disease resulted in small decrease of Aβ levels in the cerebrospinal fluid of patients with mild Alzheimer's disease (Simons *et al.*, 2002). Both studies indicate that statins can decrease Aβ levels in humans by reducing CSF and plasma Aβ.

Increasing evidence suggests that the enzymes generating Aβ, the β- and γ- secretases operate best in a high-cholesterol environment. Lipid rafts are cholesterol-enriched membrane microdomains implicated in signal transduction, protein trafficking and proteolytic processing (Simons and Toomre, 2000). Cholesterol plays a crucial role in maintaining lipid rafts in a functional state (Simons and Ikonen, 2000). Several studies demonstrated that lipid rafts are critically involved in regulating Aβ generation, and that β-secretase (BACE) is associated with rafts. Recombinant BACE expressed in three distinct cell lines has been found in lipid rafts (Riddel *et al.*, 2001). Wild-type BACE was found in rafts prepared from a human neuroblastoma cell line (SH-SY5Y), and it was shown that addition of GPI-anchor to BACE targets the enzyme exclusively to lipid raft domains. Expression of GPI-BACE drastically increased the secretion of both APPsβ and Aβ peptides (Cordy *et al.*, 2003). Production of Aβ was also increased when APP and BACE clustering was induced upon cross-linking with antibodies to cholesterol-rich membrane microdomains (Ehehalt *et al.*, 2003). Also, the γ-secretase complex was shown to be raft-associated. Consistent with this localization, it was found that gamma-secretase activity is cholesterol-dependent. Depletion of membrane cholesterol completely inhibits γ-secretase cleavage, which can be restored by cholesterol replacement (Wahrle *et al.*, 2002). Other studies showed that γ-secretase activity is present in rafts, but the membrane cholesterol content does not directly modulate the enzyme activity (Wada *et al.*, 2003). It was shown that – in contrast to β- and γ-secretases – active α-secretase ADAM-10 is not localized in lipid rafts: cholesterol depletion led to disruption of rafts and caveolae and increase in enzyme activity (Kojro *et al.*, 2001). Thus, high membrane cholesterol content favors Aβ production, and low membrane cholesterol content favors APPsα production.

5. PERSPECTIVES

The finding that neuronal overexpression of ADAM-10 in an AD mouse model prevents amyloid plaque formation and hippocampal defects, provides evidence that ADAM-10 acts as an efficient α-secretase without excluding other members of the ADAM family as α-secretases.

For AD treatment, activation of the α-secretase, especially of ADAM-10, may be particularly promising as compared to other targets, because it combines several beneficial effects: inhibition of Aβ peptide production by cleavage of APP at the α-secretase site, further reduction of the Aβ level by cleavage of soluble Aβ peptides, and, in addition, generation of neuroprotective APPsα. In spite of intense efforts, it has been difficult to find safe and selective β- and γ-secretase inhibitors, mainly resulting from the influence on other substrates. Activation of proteases degrading Aβ, like the insulin-degrading enzyme or neprilysin, represents another therapeutic approach (Leissring *et al.*, 2003). These enzymes, however, have a number of substrates, and the signaling pathways for upregulation at present are largely unknown. Cleavage of putative substrates other than APP by ADAM-10 might play a role during neural development (Hartmann *et al.*, 2002; Hattori *et al.*, 2000; Mechtersheimer *et al.*, 2001), but probably has no or only minor effects in adults. In our mouse model we found at least no indications that ADAM-10 affects Notch signaling in the brain of adult animals.

It will be important to show whether an increase of α-secretase expression and activity at later stages in life could exert such beneficial effects as shown in the animal model, since this would open approaches to a therapy of the neurodegeneration in AD.

REFERENCES

Anders, A., Gilbert, S., Garten, W., Postina, R., Fahrenholz, F., 2001, Regulation of the alpha-secretase ADAM-10 by its prodomain and proprotein convertases. *FASEB J.*, **15:** 1837-1839.

Asai, M., Hattori, C., Szabo, B., Sasagawa, N., Maruyama, K., Tanuma, S., Ishiura, S., 2003, Putative function of ADAM-9, ADAM-10, and ADAM-17 as APP alpha-secretase. *Biochem. Biophys. Res. Commun.*, **301:** 231-5.

Arribas, J., Coodly, L., Vollmer, P., Kishimoto, T. K., Rose-John, S., Massague, J., 1996, Diverse cell surface protein ectodomains are shed by a system sensitive to metalloprotease inhibitors. *J. Biol. Chem.*, **271:** 11376-82.

Avramovich, Y., Amit, T., Youdim, M. B., 2002, Non-steroidal anti-inflammatory drugs stimulate secretion of non-amyloidogenic precursor protein. *J. Biol. Chem.* **277:** 31466-31473.

Bayer, T. A., Schafer, S., Simons, A., Kemmling, A., Kamer, T., Tepests, R., Eckert, A., Schussel, K., Eikenberg, O., Sturchler-Pierrat, C., Abramowski, D., Staufenbiel, M., Multhaup, G., 2003, Dietary Cu stabilizes brain superoxide dismutase 1 activity and reduces amyloid Abeta production in APP23 transgenic mice. *Proc. Natl. Acad. Sci. USA* **100:** 14187-14192.

Black, R. A., Rauch, C. T., Kozlosky, C. J., Peschon, J. J., Slack. J. L., Wolfson, M. F., Castner, B. J., Stocking, K. L., Reddy, P., Srinivasan. S., Nelson, N., Boiani, N., Schooley, K. A., Gerhart, M., Davis, R., Fitzner, J. N., Johnson, R. S., Paxton, R. J., March, C. J., Cerretti, D. P., 1997, A metalloproteinase disintegrin that releases tumour-necrosis factor-alpha from cells. *Nature* **385:** 729-33.

Borchardt, T., Camakaris, J., Cappai, R., Masters, C. L., Beyreuther, K., Multhaup, G., 1999, Copper inhibits beta-amyloid production and stimulates the non-amyloidogenic pathway of amyloid-precursor-protein secretion. *Biochem. J.* **344:** 461-467.

Brou, C., Logeat, F., Gupta, N., Bessia, C., LeBail, O., Doedens, J. R., Cumano, A., Roux, P., Black, R. A., Israel, A., 2000, A novel proteolytic cleavage involved in Notch signaling: the role of the disintegrin-metalloprotease TACE. *Mol.Cell* **5:** 207-16.

Buxbaum, J. D., Liu, K.-N., Luo, Y., Slack. J. L., Stocking, K. L., Peschon, J. J., Johnson, R. S., Castner, B. J., Cerretti,D. P., Black,R., 1998, Evidence that tumor necrosis factor converting enzyme is involved in regulated α-secretase cleavage of the Alzheimer amyloid protein precursor. *J. Biol. Chem.* **273:** 27765-67.

Chantry, A., Gregson, N. A., Glynn, P., 1989, A novel metalloproteinase associated with brain myelin membranes. Isolation and characterization. *J. Biol. Chem.* **264:** 21603-7.

Chantry, A., Glynn, P., 1990, A novel metalloproteinase originally isolated from brain myelin membranes is present in many tissues. *Biochem J.* **268:** 245-8.

Colciaghi, F., Borroni, B., Pastorino, L., Marcello, E., Zimmermann, M., Cattabeni, F., Padovani, A., Di Luca, M., 2002, α-Secretase ADAM-10 as well as αAPPs is reduced in platelets and CSF of Alzheimer disease patients. *Mol. Med.* **8:** 67-74.

Corder, E. H., Saunders, A. M., Strittmatter, W. J., Schmechel, D. E., Gaskell, P. C., Small, G. W., Roses, A. D., Haines, J. L., Pericak-Vance, M. A., 1993, Gene dose of apolipoprotein E type 4 allele and the risk of Alzheimer's disease in late onset families. *Science* **261:** 921-3.

Cordy, J.M., Hussain, I., Dingwall, C., Hooper, N.M., Turner, A.J., 2003, Exclusively targeting beta-secretase to lipid rafts by GPI-anchor addition up-regulates beta-site processing of the amyloid precursor protein. Proc Natl Acad Sci U S A **100:** 11735-40.

Dallas, D. J., Genever, P. G., Patton, A. J., Millichip, M.I., McKie, N., Skerry, T. M., 1999, Localization of ADAM-10 and Notch receptors in bone. *Bone* **25:** 9–15.

De Strooper, B., Umans, L., Van Leuven, F., Van Den Berghe, H., 1993, Study of the synthesis and secretion of normal and artificial mutants of murine amyloid precursor protein (APP): cleavage of APP occurs in a late compartment of the default secretion pathway. *J. Cell Biol.* **121:** 295–304.

Ehehalt, R., Keller, P., Haass, C., Thiele, C., Simons, K., 2003, Amyloidogenic processing of the Alzheimer beta-amyloid precursor protein depends on lipid rafts. *J. Cell. Biol.* **160:** 113-23.

Endres, K., Anders, A., Kojro, E., Gilbert, S., Fahrenholz, F., Postina, R., 2003, Tumor necrosis factor-alpha converting enzyme is processed by proprotein-convertases to its mature form which is degraded upon phorbol ester stimulation. *Eur. J. Biochem.* **270:** 2386-93.

Fambrough, D., Pan, D., Rubin, G. M., Goodman, C. S., 1996, The cell surface metalloprotease/disintegrin Kuzbanian is required for axonal extension in Drosophila. *Proc. Natl. Acad. Sci. USA* **93:** 13233–13238.

Fassbender, K., Simons, M., Bergmann, C., Stroick, M., Lutjohann, D., Keller, P., Runz, H., Kuhl, S., Bertsch, T., von Bergmann, K., Hennerici, M., Beyreuther, K., Hartmann, T., 2001, Simvastatin strongly reduces levels of Alzheimer's disease beta -amyloid peptides Abeta 42 and Abeta 40 in vitro and in vivo. *Proc. Natl. Acad. Sci. USA.* **98:** 5856-61.

Fisher, A., Pittel, Z., Haring, R., Bar-Ner, N., Kliger-Spatz, M., Natan, N., Egozi, I., Sonego, H., Marcovitch, I., Brandeis, R., 2003, M1 muscarinic agonists can modulate some of the hallmarks in Alzheimer's disease: implications in future therapy. *J. Mol. Neurosci.* **20:** 349-356.

Friedhoff, L. T., Cullen, E. I., Geoghagen, N. S., Buxbaum, J. D., 2001, Treatment with controlled-release lovastatin decreases serum concentrations of human beta-amyloid (A beta) peptide. *Int. J. Neuropsychopharmacol.* **4:** 127-30.

Furukawa, K., Sopher, B. L., Rydel, R., Begley, J. G., Pham, D. G., Martin, G. M., Fox, M., Mattson, M. P., 1996, Increased activity- regulating and neuroprotective efficacy of α-secretase-derived secreted amyloid precursor protein conferred by a C-terminal heparin-binding domain. *J. Neurochem.* **67:** 1882-1896.

Garton, K. J., Gough, P. J., Blobel, C. P., Murphy, G., Greaves, D. R., Dempsey, P. J., Raines, E. W., 2001, Tumor necrosis factor-alpha-converting enzyme (ADAM-17) mediates the cleavage and shedding of fractalkine (CX3CL1). *J. Biol. Chem.* **276:** 37993-8001.

Gouras, G. K., Xu, H., Gross, R. S., Greenfield, J. P., Hai, B., Wang, R., Greengard, P., 2000, Testosterone reduces neuronal secretion of Alzheimer's beta-amyloid peptides. *Proc. Natl. Acad. Sci. USA* **97:** 1202-1205.

Gutwein, P., Mechtersheimer, S., Riedle, S., Stoeck, A., Gast, D., Joumaa, S., Zentgraf, H., Fogel, M., Altevogt, D. P., 2003, ADAM10-mediated cleavage of L1 adhesion molecule at the cell surface and in released membrane vesicles. *FASEB J.* **17:** 292-4.

Haass, C., Selkoe, D. J., 1993, Cellular processing of beta-amyloid precursor protein and the genesis of amyloid beta-peptide. *Cell.* **75:** 1039-1042.

Haass, C., Koo, E. H., Mellon, A., Hung, A. Y., Selkoe, D. J., 1992, Targeting of cell-surface beta-amyloid precursor protein to lysosomes: alternative processing into amyloid-bearing fragments. *Nature* **357:** 500–503.

Hartmann, D., De Strooper, B., Serneels, L., Craessaerts, K., Herreman, A., Annaert, W., Umans, L., Lubke, T., Lena, I. A., von Figura, K., Saftig, P., 2002, The disintegrin/metalloprotease ADAM-10 is essential for Notch signalling but not for α-secretase activity in fibroblasts. *Hum. Mol. Genet.* **11:** 2615-2624.

Hattori, M. ***et al.***, 2000, Regulated cleavage of a contact-mediated axon repellent. *Science* **289:** 1360–1365.

Hotoda, N., Koike, H., Sasagawa, N., Ishiura, S., 2002, A secreted form of human ADAM-9 has an alpha-secretase activity for APP. *Biochem. Biophys. Research. Commun.* **293:** 800-5.

Howard, L., Lu, X., Mitchell, S., Griffiths, S., Glynn, P., 1996, Molecular cloning of MADM: a catalytically active mammalian disintegrin-metalloprotease expressed in various cell types. *Biochem. J.* **317:** 45–50.

Hundhausen, C., Misztela, D., Berkhout, T. A., Broadway, N., Saftig, P., Reiss, K., Hartmann, D., Fahrenholz, F., Postina, R., Matthews, V., Kallen, K. J., Rose-John, S., Ludwig, A., 2003, The disintegrin-like metalloproteinase ADAM-10 is involved in constitutive cleavage of CX3CL1 (fractalkine) and regulates CX3CL1-mediated cell-cell adhesion. *Blood* **102:** 1186-95.

Ikezu, T., Trapp, B. D., Song, K. S., Schlegel, A., Lisanti, M. P., Okamoto, T., 1998, Caveolae, plasma membrane microdomains for alpha-secretase-mediated processing of the amyloid precursor protein. *J. Biol. Chem.* **273:** 10485–10495.

Izumi, Y., Hirata, M., Hasuwa, H., Iwamoto, R., Umata, T., Miyado, K., Tamai, Y., Kurisaki, T., Sehara-Fujisawa, A., Ohno, S., Mekada, E., 1998, A metalloprotease-disintegrin, MDC9/meltrin-gamma/ADAM9 and PKCdelta are involved in TPA-induced ectodomain shedding of membrane-anchored heparin-binding EGF-like growth factor. *EMBO J.* **17:** 7260-72.

Jaffe, A. B., Toran-Allerand, C. D., Greengard, P., Gandy, S. E., 1994, Estrogen regulates metabolism of Alzheimer amyloid beta precursor protein. *J. Biol. Chem.* **269:** 13065-13068.

Jick, H., Zornberg, G. L., Jick, S. S., Seshadri, S., Drachman, D. A., 2000, Statins and the risk of dementia. *Lancet* **356:** 1627-31.

Jolly-Tornetta, C., Gao, Z. Y., Lee, V. M., Wolf, B. A., 1998, Regulation of amyloid precursor protein secretion by glutamate receptors in human Ntera 2 neurons. *J. Biol. Chem.* **273:** 14015-14021.

Kang, J., Lemaire, H. G., Unterbeck, A., Salbaum, L. M., Masters, C. L., Grzeshik, K. G., Multhaup, G., Beyreuther, K., Müller-Hil, B., 1987, The precursor of Alzheimer's disease amyloid A4 protein resembles a cell surface receptor. *Nature* **325:** 733-736.

Koike, H., Tomioka, S., Sorimachi, H., Saido, T. C., Maruyama, K., Okuyama, A., Fujisawa-Sehara, A., Ohno, S., Suzuki, K.; Ishiura, S., 1999, Membrane-anchoredmetalloprotease MDC9 has an alpha-secretase activity responsible for processing the amyloid precursor protein. *Biochem. J.* **343:** 371-375.

Kojro, E., Gimpl, G., Lammich, S., Marz, W., Fahrenholz, F., 2001, Low cholesterol stimulates the nonamyloidogenic pathway by its effect on the alpha -secretase ADAM-10. *Proc. Natl. Acad. Sci. USA* **98:** 5815-20.

Kozikowski, A. P., Nowak, I., Petukhov, P. A., Etcheberrigaray, R., Mohamed, A., Tan, M., Lewin, N., Hennings, H., Pearce, L. L., Blumberg, P. M., 2003, New amide-bearing benzolactam-based protein kinase C modulators induce enhanced secretion of the amyloid precursor protein metabolite sAPPalpha. *J. Med. Chem.* **46:** 364-373.

Kuentzel, S. L., Ali, S. M., Altman, R. A., Greenberg, B. D., Raub, T. J., 1993, The Alzheimer beta-amyloid protein precursor/protease nexin-II is cleaved by secretase in a trans-Golgi secretory compartment in human neuroglioma cells. *Biochem. J.* **295:** 367–378.

Kuo, Y. M., Emmerling, M. R., Bisgaier, C. L., Essenburg, A. D., Lampert, H. C., Drumm, D., Roher, A. E., 1998, Elevated low-density lipoprotein in Alzheimer's disease correlates with brain abeta 1-42 levels. *Biochem. Biophys. Res. Commun.* **252:** 711-5.

Lammich, S., Kojro, E., Postina, R., Gilbert, S., Pfeiffer, R., Jasionowski, M., Haass, C. und Fahrenholz, F., 1999, Constitutive and regulated α-secretase cleavage of Alzheimer's amyloid precursor protein by a disintegrin metalloprotease. *Proc. Natl. Acad. Sci. USA* **96:** 3922-3927.

Lannfelt, L., Basun, H., Wahlund, L. O., Rowe, B. A., Wagner, S. L., 1995, Decreased alpha-secretase-cleaved amyloid precursor protein as a diagnostic marker for Alzheimer's disease. *Nature Med.* **1:** 829-832.

Leissring, M. A., Farris, W., Chang, A. Y., Walsh, D. M., Wu, X., Sun, X., Frosch, M. P., Selkoe, D. J., 2003, Enhanced proteolysis of beta-amyloid in APP transgenic mice prevents plaque formation, secondary pathology, and premature death. *Neuron* **40:** 1087-1093.

Lin, L., Georgievska, B., Mattsson, A., Isacson, O., 1999, Cognitive changes and modified processing of amyloid precursor protein in the cortical and hippocampal system after cholinergic synapse loss and muscarinic receptor activation. *Proc. Natl. Acad. Sci. U S.A* **96:** 12108-12113.

Loechel, F., Gilpin, B. J., Engvall, E., Albrechtsen, R., Wewer, U. M., 1998, Human ADAM 12 (meltrin alpha) is an active metalloprotease. *J. Biol. Chem.* **273:** 16993-16997.

Lum, L., Reid, M. S., Blobel, C. P., 1998, Intracellular maturation of the mouse metalloprotease disintegrin MDC15. *J. Biol. Chem.* **273:** 26236-26247.

Lunn, C. A., Fan, X., Dalie, B., Miller, K., Zavodny, P. J., Narula, S. K., Lundell, D., 1997, Purification of ADAM-10 from bovine spleen as a TNF alpha convertase. *FEBS Lett.* **400:** 333–335.

Maillet, M., Robert, S. J., Cacquevel, M., Gastineau, M., Vivien, D., Bertoglio, J., Zugaza, J. L., Fischmeister, R., Lezoualc'h, F., 2003, Crosstalk between Rap1 and Rac regulates secretion of sAPPalpha. *Nat. Cell Biol.* **5:** 633-639.

Manthey, D., Heck, S., Engert, S., Behl, C., 2001, Estrogen induces a rapid secretion of amyloid beta precursor protein via the mitogen-activated protein kinase pathway. *Eur. J. Biochem.* **268:** 4285-4291.

Marcinkiewicz, M., Seidah, N., G., 2000, Coordinated expression of β-amyloid precursor protein and the putative β- secretase BACE and α-secretase ADAM-10 in mouse and human brain. *J. Neurochem.* **75:** 2133-2143.

Maskos, K., Fernandez-Catalan, C., Huber, R., Bourenkov, G. P., Bartunik, H., Ellestad, G. A., Reddy, P., Wolfson, M. F., Rauch, C. T., Castner, B. J., Davis, R., Clarke, H. R., Petersen, M., Fitzner, J. N., Cerretti, D. P., March, C. J., Paxton, R. J., Black, R. A., Bode, W., 1998, Crystal structure of the catalytic domain of human tumor necrosis factor-alpha-converting enzyme. *Proc. Natl. Acad. Sci. USA* **95:** 3408-3412.

Mattson, M. P., Cheng, B., Culwell, A. R., Esch, F. S., Lieberburg, I., Rydel, R. E., 1993, Evidence for excitoprotective and intraneuronal calcium-regulating roles for secreted forms of the β-amyloid precursor protein. *Neuron* **10:** 243-254.

Mazzucchelli M, Porrello E, Villetti G, Pietra C, Govoni S, Racchi M (2003) Characterization of the effect of ganstigmine (CHF2819) on amyloid precursor protein metabolism in SH-SY5Y neuroblastoma cells. *J. Neural Transm.* **110:** 935-947.

Mechtersheimer, S., Gutwein, P., Agmon-Levin, N., Stoeck, A., Oleszewski, M., Riedle, S., Postina, R., Fahrenholz, F., Fogel, M., Lemmon, V., Altevogt, P., 2001, Ectodomain shedding of L1 adhesion molecule promotes cell migration by autocrine binding to integrins. *J. Cell Biol.* **155:** 661-673.

Mezaine, H., Dodart. J.-C., Mathis, C., Little, S., Clemens, J., Paul. S. M., Ungerer, A., 1998, Memory-enhancing effects of secreted forms of the β-amyloid precursor protein in normal and amnestic mice. *Proc. Natl. Acad. Sci. USA* **95:** 12683-12688.

Millichip, M. I., Dallas, D. J., Wu, E., Dale, S., McKie, N., 1998, The metallo-disintegrin ADAM-10 (MADM) from bovine kidney has type IV collagenase activity in vitro. *Biochem. Biophys. Res. Commun.* **245:** 594–598.

Moffat, S. D., Zonderman, A. B., Metter, E. J., Kawas, C., Blackman, M. R., Harman, S. M., Resnick, S. M., 2004, Free testosterone and risk for Alzheimer disease in older men. *Neurology* **62:** 188-193.

Moss, M. L., Jin, S. L., Milla, M. E., Bickett, D. M., Burkhart, W., Carter, H. L., Chen, W. J., Clay, W. C., Didsbury, J. R., Hassler, D., Hoffman, C. R., Kost, T. A., Lambert, M. H., Leesnitzer, M. A., McCauley, P., McGeehan, G., Mitchell, J., Moyer, M., Pahel, G., Rocque, W., Overton, L. K., Schoenen, F., Seaton, T., Su, J. L., Becherer, J. D., 1997, Cloning of a disintegrin metalloproteinase that processes precursor tumour-necrosis factor-alpha. *Nature* **385:** 733-6.

Mucke, L., Abraham, C. R., Masliah, E., 1996, Neurotrophic and neuroprotective effects of hAPP in transgenic mice. *Ann. N. Y. Acad. Sci.* **777:** 82-88.

Nath, D., Slocombe, P. M., Webster, A., Stephens, P. E., Docherty, A. J, Murphy, G., 2000, Meltrin gamma(ADAM-9) mediates cellular adhesion through alpha(6)beta(1)integrin, leading to a marked induction of fibroblast cell motility. *J. Cell. Sci.* **113:** 2319-28.

Nitsch, R. M., Deng, A., Wurtman, R. J., Growdon, J. H., 1997, Metabotropic glutamate receptor subtype mGluR1alpha stimulates the secretion of the amyloid beta-protein precursor ectodomain. *J. Neurochem.* **69**: 704-712.

Nitsch, R. M., Deng, M., Growdon, J. H., Wurtman, R. J., 1996, Serotonin 5-HT2a and 5-HT2c receptors stimulate amyloid precursor protein ectodomain secretion. *J. Biol. Chem.* **271:** 4188-4194.

Nitsch, R. M., Deng, M., Tennis, M., Schoenfeld, D., Growdon, J. H., 2000, The selective muscarinic M1 agonist AF102B decreases levels of total Abeta in cerebrospinal fluid of patients with Alzheimer's disease. *Ann. Neurol.* **48:** 913-918.

Nitsch, R. M., Slack, B. E., Wurtman, R. J., Growdon, J. H., 1992, Release of Alzheimer amyloid precursor derivatives stimulated by activation of muscarinic acetylcholine receptors. *Science* **258:** 304-307.

Pakaski, M., Rakonczay, Z., Kasa, P., 2001, Reversible and irreversible acetylcholinesterase inhibitors cause changes in neuronal amyloid precursor protein processing and protein kinase C level in vitro. *Neurochem. Int.* **38:** 219-226.

Pan, D., Rubin, G. M., 1997, Kuzbanian controls proteolytic processing of Notch and mediates lateral inhibition during Drosophila and vertebrate neurogenesis. *Cell* **90:** 271-80.

Pandiella, A., Massague, J., 1991, Multiple signals activate cleavage of the membrane transforming growth factor- precursor. *J. Biol. Chem.* **266:** 5769–5773.

Parkin, E. T., Trew, A., Christie, G., Faller, A., Mayer, R., Turner, A. J., Hooper, N. M., 2002, Structure-activity relationship of hydroxamate-based inhibitors on the secretases that cleave the amyloid precursor protein, angiotensin converting enzyme, CD23, and pro-tumor necrosis factor-alpha. *Biochemistry* **41:** 4972-81.

Parvathy, S., Hussain, I., Karran, E. H., Turner, A. J., Hooper, N. M., 1998, Alzheimer's amyloid precursor protein alpha-secretase is inhibited by hydroxamic acid-based zinc metalloprotease inhibitors: similarities to the angiotensin converting enzyme secretase. *Biochemistry* **37:** 1680-5.

Peschon, J. J., Slack, J. L., Reddy, P., Stocking, K. L., Sunnarborg, S. W., Lee, D. C., Russell, W. E., Castner, B. J., Johnson, R. S., Fitzner, J. N., Boyce, R. W., Nelson, N., Kozlosky, C. J., Wolfson, M. F., Rauch, C. T., Cerretti, D. P., Paxton, R. J., March, C. J., Black, R. A., 1998, An essential role for ectodomain shedding in mammalian development. *Science* **282:** 1281-4.

Postina, R., Schroeder, A., Dewachter, I., Bohl, J., Schmitt, U., Kojro, E., Prinzen, C., Endres, K., Hiemke. C., Blessing, M., Flamez, P., Dequenne, A., Godaux, E., van Leuven, F., Fahrenholz, F., 2004, A disintegrin-metalloproteinase prevents amyloid plaque formation and hippocampal defects in an Alzheimer's disease mouse model. *J. Clin. Invest.* **113:** 1456-1464.

Qi, H., Rand, M.D., Wu, X., Sestan, N., Wang, W., Rakic, P., Xu, T., Artavanis-Tsakonas, S., 1999, Processing of the notch ligand delta by the metalloprotease Kuzbanian. *Science.* **283:** 91-94.

Racchi, M., Ianna, P., Binetti, G., Trabucchi, M., Govoni, S., 1998, Bradykinin-induced amyloid precursor protein secretion: a protein kinase C-independent mechanism that is not altered in fibroblasts from patients with sporadic Alzheimer's disease. *Biochem. J.* **330:** 1271-1275.

Racchi, M., Solano, D. C., Sironi, M., Govoni, S., 1999, Activity of alpha-secretase as the common final effector of protein kinase C-dependent and -independent modulation of amyloid precursor protein metabolism. *J. Neurochem.* **72:** 2464- 70.

Refolo, L. M., Malester, B., LaFrancois, J., Bryant-Thomas, T., Wang, R., Tint, G. S., Sambamurti, K., Duff, K., Pappolla, M. A., 2000, Hypercholesterolemia accelerates the Alzheimer's amyloid pathology in a transgenic mouse model. *Neurobiol. Dis.* **7:** 321-31.

Riddell, D. R., Christie, G., Hussain, I., Dingwall, C., 2001, Compartmentalization of beta-secretase (Asp2) into low-buoyant density, noncaveolar lipid rafts. *Curr. Biol.* **11:** 1288-93.

Rio, C., Buxbaum, J. D., Peschon, J. J., Corfas, G., 2000, Tumor necrosis factor-alpha-converting enzyme is required for cleavage of erbB4/HER4. *J. Biol. Chem.* **275:** 10379-87.

Robert, S. J., Zugaza, J. L., Fischmeister, R., Gardier, A. M., Lezoualc'h, F., 2001, The human serotonin 5-HT4 receptor regulates secretion of non-amyloidogenic precursor protein. *J. Biol. Chem.* **276:** 44881-44888.

Roberts, S. B., Ripellino, J. A., Ingalls, K. M., Robakis, N. K., Felsenstein, K. M., 1994, Non-amyloidogenic cleavage of the beta-amyloid precursor protein by an integral membrane metalloendopeptidase. *J. Biol. Chem.* **269:** 3111–3116.

Roghani, M., Becherer, J. D., Moss, M. L., Atherton, R. E., Erdjument-Bromage, H., Arribas, J., Blackburn, R. K., Weskamp, G., Tempst, P., Blobel, C. P., 1999, Metalloprotease-disintegrin MDC9: intracellular maturation and catalytic activity. *J. Biol. Chem.* **274:** 3531-40.

Rooke, J., Pan, D., Xu, T., Rubin, G. M., 1996, KUZ, a conserved metalloprotease-disintegrin protein with two roles in Drosophila neurogenesis. *Science* **273:** 1227–1231.

Rosendahl, M. S., Ko, S. C., Long D. L., , Brewer, M. T., Rosenzweig, B., Hedl, E., Anderson L., Pyle, S. M., Moreland J., Meyers, M. A., Kohno, T., Lyons D., Lichenstein, H. S., 1997, Identification and characterization of a pro-tumor necrosis factor-α-processing enzyme from the ADAM family of zinc metalloproteases. *J. Biol. Chem.* **272:** 24588–24593.

Rossner, S., Beck, M., Stahl, T., Mendla, K., Schliebs, R., Bigl, V., 2000, Constitutive overactivation of protein kinase C in guinea pig brain increases alpha-secretory APP processing without decreasing beta-amyloid generation. *Eur. J. Neurosci.* **12:** 3191-3200.

Sahasrabudhe, S. R., Spruyt, M. A., Muenkel, H. A., Blume, A. J., Vitek, M. P., Jacobsen, J. S., 1992, Release of amino- terminal fragments from amyloid precursor protein reporter and mutated derivatives in cultured cells. *J. Biol. Chem.* **267:** 25602-25608.

Sambamurti, K., Shioi, J., Anderson, J. P., Pappolla, M. A., Robakis, N. K., 1992, Evidence for intracellular cleavage of the Alzheimer's amyloid precursor in PC12 cells. *J. Neurosci. Res.* **33:** 319–329.

Schlondorff, J., Becherer, J. D., Blobel, C. P., 2000, Intracellular maturation and localization of the tumour necrosis factor alpha convertase (TACE). *Biochem. J.* **347:** 131-8.

Seals, D. F., Courtneidge, S. A., 2003, The ADAMs family of metalloproteases: multidomain proteins with multiple functions. *Genes Dev.* **17:** 7-30.

Selkoe, D. J., 1996, Amyloid beta-protein and the genetics of Alzheimer's disease. *J. Biol. Chem.* **271:** 18595-18298.

Sennvik, K., Fastbom, J., Blomberg, M., Wahlund, L. O., Winblad, B., Benedikz, E., 2000, Levels of alpha- and beta-secretase-cleaved amyloid precursor protein in the cerebrospinal fluid of Alzheimer's disease patients. *Neurosc. Lett.* **278:** 169-172.

Simons, K., Ikonen, E., 2000, How cells handle cholesterol. *Science* **290:** 1721-1726.

Simons, K., Toomre, D., 2000, Lipid rafts and signal transduction. *Nat. Rev. Mol. Cell Biol.* **1:** 31-39.

Simons, M., De Strooper, B., Multhaup, G., Tienari, P. J., Dotti, C. G., Beyreuther, K., 1996, Amyloidogenic processing of the human amyloid precursor protein in primary cultures of rat hippocampal neurons. *J. Neurosci.* **16:** 899–908.

Simons, M., Schwarzler, F., Lutjohann, D., von Bergmann, K., Beyreuther, K., Dichgans, J., Wormstall, H., Hartmann, T., Schulz, J. B., 2002, Treatment with simvastatin in normocholesterolemic patients with Alzheimer's disease: a 26-week randomized, placebo-controlled, double-blind trial. *Ann. Neurol.* **52:** 346-350.

Sisodia, S. S., 1992, β-amyloid precursor protein cleavage by a membrane-bound protease. *Proc. Natl. Acad. Sci. USA* **89:** 6075-6079.

Six, E., Ndiaye, D., Laabi, Y., Brou, C., Gupta-Rossi, N., Israel, A., Logeat, F., 2003, The Notch ligand Delta1 is sequentially cleaved by an ADAM protease and gamma-secretase. *Proc. Natl. Acad. Sci. USA* **100:** 7638-43.

Skovronsky, D. M., Fath, S., Lee, V. M., Milla, M. E., 2001, Neuronal localization of the TNFalpha converting enzyme (TACE) in brain tissue and its correlation to amyloid plaques. *J. Neurobiol.* **49:** 40-6.

Slack, B. E., Breu, J., Muchnicki, L., Wurtman, R. J., 1997, Rapid stimulation of amyloid precursor protein release by epidermal growth factor: role of protein kinase C. *Biochem. J.* **327:** 245-249.

Solano, D.C., Sironi, M., Bonfini, C., Solerte, S.B., Govoni, S., Racchi, M., 2000, Insulin regulates soluble amyloid precursor protein release via phosphatidyl inositol 3 kinase-dependent pathway. *FASEB J.* **14:** 1015-1022

Srour, N., Lebel, A., McMahon, S., Fournier, I., Fugere, M., Day, R., Dubois, C. M., 2003, TACE/ADAM-17 maturation and activation of sheddase activity require proprotein convertase activity. *FEBS Lett.* **554:** 275-83.

Tomita, S., Kirino, Y., Suzuki, T., 1998, Cleavage of Alzheimer's amyloid precursor protein (APP) by secretases occurs after O-glycosylation of APP in the protein secretory pathway. Identification of intracellular compartments in which APP cleavage occurs without using toxic agents that interfere with protein metabolism. *J. Biol. Chem.* **273:** 6277–6284.

Tyler, S. J., Dawbarn, D., Wilcock, G. K., Allen, S. J., 2002, Alpha- and beta-secretase: profound changes in Alzheimer's disease. *Biochem. Biophys. Res. Commun.* **299:** 373-6.

Wada, S., Morishima-Kawashima, M., Qi, Y., Misono, H., Shimada, Y., Ohno-Iwashita, Y., Ihara, Y., 2003, Gamma-secretase activity is present in rafts but is not cholesterol-dependent. *Biochemistry.* **42:** 13977-86.

Wahrle, S., Das, P., Nyborg, A. C., McLendon, C., Shoji, M., Kawarabayashi, T., Younkin, L. H., Younkin, S. G., Golde, T. E., 2002, Cholesterol-dependent gamma-secretase activity in buoyant cholesterol-rich membrane microdomains. *Neurobiol. Dis.* **9:** 11-23.

Weidemann, A., König, G., Bunke, D., Fischer, P., Salbaum, J., Masters, C. L., Beyreuther, K., 1989, Identification, biogenesis, and localization of precursors of Alzheimer's disease A4 amyloid protein. *Cell* **57:** 115-126.

Wolozin, B., Kellman, W., Ruosseau, P., Celesia, G. G., Siegel, G., 2000, Decreased prevalence of Alzheimer disease associated with 3-hydroxy-3-methyglutaryl coenzyme A reductase inhibitors. *Arch. Neurol.* **57:** 1439-43.

Weskamp, G., Cai, H., Brodie, T. A., Higashyama, S., Manova, K., Ludwig, T., Blobel, C. P., 2002, Mice lacking the metalloprotease-disintegrin MDC9 (ADAM-9) have no evident major abnormalities during development or adult life. *Mol. Cell. Biol.* **22:** 1537-44.

Wu, E., Croucher, P. I., McKie, N., 1997, Expression of members of the novel membrane linked metalloproteinase family ADAM in cells derived from a range of hematological malignancies. *Biochem. Biophys. Res. Commun.* **235:** 437-42.

Xu, H., Gouras, G. K., Greenfield, J. P., Vincent, B., Naslund, J., Mazzarelli, L., Fried, G., Jovanovic, J. N., Seeger, M., Relkin, N. R., Liao, F., Checler, F., Buxbaum, J. D., Chait, B. T., Thinakaran, G., Sisodia, S. S., Wang, R., Greengard, P., Gandy, S., 1998, Estrogen reduces neuronal generation of Alzheimer beta-amyloid peptides. *Nat. Med.* **4:** 447-451.

Youdim, M. B., Amit, T., Bar-Am, O., Weinstock, M., Yogev-Falach, M., 2003, Amyloid processing and signal transduction properties of antiparkinson-antialzheimer neuroprotective drugs rasagiline and TV3326. *Ann. N. Y. Acad. Sci.* **993:** 378-386.

Zhong, Z., Higaki, J., Murakami, K., Wang., Y., Catalano, R., Quon, D., Cordell, B., 1994, Secretion of beta-amyloid precursor protein involves multiple cleavage sites. *J. Biol. Chem.* **269:** 627–632.

Chapter 6

Amyloid β Degradation: A Challenging Task for Brain Peptidases

Laura Morelli, Ayelén Bulloj, María Celeste Leal and Eduardo M. Castaño
Instituto de Química y Fisicoquímica Biológicas (IQUIFIB), CONICET. Cátedra de Química Biológica Patológica, Departamento de Química Biológica, Facultad de Farmacia y Bioquímica, Universidad de Buenos Aires, Junin 956, C113AAD, Buenos Aires, Argentina

Abstract: Amyloid β (Aβ) accumulates in the neuropil and within the walls of cerebral vessels in association with normal aging, dementia or stroke. Aβ is released from its precursor protein as soluble monomeric species yet, under pathological conditions, it self-aggregates to form soluble oligomers or insoluble fibrils that may be toxic to neurons and vascular cells. Aβ levels could be lowered by inhibiting its generation or by promoting its clearance by transport or degradation. Here we will summarize recent findings on brain proteases capable of degrading Aβ, with a special focus on those enzymes for which there is genetic, transgenic or biochemical evidence supporting a role in the proteolysis of Aβ *in vivo*.

Key words: Amyloid β, dementia, stroke, amyloidoses, Alzheimer's disease, brain proteases, insulin degrading enzyme.

1. INTRODUCTION

The progressive accumulation of amyloid β peptide (Aβ) in the form of soluble oligomers or insoluble fibrils is part of the normal process of brain aging (Glenner and Wong, 1984; Coria *et al.*, 1987; Wang *et al.*, 1999). Such accumulation reaches very high levels in the neuropil and cerebral microvessels in several neuropathological conditions characterized by dementia or stroke. These include Alzheimer's disease (AD), Down's syndrome beyond the age of 35 years, sporadic cerebral amyloid angiopathy (SCAA) and dementia pugilistica (reviewed in Morelli *et al.*, 2002). The development of these sporadic disorders may be strongly influenced by genetic and environmental factors such as the apolipoprotein E (apoE) genotype (reviewed in Roses, 1997), chronic trauma or cholesterol metabolism (Hartman *et al.*, 2002; Hartmann, 2001; Pappolla *et al.*, 2003). In addition, the Aβ-associated disorders comprise rare genetic diseases known as familial early-onset AD, hereditary cerebral haemorrhage with amyloidosis (HCHWA) of Dutch and Italian types and dementia with amyloid angiopathy of Flemish, Arctic or Iowa origin. This group of autosomal dominant disorders is associated with mutations in the amyloid β-precursor protein (APP) and Presenilin (PS)1/2 genes (reviewed in Rocchi *et al.*, 2003; Miravalle *et al.*, 2000). From the study of APP and PS mutations using *in vitro* (synthetic peptides, cell cultures) and *in vivo* (transgenic mice) models, three basic mechanisms to understand Aβ accretion in the brain have emerged. The first mechanism is related to an increase in the rate of total Aβ production, as described with the Swedish double mutation (NL-APP) at the β-secretase site and with the Flemish mutation (A692G) (Citron *et al.*, 1992; De Jonghe *et al.*, 1998). While the former seems to accelerate the generation of the N-terminus of Aβ by β-secretase, the latter may impair the non-amyloidogenic α-secretase cleavage between positions 16-17 of Aβ. A second mechanism is the relative increase in the production of "long Aβ" species ending at Ala42 or Thr43 that are more hydrophobic and tend to aggregate more rapidly than "short Aβ" ending at Val40. This shift in the processing of Aβ C-terminus is related to a group of APP mutations close to the γ-secretase site and with most of the PS1/2 mutations studied so far (Suzuki *et al.*, 1994; Citron *et al.*, 1997). The third mechanism is associated with mutations that are clustered in the middle region of Aβ and result in the generation of Aβ variants with a high fibrillogenic potential under physiologic pH and ionic strength, such as the Dutch (E22Q), Arctic (E22G) and Iowa (D23N) species (reviewed in Miravalle *et al.*, 2000; Van Nostrand *et al.*, 2001). However, none of these mechanisms seem to participate in the accumulation of cerebral Aβ in the vast majority of cases, such as in sporadic AD, SCAA or normal aging, in which mutations are not in play. Recent studies in transgenic mice carrying human APP mutants suggest that the removal of Aβ from the brain is a very active process, raising the possibility that a defective clearance of Aβ may contribute to its accumulation under physiological and pathological situations (Dewachter *et al.*, 2000). After being released by secretases in the CNS, soluble Aβ can follow three possible main pathways (Figure 1): 1) it can be transported to the systemic circulation along perivascular drainage channels *via* the CSF or directly, across the blood brain barrier (Shibata *et al.*, 2000; Kuo *et al.*, 2001), 2) it can undergo extracellular or intracellular proteolytic degradation, and 3) it can self-assemble, aggregate and deposit in the form of brain amyloid plaques or amyloid angiopathy. Since this third pathway is critically dependent on concentration, minor flaws in the two mechanisms of clearance (transport and proteolysis) may rapidly shift the equilibrium to Aβ oligomerization. In this Chapter we will briefly review the recent advances on the major proteases involved in Aβ clearance, their possible participation in disease and their potential significance from a therapeutic perspective in cerebral Aβ amyloidoses.

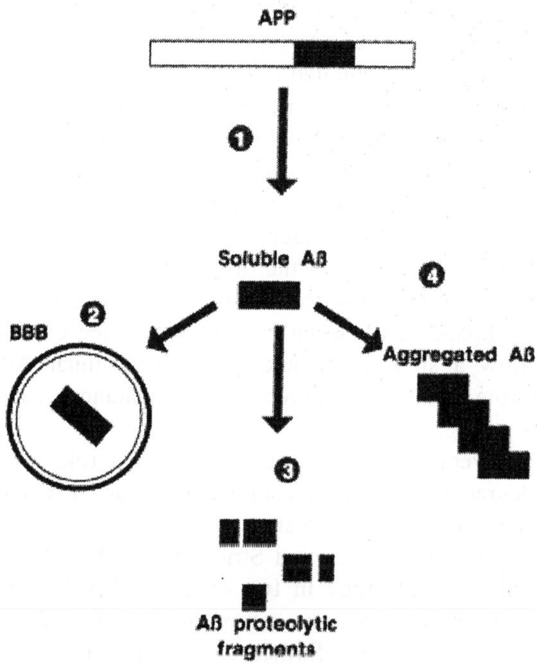

Figure 1. Schematic representation of Aβ pathways in the brain. 1) Aβ is released from APP by β and γ secretases; 2) soluble Aβ can be transported to the systemic circulation across the blood-brain-barrier (BBB) or along perivascular drainage pathways; 3) Aβ can be degraded by proteases in the intracellular or extracellular compartments. 4) Aβ that escapes from pathways 2 and 3 accumulates as aggregated species.

2. Aβ DEGRADATION PATHWAYS

The proteolytic degradation of soluble Aβ in the brain is believed to take place both in intracellular and extracellular compartments. Various mechanisms of Aβ internalization into neurons and glial cells have been described, including non-saturable uptake as well as receptor-mediated endocytosis by the serpin-enzyme complex, the LDL-receptor related proteins (LRP-1 and LRP-2), scavenger receptors and the receptor for end glycation products (RAGE) (reviewed in Morelli *et al.*, 2002). The LRP pathway mediates the internalization of several ligands including apolipoprotein E, α-2 macroglobulin and APP, all related to the pathogenesis of AD, with LRP-1 being expressed mainly in neurons and LRP-2 in choroids plexus and ependymal cells. As a result of internalization, Aβ may be degraded by lysosomal proteases including cathepsins D, B and S, which have been shown to hydrolyze Aβ *in vitro* (Hamazaki, 1996; Liuzzo *et al.*, 1999; Gupta-Bansal *et al.*, 1995). In support of this mechanism, chloroquine treatment of cultured cells inhibited the degradation of internalized Aβ by the LRP receptor and co-infusion of synthetic Aβ with leupeptin in the rat brain resulted in the intra and extracellular accumulation of Aβ (Hammad *et al.*, 1997; Frautschy *et al.*, 1998). In AD brains, the high expression of lysosomal hydrolases in vulnerable neurons and extracellularly, in close association with Aβ deposits suggests a significant activation of the lysosomal system in the course of the disease (Cataldo and Nixon, 1990). Moreover, mice with an 80% reduction in LRP levels due to genetic disruption of its ligand RAP, when crossed with mice expressing mutant human APP show a significant increase in extracellular Aβ deposition, reinforcing the role of the LRP pathway in Aβ clearance (Van Uden *et al.*, 2002).

In addition, several lines of evidence support a role for the plasmin system in the degradation of Aβ. Plasminogen, tissue (tPA) and urokinase-like plasminogen activators (uPA) are synthesized by neurons, with high levels in the hippocampus (Chen and Strickland, 1997). Plasmin degrades Aβ *in vitro*, and a recent study in two models of APP transgenic mice showed that the plasmin system was compromised long before the onset of detectable amyloid deposition, further suggesting the importance of a degradative failure in this process (Melchor *et al.*, 2003). Moreover, an important reduction in the levels of plasmin and plasminogen found in the neocortex and hippocampus of AD patients raises the possibility that an initial up-regulation of the system may become exhausted in the course of the disease (Ledesma *et al.*, 2000). Other peptidases that may participate in the extracellular degradation of Aβ are matrix metalloproteinases (MMPs). MMP-9 has been detected in pyramidal neurons in human hippocampus, is found in neuritic plaques and seems to be overexpressed in AD brains as compared to controls (Backstrom *et al.*, 1996). MMP-9 can degrade Aβ *in vitro* and species ending at Gly_{37}, consistent with its activity, have been found in amyloid extracted from leptomeningeal and cortical vessels. Active forms of MMP-2 are capable of degrading Aβ1-40 and Aβ1-42 purified from vascular amyloid and treatment of primary hippocampal cultures with synthetic Aβ increased the production of MMP-2, MMP-9 and MMP-3 (Deb *et al.*, 1996; 1999).

In addition to the proteolytic systems mentioned above, that may participate in the degradation of Aβ under pathological conditions, recent compelling evidence indicates that a group of zinc-metallopeptidases including neprilysin, endothelin-converting enzymes and insulin-degrading enzyme may regulate the steady state levels of Aβ in the mammalian brain (Table I).

Table I. Major Aβ-degrading proteases

Protease	Type	Cellular localization	Overexpression (transfected cells)	Knock out mice
Neprilysin	Neutral, Zn-metallo	Plasma membrane	↓ Aβ (EC,IC)	↑ Aβ* (gene-dose effect)
Endothelin - Converting Enzymes	Neutral / acidic Zn-metallo	Plasma membrane, Internal membranes	↓ Aβ (EC)	↑ Aβ* (gene-dose effect)
Insulin-Degrading Enzyme	Neutral, Zn-metallo	Cytosol, Membranes, Peroxisomes, Secreted	↓ Aβ (EC,IC)	↑ Aβ* (gene-dose effect)

EC, extracellular; IC, intracellular, * As determined in brain tissue homogenates

3. NEPRILYSIN

Neprilysin is also known as neutral endopeptidase (NEP), EC3.4.24.11, enkephalinase, CD10 and common acute lymphoblastic leukaemia antigen (CALLA). NEP is a widely expressed zinc metallopeptidase with a HEXXH active motif that is strongly inhibited by thiorphan and phosphoramidon, a property shared with other members of the M13 subfamily (reviewed in Turner and Tanzawa, 1997). NEP is a type II integral membrane protein with a large lumenal-extracellular domain that contains the active site (Oefner *et al.*, 2000). Such topology allows NEP to hydrolyse several peptides at the cell surface, including enkephalins, substance P, atrial natriuretic peptide, somatostatin, endothelin and insulin B chain. Within the CNS, NEP predominates in synaptic terminals of the neuropil in which it may participate in terminating the action of neuropeptides (Schwartz *et al*, 1980; Turner *et al.*, 2001). Early reports on the possible relationship of NEP with Aβ metabolism included the staining of AD senile plaques with a monoclonal anti-NEP and the degradation of synthetic Aβ1-40 *in vitro* by NEP purified from rabbit kidney (Sato *et al.*, 1991; Howell *et al.*, 1995). More recently, it was shown that synthetic radiolabeled Aβ1-42 injected into the rat hippocampus was degraded by a thiorphan-sensitive NEP-like activity. Remarkably, the continuous infusion in these animals of thiorphan for several weeks resulted in the accumulation of endogenous Aβ42 in the form of Congo red-negative deposits that resembled diffuse plaques in the cortex and hippocampus (Iwata *et al.*, 2000). Mice with disruption of the NEP gene have a lower capability in degrading injected Aβ in a gene-dose dependent manner. Moreover, the accumulation of endogenous Aβ is highest in the hippocampus and lowest in the cerebellum, in parallel with the severity of pathology in AD and transgenic mice models of AD (Iwata *et al.*, 2001). In the human brain, NEP distribution shows an inverse correlation with the vulnerability to Aβ deposition, including a low expression in hippocampal neurons and smooth muscle cells in the microvasculature affected with amyloid angiopathy (Yasojima *et al.*, 2001; Carpentier *et al.*, 2002). The potential anti-amyloidogenic effects of NEP *in vivo* has been recently assessed by the injection of a lentiviral vector expressing this gene into the CNS of APP transgenic mice that develop amyloid deposition. A higher expression of NEP in the injected hemi-brain correlated with smaller plaques, a reduction of approximately 50% in Aβ burden and a mild improvement in MAP-2 dendritic staining that suggested less neuronal damage (Marr *et al.*, 2003).

4. ENDOTHELIN-CONVERTING ENZYMES

Endothelin-converting enzymes (ECEs) are type II integral membrane zinc metalloproteases belonging like NEP, to the M13 subfamily of mammalian neutral endopeptidases. One major physiologic role of ECEs is the cleavage of big endothelin to generate the potent vasoconstrictor peptides known as endothelins (Turner *et al.*, 1997). Several isoforms of ECEs have been identified, namely ECE-1, ECE-2, and ECE-3. Within each of the first two groups, several sub-isoforms derived from the splicing of single genes have also been determined (reviewed in D'Orleans-Juste *et al.*, 2003). ECE-1 isoforms are mainly located on the plasma membrane and widely expressed in endothelial cells and non-vascular cells, including neurons and astrocytes in the CNS. ECE-2 isoforms, that predominate in intracellular membranes, also have a ubiquitous expression with high levels in the brain, (reviewed in Davenport *et al.*, 2000; Eckman *et al.*, 2001). Recently, Eckman *et al.* (2001) reported that treatment of cell lines that express ECE with phosphoramidon caused a 2-3-fold increase in extracellular Aβ steady-state levels and that overexpression of ECE-1 in CHO cells, which lack endogenous ECE activity, reduced dramatically extracellular Aβ concentration. Moreover, a recombinant soluble form of ECE was capable of cleaving synthetic Aβ1-40 at the amino- side of Leu_{17}, Val_{18} and Phe_{19}, consistent with the known preference of ECEs for hydrophobic residues at P1'. ECE-1 homozygous knock out mice have severe cardiac and cranial abnormalities that result in embryonic lethality, however, ECE-1 (+/-) and ECE-2 (-/-) animals are viable and healthy. The study of mice with these genotypes at 3-4 weeks of age showed a mild (20-30%) though significant increase in Aβ levels in a soluble fraction of brain homogenates. Interestingly, ECE-2 deficiency showed a gene-dose effect on Aβ accumulation (Eckman *et al.*, 2003). These results strongly support a key participation of ECEs in the physiological catabolism of Aβ in the mammalian brain.

5. INSULIN-DEGRADING ENZYME

Insulin-degrading enzyme (IDE) or Insulysin is a thiol zinc-metallopeptidase that belongs to the "inverzincin" family, name that describes an inverted active sequence HXXEH as compared to the typical zinc peptidases such as thermolysin, NEP or ECEs (Becker and Roth, 1992). IDE is highly conserved in evolution and has a ubiquitous expression including high levels in testis, ovary, kidney, liver, red cells, pancreas, muscle and brain (Kuo *et al.*, 1993). IDE has been implicated in broad physiological roles, such as cellular growth and differentiation, the modulation of proteasomal activity and steroid signaling. Although it is mainly found in the cystosol and peroxisomes, IDE has been also located in active form at the surface of many cell types, including neurons (reviewed in Duckworth *et al.*, 1998; Vekrellis *et al.*, 2000). Moreover, a secreted form of the protease has been found in the conditioned media of different cell lines and primary cultures (Safavi *et al.*, 1996; Gasparini *et al.*, 2001). The biochemistry of the association of IDE to membranes and the mechanism of its secretion remain to be studied since IDE seems not to have a typical signal peptide and no isoforms or post-translational modifications have been described so far.

5.1 IDE: an amyloid-degrading protease

IDE is known to degrade several peptides, many of which have a propensity to form amyloid fibrils *in vitro* and *in vivo* including insulin, glucagon, amylin, atrial natriuretic peptide, calcitonin and Aβ (reviewed in Duckworth *et al.*, 1998). Although IDE has no sequence specifity, it shows some preference for basic or hydrophobic residues on the carboxyl-side of the scissile bonds, suggesting a recognition of secondary or tertiary structure shared by amyloid-forming peptides. Aβ degradation by IDE was first shown with synthetic peptides and purified protease and subsequently demonstrated in a variety of cultured cells and brain homogenates (Vekrellis *et al.*, 2000; Kurochkin and Goto, 1994; Sudoh *et al.*, 2002; McDermott and Gibson, 1997; Perez *et al.*, 2000). With regard to the effect of Aβ oligomerization upon IDE activity and specificity, it has been proposed that IDE is capable of degrading monomeric Aβ more efficiently than oligomeric Aβ species, which are thought to be toxic to neurons or vascular cells (Walsh *et al.*, 2002). We have recently shown, in support of such notion, that a recombinant rat IDE degraded very efficiently Aβ monomers while SDS-resistant dimers were almost completely resistant to proteolysis (Figure 2).

IDE may recognize a motif on monomeric Aβ within positions 18-22, a hydrophobic stretch that is critical for amyloid formation (McDermott and Gibson, 1997; Soto *et al.*, 1995). It is possible that when these residues participate in peptide self-assembly, they are no longer accessible to IDE catalytic site. To confirm the effect of Aβ dimerization upon proteolysis, the study of peptide species isolated under native conditions is needed, yet, the possibility that Aβ dimers exist at physiological concentrations in the brain (Garzon-Rodriguez *et al.*, 1997) raises an important question about Aβ catabolism with potential therapeutic implications.

Recently, mice with targeted deletions of the IDE gene have been generated and characterized (Farris *et al.*, 2003; Miller *et al.*, 2003). Brain membranes and soluble fractions isolated from IDE (-/-) animals showed a significant decrease in synthetic Aβ degradation. Moreover, endogenous brain levels of Aβ were increased, with an inverse correlation with IDE activity (Miller *et al.*, 2003). Interestingly, the loss of activity of IDE in this animal model also resulted in the accumulation of an unphosphorylated form of the carboxyl-terminal domain of APP, hyperinsulinemia and glucose intolerance (Farris *et al.*, 2003).

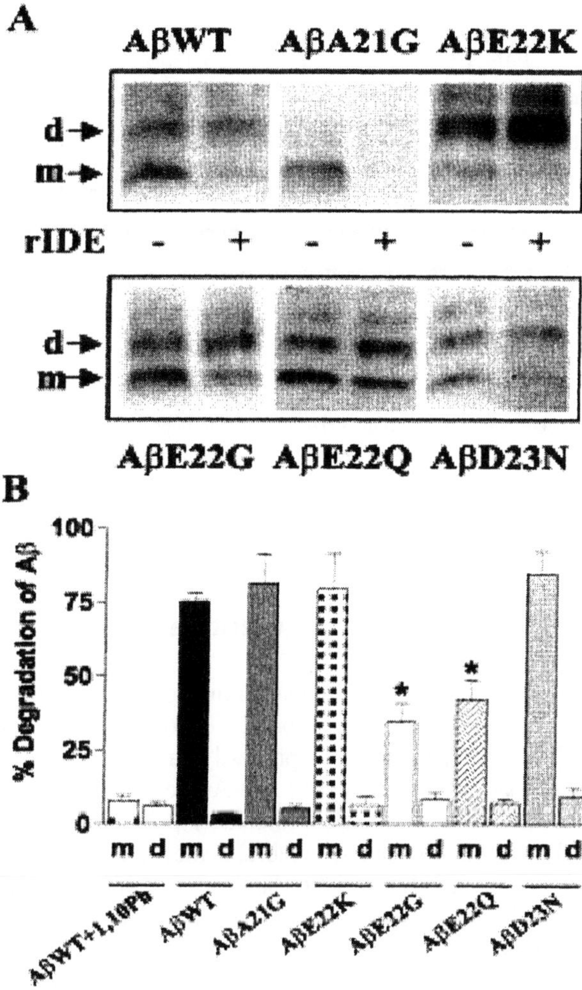

Figure 2. Degradation of Aβ variants by recombinant rat IDE. Panel A, western blot analysis with monoclonal 6E10 of degradation of AβWT and Aβ genetic variants by rIDE. m, Aβ monomers; d, Aβ dimers. Panel B, densitometric quantitation of the immunoreactivity of Aβ as in panel A. m, monomers; d, dimers. Bars represent the mean ± SE of three independent experiments. *, $p<0.01$ (Student's *t* test) as compared with AβWT. From Morelli *et al.*, 2003, by permission of the American Society for Biochemistry and Molecular Biology, Inc.

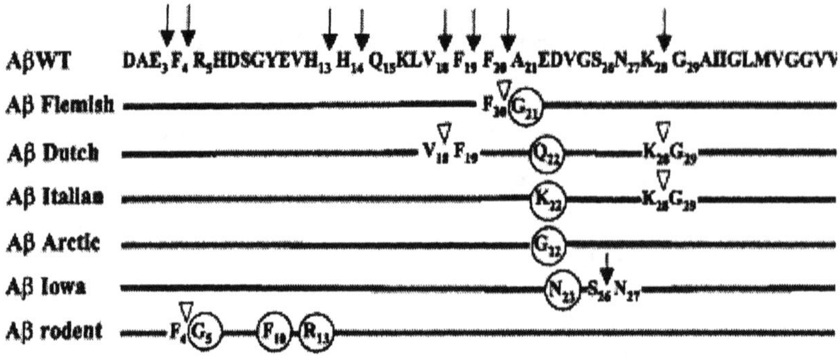

Figure 3. Schematic representation of cleavage sites within AβWT, Aβ rodent, and Aβ human genetic variants by rIDE. *Arrows* show the peptide bonds hydrolysed by rIDE. *Open arrowheads* (∇) indicate loss of cleavage under the experimental conditions tested. The *lines* represent identity of amino acid sequence and cleavage sites. Amino acid substitutions in each Aβ variant are shown in *circles.* From Morelli *et al.*, 2003, by permission of the American Society for Biochemistry and Molecular Biology, Inc.

5.2 IDE degrades Aβ genetic variants

A group of autosomal dominant mutations in the APP gene result in amino acid substitutions at positions 21, 22 or 23 of Aβ sequence. Although these Aβ variants present with a primarily vascular deposition, they translate into different clinical phenotypes. Aβ Arctic (E22G) and Aβ Iowa (D23N) are characterized by presenile dementia, Aβ Flemish (A21G) is associated with early onset dementia and cerebral hemorrhage, while Aβ Dutch (E22Q) and Aβ Italian (E22K) variants have a predominant vascular phenotype characterized by massive strokes (reviewed in Miravalle *et al.*, 2000). We have recently studied the degradation of Aβ WT and Aβ variants by recombinant IDE. Monomers of AβWT, AβA21G, AβE22K and AβD23N were readily degraded with an apparent similar efficiency. In contrast, AβE22Q and AβE22G monomers were substantially more resistant to proteolysis by rIDE (Figure 2). More studies are needed to clarify the mechanism of such resistance, including the influence of Aβ oligomerization and the kcat/Km for each Aβ species at low concentrations to minimize aggregation. Interestingly, amino acid substitutions did not affect substantially IDE specificity, with the exception of minor changes in the sites of cleavage in the Flemish and Iowa variants (Figure 3). These results support the concept that up-regulation of IDE activity may promote the removal of soluble Aβ in hereditary forms of Aβ diseases.

5.3 A pathogenic role of IDE in AD?

An early study by our group suggested that IDE activity was significantly reduced in a soluble fraction obtained from the temporal cortex of AD patients as compared to age-matched controls (Perez *et al.*, 2000). Subsequently, Cook *et al.* (2003) have shown that a decrease in IDE expression was more pronounced in the hippocampus in late-onset AD patients carrying the APOE ε4 allele. Genetic linkage studies indicate that a locus on chromosome 10q within 195 kilobases of the IDE gene is associated with late-onset Alzheimer's disease (LOAD) (Bertram *et al.*, 2000; Myers *et al.*, 2000; Ertekin-Taner *et al.*, 2000; Abraham *et al.*, 2001). So far, the analysis of all the coding exons, untranslated regions, and two intronic polymorphisms of the IDE gene has failed to reveal association with AD (Prince *et al.*, 2003; Boussaha *et al.*, 2002). However, a more recent study of five IDE polymorphisms stratified by APOE genotype raised the possibility that the lack of a minor IDE haplotype may be predictive of AD in APOE ε4-negative individuals (Edland *et al.*, 2003). Although these results need confirmation with larger numbers of AD patients and controls, they raise the interesting possibility that IDE expression levels are related to the APOE genotype. Moreover, together with biochemical and functional data, they suggest that IDE may participate in the pathogenesis of LOAD.

6. FUTURE PERSPECTIVES

The recent and significant advances on our knowledge about Aβ degradation raise the possibility that in the near future brain proteases may be manipulated to promote Aβ clearance. However, several aspects of Aβ metabolism remain to be clarified, both in normal conditions and disease. It is yet not clear which is the main compartment for Aβ degradation in the brain and whether the initial Aβ oligomerization takes place intra or extracellularly. The apparent redundancy in Aβ proteolysis suggests that the process is ubiquitous, with Aβ degradation taking place, at a given time, in different compartments and by different peptidases. Yet, protease efficiency may be strongly dependent upon Aβ oligomerization. This process is influenced, in turn, by the primary structure of the peptide, postranslational modifications, or interaction of Aβ with other proteins. Moreover, several aspects related to protease manipulation need to be carefully addressed in animal models. The list includes the impact on basic aspects of brain physiology such as insulin signaling, neuropeptide transmission, blood flow, clot formation or extracellular matrix integrity. As an example, fibrinolytic therapy of myocardial infarction with tPA has been associated with cerebral hemorrhage in patients with SCAA (Wijdicks and Jack, 1993), thus, the pharmacological activation of plasmin may be particularly problematic in this group of patients. Several proteolytic systems seem to be chronically stimulated in AD brains and therefore, increasing their expression or activity may have little impact on Aβ metabolism. On the other hand, the promotion of the activity of proteases that are thought to be decreased in AD such as NEP or IDE may have profound effects on the steady-state levels of soluble Aβ. With the transgenic animal models currently available, the proteases that degrade Aβ can be up-modulated and the consequences upon Aβ levels and its purported toxicity assessed *in vivo* (Marr *et al.*, 2003). These experiments may be decisive to clarify the potential of treatments based on Aβ degradation for the management of AD and related disorders.

ACKNOWLEDGEMENTS

This work was supported by a grant from the Alzheimer's Association and Beca Carrillo-Oñativia (to E.M.C) and from the Universidad de Buenos Aires (B608, to L.M.)

REFERENCES

Abraham, R., Myers, A., Wavrant-DeVrieze, F., Hamshere, M.L., Thomas, H.V., Marshall, H., Compton, D., Spurlock, G., Turic, D., Hoogendoorn, B., Kwon, J.M., Petersen, R.C., Tangalos, E., Norton, J., Morris, J.C., Bullock, R., Liolitsa, D., Lovestone, S., Hardy, J., Goate, A., O'Donovan, M., Williams, J., Owen, M.J., and Jones, L., 2001, Substantial linkage disequilibrium across the insulin-degrading enzyme locus but no association with late-onset Alzheimer's disease. *Hum. Genet.* **109:** 646-652.

Backstrom, J.R., Lim, G.P., Cullen, M.J., and Tokes, Z.A., 1996, Matrix metalloproteinase 9 (MMP-9) is synthesized in neurons of the human hippocampus and is capable of degrading the amyloid β peptide (1-40). *J. Neurosci.* **16:** 7910-7919.

Becker, A.B., and Roth, R.A., 1992, An unusual active site identified in a family of zinc metalloendopeptidases. *Proc. Natl. Acad. Sci. USA.* **89:** 3835-3839.

Bertram, L., Blacker, D., Mullin, K., Keeney, D., Jones, J., Basu, S., Yhu, S., McInnis, M.G., Go, R.C., Vekrellis, K., Selkoe, D.J., Saunders, A.J., and Tanzi, R.E., 2000, Evidence for genetic linkage of Alzheimer's disease to chromosome 10q. *Science* **290:** 2302-2303.

Boussaha, M., Hannequin, D., Verpillat, P., Brice, A., Frebourg, T., and Campion, D., 2002, Polymorphisms of insulin degrading enzyme gene are not associated with Alzheimer's disease. *Neurosci. Lett.* **329:** 121-123.

Carpentier, M., Robitaille, Y., DesGroseillers, L., Boileau, G., and Marcinkiewicz, M., 2002, Declining expression of neprilysin in Alzheimer disease vasculature: possible involvement in cerebral amyloid angiopathy. *J. Neuropathol. Exp. Neurol.* **61:** 849-856.

Cataldo, A.M., and Nixon, R.A., 1990, Enzymatically active lysosomal proteases are associated with amyloid deposits in Alzheimer brain. *Proc. Natl. Acad. Sci. USA.* **87:** 3861-3865.

Chen, Z.L., and Strickland, S., 1997, Neuronal death in the hippocampus is promoted by plasmin-catalyzed degradation of laminin. *Cell.* **91:** 917-925.

Citron, M., Oltersdorf, T., Haass, C., McConlogue, L., Hung, A.Y., Seubert, P., Vigo-Pelfrey, C., Lieberburg, I.,and Selkoe, D.J., 1992, Mutation of the β-amyloid precursor protein in familial Alzheimer's disease increases β-protein production. *Nature.* **360:** 672-674.

Citron, M., Westaway, D., Xia, W., Carlson, G., Diehl,T., Levesque, G., Johnson-Wood, K., Lee, M., Seubert, P., Davis, A., Kholodenko, D., Motter, R., Sherrington, R., Perry, B., Yao, H., Strome, R., Lieberburg, I., Rommens, J., Kim, S., Schenk, D., Fraser, P., St George Hyslop, P., and Selkoe, D.J., 2003, Mutant presenilins of Alzheimer's disease increase production of 42-residue amyloid β-protein in both transfected cells and transgenic mice. *Nat. Med.* **3:** 67-72.

Cook, D.G., Leverenz, J.B., McMillan, P.J., Kulstad, J.J., Ericksen, S., Roth, R.A., Schellenberg, G.D., Jin, L.W., Kovacina, K.S., and Craft, S., 2003, Reduced hippocampal insulin-degrading enzyme in late-onset Alzheimer's disease is associated with the apolipoprotein E-ε4 allele. *Am. J. Pathol.* **162:** 313-319.

Coria, F., and Castaño, E.M., 1987, Frangione B. Brain amyloid in normal aging and cerebral amyloid angiopathy is antigenically related to Alzheimer's disease β-protein. *Am .J. Pathol.* **129:** 422-428.

Davenport, A.P., and Kuc, R.E., 2000, Cellular expression of isoforms of endothelin-converting enzyme-1 (ECE-1c, ECE-1b and ECE-1a) and endothelin-converting enzyme-2. *J. Cardiovasc. Pharmacol.* **36:** S12-S14.

De Jonghe, C., Zehr, C., Yager, D., Prada, C.M., Younkin, S., Hendriks, L., Van Broeckhoven, C., and Eckman, C.B., 1998, Flemish and Dutch mutations in amyloid precursor protein have different effects on amyloid β secretion. *Neurobiol. Dis.* **5:** 281-286.

Deb, S., and Gottschall, P.E., 1996, Increased production of matrix metalloproteinases in enriched astrocyte and mixed hippocampal cultures treated with β-amyloid peptides. *J. Neurochem.* **66:** 1641-1647.

Deb, S., Zhang, J.W., and Gottschall, P.E., 1999, Activated isoforms of MMP-2 are induced in U87 human glioma cells in response to β-amyloid peptide. *J. Neurosci. Res.* **55:** 44-53.

Dewachter, I., Van Dorpe, J., Smeijers, L., Gilis, M., Kuiperi, C., Laenen, I., Caluwaerts, N., Moechars, D., Checler, F., Vanderstichele, H., and Van Leuven, F., 2000, Aging increased amyloid β peptide and caused amyloid plaques in brain of old APP/V717I transgenic mice by a different mechanism than mutant presenilin 1. *J. Neurosci.* **20:** 6452-6458.

D'Orleans-Juste, P., Plante, M., Honore, J.C., Carrier, E., and Labonte, J., 2003, Synthesis and degradation of endothelin-1. *Can. J. Physiol. Pharmacol.* **81:** 503-510.

Duckworth, W.C., Bennett, R.G., and Hamel, F.G., 1998, Insulin degradation: progress and potential. *Endocr. Rev.* **19:** 608-624.

Eckman, E.A., Reed, D.K., and Eckman, C.B., 2001, Degradation of the Alzheimer's amyloid β peptide by endothelin-converting enzyme. *J. Biol. Chem.* **276:** 24540-24548.

Eckman, E.A., Watson, M., Marlow, L., Sambamurti, K., and Eckman, C.B., 2003, Alzheimer's disease β-amyloid peptide is increased in mice deficient in endothelin-converting enzyme. *J. Biol. Chem.* **278:** 2081-2084.

Edland, S.D., Wavrant-De Vriese, F, Compton, D., Smith, G.E., Ivnik, R., Boeve, B.F., Tangalos, E.G., and Petersen, R.C., 2003, Insulin degrading enzyme (IDE) genetic variants and risk of Alzheimer's disease: evidence of effect modification by apolipoprotein E (APOE). *Neurosci. Lett.* **345:** 21-24.

Ertekin-Taner, N., Graff-Radford, N., Younkin, L.H., Eckman, C., Baker, M., Adamson, J., Ronald, J., Blangero, J., Hutton, M., and Younkin, S.G., 2000, Linkage of plasma Aβ42 to a quantitative locus on chromosome 10 in late-onset Alzheimer's disease pedigrees. *Science.* **290:** 2303-2304.

Farris, W., Mansourian, S., Chang, Y., Lindsley, L., Eckman, E.A., Frosch, M.P., Eckman, C.B., Tanzi, R.E., Selkoe, D.J., and Guenette, S., 2003, Insulin-degrading enzyme regulates the levels of insulin, amyloid β-protein, and the β-amyloid precursor protein intracellular domain in vivo. *Proc. Natl. Acad. Sci. U S A.* **100:** 4162-4167.

Frautschy, S.A., Horn, D.L., Sigel, J.J., Harris-White, M.E., Mendoza, J.J., Yang ,F., Saido, T.C., and Cole, G.M., 1998, Protease inhibitor coinfusion with amyloid β-protein results in enhanced deposition and toxicity in rat brain. *J. Neurosci.* **18:** 311-8321.

Garzon-Rodriguez, W., Sepulveda-Becerra, M., Milton, S., and Glabe, C.G., 1997, Soluble amyloid Aβ-(1-40) exists as a stable dimer at low concentrations. *J. Biol. Chem.* **272:** 21037-21044.

Gasparini, L., Gouras, G.K., Wang, R., Gross, R.S., Beal, M.F., Greengard, P., and Xu, H., 2001, Stimulation of β-amyloid precursor protein trafficking by insulin reduces intraneuronal β-amyloid and requires mitogen-activated protein kinase signaling. *J. Neurosci.* **21:** 2561-2570.

Glenner, G.G., and Wong, C.W., 1984, Alzheimer's disease and Down's syndrome: sharing of a unique cerebrovascular amyloid fibril protein. *Biochem. Biophys. Res. Commun.* **122:** 1131-1135.

Gupta-Bansal, R., Frederickson, R.C., and Brunden, K.R., 1995, Proteoglycan-mediated inhibition of Aβ proteolysis. A potential cause of senile plaque accumulation. *J. Biol. Chem.* **270:** 18666-18671

Hamazaki, H., 1996, Cathepsin D is involved in the clearance of Alzheimer's β-amyloid protein. *FEBS Lett.* **396:** 139-142.

Hammad, S.M., Ranganathan, S., Loukinova, E., Twai, W.O., and Argraves, W.S., 1997, Interaction of apolipoprotein J-amyloid β peptide complex with low density lipoprotein receptor-related protein -2/megalin. A mechanism to prevent pathological accumulation of amyloid β peptide. *J. Biol. Chem.* **272:** 18644-18649.

Hartman, R.E., Laurer, H., Longhi, L., Bales, K.R.Paul, S.M., McIntosh, T.K., and Holtzman, D.M., 2002, Apolipoprotein E4 influences amyloid deposition but not cell loss after traumatic brain injury in a mouse model of Alzheimer's disease. *J. Neurosci.* **22:** 10083-10087.

Hartmann, T., 2001, Cholesterol, A and Alzheimer's disease. *Trends Neurosci.* **24**: S45-S48.

Howell, S., Nalbantoglu, J., and Crine, P., 1995, Neutral endopeptidase can hydrolyze - amyloid β (1-40) but shows no effect on β-amyloid precursor protein metabolism. *Peptides.* **16:** 647-652.

Iwata, N., Tsubuki, S., Takaki, Y., Shirotani, K., Lu, B., Gerard, N.P., Gerard,C., Hama, E., Lee, H.J., and Saido, T.C., 2001, Metabolic regulation of brain Aβ by neprilysin. *Science* **292:** 1550-1552.

Iwata, N., Tsubuki, S., Takaki, Y., Watanabe, K., Sekiguchi, M., Hosoki, E., Kawashima-Morishima, M., Lee, H.J., Hama, E., Sekine-Aizawa, Y., and Saido, T.C., 2000, Identification of the major Aβ (1-42) degrading catabolic pathway in brain parenchyma: suppression leads to biochemichal and pathological deposition. *Nat. Med.* **6:** 143-150.

Kuo, W.L., Montag, A.G., and Rosner, M.R., 1993, Insulin-degrading enzyme is differentially expressed and developmentally regulated in various rat tissues. *Endocrinology.* **132:** 604-611.

Kuo, Y.M., Beach, T.G., Sue, L.I., Scott, S., Layne, K.J., Kokjohn, T.A., Kalback, W.M., Luehrs, D.C., Vishnivetskaya, T.A., Abramowski, D., Sturchler-Pierrat, C., Staufenbiel, M., Weller, R.O., and Roher, A.E., 2001, The evolution of Aβ peptide burden in the

APP23 transgenic mice: implications for Aβ deposition in Alzheimer disease. *Mol. Med.* **7:** 609-618.

Kurochkin, I.V., and Goto, S., 1994, Alzheimer's β-amyloid peptide specifically interacts with and is degraded by insulin degrading enzyme. *FEBS Lett.* **345:** 33-37.

Ledesma, M.D., Da Silva, J.S., Crassaerts, K., Delacourte, A., De Strooper, B., and Dotti, C.G., 2000, Brain plasmin enhances APP α-cleavage and Aβ degradation and is reduced in Alzheimer 's disease brains. *EMBO reports* **1:** 530-535.

Liuzzo,J.P., Petanceska, S.S., and Devi, L.A., 1999, Neurotrophic factors regulate cathepsin S in macrophages and microglia: a role in the degradation of myelin basic protein and amyloid β peptide. *Mol. Med.* **5:** 334-343.

Marr, R.A., Rockenstein, E., Mukherjee, A., Kindy, M.S., Hersh, L.B., Gage, F.H., Verma, and I.M., Masliah, E., 2003, Neprilysin gene transfer reduces human amyloid pathology in transgenic mice. *J. Neurosci.* **23:** 1992-1996.

McDermott, J.R., and Gibson, A.M., 1997, Degradation of Alzheimer's β-amyloid protein by human and rat brain peptidases: involvement of insulin-degrading enzyme. *Neurochem. Res.* **22:** 49-56.

Melchor, J.P., Pawlak, R., and Strickland, S., 2003, The tissue plasminogen activator-plasminogen proteolytic cascade accelerates amyloid-β degradation and inhibits Aβ-induced neurodegeneration. *J. Neurosci.* **23:** 8867-8871.

Miller, B.C., Eckman, E.A., Sambamurti, K., Dobbs, N., Chow, K.M., Eckman, C.B., Hersh, L.B., and Thiele, D.L., 2003, Amyloid-β peptide levels in brain are inversely correlated with insulysin activity levels in vivo. *Proc. Natl. Acad. Sci.USA.* **100:** 6221-6226.

Miravalle, L., Tokuda, T., Chiarle, R., Giaccone, G., Bugiani, O., Tagliavini, F., Frangione, B., and Ghiso, J., 2000, Substitutions at codon 22 of Alzheimer's Aβ peptide induce diverse conformational changes and apoptotic effects in human cerebral endothelial cells. *J. Biol. Chem.* **275:** 27110-27116.

Morelli, L., Llovera, R., Gonzalez, S.A., Affranchino, J.L., Prelli, F., Frangione, B., Ghiso, J., and Castaño, E.M., 2003, Differential degradation of amyloid β genetic variants associated with hereditary dementia or stroke by insulin-degrading enzyme. *J. Biol. Chem.* **278:** 23221-23226.

Morelli, L., Llovera, R., Ibendahl, S., and Castaño, E.M., 2002, The degradation of amyloid β as a therapeutic strategy in Alzheimer's disease and cerebrovascular amyloidoses. *Neurochem. Res.* **27:** 1387-1399.

Myers, A., Holmans, P., Marshall, H., Kwon, J., Meyer, D., Ramic, D., Shears, S., Booth, J., DeVrieze, F.W., Crook, R., Hamshere, M., Abraham, R., Tunstall, N., Rice, F., Carty, S., Lillystone, S., Kehoe, P., Rudrasingham, V., Jones, L., Lovestone, S., Perez-Tur, J., Williams, J., Owen, M.J., Hardy, J., and Goate, A.M., 2000. Susceptibility locus for Alzheimer's disease on chromosome 10. *Science* **290:** 2304-2305.

Oefner, C., D'Arcy, A., Hennig, M., Winkler, F.K., and Dale, G.E., 2000, Structure of human neutral endopeptidase (Neprilysin) complexed with phosphoramidon. *J. Mol. Biol.* **296:** 341-349.

Pappolla, M.A., Bryant-Thomas, T.K., Herbert, D., Pacheco, J., Fabra Garcia, M., Manjon ,M., Girones ,X., Henry, T.L., Matsubara, E., Zambon, D., Wolozin, B., Sano, M., Cruz-Sanchez, F.F., Thal, L.J., Petanceska, S.S., and Refolo, L.M., 2003, Mild hypercholesterolemia is an early risk factor for the development of Alzheimer amyloid pathology. *Neurology.* **61:** 199-205.

Perez, A., Morelli, L., Cresto, J.C., and Castaño, E.M., 2000, Degradation of soluble amyloid β-peptides 1-40, 1-42, and the Dutch variant 1-40Q by insulin degrading enzyme from Alzheimer disease and control brains. *Neurochem. Res.* **25:** 247-255.

Prince, J.A., Feuk, L., Gu, H.F., Johansson, B., Gatz, M., Blennow, K., and Brookes, A.J., 2003, Genetic variation in a haplotype block spanning IDE influences Alzheimer disease. *Hum. Mutat.* **22:** 363-371.

Rocchi, A., Pellegrini, S., Siciliano, G., and Murri, L., 2003, Causative and susceptibility genes for Alzheimer's disease: a review. *Brain Res. Bull.* **61:** 1-24.

Roses, A.D., 1997, A model for susceptibility polymorphisms for complex diseases: apolipoprotein E and Alzheimer disease. *Neurogenetics.* **1:** 3-11.

Safavi, A., Miller, B.C., Cottam, L., and Hersh, L.B., 1996, Identification of γ-endorphin-generating enzyme as insulin-degrading enzyme. *Biochemistry* **35:** 14318-14325.

Sato, M., Ikeda, K., Haga, S., Allsop, D., and Ishii, T., 1991, A monoclonal antibody to common acute lymphoblastic leukemia antigen immunostains senile plaques in the brains of patients with Alzheimer′ disease. *Neurosci. Lett.* **121:** 271-273.

Schwartz, J.C., De La Baume, S., Malfroy, B., Patey, G., Perdrisot, R., Swerts, J.P., Fournie-Zaluski , M.C., Gacel, G., and Roques, B.P., 1980, "Enkephalinase", a newly characterized dipeptidyl carboxypeptidase: properties and possible role in enkephalinergic transmission. *Intl. J. Neurol.* **14:** 195-204.

Shibata, M., Yamada, S., Kumar, S.R., Calero, M., Bading, J., Frangione, B., Holtzman, D.M., Miller, C.A., Strickland, D.K., Ghiso, J., and Zlokovic, B.V., 2000, Clearance of Alzheimer's amyloid-β (1-40) peptide from brain by LDL receptor-related protein-1 at the blood-brain barrier. *J. Clin. Invest.* **106:** 1489-1499.

Soto, C., Castaño, E.M., Frangione, B., and Inestrosa, N.C., 1995, The α-helical to β-strand transition in the amino-terminal fragment of the amyloid β-peptide modulates amyloid formation. *J. Biol. Chem.* **270:** 3063-3067.

Sudoh, S., Frosch, M.P., and Wolf, B.A., 2002, Differential effects of proteases involved in intracellular degradation of amyloid β-protein between detergent-soluble and insoluble pools in CHO-695 cells. *Biochemistry* **41:** 1091-1099.

Suzuki, N., Cheung, T.T., Cai, X.D., Odaka, A., Otvos, L. Jr., Eckman, C., Golde, T.E., and Younkin, S.G., 1994, An increased percentage of long amyloid β protein secreted by familial amyloid β protein precursor (βAPP717) mutants. *Science* **264:** 1336-1340.

Turner, A.J., Isaac, R.E., and Coates, D., 2001, The neprilysin (NEP) family of zinc metalloendopeptidases: genomics and function. *BioEssays* **23:** 261-269.

Turner, A.J., and Tanzawa, K., 1997, Mammalian membrane metallopeptidases: NEP, ECE, KELL, and PEX. *FASEB J.* **11:** 355-364.

Van Nostrand, W.E., Melchor, J.P., Cho, H.S., Greenberg, S.M., and Rebeck, G.W., 2001, Pathogenic effects of D23N Iowa mutant amyloid β-protein. *J. Biol. Chem.* **276:** 32860-32866.

Van Uden, E., Mallory, M., Veinbergs, I., Alford, M., Rockenstein, E., and Masliah, E., 2002, Increased extracellular amyloid deposition and neurodegeneration in human amyloid precursor protein transgenic mice deficient in receptor-associated protein. *J. Neurosci.* **22:** 9298-9304.

Vekrellis, K., Ye, Z., Qiu ,W.Q., Walsh, D., Hartley, D., Chesneau, V., Rosner, M.R., and Selkoe, D.J., 2000, Neurons regulate extracellular levels of amyloid β-protein via proteolysis by insulin-degrading enzyme. *J. Neurosci.* **20:** 1657-1665.

Walsh, D.M., Klyubin, I., Fadeeva, J.V., Cullen, W.K., Anwyl, R., Wolfe, M.S., Rowan, M.J., and Selkoe, D.J., 2002, Naturally secreted oligomers of amyloid β protein potently inhibit hippocampal long-term potentiation in vivo. *Nature* **416:** 535-539.

Wang, J., Dickson, D.W., Trojanowski, J.Q., and Lee, V.M., 1999, The levels of soluble versus insoluble brain Aβ distinguish Alzheimer's disease from normal and pathologic aging. *Exp. Neurol.* **158:** 328-337.

Wijdicks, E.F., and Jack, C.R. Jr., 1993, Intracerebral hemorrhage after fibrinolytic therapy for acute myocardial infarction. *Stroke.* **24:** 554-557.

Yasojima, K., Haruhiko, A., McGeer, E.G., and McGeer, P.L., 2001, Reduced neprilysin in high plaque areas of Alzheimer brain: a possible relationship to deficient degradation of β-amyloid peptide. *Neurosci. Lett.* **297:** 97-100.

Chapter 7

The Protective Role of Vitamin E in Vascular Amyloid β-Mediated Damage

Francisco José Muñoz, Montserrat Solé and Mireia Coma
Unitat de Senyalització Cel lular, Departament de Ciències Experimentals i de la Salut, Universitat Pompeu Fabra (UPF), 08003-Barcelona, Spain

Abstract: Amyloid β peptide (Aβ) accumulation produces the senile plaques in the brain parenchyma characteristic of Alzheimer's Disease (AD) and the vascular deposits of Cerebral Amyloid Angiopathy (CAA). Oxidative stress is directly involved in Aβ-mediated cytotoxicity and antioxidants have been reported as cytoprotective in AD and CAA. Vitamin E has antioxidant and hydrophobic properties that render this molecule as the main antioxidant present in biological membranes, preventing lipid peroxidation, carbonyl formation and inducing intracellular modulation of cell signalling pathways. Accordingly, vascular damage produced by Aβ and prooxidant agents can be decreased or prevented by vitamin E. The protective effect of vitamin E against Aβ cytotoxicity in vascular cells in comparison to the neuronal system is reviewed in this chapter.

Key words: Amyloid β-peptide, Cerebral Amyloid Angiopathy, vitamin E, vascular cells, antioxidants, oxidative stress.

1. VASCULAR AMYLOIDOSIS

The different systemic diseases generally termed *amyloidosis* are all characterized by the misfolding of proteins into β-pleated sheet-rich structures that in turn leads to the aggregation and fibrillogenesis of the proteins, triggering pathological processes in the tissues. The most representative pathologies affecting the brain vessels are those produced by the aggregation of cystatin C (hereditary cerebral hemorraghe with amyloidosis of the Icelandic type; HCHWA-I), transthyretin (familial transthyretin amyloidosis; TTR), gelsolin (familial amyloidosis of the Finnish type; FAF), prion protein (Gerstmann-Sträussler-Scheinker syndrome; GSS), ABri (familial British dementia; FBD), ADan (familial Dannish dementia) and the amyloid β-peptide (Alzheimer disease and cerebral amyloid angiopathy; AD and CAA).

2. CEREBRAL AMYLOID ANGIOPATHY

CAA is present in most cases of AD and it is characterized by the deposition of amyloid β-peptide (Aβ) in the media and adventitia of both leptomeningeal arteries and intracortical arterioles and capillaries (Figure 1), and less frequently in veins (Vinters *et al.*, 1988; Calhoun *et al.*, 1999). These vascular deposits are mostly composed by $A\beta_{1\text{-}40}$ wild type (Castaño *et al.*, 1996) and produce degeneration of vascular smooth muscle cells (VSMCs) from the media (Wisniewski and Wegiel, 1994; Zhang *et al.*, 1998) and endothelial cells from the intima (Wisniewski *et al.*, 1992; Kalaria, 1997). A variant of CAA with an early onset of the disease is hereditary cerebral haemorrhage with amyloidosis of the Dutch type (HCHWA-D). HCHWA-D is caused by Aβ-encoding gene point mutation which produces substitution of Glu→Gln at the position 22 (Levy *et al.*, 1990) resulting in a peptide with increased ability to form amyloid fibrils (Wisniewski *et al.*, 1991). Although HCHWA-D patients show diffuse amyloid deposition in the brain parenchyma, the main hallmarks of AD (mature senile plaques and neurofibrillary tangles) are not observed (Timmers *et al.*, 1990; Maat-Schieman *et al.*, 1994). In both CAA and HCHWA-D, the vascular amyloid deposits contain extracellular matrix molecules and other common components of senile plaques from the neuropil of AD patients (Snow *et al.*, 1988; Verbeek *et al.*, 1998; Mesulam *et al.*, 1992; Van Duinen *et al.*, 1995). Significantly, as in senile plaques (Vehmas et al., 2003), the vascular deposits show the presence of reactive glia (Uchihara et al., 1997).

3. ORIGIN OF VASCULAR Aβ

The amyloid precursor protein (APP) is present in VSMCs, pericytes and endothelial cells (Schmechel *et al.*, 1988; Shoji *et al.*, 1990; Tagliavini et al., 1990; Wisniewski and Wegiel, 1994), but the origin of vascular Aβ deposits

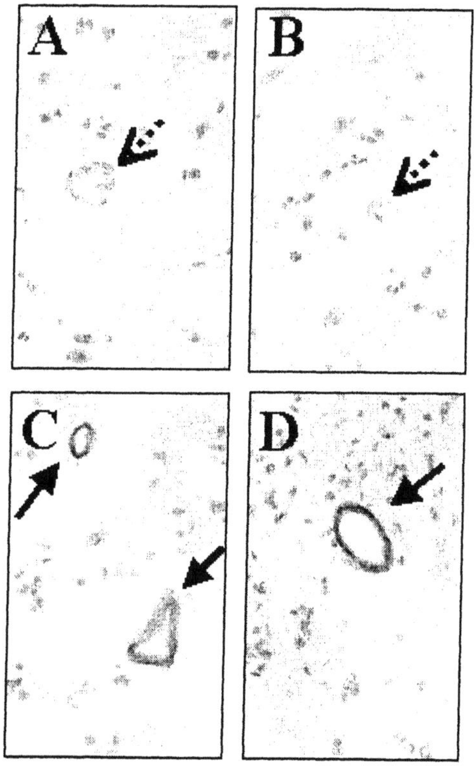

Figure 1. Amyloid deposits identified by Congo red staining in brain samples from the frontal cortex of control (A,B) and AD patients (C,D) with AD in the VI stage. Arrows show the blood vessels. The samples from AD patients show that most of the blood vessels are Congo red-positive.

is controversial (Weller *et al.*, 1998). VSMCs are able to produce Aβ (Frackowiak *et al.*, 1995), which has been identified even in intracellular compartments of VSMCs (Mazur-Kolecka *et al.*, 1995; Wisniewski *et al.*, 2000). Nevertheless, transgenic mice overexpressing neuronal mutated APP (Dutch-Iowa-Swedish mutations) also develop CAA (Davis *et al.*, 2004), indicating that neuronal Aβ is deposited in the vessels. It could be due to the flux from the neuronal Aβ drainage, but the contribution of VSMCs to the vascular Aβ deposits could be also relevant by producing seeds for the fibrillation of the Aβ coming from the neurons. It could be due to the flux from the neuronal Aβ drainage, but the contribution of VSMCs to the vascular Aβ deposits has been demonstrated in cell cultures (Frakcowiak *et al.*, 2004) and human brain vessel cultures (Mazur-Kolecka *et al.*, 2004). Moreover, if neurons were the producers of all the vascular Aβ, a gradient of immature Aβ deposits should be shown from the parenchyma to the vessels and it does not occur in arteries or small arterioles, where Aβ deposits have been found. Nevertheless, such a gradient is shown in the proximity of veins, probably corresponding to the clearance of neuronal Aβ. These findings suggest that the vascular Aβ is produced by both type of cells, neurons and VSMCs, and that VSMCs play a key role in the Aβ secretion.

4. Aβ EFFECTS ON THE VESSELS

CAA is characterized by the degeneration of VSMCs and endothelial cells (Miyakawa *et al*, 1997; Kalaria, 1997). Aβ deposits are present in the tunica media of large vessels at early stages of AD. VSMCs close to Aβ deposits have swollen nuclei and express the proliferating cell nuclear antigen. When the amyloidosis is at advanced stages, the tunica media is replaced by amyloid deposits and VSMCs degenerate becoming scarce (Wisniewscki *et al.*, 2000). Thus, it has been reported that there is an increase in the number of apoptotic VSMCs and endothelial cells in AD (De la Monte *et al.*, 2000). A direct toxic effect of Aβ on VSMCs *in vitro* has been also demonstrated (Davis and Van Nostrand, 1996; Muñoz *et al.*, 2002).

At the functional level, Aβ enhances the vessel contraction (Crawford *et al.*, 1998; Suo *et al.*, 2000) and decreases the endothelium-dependent vasodilatation (Thomas *et al.*, 1997) despite the increased nitric oxide (NO) production by AD endothelial cells (Grammas *et al.*, 2000). The lack of vasodilatatory properties of NO may be due to the sequestration of NO in a pro-oxidant environment to produce peroxynitrite, a powerful oxidant produced from the reaction of superoxide anion ($O_2^{\cdot-}$) and NO. There is also evidence of endothelial cell degeneration in CAA (Miyakawa *et al.*, 1997), which produces blood vessel damage and increased permeability of the blood-brain barrier (Wisniewski *et al.*, 2000).

5. OXIDATIVE STRESS IN THE ETIOLOGY OF AD

Oxidative stress could be involved in the development of AD since it has been demonstrated that the expression and activity of BACE, the proposed β-secretase for APP, is increased by oxidative stress (Tamagno *et al.*, 2002), and that oxidative stress enhances the production of Aβ (Frederikse *et al.*,

1996). Moreover, homocysteine, a well-known risk factor for atherogenic damage and vascular disease, has been also proposed as a risk factor for AD (Seshadri *et al.*, 2002), and the deleterious effect of homocysteine on blood vessels may be mediated by oxidative stress (Perna *et al.*, 2003). Thus the putative role of homocysteine in the development of AD could be related to a pro-oxidant activity yielding to an increase in the production of Aβ, and/or because homocysteine increases Aβ-induced cytotoxicity (Mok *et al.*, 2002).

6. OXIDATIVE STRESS IN Aβ-MEDIATED CYTOTOXICITY

The existence of a specific receptor mediating the cytotoxicity induced by Aβ has been proposed, but none of the putative candidates can explain all the cytotoxic effects observed. There is an increased amount of experimental and histopathological evidence suggesting that oxidative stress plays a key role in Aβ-mediated cytotoxicity (Behl *et al.* 1994; Miranda *et al.*, 2000; Muñoz *et al.*, 2002).

Aβ generates hydrogen peroxide (H_2O_2) through metal ion reduction (Huang *et al.*, 1999) and is able to increase the free radical generation by metals such as iron, copper and zinc (Bondy *et al.*, 1998), which are highly concentrated within the core and periphery of Aβ deposits (Lovell *et al.*, 1998). The oxidative damage of proteins generates an increase in carbonyl groups (Stadtman, 1990). Carbonyl residues and protein nitration in AD brain may come from the action of peroxynitrites (Smith *et al.*, 1997). Proteins can also be modified by non-enzymatic reaction with monosaccarides, such as the Maillard reaction, when irreversible advanced glycation end products (AGEs) are formed, concomitant with hydroxyl radical ($HO^{\cdot}$) generation (Münch *et al.*, 1997a). It has been demonstrated *in vitro* that Aβ induces lipoperoxidation of membranes (Koppal *et al.*, 1998, Mark *et al.*, 1997) leading to the disruption of the physiological signalling pathways (Kelly *et al.*, 1996). Moreover, it impairs the function of membrane-regulatory proteins, including cation transport ATPases (Mark *et al.*, 1995). The impairment of ATPase function means that the intracellular calcium cannot be pumped out. The progressive cytoplasmic accumulation of calcium and the oxidative damage on mitochondria and nucleic acid (Gabbita *et al.*, 1998; Nunomura *et al.*, 1999) trigger cellular apoptotic mechanisms. Therefore, extracellular Aβ produces a cascade of ROS which induces intracellular damage (Figure 2).

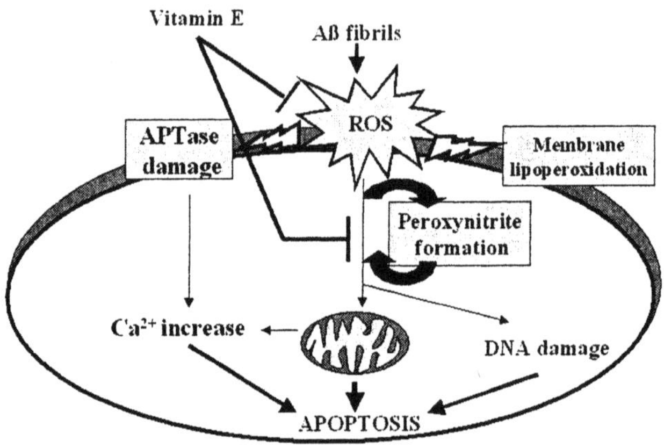

Figure 2. Aβ produces oxidative damage leading to the cell death. Reactive oxygen species (ROS) induce peroxidation with damage to both membrane lipid and protein. This event disrupts cell homeostasis. The intracellular ROS cascade is increased by the formation of peroxynitrites. Mitochondrial oxidative damage is triggering apoptotic pathways by the release of cytochrome C and the enhancement of intracellular calcium, which even activates endonucleases. Vitamin E can prevent the cell death by inhibiting the damage from extracellular ROS in the membrane and in the intracellular ROS cascade.

7. INTRACELLULAR SIGNALLING PATHWAYS INVOLVED IN THE OXIDATIVE DAMAGE

The mitogen-activated protein kinase (MAPK) pathways are involved in the deleterious effect of Aβ, but the specific role of the extracellular signal-regulated kinase (ERK), c-Jun NH_2-terminal protein kinase (JNK) and p38/RK/MpK2/CSBP kinases in the Aβ toxicity is controversial (Zhu *et al.*, 2002).

Low levels of radical/reactive oxygen species (ROS) play an important role in normal cell proliferation (Burdon, 1995; Benhar *et al.*, 2002) and regulate cellular signalling by the activation of MAPKs leading to induction of gene expression to protect cells. But at high concentrations, these agents activate ERK2 and JNK and ICE/Ced-3 caspase pathway inducing apoptosis

(Kong et al., 1998). Other authors propose that low concentrations of H_2O_2 activates phosphatidylinositol-3-kinase (PI-3K) giving an increase in the cell survival by the activation of c-AMP response element binding protein (CREB) throughout the action of ERK1/2 and Akt/PKB pathways, while high concentrations of H_2O_2 are proapoptotic throughout the activation of JNK/c-jun cascade in cortical neurons (Crossthwaite et al., 2002). In cardiac myocytes two of the MAPK pathways, JNK and p38 are reported as proapoptotic, whereas ERK pathway is considered antiapoptotic (Aikawa et al., 1997; Turner et al., 1998).

Regarding Aβ, it triggers the JNK/c-jun cascade (Figure 3) producing cell death and increasing the expression of proapoptotic molecules in neurons (Okazawa and Estus, 2002). Therefore, the inhibition of JNK protects PC12 cells against Aβ-mediated cytotoxicity (Troy *et al.*, 2001). The involvement of JNK and p38 pathways in AD has been also demonstrated in an animal model. Thus, both MAPK pathways are activated in the cerebral cortex of a double transgenic mice for mutant APP (Swedish mutations) and mutated presenilin-1 (P264L), which produces a dramatic increase in the production and the consequent aggregation of Aβ (Savage *et al.*, 2002). All these results show that firstly the activation of MAPK pathways rendering apoptosis or protection depends on cell type and the concentration of the prooxidant agent. Secondly, the specific role of ERK1/2 in oxidative stress/AD/CAA is the most controversial but JNK and p38 appears to be directly involved in the cell damage.

On the other hand, oxidative stress produces the activation of redox-sensitive transcription factors, such as nuclear factor-κβ (NF-κβ) (Piette *et al.*, 1997) and activator protein-1 (AP-1) (Lo *et al.*, 1996; Vollgraf *et al.*, 1999) triggering apoptosis or inducing the protection of the cells (Bossy-Wetzel et al., 1997). AP-1 is a protein complex containing Jun and Fos proteins or Jun dimers (Gass and Herdegen, 1995), and the activation of the migration to the nucleus of AP-1 and NF-κβ is mainly controlled by JNK and p38 pathways (Behrens *et al.*, 1999). These mechanisms have been demonstrated to occur under the effect of Aβ in neurons (Kaltschmidt *et al.*, 1997; Mattson *et al.*, 1997), and in vascular cells with different pro-oxidant insults (Yin *et al.*, 2002; Wang *et al.*, 2002; Robbesyn *et al.*, 2003)

8. INTRACELLULAR ANTIOXIDANT DEFENCES

The cellular mechanism of protection against oxidative stress is constituted by different intacellular enzymes, mainly catalase, superoxide

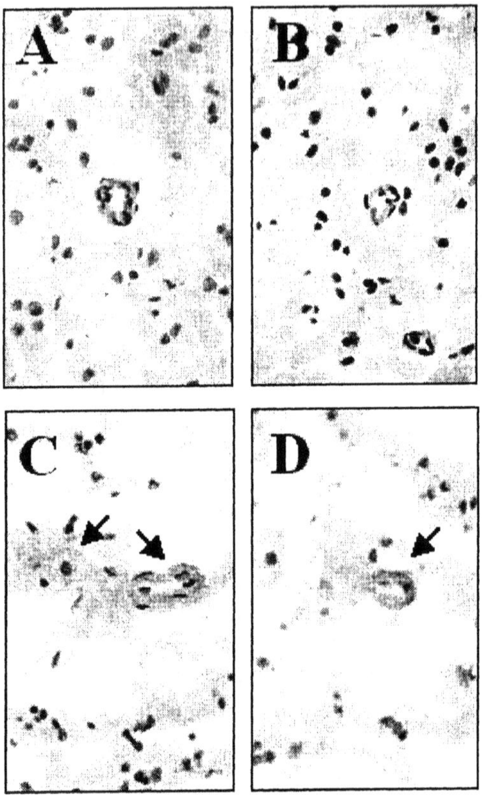

Figure 3. c-jun activation identified by immunohistochemical staining with peroxidase in brain samples from the frontal cortex of control (A, B) and AD patients (C, D) with AD in the VI stage. Asterisks indicate the presence of vessels. Arrows show the positive areas for c-jun activation into the vessels and the brain parenchyma. Positive areas for c-jun correlate with AD brain vessels and their periphery in all the samples analysed from AD patients.

dismutase (SOD), thioredoxin, the peroxiredoxins and the enzymes related to the glutathione (GSH) pathway. GSH is a tripeptide formed by glutamate, glycine and cysteine, and its antioxidant properties depends on the thiol group of the molecule. GSH-peroxidase is considered one of the most important enzymes involved in the hydrolysis of peroxides in the brain. Furthermore, different neuronal cell lines showed resistance against Aβ-mediated cytotoxicity which was directly proportional to the levels of GSH-peroxidase (Calderón *et al.*, 1999). The relevance of the protective role of antioxidants in the vasculature was first evidenced by the observation that Aß-induced endothelial damage is prevented by the enzyme SOD (Thomas *et al.*, 1996; 1997; Crawford *et al.*, 1997). These results are in agreement with a direct effect of $O_2^{\cdot -}$ in Aβ-mediated cytotoxicity in vascular cells, as it was demonstrated by the measurement of the dihydroethidium fluorescence, an indicator of ROS, in VSMCs and endothelial cells challenged by Aβ (Muñoz *et al.*, 2002).

9. PROTECTION BY ANTIOXIDANT MOLECULES

Antioxidants such as vitamin E, 17β-estradiol or melatonin have demonstrated protective properties on neuronal cells against the Aβ-mediated cytotoxicity (Behl, 2000; 2002; Behl *et al.*, 1992; 1997; Mattson and Goodman, 1995; Pappolla *et al.*, 1997; Bonnefont *et al.*, 1998), (see also Chapter 3). Vitamin E can protect VSMCs and endothelial cells against alcohol, which may induce oxidative stress (Altura and Gebrewold, 1996), and against Aβ-mediated cytotoxicity (Muñoz *et al.*, 2002). Vitamin C, which shares anti-oxidant properties (Podmore *et al.*, 1998), prevents β-amyloid-induced intracellular calcium increase and cell death in PC12 cells (Yallampalli *et al.*, 1998). However other authors have not found any protection with antioxidants such as vitamin E, trolox (a hydrosoluble form of vitamin E), vitamin C or N-acetyl-L-cysteine (NAC) on neuronal cells (Lockhart *et al.*, 1994; Pike *et al.*, 1997). This lack of protection could be due to the experimental procedures. The steroidal hormone 17β-estradiol has also been proposed to play a key role in the prevention or retardation of AD related pathologies, since women treated with estrogen replacement therapy showed a lower prevalence of AD (Tang *et al.*, 1996; Kawas *et al.*, 1997) due to the pleiotropic effects of 17β-estradiol (Behl, 2002).

10. PROTECTION BY VITAMIN E

Vitamin E was discovered by Evans and Bishop in 1922. Vitamin E is a term which includes a group of tocopherols and tocotrienols, both having four isomers (alpha, beta, gamma and delta). The alpha-tocopherol is the most active in humans because of the high affinity of the tocopherol transporter protein (TTP) for this molecule (Hosomi *et al.*, 1997). The relevance of this transporter is shown when there are mutations in the TTP gene. It produces a reduction of alpha-tocopherol in plasma and tissues yielding to ataxia with vitamin E deficiency (Ben Hamida *et al.*, 1993). When considering the vitamin E distribution in brain, there are no specific areas of the brain or spinal cord that are richer in vitamin E than others, however, the uptake of vitamin E is considerably high in the cerebellum (Vatassery, 1992).

Due to the antioxidant and hydrophobic characteristics of vitamin E, it is the main antioxidant present in biological membranes (Perly *et al.*, 1985), preventing lipid peroxidation (Halliwell and Gutteridge, 1984) by trapping the peroxyl radicals (Naiki *et al.*, 1998). Vitamin E has protective properties on neuronal cells against the Aβ-mediated cytotoxicity (Behl *et al.*, 1992). These neuronal protective properties have been demonstrated even in synaptosomas challenged with Aβ (Koppal *et al.*, 1998). Regarding the role of vitamin E on blood vessels (Figure 4), it has been reported that alpha-tocopherol can protect VSMCs and endothelial cells against alcohol (Altura and Gebrewold, 1996) and against Aβ-mediated cytotoxicity even when both types of cells were challenged with the Dutch variant of Aβ, which is more toxic for vascular cells than the wild type Aβ (Muñoz *et al.*, 2002).

The protective effect of vitamin E against oxidative stress is not just due to the free radical scavenging activity of the molecule but also due to the modulation of signalling pathways. In fact, vitamin E has been reported to protect against oxidative stress by decreasing JNK activity and increasing the ERK activity in cardiac myocytes (Qin *et al.*, 2003), and inhibiting caspase-3 activation in vascular endothelial cells (Uemura *et al.*, 2002).

Moreover, in HeLa cells, pretreatment with free radical scavengers NAC, GSH or vitamin E, inhibited JNK pathway activation by prooxidant agents (Kong *et al.*, 1998). Furthermore, the protective roles of vitamin E could also be related to other intracellular effects such as the activation of PP2A, the inhibition of alpha-PKC in VSMCs (Ricciarelli *et al.*, 1998), the inhibition of the production of eicosanoids (Pratico *et al.*, 1998; Jialial *et al.*, 2001; Lee *et al.*, 1999) or the inhibition of inducible NO synthase (iNOS) (Badger *et al.*, 2000; Guan *et al.*, 1998) which leads to a decrease in the protein peroxynitration. On the other hand, vitamin E protects neurons and vascular cells against oxidative cell death *in vitro* by the activation of NF-κß (Behl, 2000; Li-Weber *et al.*, 2002). Alpha-tocopherol also induces the expression of connective tissue growth factor (CTGF) in VSMCs in a PKC-independent pathway (Villacorta *et al.*, 2003) and increases the synthesis of alpha-tropomyosin in VSMCs (Aratri *et al.*, 1999), suggesting an improvement in the VSMC function in the vessels. Moreover, vitamin E can be a vasodilatatory agent since thrombin-mediated PKC activation and endothelin secretion are inhibited by alpha-tocopherol in endothelial cells (Martin-Nizard *et al.*, 1998). A clinical trial confirmed the positive role of vitamin E in preventing AD in ageing people (Sano *et al.*, 1997). There are also reports suggesting that use of high vitamin E and vitamin C supplements may decrease the risk of AD (Morris *et al.*, 1998). Other studies suggest that vitamin E could be also protective in vascular disease (*reviewed by*

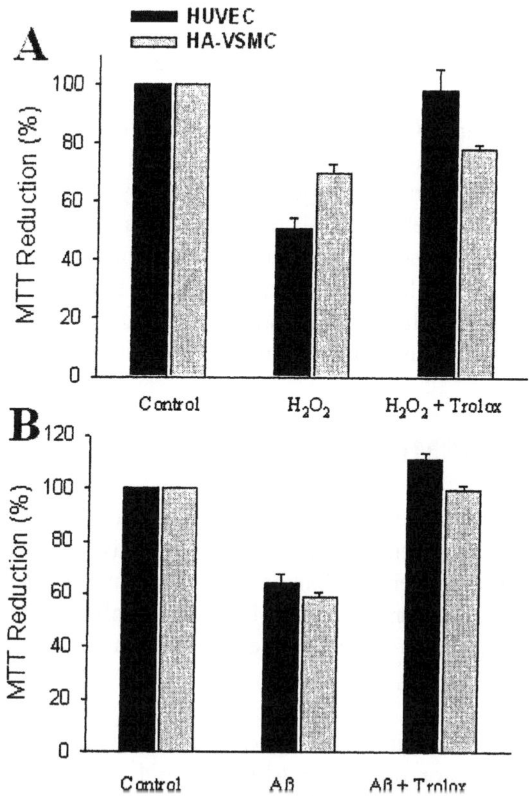

Figure 4. The water-soluble analogue of Vitamin E (Trolox) protects human umbilical vein endothelial cells (HUVEC) and human aortic vascular smooth muscle cells (HA-VSMC) from H_2O_2 and Dutch $A\beta_{1\text{-}40}$ fibrils cytotoxicity. Representative experiments were performed in quadruplicate. Cells were challenged with 4 μM H_2O_2 in HUVEC and 50 μM H_2O_2 in HA-VSMCs (A), and 0.25 μM Aβ in HUVEC and 0.125 μM Aβ in HA-VSMCs (B). For the protection studies, cells were treated with 500 μM Trolox. Cell viability was evaluated by MTT reduction after 24h of incubation. Control cells were assumed to have 100% of viability.

Steinberg, 1995; Gotto, 2003). In addition, there are experimental data demonstrating that treatment with antioxidants such as idebenone and alpha-tocopherol prevents learning and memory deficits caused by Aβ in rats (Yamada *et al.*, 1999). Furthermore, decreasing serum levels of vitamin E were associated with poor memory performance in older people (Perkins *et al.*, 1999), however this effect was not found in rats (Ichitani *et al.*, 1992).

11. CONCLUSIONS

Oxidative stress is directly involved in Aβ-mediated cytotoxicity. Thus, free radical scavengers and antioxidants are considered key pharmacological tools against AD and CAA. Vascular damage produced by Aβ and prooxidant agents can also be decreased or avoid by vitamin E. These findings suggest that vitamin E is a good biological cytoprotective agent, but more work is needed to elucidate all the intracellular mechanism triggered by vitamin E and the changes in the expression of specific protective genes such antiapoptotic molecules. Moreover, considering that Aβ production has been related directly with oxidative stress, vitamin E could prevent the triggering of AD by decreasing the production of Aβ in vascular cells.

ACKNOWLEDGEMENTS

We acknowledge Dr. Elena Hidalgo and Dr. Miguel A. Valverde for the very fruitful critical suggestions on the manuscript. This work was supported by grants from FIS (Ministerio de Sanidad, Spain; grant nº 01-1029; and Red HERACLES) and Fundación Domingo Martínez. We also thank the Banc de Teixits Neurologics, Universitat de Barcelona-Hospital Clínic for providing brain samples.

Abbreviations: AD, Alzheimer's disease; AGEs, advanced glycation end products; AP-1, activator protein-1; APP, amyloid precursor protein; Aß, amyloid ß-peptide; CAA, cerebral amyloid angiopathy; CREB, c-AMP response element binding protein; CTGF, connective tissue growth factor; ERK, extracellular signal-regulated kinase; GSH, glutathione; HCHWA-D, hereditary cerebral haemorrhage with amyloidosis of the Dutch type; iNOS, inducible NO synthase; JNK, c-Jun NH_2-terminal protein kinase; MPAKs, mitogen-activated protein kinases; MTT, 3-(4,5-dimethylthiazol-2-yl)-2,5-diphenyltetrazolium bromide; NAC, N-acetyl-L-cysteine; NF-κβ, nuclear factor-κβ; NO, nitric oxide; PI-3K, phosphatidylinositol-3-kinase; ROS, radical oxygen species; SOD, superoxide dismutase; TTP, tocopherol transporter protein; VSMCs, vascular smooth muscle cells.

REFERENCES

Aikawa, R., Komuro, I., Yamazaki, T., Zou, Y., Kudoh, S., Tanaka, M., Shiojima, I., Hiroi, Y., and Yazaki, Y., 1997, Oxidative stress activates extracellular signal-regulated kinases through Src and Ras in cultured cardiac myocytes of neonatal rats. *J. Clin. Invest.* **100:** 1813-1821.

Altura, B.M., Gebrewold, A., 1996, Alpha-tocopherol attenuates alcohol-induced cerebral vascular damage in rats: possible role of oxidants in alcohol brain pathology and stroke. *Neurosci Lett.* **220**: 207-210.

Aratri, E., Spycher, S.E., Breyer, I., Azzi, A., 1999, Modulation of alpha-tropomyosin expression by alpha-tocopherol in rat vascular smooth muscle cells. *FEBS Lett.* **447**: 91-94.

Badger, A.M., 2000, Differential effects of SB 242235, a selective p38 mitogen-activated protein kinase inhibitor on IL-1 treated bovine and human cartilage/chondrocyte cultures. *Osteoarthr. Cartil.* **8**: 434-443.

Ben Hamida, C., Doerflinger, N., Belal, S., Linder, C., Reutenauer, L., Dib, C., Gyapay, G., Vignal, A., Le Paslier, D., Cohen, D., Pandolfo, M., Mokini, V., Novelli, G., Hentati, F., Ben Hamida, M., Mandel, J. L., and Koenig, M., 1993, Localization of Friedreich ataxia phenotype with selective vitamin E deficiency to chromosome 8q by homozygosity mapping. *Nat. Genet.* **5**:195-200.

Behl, C., 2002, Oestrogen as a neuroprotective hormone. *Nat. Neurosci.* **3**: 433-442.

Behl, C., 2000, Vitamin E protects neurons against oxidative cell death in vitro more effectively than 17-beta estradiol and induces the activity of the transcription factor NF-kappaB,. *J. Neural Transm.*, **107**: 393-407.

Behl, C., Davis, J., Cole, G.M., and Schubert, D., 1992, Vitamin E protects nerve cells from amyloid beta protein toxicity. *Biochem. Biophys. Res. Commun.* **186**: 944-950.

Behl, C., Davis, J., Lesley, R., and Schubert, D., 1994, Hydrogen peroxide mediates amyloid beta protein toxicity. *Cell* **77**: 817-827.

Behrens, A., Sibila, M., and Wagner, E.F., 1999, Amino-terminal phosphorylation of c-Jun regulates stress-induced apoptosis and cellular proliferation. *Nat. Genet.* **21**: 326-329.

Ben Hamida, C., Doerflinger, N., Belal, S., Linder, C., Reutenauer, L., Dib, C., Gyapay, G., Vignal, A., Le Paslier, D., Cohen, D., Pandolfo, M., Mokini, V., Novelli, G., Hentati, F., Ben Hamida, M., Mandel, J. L., and Koenig, M., 1993, Localization of Friedreich ataxia phenotype with selective vitamine deficiency to chromosome 8q by homozygosity mapping. *Nat. Genet.* **5**: 195-200.

Benhar, M., Engelberg, D., and Levitzki, A., 2002, ROS, stress-activated kinases and stress signaling in cancer. *EMBO Rep.* **3**: 420-425.

Bondy, S.C., Guo-Ross, S.X., and Truong, A.T., 1988, Promotion of transition metal-induced reactive oxygen species formation by beta-amyloid. *Brain Res.* **799**: 91-96.

Bonnefont, A.B., Muñoz, F.J., and Inestrosa, N.C., 1998, Estrogen protects neuronal cells from the cytotoxicity induced by acetylcholinesterase-amyloid complexes. *FEBS Lett.* **441**: 220-224.

Bossy-Wetzel, E., Bakiri, L., and Yaniv, M., 1997, Induction of apoptosis by the transcription factor c-Jun. *EMBO J.* **16**: 1695-1709.

Burdon, R.H., 1995, Superoxide and hydrogen peroxide in relation to mammalian cell proliferation. *Free Radic. Biol. Med.* **18**: 775-794.

Calderón, F.H., Bonnefont, A., Muñoz, F.J., Fernández, V., Videla, L.A., and Inestrosa, N.C., 1999, PC12 and neuro 2a cells have different susceptibilities to acetylcholinesterase-amyloid complexes, amyloid25-35 fragment, glutamate, and hydrogen peroxide. *J. Neurosci. Res.* **56**: 620-631.

Calhoun, M.E., Burgermeister, P., Phinney, A.L., Stalder, M., Tolnay, M., Wiederhold, K.H., Abramowski, D., Sturchler-Pierrat, C., Sommer, B., Staufenbiel, M., and Jucker, M., 1999, Neuronal overexpression of mutant amyloid precursor protein results in prominent deposition of cerebrovascular amyloid. Proc. Natl. Acad. Sci. USA. **96:** 14088-14093.

Castaño, E.M., Prelli, F., Soto, C., Beavis, R., Matsubara, E., Shoji, M., and Frangione, B., 1996, The length of amyloid-beta in hereditary cerebral hemorrhage with amyloidosis,

Dutch type. Implications for the role of amyloid-beta 1-42 in Alzheimer's disease. *J. Biol. Chem.* **271**: 32185-32191.

Crawford, F., Suo, Z., Fang, C., and Mullan, M., 1998, Characteristics of the in vitro vasoactivity of beta-amyloid peptides. *Exp. Neurol.* **150**: 159-168.

Crawford, F., Suo, Z., Fang, C., Sawar, A., Su, G., Arendash, G., and Mullan, M., 1997, The vasoactivity of A beta peptides. *Ann. N. Y. Acad. Sci.* **826**: 35-46.

Crossthwaite, A.J., Hasan, S., and Williams, R.J., 2002, Hydrogen peroxide-mediated phosphorylation of ERK1/2, Akt/PKB and JNK in cortical neurones: dependence on Ca(2+) and PI3-kinase. *J. Neurochem.* **80**: 24-35.

Davis, J., and Van Nostrand, W.E., 1996, Enhanced pathologic properties of Dutch-type mutant amyloid beta-protein. *Proc. Natl. Acad. Sci. USA.* **93**: 2996-3000.

De la Monte, S.M., Sohn, Y.K., Etienne, D., Kraft, J., and Wands, J.R., 2000, Role of aberrant nitric oxide synthase-3 expression in cerebrovascular degeneration and vascular-mediated injury in Alzheimer's disease. *Ann. N. Y. Acad. Sci.* **903**: 61-71.

Evans, H.M., and Bishop, K.S., 1922, On the existence of a hitherto unrecognised dietary factor essential for reproduction. *Science* **56**: 650-651.

Frackowiak, J., Mazur-Kolecka, B., Wisniewski, H.M., Potempska, A., Carroll, R.T., Emmerling, M.R., and Kim, K.S., 1995, Secretion and accumulation of Alzheimer's beta-protein by cultured vascular smooth muscle cells from old and young dogs. *Brain Res.* **676**: 225-230.

Frackowiak, J., Sukontasup, T., Potempska, A., and Mazur-Kolecka, B., 2004, Lysosomal deposition of Abeta in cultures of brain vascular smooth muscle cells is enhanced by iron. *Brain Res.* **1002**: 67-75.

Frederikse, P.H., Garland, D., Zigler, J.S.Jr, and Piatigorsky, J., 1996, Oxidative stress increases production of beta-amyloid precursor protein and beta-amyloid (Abeta) in mammalian lenses, and Abeta has toxic effects on lens epithelial cells. *J. Biol. Chem.* **271**: 10169-10174.

Gabbita, S.P., Lovell, M.A., and Markesbery, W.R., 1998, Increased nuclear DNA oxidation in the brain in Alzheimer's disease,.*J. Neurochem.* **71**: 2034-2040.

Gass, P., and Herdegen, T., 1995, Neuronal expression of AP-1 proteins in excitotoxic neurodegenerative disorders and following nerve fiber lesions. *Prog. Neurobiol.* **47**: 257-290.

Gotto, A.M., 2003, Antioxidants, statins, and atherosclerosis. *J. Am. Coll. Cardiol.* **41**: 1205-1210.

Grammas, P., Reimann-Philipp, U., and Wegiel, P.H., 2000, Cerebrovasculature-mediated neuronal cell death. *Ann. N. Y. Acad. Sci.* **903**: 55-60.

Guan, Z.H., Buckman, S.Y., Pentland, A.P., Templeton, D.J., and Morrison, A.R., 1998, Induction of cyclo-oxygenase-2 by the activated MEKK1-/SEK1/MKK4-/p38 mitogen-activated protein kinase pathway. *J. Biol. Chem.* **273**: 12901-12908.

Halliwell, B., and Gutteridge, J.M., 1984, Free radicals, lipid peroxidation, and cell damage. *Lancet* **2**: 1095.

Hosomi, A., Arita, M., Sato, Y., Kiyose, C., Ueda, T., Igarashi, O., Arai, H., and Inoue, K., 1997, Affinity for alpha-tocopherol transfer protein as a determinant of the biological activities of vitamin E analogs. *FEBS Lett.* **409**: 105-108.

Huang, X., Atwood, C.S., Hartshorn, M.A., Multhaup, G., Goldstein, L.E., Scarpa, R.C., Cuajungco, M.P., Gray, D.N., Lim, J., Moir, R.D., Tanzi, R.E., and Bush, A.I., 1999, The A beta peptide of Alzheimer's disease directly produces hydrogen peroxide through metal ion reduction. *Biochemistry.* **38**: 7609-7616.

Ichitani, Y., Okaichi, H., Yoshikawa, T., and Ibata, Y., 1992, Learning behaviour in chronic vitamin E-deficient and -supplemented rats: radial arm maze learning and passive avoidance response. *Behav. Brain Res.*, **51**: 157-164.

Jialal, I., Devaraj, S., and Kaul, N., 2001, The effect of alpha-tocopherol on monocyte proatherogenic activity. *J. Nutr.*, **131**: 389S-394S.

Kalaria, R.N., 1997, Cerebrovascular degeneration is related to amyloid-beta protein deposition in Alzheimer's disease. *Ann. N.Y.Acad. Sci.* **826**: 263-271.

Kaltschmidt, B., Uherek, M., Volk, B., Baeuerle, P.A., and Kaltschmidt, C., 1997, Transcription factor NF-kappaB is activated in primary neurons by amyloid beta peptides and in neurons surrounding early plaques from patients with Alzheimer disease. *Proc. Natl. Acad. Sci. USA.* **94**: 2642-2647.

Kawas, C., Resnick, S., Morrison, A., Brookmeyer, R., Corrada, M., Zonderman, A., Bacal, C., Lingle, D.D., and Metter, E., 1997, A prospective study of estrogen replacement therapy and the risk of developing Alzheimer's disease: the Baltimore Longitudinal Study of Aging. *Neurology.* **48**: 1517-1521.

Kelly, J.F., Furukawa, K., Barger, S.W., Rengen, M.R., Mark, R.J., Blanc, E.M., Roth G.S., and Mattson, M.P., 1996, Amyloid beta-peptide disrupts carbachol-induced muscarinic cholinergic signal transduction in cortical neurons. *Proc. Natl. Acad. Sci. USA.* **93**: 6753-6758.

Kong, A.N., Yu, R., Lei, W., Mandlekar, S., Tan, T.H., and Ucker, D.S., 1998, Differential activation of MAPK and ICE/Ced-3 protease in chemical-induced apoptosis. The role of oxidative stress in the regulation of mitogen-activated protein kinases (MAPKs) leading to gene expression and survival or activation of caspases leading to apoptosis. *Restor. Neurol. Neurosci.* **12**: 63-70.

Koppal, T., Subramaniam, R., Drake, J., Prasad, M.R., Dhillon, H., and Butterfield, D.A., 1998, Vitamin E protects against Alzheimer's amyloid peptide (25-35)-induced changes in neocortical synaptosomal membrane lipid structure and composition. *Brain Res.* **786**: 270-273.

Lee, I.K., Koya, D., Ishi, H., Kanoh, H., and King, G.L., 1999, d-Alpha-tocopherol prevents the hyperglycemia induced activation of diacylglycerol (DAG)-protein kinase C (PKC) pathway in vascular smooth muscle cell by an increase of DAG kinase activity. *Diabetes Res. Clin. Pract.* **45**: 183-190.

Levy, E., Carman, M.D., Fernandez-Madrid, I.J., Power, M.D., Lieberburg, I., Van Duinen, S.G., Bots, G.T., Luyendijk, W., and Frangione, B., 1990, Mutation of the Alzheimer's disease amyloid gene in hereditary cerebral hemorrhage, Dutch type. *Science* **248**: 1124-1126.

Li-Weber, M., Weigand, M.A., Giaisi, M., Suss, D., Treiber, M.K., Baumann, S., Ritsou, E., Breitkreutz, R., and Krammer, P.H., 2002, Vitamin E inhibits CD95 ligand expression and protects T cells from activation-induced cell death. *J. Clin. Invest.* **110**: 681-690.

Lo, Y.Y.C., Wong, J.M.S., and Cruz, T.F., 1996, Reactive oxygen species mediate cytokine activation of c-jun NH2-terminal kinases,*J. Biol. Chem.* **271**: 15703-15707.

Lockhart, B.P., Benicourt, C., Junien J.L., and Privat, A., 1994, Inhibitors of free radical formation fail to attenuate direct beta-amyloid25-35 peptide-mediated neurotoxicity in rat hippocampal cultures. *J. Neurosci. Res.* **39**: 494-505.

Lovell, M.A., Robertson, J.D., Teesdale, W.J., Campbell, J.L., and Markesbery, W.R., 1998, Copper, iron and zinc in Alzheimer's disease senile plaques. *J. Neurol. Sci.* **158**: 47-52.

Maat-Schieman, M.L.C., Radder, C.M., van Duinen, S.G., Haan, J., and Roos, R.A.C., 1994, Hereditary cerebral haemorrhage with amyloidosis (Dutch): a model for congophilic plaque formation without neurofibrillary pathology. *Acta Neuropathol.*, **88**: 371-378.

Maat-Schieman, M.L.C., van Duinen, S.G., Haan, J., and Roos, R.A.C., 1992, Morphology of cerebral plaque-like lesions in hereditary cerebral haemorrhage with amyloidosis (Dutch). *Acta Neuropathol.* **84**: 674-679.

Mark, R.J., Hensley, K., Butterfield, D.A., and Mattson, M.P., 1995, Amyloid beta-peptide impairs ion-motive ATPase activities: evidence for a role in loss of neuronal Ca2+ homeostasis and cell death. *J. Neurosci.* **15**: 6239-6249.

Mark, R.J., Pang, Z., Geddes, J.W., Uchida, K., and Mattson, M.P., 1997, Amyloid beta-peptide impairs glucose transport in hippocampal and cortical neurons: involvement of membrane lipid peroxidation. *J Neurosci.* **17**: 1046-1054.

Martin-Nizard, F., Boullier, A., Fruchart, J.C., and Duriez, P., 1998, Alpha-tocopherol but not beta-tocopherol inhibits thrombin-induced PKC activation and endothelin secretion in endothelial cells. *J. Cardiovasc. Risk.* **5**: 339-345.

Mattson, M.P., 1997, Cellular actions of beta-amyloid precursor protein and its soluble and fibrillogenic derivatives. *Physiol. Rev.* **77**: 1081-1132.

Mattson, M.P., and Goodman, Y., 1995, Different amyloidogenic peptides share a similar mechanism of neurotoxicity involving reactive oxygen species and calcium. *Brain Res.* **676**: 219-224.

Mazur-Kolecka, B., Frackowiak, J., and Wisniewski, H.M., 1995, Apolipoproteins E3 and E4 induce, and transthyretin prevents accumulation of the Alzheimer's beta-amyloid peptide in cultured vascular smooth muscle cells. *Brain Res.* **698**: 217-222.

Mesulam, M.-M., Carson, K., Price, B., and Geula, C., 1992, Cholinesterases in the amyloid angiopathy of Alzheimer's disease. *Ann. Neurol.* **31**: 565-569.

Miranda, S., Opazo, C., Larrondo, L.F., Muñoz, F.J., Ruiz, F., Leighton, F., Inestrosa, N.C., 2000, The role of oxidative stress in the toxicity induced by amyloid beta-peptide in Alzheimer's disease. *Prog. Neurobiol.* **62**: 633-648.

Miyakawa, T., Katsuragi, S., Higuchi, Y., Yamashita, K., Kimura, T., Teraoka, K., Ono, T., and Ishizuka, K., 1997, Changes of microvessels in the brain with Alzheimer's disease. *Ann. N. Y. Acad. Sci.* **826**: 428-432.

Mok, S.S., Turner, B.J., Beyreuther, K., Masters, C.L., Barrow, C.J., and Small, D.H., 2002, Toxicity of substrate-bound amyloid peptides on vascular smooth muscle cells is enhanced by homocysteine. *Eur. J. Biochem.* **269**: 3014-3022

Morris, M.C., Beckett, L.A., Scherr, P.A., Hebert, L.E., Bennett, D.A., Field, T.S., and Evans, D.A., 1998, Vitamin E and vitamin C supplement use and risk of incident Alzheimer disease. *Alzheimer Dis. Assoc. Disord.* **12**: 121-126.

Muñoz, F.J., Opazo, C., Gil-Gómez, G., Tapia, G., Fernández, V., Valverde, M.A., Inestrosa, N.C., 2002, Vitamin E but not 17beta-estradiol protects against vascular toxicity induced by beta-amyloid wild type and the Dutch amyloid variant. *J. Neurosci.* **22**: 3081-3089.

Mazur-Kolecka, B., Kowal, D., Sukontasup, T., Dickson, D., and Frackowiak, J., 2004, The effect of oxidative stress on amyloid precursor protein processing in cells engaged in ß-amyloidosis is related to apolipoprotein E genotype. *Acta Neuropathol. (Berl.)* In press.

Naiki, H., Hasegawa, K., Yamaguchi, I., Nakamura, H., Gejyo, F., and Nakakuki, K., 1998, Apolipoprotein E and antioxidants have different mechanisms of inhibiting Alzheimer's beta-amyloid fibril formation in vitro. *Biochemistry* **37**: 17882-17889.

Nunomura, A., Perry, G., Hirai, K., Aliev, G., Takeda, A., Chiba, S., and Smith, M.A., 1999, Neuronal RNA oxidation in Alzheimer's disease and Down's syndrome. *Ann. N. Y. Acad. Sci.* **893**: 362-364.

Okazawa, H., and Estus, S., 2002, The JNK/c-Jun cascade and Alzheimer's disease. *Am. J. Alzheimers Dis. Other Demen.* **17**: 79-88.

Pappolla, M.A., Sos, M., Omar, R.A., Bick, R.J., Hickson-Bick, D.L., Reiter, R.J., Efthimiopoulos, S., and Robakis, N.K., 1997, Melatonin prevents death of neuroblastoma cells exposed to the Alzheimer amyloid peptide. *J. Neurosci.* **17**: 1683-1690.

Perna, A.F., Ingrosso, D., and De Santo, N.G., 2003, Homocysteine and oxidative stress, *Amino Acids,* **25**: 409-417.

Perly, B., Smith, I.C., Hughes, L., Burton, G.W., and Ingold, K.U., 1985, Estimation of the location of natural alpha-tocopherol in lipid bilayers by 13C-NMR spectroscopy. *Biochim. Biophys. Acta.* **819**:131-135.

Piette, J., Piret, B., Bonizzi, G., Schoonbroodt, S., Merville, M.P., Legrand-Poels, S., and Bours, V., 1997, Multiple redox regulation in NF-kappa beta transcription factor activation. *J. Biol. Chem.* **378**: 1237-1245.

Pike, C.J., Ramezan-Arab, N., and Cotman, C.W., 1997, Beta-amyloid neurotoxicity in vitro: evidence of oxidative stress but not protection by antioxidants. *J. Neurochem.* **69**: 1601-1611.

Podmore, I.D., Griffiths, H.R., Herbert, K.E., Mistry, N., Mistry, P., and Lunec, J., 1998, Vitamin C exhibits pro-oxidant properties. *Nature* **392**: 559.

Pratico, D., Tangirala, R.K., Rader, D.J., Rokach, J., and Fitzgerald, G.A., 1998, Vitamin E suppresses isoprostane generation in vivo and reduces atherosclerosis in ApoE-deficient mice. *Nat. Med.* **4**: 1189-1192.

Qin, F., Shite, J., and Liang, C., 2003, Antioxidants attenuate myocyte apoptosis and improve cardiac function in CHF: association with changes in MAPK pathways. *Am. J. Physiol. Heart Circ. Physiol.* **285**: 822-832.

Ricciarelli, R., Tasinato, A., Clement, S., Ozer, N.K., Boscoboinik, D., and Azzi, A., 1998, alpha-Tocopherol specifically inactivates cellular protein kinase C alpha by changing its phosphorylation state. *Biochem. J.* **334**: 243-249.

Robbesyn, F., Garcia, V., Auge, N., Vieira, O., Frisach, M.F., Salvayre, R., and Negre-Salvayre, A., 2003, HDL counterbalance the proinflammatory effect of oxidized LDL by inhibiting intracellular reactive oxygen species rise, proteasome activation, and subsequent NF-kappaB activation in smooth muscle cells. *FASEB J.* **17**: 743-745.

Sano, M., Ernesto, C., Thomas, R.G., Klauber, M.R., Schafer, K., Grundman, M., Woodbury, P., and Growdon, J., 1997, A controlled trial of selegiline, alpha-tocopherol, or both as treatment for Alzheimer's disease. The Alzheimer's Disease Cooperative Study. *N. Engl. J. Med.* **336**: 1216-1222.

Savage, M.J., Lin, Y.G., Ciallella, J.R., Flood, D.G., and Scott, R.W., 2002, Activation of c-Jun N-terminal kinase and p38 in an Alzheimer's disease model is associated with amyloid deposition. *J. Neurosci.* **22**: 3376-3385.

Schmechel, D.E., Goldgaber, D., Burkhart, D.S., Gilbert, J.R., Gadjuseck, D.C., and Roses, A.D., 1988, Cellular localization of messenger RNA encoding amyloid-beta-protein in normal tissue and in Alzheimer disease. *Alz. Dis. Assoc. Dis.* **2**: 96-111.

Seshadri, S., Beiser, A., Selhub, J., Jacques, P.F., Rosenberg, I.H., D'Agostino, R.B., Wilson, P.W., and Wolf, P.A., 2002, Plasma homocysteine as a risk factor for dementia and Alzheimer's disease. *N. Engl. J. Med.* **346**: 476-483.

Shoji, M., Hirai, S., Harigaya, Y., Kawarabayashi, T., and Yamaguchi, H., 1990, The amyloid beta-protein precursor is localized in smooth muscle cells of leptomeningeal vessels. *Brain Res.* **530**: 113-116.

Smith, M.A., Richey Harris, P.L., Sayre, L.M., Beckman, J.S., and Perry, G., 1997, Widespread peroxynitrite-mediated damage in Alzheimer's disease. *J. Neurosci.* **17**: 2653-2657.

Snow, A.D., Mar, H., Nochlin, D., Kimata, K., Kato, M., Suzuki, S., Hassell, J., and Wight, T.N., 1988, The presence of heparan sulfate proteoglycans in the neuritic plaques and congophilic angiopathy in Alzheimer's disease,.*Am. J. Pathol.* **133**: 456-463.

Stadtman, E.R., 1990, Metal ion-catalyzed oxidation of proteins: biochemical mechanism and biological consequences. *Free Radic. Biol. Med.* **9**: 315-325.

Steinberg, D., 1995, Clinical trials of antioxidants in atherosclerosis: are we doing the right thing? *Lancet* **346**: 36-38.

Suo, Z., Su, G., Placzek, A., Kundtz, A., Humphrey, J., Crawford, F., and Mullan, M., 2000, A beta vasoactivity in vivo. *Ann. N. Y. Acad. Sci.* **903**: 156-163.

Tagliavini, F., Ghiso, J., Timmers, W.F., Giaccone, G., Bugiani, O., and Frangione, B., 1990, Coexistence of Alzheimer's amyloid precursor protein and amyloid protein in cerebral vessel walls. *Lab. Invest.* **62**: 761-767.

Tamagno, E., Bardini, P., Obbili, A., Vitali, A., Borghi, R., Zaccheo, D., Pronzato, M.A., Danni, O., Smith, M.A., Perry, G., and Tabaton, M., 2002, Oxidative stress increases expression and activity of BACE in NT2 neurons. *Neurobiol. Dis.* **10**: 279-288.

Tang, M.X., Jacobs, D., Stern, Y., Marder, K., Schofield, P., Gurland, B., Andrews, H., and Mayeux, R., 1996, Effect of oestrogen during menopause on risk and age at onset of Alzheimer's disease. *Lancet* **348**: 429-432.

Thomas, T., McLendon, C., Sutton, E.T., and Thomas, G., 1997, Beta-Amyloid-induced cerebrovascular endothelial dysfunction. *Ann. N. Y. Acad. Sci.* **826**: 447-451.

Thomas, T., Sutton, E.T., Hellermann, A., and Price, J.M., 1997, Beta-amyloid-induced coronary artery vasoactivity and endothelial damage. *J. Cardiovasc. Pharmacol.* **30**: 517-522.

Timmers, W.F., Tagliavini, F., Haan, J., and Frangione, B., 1990, Parenchymal preamyloid and amyloid deposits in the brains of patients with hereditary cerebral hemorrhage with amyloidosis-Dutch type. *Neurosci. Lett.*, **118**: 223-226.

Troy, C.M., Rabacchi, S.A., Xu, Z., Maroney, A.C., Connors, T.J., Shelanski, M,L., and Greene, L.A., 2001, Beta-Amyloid-induced neuronal apoptosis requires c-Jun N-terminal kinase activation, *J. Neurochem.* **77**:157-164.

Turner, N.A., Xia, F., Azhar, G., Zhang, X., Liu, L., and Wei, J., 1998, Oxidative stress induces DNA fragmentation and caspase activation via the c-Jun NH2-terminal kinase pathway in H9c2 cardiac muscle cells. *J.Mol.Cell Cardiol.* **30**: 1789-1801.

Uchihara, T., Akiyama, H., Kondo, H., and Ikeda, K., 1997, Activated microglial cells are colocalized with perivascular deposits of amyloid-beta protein in Alzheimer's disease brain. *Stroke* **28**: 1948-1950.

Uemura, M., Manabe, H., Yoshida, N., Fujita, N., Ochiai, J., Matsumoto, N., Takagi, T., Naito, Y., and Yoshikawa, T., 2002, Alpha-tocopherol prevents apoptosis of vascular endothelial cells via a mechanism exceeding that of mere antioxidation. *Eur. J. Pharmacol.* **456**: 29-37.

Van Dorpe, J., Smeijers, L., Dewachter, I., Nuyens, D., Spittaels, K., Van Den Haute, C., Mercken, M., Moechars, D., Laenen, I., Kuiperi, C., Bruynseels, K., Tesseur, I., Loos, R., Vanderstichele, H., Checler, F., Sciot, R., Van Leuven, F., 2000, Prominent cerebral amyloid angiopathy in transgenic mice overexpressing the london mutant of human APP in neurons. *Am. J. Pathol.* **157**: 1283-1298.

Van Duinen, S.G., Maat-Schieman, M.L., Bruijn, J.A., Haan, J., and Roos, R.A., 1995, Cortical tissue of patients with hereditary cerebral hemorrhage with amyloidosis (Dutch) contains various extracellular matrix deposits. Lab. Invest. **73**: 183-189.

Vatassery, G.T., 1992, Vitamin E. Neurochemistry and implications for neurodegeneration in Parkinson's disease. *Ann. N. Y. Acad. Sci.* **669**: 97-109;

Vehmas, A.K., Kawas, C.H., Stewart, W.F., and Troncoso, J.C., 2003, Immune reactive cells in senile plaques and cognitive decline in Alzheimer's disease. *Neurobiol. Aging* **24**: 321-331.

Verbeek, M.M., Otte-Holler, I., Veerhuis, R., Ruiter, D.J., and De Waal, R.M., 1998, Distribution of A-beta-associated proteins in cerebrovascular amyloid of Alzheimer's disease. *Acta Neuropathol. (Berl)* **96**: 628-636.

Villacorta, L., Graca-Souza, A.V., Ricciarelli, R., Sing., J.M., and Azzi, A., 2003, Alpha-tocopherol induces expression of connective tissue growth factor and antagonizes tumor necrosis factor-alpha-mediated downregulation in human smooth muscle cells. *Circ. Res.* **92**: 104-110.

Vinters, H.V., Pardrigde, W.M., Secor, D.L., and Ishii, N., 1988, Immunohistochemical study of cerebral amyloid angiopathy. *Am. J. Pathol.* **133**: 150-162.

Vollgraf, U., Wegner, M., and Richter-Landsberg, C., 1999 Activation of AP-1 and Nuclear Factor-kb transcription factors is involved in hydrogen peroxide-induced apoptotic cell death of oligodendrocytes, *J. Neurochem.*, **73**: 2501-2509.

Wang, S., Kotamraju, S., Konorev, E., Kalivendi, S., Joseph, J., and Kalyanaraman, B., 2002, Activation of nuclear factor-kappaB during doxorubicin-induced apoptosis in endothelial cells and myocytes is pro-apoptotic: the role of hydrogen peroxide, *Biochem. J.*, **367**: 729-740.

Weller, R.O., Massey, A., Newman, T.A., Hutchings, M., Kuo, Y.M., and Roher, A.E., 1998, Amyloid β accumulates in putative interstitial fluid drainage pathways in Alzheimer's disease, *Am. J. Pathol.*, **153**: 725-733.

Wisniewski, T., Ghiso, J., and Frangione, B., 1991, Peptides homologous to the amyloid protein of Alzheimer's disease containing a glutamine for glutamic acid substitution have accelerated amyloid fibril formation, *Biochem. Biophys. Res. Commun.*, **179**: 1247-54.

Wisniewski, H.M., and Wegiel, J., 1994, Beta-amyloid formation by myocytes of leptomeningeal vessels, *Acta Neuropathol. (Berl)*, **87**: 233-241.

Wisniewski, H.M., Wegiel, J., Vorbrodt, A.W., Mazur-Kolecka, B., and Frackowiak, J., 2000, Role of perivascular cells and myocytes in vascular amyloidosis, *Ann.N.Y.Acad. Sci.*, **903**: 6-18.

Wisniewski, H.M., Wegiel, J., Wang, K.C., and Lach, B., 1992, Ultrastructural studies of the cells forming amyloid in the cortical vessel wall in Alzheimer's disease, *Acta Neuropathol. (Berl)*, **84**: 117-127.

Yallampalli, S., Micci, M.A., and Taglialatela, G., 1998, Ascorbic acid prevents beta-amyloid-induced intracellular calcium increase and cell death in PC12 cells. *Neurosci. Lett.*, **251**:105-108.

Yamada, K., Tanaka, T., Han, D., Senzaki, K., Kameyama, T., and Nabeshima, T., 1999, Protective effects of idebenone and alpha-tocopherol on beta-amyloid-(1-42)-induced learning and memory deficits in rats: implication of oxidative stress in beta-amyloid-induced neurotoxicity in vivo, *Eur. J. Neurosci.*, **11**: 83-90.

Yin, K.J., Lee, J.M., Chen, S.D., Xu, J., and Hsu, C.Y., 2002, Amyloid-beta induces Smac release via AP-1/Bim activation in cerebral endothelial cells, *J. Neurosci.*, **22**: 9764-9770.

Zhang, W.W., Lempessi, H., and Olsson, Y., 1998, Amyloid angiopathy of the human brain: immunohistochemical studies using markers for components of extracellular matrix, smooth muscle actin and endothelial cells, *Acta Neuropathol. (Berl)*, **96**: 558-563.

Zhu, X., Lee, H.G., Raina, A.K., Perry, G., and Smith, M.A., 2002, The role of mitogen-activated protein kinase pathways in Alzheimer's disease, *Neurosignals*, **11**: 270-281.

Chapter 8

Amyloid Accumulation and Pathogensis of Alzheimer's Disease: Significance of Monomeric, Oligomeric and Fibrillar Aβ

Charles C. Glabe
University of California, Irvine. Irvine, CA 92697, USA

Abstract: This chapter reviews recent findings that indicate that soluble amyloid oligomers may represent the primary pathological species in Alzheimer's and other degenerative diseases. Various amyloids share a number of common properties, including their structures and pathways for fibril formation and accumulation in disease. Recent findings suggest that toxic amyloid oligomers share a common structure, suggesting that they also share a common mechanism of pathogenesis

Key words: Amyloid, Aβ, oligomer, fibril, pathogenic mechanisms

1. INTRODUCTION

The accumulation of amyloids is increasingly recognized as a key feature of the pathogenesis of a variety of diseases, including Alzheimer's, Parkinson's, Huntington's diseases and type II diabetes. The amyloid deposits present in these diseases characteristically consists of a particular protein or proteolytic fragment of protein, although it has been increasingly recognized that more than one type of amyloid may accumulate in a disease. Regardless of the protein sequence, the amyloid fibrils have a characteristic "crossed β-sheet" structure and share a common aggregation pathway involving the misfolding or partial denaturation of the protein, adoption of β-sheet structure and oligomerization to form soluble aggregation intermediates on the pathway to fibril formation. Amyloids are deleterious to cells *in vitro* and *in vivo*, suggesting that they play a causal role in pathogenesis. Recent evidence suggests that amyloid oligomers and fibrils have a common structure, suggesting that they all may share a common primary mechanism of pathogenesis.

2. KEY EVENT: PROTEIN MISFOLDING

Amyloidoses are fundamentally diseases of protein misfolding. The propensity to misfold may be enhanced by mutation, environmental factors or posttranslational modification and some proteins are "natively misfolded" or conformationally mutable. Cells contain elaborate machinery, like chaperones to ensure that proteins fold properly and selective degradation systems, such as the proteasome to dispose of misfolded proteins. What distinguishes amyloids from other misfolded proteins appears to be that the ability to aggregate into higher order oligomeric and fibrillar structures allows them to evade these quality control systems. Inhibition of the proteasome (Waelter *et al.*, 2001) or lysosomal degradation systems (Bendiske and Bahr, 2003; Hajimohammadreza *et al.*, 1994) can lead to the accumulation of misfolded amyloidogenic proteins and peptides. Amyloid aggregates also inhibit the proteasome, perhaps by adopting stably folded structures that cannot be unfolded to allow transport into the central catalytic pore of the enzyme complex (Gregori *et al.*, 1995; Keck *et al.*, 2003). The adoption of a stable aggregated β-sheet structure my be a critical initiating or nucleating event in the formation of amyloid seeds that can grow and propagate because the amyloid structure is relatively resistant to degradation *in vivo*. The formation of these stable nuclei is dependent on the concentration of aggregation competent misfolded proteins and this may explain why a small increase in the concentration of misfolded proteins can dramatically accelerate onset of disease.

3. COMMON STRUCTURES

Many proteins assemble into amyloid fibrils, even ones that are not disease related. Amyloid fibrils are approximately 6-10 nm diameter unbranched fibrils characterized by a "crossed β-sheet" structure in which the polypeptide chain is oriented perpendicular to the fiber axis with the hydrogen bonding parallel to the axis. The ability to form amyloids is not limited to disease-related proteins (Dobson, 1999). Rather, it appears to be a fundamental polymer motif of the polypeptide backbone, although it is not commonly observed in native proteins. The observation that a number of different polyamino acids form crossed β-sheet amyloid fibers indicates that specific sequences or side chain interactions is not required for amyloid formation and suggests that the fundamental tendency to form amyloids is overcome by the propensity of specific sequences to adopt other stable structures (Fandrich and Dobson, 2002). Amyloid fibrils have a number of common properties that may be a reflection of their underlying common structure. They display characteristic tinctorial properties, binding Congo red and thioflavinee dyes (LeVine, 1993). A wide variety of different amyloid fibrils have also been reported to be recognized by a conformation dependent monoclonal antibody, independently of their sequence. More recent structural analysis suggests that they have a common parallel, in-register organization that may be characteristic of their structure (Balbach *et al.*, 2002; Benzinger *et al.*, 1998; Der-Sarkissian *et al.*, 2003; Torok *et al.*, 2002). The β sheets of Aβ40 and Aβ42 amyloid fibrils are formed from parallel β strands (Antzutkin *et al.*, 2002), while fibrils formed from smaller peptide segments, such as Aβ11-25 are antiparallel (Petkova *et al.*, 2004). Although the spacing is different in the structures formed from smaller fragments, the side chains are still in register along the strands that are facing in the same direction.

4. COMMON PATHWAYS OF AGGREGATION

Amyloid β aggregation/oligomerization pathways were among the earliest characterized because the ability to chemically synthesize large amounts of the naturally occurring peptide provided ample biochemical material for study. These early studies established that peptide concentration, low pH, time of incubation and the length of the carboxyl terminus are important parameters in determining the rate of fibril assembly (Burdick *et al.*, 1992). Some of this early work reported conflicting results and even variable results between different lots of peptide or peptide preparation, but more recent analysis indicates that much of this variation can be traced to differences in the solvents used and conformationally heterogeneous samples that retain a considerable amount of "conformational history" after purification and lyophilization (Shen and Murphy, 1995; Zagorski *et al.*, 1999). This variation can be largely eliminated by starting with stock solutions of fully solvated random coil peptide, such as peptides dissolved in hexafluoroisopropanol (Dahlgren *et al.*, 2002; Kayed *et al.*, 2003; Zagorski *et al.*, 1999).

In addition to fibrils and monomer, quasi-stable intermediate aggregates ranging from dimers up to particles of a million Da or greater have been observed by a variety of techniques, including SDS gel electrophoresis

(Burdick *et al.*, 1992), gel filtration (Hilbich *et al.*, 1991; Soreghan *et al.*, 1994), fluorescence resonance energy transfer (Garzon-Rodriguez *et al.*, 1997), light scattering (Walsh *et al.*, 1997), and atomic force microscopy (Harper *et al.*, 1997b). In particular, soluble spherical aggregates of 2-20 nm have been observed for many different types of amyloids by electron and atomic force microscopy (Anguiano *et al.*, 2002; Harper *et al.*, 1997b; Lashuel *et al.*, 2002). These spherical oligomers appear at intermediate times of incubation and disappear as fibrils appear, suggesting that they represent intermediates in the pathway of fibril formation. Although these various aggregates have been observed by many different laboratories and under different conditions, the relationships of these intermediates, their roles if fibril growth and elongation and relationship to disease pathogenesis are only now being clarified.

One particular hypothetical pathway for Aβ fibril assembly is shown in Figure 1. The earliest event in the process of aggregation is the formation of dimers (Hilbich *et al.*, 1991; Soreghan *et al.*, 1994). Upon further incubation increasing amounts of higher order aggregates form as the amount of dimer decreases (Hartley *et al.*, 1999). These high molecular weight aggregates have apparent sizes ranging from approximately (Lomakin *et al.*, 1997; Soreghan *et al.*, 1994) 10^5 to 10^6 Da with an average size corresponding to approximately 24 monomers. The high molecular weight Aβ oligomers appear as spherical particles of approximately 3 nm in diameter (Anguiano *et al.*, 2002; Harper *et al.*, 1997b; Lashuel *et al.*, 2002) and have the characteristics of a protein micelle (Soreghan *et al.*, 1994). Aβ is amphipathic and lowers the surface tension of water. The formation of the high molecular weight aggregates displays a critical concentration of approximately 25 μM and the formation of the high MW aggregates is correlated with the formation of a new hydrophobic environment into which hydrophobic fluorescent dyes partition (LeVine, 2002; Soreghan *et al.*, 1994). At longer times of incubation, the spherical oligomers appear to co-aggregate to form curvilear fibrils with a characteristic beaded appearance that have been termed "protofibrils" (Harper *et al.*, 1997a; Hartley *et al.*, 1999). The spherical oligomers and protofibrils appear to be intermediates in the pathway of fiber formation because they disappear at mature fibrils accumulate and the rate of monomer dissociation from them is too slow to account for fibril growth (Kayed *et al.*, 2003). The transition from spherical oligomers and protofibrils to mature fibrils appears to involve a major conformational change, because fluorescence quenching analysis indicates that the carboxyl terminus is highly shielded from the aqueous solvent in the soluble oligomeric state, whereas it is exposed to the solvent in the fibrillar state (Garzon-Rodriguez *et al.*, 2000). Once the amyloid fibril lattice has been established, it can grow by the addition of monomer onto the ends of the fibrils (Esler *et al.*, 2000; Tseng *et al.*, 1999).

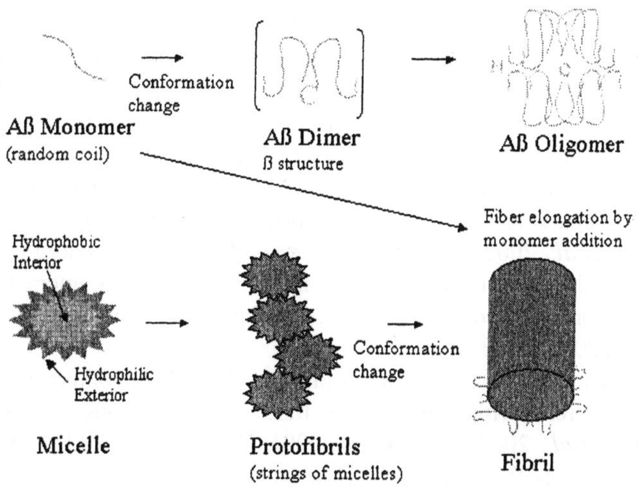

Figure 1. Pathway of Aβ aggregation and fibril formation. Top row: Aβ is derived from a segment of APP that is believed to be predominantly alpha helical in its native state. Monomeric Aβ is largely random coil in solution. The formation of Aβ dimers is coincident with the adoption of partial β structure. The Aβ dimers assemble into higher order aggregates or oligomers. Bottom row: The higher order Aβ aggregates are protein micelles with a hydrophobic core and a polar exterior that appear as 3 nm spherical particles in the atomic force or electron microscope. At later times, these spherical particles appear to form curvilinear strings or "protofibrils". The protofibrils undergo a conformation change to form the straight, unbranched mature fibrils. Once the amyloid fibril lattice has been established, the fibril can grow by the addition of Aβ monomer or dimer at the ends of the fibril

5. COMMON MECHANISMS

The concept that amyloid oligomers rather than mature fibrils represent the primary toxic species arose as a consequence of the inadequacy of fibril accumulation to account for pathogenesis. Terry and coworker have long noted that there is a poor correlation between amyloid deposits and dementia or disease progression (Terry, 1996). Non-demented control individuals occasionally have high levels of AD pathology as determined by the presence of amyloid plaques. Moreover, there are reports of individuals bearing FAD PS mutations that had massive amounts of "cotton wool" amyloid deposits and these individuals were not demented (Houlden *et al.*, 2000; Verkkoniemi *et al.*, 2000). Alzheimer's disease brain also contains soluble Aβ oligomers that are typically defined as the fraction of peptide not sedimenting at 100,000 x g or more (Gong *et al.*, 2003; Kuo *et al.*, 1996; Pitschke *et al.*, 1998). This soluble population of Aβ correlates better with dementia than the fibrillar Aβ (Lue *et al.*, 1999; McLean *et al.*, 1999). In some transgenic mouse models of AD, neurological deficits precede the deposition of significant amounts of Aβ, suggesting that the onset of pathology occurs prior to amyloid fibril deposition (Hsia *et al.*, 1999; Westerman *et al.*, 2002). Because of these problems, the amyloid hypothesis has been modified to include the possibility that soluble oligomers of Aβ and not the fibrillar amyloid deposits represent the primary toxic species (Hardy and Selkoe, 2002; Klein *et al.*, 2001).

Numerous *in vitro* studies have indicated that Aβ oligomers are toxic to cells or interfere with the normal function of neurons (reviewed in: (Kirkitadze *et al.*, 2002; Klein *et al.*, 2001; Walsh *et al.*, 2002). Mounting evidence indicates that soluble amyloid oligomers are generally toxic for a wide variety of disease related amyloids (Caughey and Lansbury, 2003) and indeed soluble oligomers are intrinsically toxic even when they are formed by proteins that are not disease related (Bucciantini *et al.*, 2002). Although the idea that soluble Aβ oligomers may play a primary role in AD pathogens is attractive, the evidence in support of this is largely correlative and derived from *in vitro* toxicity studies. In order to evaluate the role of soluble Aβ oligomers in diseased human brain, we would need some way of distinguishing them from soluble Aβ monomer, APP and Aβ fibrils. All of these species contain the same amino acid sequence, but differ in conformation, so a conformation-dependent antibody that distinguishes between the different forms would be very useful. We produced a polyclonal antibody that is specific for soluble Aβ oligomers and does not recognize Aβ monomer, APP or Aβ fibrils (Kayed *et al.*, 2003). The experimental approach was to immunize rabbits with a covalently constrained mimic of Aβ in same the conformation as it exists in soluble spherical oligomers. The Aβ peptide in the molecular mimic antigen is covalently coupled to the surface of colloidal gold particles *via* its carboxyl terminus to mimic the overall organization that exists in soluble oligomers (Figure 1). The covalent attachment of Aβ prevents the dissociation of Aβ to monomers and the conformation change that is required for the transition to mature fibrils, locking Aβ in the oligomeric conformation. The resulting immune serum is remarkably specific for spherical oligomers and protofibrils as determined by dot blot and ELISA assays. Even after immunization more than 12 times over a period of 9 months, no reactivity against Aβ monomer or fibrils was detected. The oligomer-specific epitope in Aβ 42 appears after approximately 6 hr of incubation, reaches a maximum at approximately 24-48 hours and disappears coincident with the formation of mature fibrils. The smallest size oligomer recognized by the antibody has the approximate size of an octamer (Kayed *et al.*, 2003).

Surprisingly the structure of soluble oligomers from different types of amyloidogenic proteins is much more similar than their obvious morphological similarity suggests. The antibody specific for Aβ oligomers also recognizes soluble oligomers from a wide variety of amyloid forming peptides and proteins equally well (Kayed *et al.*, 2003). As for Aβ, the monomeric and fibrillar forms of these amyloids are not recognized by the antibody. This indicates that the antibody recognizes an epitope that is independent of the amino acid sequence and is shared in common among all types of amyloids examined. The anti-oligomer antibody also blocks the toxicity of all types of soluble oligomers *in vitro*. Since the oligomers share a common structure and they are all intrinsically toxic to cells, this suggests that mechanism of toxicity is also shared in common among the various amyloid-related degenerative diseases.

The availability of an antibody that can distinguish the conformations of amyloids in the soluble oligomeric state from the native conformation, random coil monomer and fibrillar states facilitates an examination of the localization of oligomeric amyloid in AD brain. The anti-oligomer antibody detects a unique subset of amyloid deposits in AD brain and does not co-localize with thioflavine-S positive fibrillar amyloid plaques (Kayed *et al.*, 2003). The staining is predominantly extracellular and does not appear to be preferentially localized with any cell type. Occasionally, anti-oligomer immunoreactivity is observed in the vicinity of large diffuse amyloid deposits stained with the Aβ 10-17 specific antibody, 6E10, but the staining is distinct and does not overlap. Diffuse amyloid deposits are generally regarded as non-fibrillar because of the lack of thioflavine-S staining and the absence of fibrils upon ultrastructural examination. The fact that the majority of these diffuse Aβ deposits are not labeled by anti-oligomer antibody indicates that they are conformationally distinct from both the thioflavine-S positive fibrillar neuritic and cored plaques and the soluble oligomer deposits identified by the anti-oligomer antibody. Because soluble oligomers appear to be an intermediate on the pathway to fibril formation *in vitro*, this suggests that the anti-oligomer immunoreactive deposits may represent an early stage of amyloid deposition and perhaps the nidus for the formation of fibrillar amyloid deposits. The relationship between the diffuse plagues and oligomeric deposits remains a mystery.

An important issue that remains to be addressed is whether the oligomeric amyloid deposits precede the appearance of fibrillar and diffuse and fibrillar amyloid deposits and whether they are intimately associated with imitation of dementia and cognitive decline. In an initial study of soluble PBS lysates of frontal cortex from 27 AD patients, 2 patients with mild Braak changes (MBC) and 5 age matched controls, indicated that soluble oligomers are greatly reduced or absent in control brains, while they are detected in both MBC samples and the majority, but not all of the AD brain samples. The absence of detectable soluble oligomers in some of the AD cases is consistent with the idea that soluble oligomers represent a transient intermediate in the formation of fibrillar deposits and that they are not continuously produced during the progression of disease. It remains to be established whether their presence is a good predictor of disease onset or progression.

If soluble oligomers have a common mechanism of toxicity, it predicts that they would act on the same primary target. This restricts the number of potential targets to ones that are plausible for all of the different types of soluble amyloid oligomers. Since some amyloids are cytosolic while others are lumenal or extracellular, this suggests that the primary target of oligomers must be accessible to both compartments. The most obvious target that meets this criterion is the plasma membrane that separates the two compartments. Amyloid oligomers, such as Aβ, alpha synuclein and IAPP have been widely reported to disrupt membranes or to form pores or channels in membranes (Caughey and Lansbury, 2003; Kagan *et al.*, 2002), *see also* Chapter 1. The disruption of intracellular Ca homeostasis that has been reported as a common component of amyloid toxicity could be directly related to membrane perturbations that increase membrane permeability (Mattson, 2002). A wide variety of transmembrane signalling processes and the production of reactive oxygen species could also be directly related to membrane perturbation by soluble oligomers. Even though soluble oligomers may not be acutely toxic *in vivo* as they are *in vitro*, the chronic leakage of ions across the plasma membrane may be sufficient to disrupt normal neuronal function and serve as a source of chronic stress in maintaining a normal membrane potential.

REFERENCES

Anguiano, M., Nowak R.J., and Lansbury, P.T.Jr, 2002, Protofibrillar islet amyloid polypeptide permeabilizes synthetic vesicles by a pore-like mechanism that may be relevant to type II diabetes. *Biochemistry* **41:** 11338-11343.

Antzutkin, O.N., Leapman, R.D., Balbach, J.J., and Tycko, R., 2002, Supramolecular structural constraints on Alzheimer's beta-amyloid fibrils from electron microscopy and solid-state nuclear magnetic resonance. *Biochemistry* **41:** 15436-15450.

Balbach, J.J., Petkova, A.T., Oyler, N.A., Antzutkin, O.N., Gordon, D.J., Meredith, S.C., and Tycko, R., 2002, Supramolecular Structure in Full-Length Alzheimer's beta-Amyloid Fibrils: Evidence for a Parallel beta-Sheet Organization from Solid- State Nuclear Magnetic Resonance. *Biophys. J.* **83:** 1205-1216.

Bendiske, J., and Bahr, B.A., 2003, Lysosomal activation is a compensatory response against protein accumulation and associated synaptopathogenesis--an approach for slowing Alzheimer disease? *J. Neuropathol.. Ex..p Neurol.* **62:** 451-463.

Benzinger, T.L., Gregory, D.M., Burkoth, T.S., Miller-Auer, H., Lynn, D.G., Botto, R.E., and Meredith, S.C., 1998, Propagating structure of Alzheimer's beta-amyloid(10-35) is parallel beta-sheet with residues in exact register. *Proc. Natl. Acad. Sci. USA* **95:** 13407-13412.

Bucciantini, M., Giannoni, E., Chiti, F., Baroni, F., Formigli, L., Zurdo, J., Taddei, N., Ramponi, G., Dobson, C.M. ,and Stefani,. M., 2002, Inherent toxicity of aggregates implies a common mechanism for protein misfolding diseases. *Nature* **416:** 507-511.

Burdick, D., Soreghan, B., Kwon, M., Kosmoski, J., Knauer, M., Henschen, A., Yates, J., Cotman, C., and Glabe. C., 1992, Assembly and aggregation properties of synthetic Alzheimer's A4/beta amyloid peptide analogs. *J. Biol. Chem.* **267:** 546-554.

Caughey, B., and Lansbury, P.T., 2003, Protofibrils, pores, fibrils, and neurodegeneration: separating the responsible protein aggregates from the innocent bystanders. *Ann. Rev. Neurosci.* **26:** 267-298.

Dahlgren, K.N., Manelli, A.M., Stine, W.B.Jr., Baker, L.K., Krafft, G.A., and LaDu, M.J.. 2002, Oligomeric and fibrillar species of amyloid-beta peptides differentially affect neuronal viability. *J. Biol. Chem.* **277:** 32046-32053.

Der-Sarkissian, A., Jao, C.C., Chen, J., and Langen, R., 2003, Structural organization of alpha-synuclein fibrils studied by site-directed spin labeling. *J Biol Chem.* 278:37530-5.

Dobson, C.M., 1999.,Protein misfolding, evolution and disease. *TIBS.* **24:** 3293-32.

Esler, W.P., Stimson, E.R., Jennings, J.M., Vinters, H.V., Ghilardi, J.R.,Lee, J.P., Mantyh, P.W., and Maggio, J.E., 2000, Alzheimer's disease amyloid propagation by a template-dependent dock-lock mechanism. *Biochemistry* **39:** 6288-6295.

Fandrich, M., and Dobson, C.M., 2002, The behaviour of polyamino acids reveals an inverse side chain effect in amyloid structure formation. *EMBO J.* **21:** 5682-5690.

Garzon-Rodriguez, W., Sepulveda-Becerra, M., Milton, S., and Glabe, C.G., 1997, Soluble amyloid Abeta-(1-40) exists as a stable dimer at low concentrations. *J Biol Chem.* **272:** 21037-21044.

Garzon-Rodriguez, W., Vega, A., Sepulveda-Becerra, M., Milton, S., Johnson, D.A., Yatsimirsky, A.K., and Glabe, C.G., 2000, A conformation change in the carboxyl terminus of Alzheimer's Abeta (1-40) accompanies the transition from dimer to fibril as revealed by fluorescence quenching analysis. *J. Biol. Chem.* **275:** 22645-22649.

Gong, Y., Chang, L., Viola, K.L, .Lacor, P.N., Lambert, M.P., Finch, C.E., Krafft, G.A., and Klein, W.L., 2003, Alzheimer's disease-affected brain: presence of oligomeric A beta ligands (ADDLs) suggests a molecular basis for reversible memory loss. *Proc. Natl. Acad. Sci. U S A.* **100:** 10417-10422.

Gregori, L., Fuchs, C., Figueiredo-Pereira, M.E., Van Nostrand, W.E., and Goldgaber, D., 1995, Amyloid beta-protein inhibits ubiquitin-dependent protein degradation in vitro. *J. Biol. Chem.* **270:** 19702-19708.

Hajimohammadreza, I., Anderson, V.E., Cavanagh, J.B., Seville, M.P., Nolan, C.C., Anderton, B.H., and Leigh, P.N., 1994, beta-Amyloid precursor protein fragments and

lysosomal dense bodies are found in rat brain neurons after ventricular infusion of leupeptin. *Brain Res.* **640:** 25-32.
Hardy, J., and Selkoe. D.J., 2002, The amyloid hypothesis of Alzheimer's disease: progress and problems on the road to therapeutics. *Science* **297:** 353-356.
Harper, J.D., Lieber, C.M., and. Lansbury, P.T. 1997a, Atomic force microscopic imaging of seeded fibril formation and fibril branching by the Alzheimer's disease amyloid-beta protein. *Chem. Biol.* **4:** 951-959.
Harper, J.D., Wong, S.S., Lieber, C.M., and Lansbury, P.T., 1997b, Observation of metastable Abeta amyloid protofibrils by atomic force microscopy. *Chem Biol.* **4:** 119-125.
Hartley, D.M., Walsh, D.M, Ye, C.P., Diehl, T., Vasquez, S., Vassilev, P.M., Teplow D.B., and Selkoe. D.J., 1999, Protofibrillar intermediates of amyloid beta-protein induce acute electrophysiological changes and progressive neurotoxicity in cortical neurons. J. *Neurosci.* **19:** 8876-8884.
Hilbich, C., Kisters-Woike, B., Reed, J., Masters, C.L., and Beyreuther, K., 1991, Aggregation and secondary structure of synthetic amyloid beta A4 peptides of Alzheimer's disease. *J. Mol. Biol.* **218:** 149-163.
Houlden, H., Baker, M., McGowan, E., Lewis, P., Hutton, M., Crook, R., Wood, N.W., Kumar-Singh, S., Geddes, J., Swash, M., Scaravilli, F., Holton, J.L., Lashley, T., Tomita, T., Hashimoto, T., Verkkoniemi, A., Kalimo, H., Somer, M., Paetau, A., Martin, J.J., Van Broeckhoven, C., Golde, T., Hardy, J., Haltia, M., and Revesz, T., 2000, Variant Alzheimer's disease with spastic paraparesis and cotton wool plaques is caused by PS-1 mutations that lead to exceptionally high amyloid-beta concentrations. *Ann. Neurol.* **48:** 806-808.
Hsia, A.Y., Masliah, E., McConlogue, L., Yu, G.Q., Tatsuno, G., Hu, K., Kholodenko, D., Malenka, R.C., Nicoll, R.A., and Mucke, L., 1999, Plaque-independent disruption of neural circuits in Alzheimer's disease mouse models. *Proc. Natl. Acad. Sci. USA* **96:** 3228-3233.
Kagan, B.L., Hirakura, Y., Azimov, R., Azimova, R., and Lin M.C., 2002, The channel hypothesis of Alzheimer's disease: current status. *Peptides.* **23:** 1311-1315.
Kayed, R., Head, E., Thompson, J.L., McIntire, T.M., Milton, S.C., Cotman, C.W., and Glabe, C.G., 2003, Common structure of soluble amyloid oligomers implies common mechanism of pathogenesis. *Science* **300:** 486-489.
Keck, S., Nitsch, R., Grune, T., and Ullrich, O., 2003, Proteasome inhibition by paired helical filament-tau in brains of patients with Alzheimer's disease. *J. Neurochem.* **85:** 115-122.
Kirkitadze, M.D., Bitan, G., and Teplow. D.B., 2002, Paradigm shifts in Alzheimer's disease and other neurodegenerative disorders: the emerging role of oligomeric assemblies. *J. Neurosci. Res.* **69:** 567-577.
Klein, W.L., Krafft, G.A., and Finch, C.E., 2001, Targeting small Abeta oligomers: the solution to an Alzheimer's disease conundrum? *Trends Neurosci.* **24:** 219-224.
Kuo, Y.M., Emmerling, M.R., Vigo-Pelfrey, C., Kasunic, T.C., Kirkpatrick, J.B., Murdoch, G.H., Ball, M.J., and Roher, A.E., 1996, Water-soluble Abeta (N-40, N-42) oligomers in normal and Alzheimer disease brains. *J. Biol. Chem.* **271:** 4077-4081.
Lashuel, H.A., Hartley, D., Petre, B.M., Walz, T., and Lansbury, P.T.Jr., 2002, Neurodegenerative disease: amyloid pores from pathogenic mutations. *Nature* **418:** 291.
LeVine, H., 2002, 4,4(')-Dianilino-1,1(')-binaphthyl-5,5(')-disulfonate: report on non- beta-sheet conformers of Alzheimer's peptide beta(1-40). *Arch. Biochem, Biophys.* **404:** 106-115.
LeVine, H.D., 1993, Thioflavinee T interaction with synthetic Alzheimer's disease beta-amyloid peptides: detection of amyloid aggregation in solution. *Protein Sci.* **2:** 404-410.

Lomakin, A., Teplow, D.B., Kirschner, D.A., and Benedek, G.B., 1997. Kinetic theory of fibrillogenesis of amyloid beta-protein. *Proc. Natl. Acad. Sci. USA* **94:** 7942-7947.

Lue, L.F., Kuo, Y.M., Roher, A.E., Brachova, L., Shen, Y., Sue, L., Beach, T., Kurth, J.H., Rydel, R.E., and Rogers, J., 1999, Soluble amyloid beta peptide concentration as a predictor of synaptic change in Alzheimer's disease. *Amer. J. Pathol.* **155:** 853-862.

Mattson, M.P., 2002, Oxidative stress, perturbed calcium homeostasis, and immune dysfunction in Alzheimer's disease. *J. Neurovirol.* **8:** 539-550.

McLean, C.A., Cherny, R.A., Fraser, F.W., Fuller, S.J., Smith, M..J., Beyreuther, K., Bush, A.I., and Masters. C.L., 1999, Soluble pool of Abeta amyloid as a determinant of severity of neurodegeneration in Alzheimer's disease. *Ann. Neurol.* **46:** 860-866.

Petkova, A.T., Buntkowsky, G., Dyda, F., Leapman, R.D., Yau, W.M., and Tycko, R., 2004, Solid State NMR Reveals a pH-dependent Antiparallel beta-Sheet Registry in Fibrils Formed by a beta-Amyloid Peptide. *J. Mol. Biol.* **335:** 247-260.

Pitschke, M., Prior, R., Haupt, M., and Riesner, D., 1998, Detection of single amyloid beta-protein aggregates in the cerebrospinal fluid of Alzheimer's patients by fluorescence correlation spectroscopy [see comments]. *Nature Med.* **4:** 832-834.

Shen, C.L., and Murphy, R.M., 1995, Solvent effects on self-assembly of beta-amyloid peptide. *Biophys. J.* **69:** 640-651.

Soreghan, B., Kosmoski, J., and Glabe, C., 1994, Surfactant properties of Alzheimer's A beta peptides and the mechanism of amyloid aggregation. *J. Biol. Chem.* **269:** 28551-28554.

Terry, R.D., 1996, The pathogenesis of Alzheimer disease: an alternative to the amyloid hypothesis. *J. Neuropathol Exp. Neurol.* **55:** 1023-1025.

Torok, M., Milton, S., Kayed, R., Wu, P., McIntire, T., Glabe, C.C., and Langen, R., 2002, Structural and dynamic features of Alzheimer's Abeta peptide in amyloid fibrils studied by site-directed spin labeling. *J. Biol. Chem.* **277:** 40810-40815.

Tseng, B.P., Esler, W.P., Clish, C.B., Stimson, E.R., Ghilardi, J.R., Vinters, H.V., Mantyh, P.W., Lee, J.P., and Maggio, J.E., 1999, Deposition of monomeric, not oligomeric, Abeta mediates growth of Alzheimer's disease amyloid plaques in human brain preparations. *Biochemistry* **38:** 10424-10431.

Verkkoniemi, A., Somer, M., Rinne, J.O., Myllykangas, L., Crook, R., Hardy, J., Viitanen, M., Kalimo, H., and Haltia, M., 2000, Variant Alzheimer's disease with spastic paraparesis: clinical characterization. *Neurology* **54:** 1103-1109.

Waelter, S., Boeddrich, A., Lurz, R., Scherzinger, E., Lueder, G., Lehrach, H., and Wanker, E.E., 2001, Accumulation of mutant huntingtin fragments in aggresome-like inclusion bodies as a result of insufficient protein degradation. *Mol. Biol. Cell* **12:** 1393-1407.

Walsh, D.M., Klyubin, I., Fadeeva, J.V., Rowan, M.J., and Selkoe, D.J., 2002, Amyloid-beta oligomers: their production, toxicity and therapeutic inhibition. *Biochem. Soc. Trans.* **30:** 552-557.

Walsh, D.M., Lomakin, A., Benedek, G.B., Condron, M.M., and Teplow, D.B., 1997, Amyloid beta-protein fibrillogenesis. Detection of a protofibrillar intermediate. *J. Biol. Chem.* **272:** 22364-22372.

Westerman, M.A., Cooper-Blacketer, D., Mariash, A., Kotilinek, L., Kawarabayashi, T., Younkin, L.H., Carlson, G.A., Younkin, S.G., and Ashe, K.H., 2002, The relationship between Abeta and memory in the Tg2576 mouse model of Alzheimer's disease. *J. Neurosci.* **22:** 1858-1867.

Zagorski, M.G., Yang, J., Shao, H., Ma, K., Zeng, H., and Hong, A., 1999, Methodological and chemical factors affecting amyloid beta peptide amyloidogenicity. *Meth. Enzymol.* **309:** 189-204.

Chapter 9

Cholesterol and Amyloid β Fibrillogenesis

Katsuhiko Yanagisawa
National Institute for Longevity Sciences, 474-8522 Obu, Japan

Abstract: Evidence is accumulating to suggest that cholesterol is a potent risk factor for the development of Alzheimer's disease. An increase in cholesterol level in neuronal membranes may facilitate the generation and aggregation of the amyloid β-protein (Aβ). Our results and those of other groups suggest that cholesterol has both direct and indirect effects of acceleration of Aβ fibrillogenesis. A novel concept of cholesterol neurobiology is necessary to elucidate the mechanism underlying cholesterol-dependent Aβ pathology.

Key words: Amyloid β-protein, fibrillogenesis, cholesterol, GM1 ganglioside, lipid raft, senile plaque, aging, apolipoprotein

1. INTRODUCTION

Amyloid β-protein (Aβ) deposition in the brain is one of the fundamental processes in the development of Alzheimer's disease (AD); however, the molecular mechanism underlying the conversion of Aβ from its nontoxic soluble form to its toxic aggregated form remains to be determined. Aβ, a naturally occurring proteolytic (β-secretase) derivative of the amyloid precursor protein (APP), is believed to self-aggregate. Indeed, Aβ forms amyloid fibrils following its incubation *in vitro* at concentrations higher than those in physiological biological fluids. However, Aβ aggregation *in vitro* can be significantly suppressed by subjecting an Aβ solution to "de-seeding". In the case of familial AD, the expression of genes responsible for AD is likely to enhance Aβ deposition through accelerated Aβ generation.

However, no evidence has ever been reported to indicate that Aβ generation is perturbed in sporadic AD, a major form of the disease. Thus, it is reasonable to assume that Aβ deposition in the brain is induced by not only an increase in Aβ concentration in cells and/or extracellular space but also as yet unknown environmental factors. Recently, evidence is accumulating to suggest that cholesterol is a potential risk factor for AD development (Simons *et al.*, 2001; Puglielli *et al.*, 2003). In regard to the possible role(s) of cholesterol in the acceleration of Aβ deposition *in vivo*, several investigators have reported that processes in Aβ generation, including APP cleavage by β- and γ-secretases, can be facilitated by an increase in cellular cholesterol concentration (Wahrle *et al.*, 2002; Fassbender *et al.*, 2001; Frears *et al.*,1999; Simons *et al.*, 1998). Alternatively, we and other groups have suggested that Aβ fibrillogenesis is also modulated through direct and indirect interactions with cellular lipids including cholesterol (McLaurin *et al.*, 2002; Yamazaki *et al.*, 2001; Yip *et al.*, 2001; Kakio *et al.*, 2001; Mizuno *et al.*, 1999; Mizuno *et al.*, 1998).

2. CHOLESTEROL AS A RISK FACTOR FOR AD DEVELOPMENT

Previous studies revealed that a number of genetic risk factors are involved in AD development. Notably, among them, there are many genes that encode proteins directly involved in the regulation of lipid metabolism. Apolipoprotein E (ApoE) is a major lipoprotein in the central nervous system. Moreover, it is generally accepted that the genetic polymorphism of the *ApoE* gene is closely related to the prevalence of AD (Corder *et al.*, 1993; Poirier *et al.*, 1993; Saunders *et al.*, 1993; Strittmatter *et al.*, 1993). Much effort has been exerted to elucidate the pathogenic effect of one of ApoE isoforms, ApoE4, the gene product of ApoE allele ε4, the presence of which is a strong risk factor for AD. However, it still remains to be determined how ApoE4 accelerates the progression of AD at molecular and cellular levels. We have been attempting to clarify this question from a viewpoint of the physiological function of ApoE, that is, regulation of cholesterol metabolism in the central nervous system (Michikawa and Yanagisawa, 1998). We have reported that ApoE regulates the turnover of cholesterol and other lipids through the modulation of influx and efflux of these lipids in an isoform-dependent manner (Gong *et al.*, 2002; Michikawa *et al.*, 2000). Recently, we have also found that ApoE modulates the cholesterol distribution in neuronal membranes as discussed below (Hayashi *et al.*, 2002).

Since the relationship between the presence of ApoE allele ε4 and the prevalence of AD has been confirmed in various ethnic groups, a number of investigators have paid attention to the question as to whether cholesterol metabolism is perturbed in the central nervous system of individuals with AD. Jarvik and colleagues reported that the role of ApoE allele ε4 in the acceleration of AD development is dependent on serum total cholesterol (TC) level as well as age and sex (Jarvik *et al.*, 1995). Additionally, Notkola *et al.* (1998) performed a longitudinal study and concluded that an increase in serum TC level in midlife can be a risk factor for the development of late-life AD. This possibility was supported by the results of Kivipelto *et al.* (2001) who also confirmed that there was a relationship between the level of midlife hypercholesterolemia and the prevalence of late-life mild cognitive impairment (MCI), a putative preclinical stage of AD. Recently, two groups have independently performed retrospective epidemiological studies and reported that statin, an inhibitor of 3-hydroxy-3-methylglutaryl coenzyme A reductase, a key enzyme in cholesterol synthesis, has a beneficial inhibitory effect on suppression of AD development (Jick *et al.*, 2000; Wolozin *et al.*, 2000), *see also* Chapter 19.

Previous studies using animal models of AD also suggested that cholesterol can be a risk factor for AD development. Sparks *et al.* (1994) reported that a high-cholesterol diet potentially induces an increase in Aβ immunoreactivity in the brain (Sparks *et al.*, 1994). However, it remains to be clarified how the oral intake of cholesterol can alter the metabolism of lipids and proteins in the central nervous system. Although the accelerated Aβ deposition in the brain induced by high-cholesterol diet requires an explanation, the result of Sparks and collelagues has been supported by recent studies by Refolo *et al.* (2000) who reported that hypercholesterolemia induced by high-cholesterol diet accelerated the Aβ deposition in the brain of APP-transgenic mouse model of AD. They also reported that a cholesterol-lowering drug that inhibits 7-dehydrocholesterol-reductase, the enzyme working upstream of cholesterol biosynthesis, can reduce Aβ deposition in the brain of the same transgenic mouse model (Refolo *et al.*, 2001). These lines of evidence suggest that cholesterol can be transported from plasma to the brain and alter Aβ pathology.

3. CHOLESTEROL AND Aβ FIBRILLOGENESIS

3.1 Direct effect

The interaction of Aβ with plasma membranes is an important subject that needs elucidation. Wood and his colleagues previously examined the possibility of the direct binding of Aβ to lipids, including cholesterol, phosphatidylcholine and fatty acids (Avdulov *et al.*, 1997). They found that Aβ binding to lipids is dependent on the aggregation state of Aβ, that is, monomeric Aβ does not bind to lipids and only aggregated Aβ binds to lipids in a lipid class-dependent manner. Alternatively, McLaurin and her colleagues investigated whether Aβ-lipid interaction is modulated by cholesterol level in membranes using various techniques (Yip *et al.*, 2001). They reported that the extent of Aβ binding to the surface of cultured PC 12 cells decreases with increasing cholesterol content in the cells and that cholesterol-depleted cells had a higher extent of Aβ-cell surface binding. Müeller and his colleagues found that cholesterol content in neuronal membranes negatively correlates with the membrane perturbing effect of Aβ (Kirsch *et al.*, 2002). Taken all together, it is highly likely that Aβ interaction with neuronal membrane is significantly modulated by cholesterol content in such membranes. However, in regard to the potency of cholesterol to accelerate or inhibit Aβ fibrillogenesis, previous studies have caused controversy. Harris investigated how Aβ fibrillogensis *in vitro* is modulated in the presence of cholesterol by transmission electron microscopy using aqueous suspensions of microcrystalline cholesterol and other lipids (Harris, 2002). He found that the marked acceleration of Aβ aggregation is induced in the presence of cholesterol and cholesterol-containing liposomes. Furthermore, Harris suggested that the potentiation of Aβ polymerization is likely to be induced by the hydrophobic interaction between the amino acid side chains of Aβ and the tetracyclic sterol nucleus. In contrast, McLaurin and her colleagues suggested on the basis of the result obtained from *in situ* atomic force microscopy and *ex situ* electron microscopy that bilayer cholesterol content in total brain lipid extracts inversely correlates with the extent of Aβ insertion into the membranes and subsequent Aβ fibrillogenesis on the membrane surface (Yip *et al.*, 2001). Sui and his colleagues focused on Aβ insertion into the membranes and they suggested that the extent of Aβ insertion into lipid bilayers rather increases as cholesterol content increases and such insertion subsequently inhibits Aβ fibrillogenesis on the membrane surface (Ji *et al.*, 2002). Thus, although much effort has been exerted in elucidating the direct effect of cholesterol content on Aβ fibrillogenesis, the results reported to date are still controversial. It is very difficult at this point to conclude whether an increase in the cholesterol content of neuronal membranes directly accelerates or inhibits Aβ fibrillogenesis.

3.2 Indirect effect

3.2.1 Identification of GM1 ganglioside-bound Aβ

To elucidate the molecular mechanism underlying the initiation of Aβ aggregation in the brain, we previously attempted to identify an Aβ species that initially accumulates in the brains of AD patients (Yanagisawa *et al.*, 1995). We performed sucrose density gradient fractionation of cerebral cortices. We prepared a fraction with the density of the plasma membrane (membrane fraction), which was not likely to contain aggregated Aβ, as well as a fraction with the protein density (amyloid fraction), which was enriched with amyloid cores of senile plaques. We performed Western blot analyses of such fractions using various anti-Aβ monoclonal antibodies, including antibodies specific to the carboxyl- and amino-terminals, and midportion of Aβ. On the blot of the fractions prepared from AD and Down's syndrome (DS) patients in the advanced stage of AD, strong immunoreactivities of those antibodies were observed in the amyloid fraction with a molecular weight of 4 KD. No Aβ immunoreactivity was detected in any fractions prepared from control non-demented individuals, whose cerebral cortical sections did not show Aβ deposition. In contrast, a unique Aβ immunoreactivity was observed in the membrane fractions prepared from the brains which exhibited early pathological changes of AD, including diffuse amyloid plaques. Briefly, Aβ was only detected by antibodies specific to the carboxyl- and amino-terminals of Aβ but not by antibodies specific to the midportion of Aβ. Aβ also showed smearing on the blot with retarded mobility on the gel compared with synthetic Aβ. To determine that the generation of this unique Aβ species is associated with the early pathological stage of AD, we examined a number of brains, including those of AD and DS patients, and nondemented individuals. We performed quantitative densitoscanning of the Western blots and neuropathological analyses of cerebral cortices. The Aβ immunoreactivity ratio of the amyloid fraction to the membrane fraction was determined for each case. The level of diffuse plaque formation was used as a marker of the early stage of AD pathology whereas the level of neurofibrillary tangle formation as that of the advanced stage of AD pathology. From this assessment, the Aβ imunoreactivity ratio of the amyloid fraction to the membrane fraction is positively and negatively correlates with the level of neurofibrillary tangle formation and the level of diffuse plaque formation, respectively. These results suggest that membrane-bound Aβ is preferably generated in the early stage of AD.

Based on the unique molecular features of membrane-bound Aβ, including the absence of the immunoreactivity of anti-Aβ antibodies, which were specific to the midportion of Aβ, smearing and retarded mobility on the gel, we hypothesized that a small molecule, such as a lipid, is bound to Aβ. To examine this possibility, we performed delipidation using various organic solvents, including methanol, ethanol, ether, and acetone. Interestingly, only treatment with methanol markedly altered the features of Aβ, that is, immunoreactivity of the antibody specific to the midportion of Aβ was recovered, and the smearing and retarded mobility of Aβ disappeared after methanol treatment. Since a lipid with a charge is sensitive to methanol, we expected that ganglioside is bound to Aβ in the membrane fraction. To confirm this possibility, we incubated the blot of the membrane fraction with cholera toxin, a specific ligand to GM1 ganglioside. Notably, Aβ in the membrane fraction was labeled with cholera toxin. This binding of cholera toxin to Aβ was lost following methanol treatment. Alternatively, to determine the molecular species of Aβ, we performed Western blot analysis using antibodies specific to Aβ40 and Aβ42. Aβ in the membrane fraction was only recognized by the antibody specific to Aβ42. To determine the amino terminus of the Aβ, we performed mass spectrometry following peptidase treatment. Based on the results of these experiments, we concluded that the Aβ species that accumulated in the membrane fraction is Aβ1-42. To further confirm that Aβ1-42 is generated in the membrane fraction of the brain, we attempted to immunoprecipitate Aβ from the membrane fraction prepared from the AD brain. In this experiment, Aβ1-42 in the GM1 ganglioside-bound form was successfully precipitated (Yanagisawa *et al.*, 1998). Interestingly, in the experiment of immunoprecipitation, the antibody specific to the midportion of Aβ failed to precipitate Aβ from the membrane fraction. This result supports our hypothesis that Aβ adopts an altered conformation through binding to GM1 ganglioside. To further confirm the alteration of Aβ conformation through its binding to GM1 ganglioside, we attempted to generate a monoclonal antibody specific to the GM1 ganglioside-bound form of Aβ (GAβ) (Yanagisawa *et al.*, 1997). Because of the limitated amounts of Aβ obtained from cerebral cortices, we performed *in vitro* immunization. The IgM monoclonal antibody (4396) was obtained by this method. MAb 4396 specifically recognized the Aβ species in the membrane fractions prepared from the brains which exhibited early Aβ and the Aβ recovered in the amyloid fraction prepared from AD brains. pathological changes of AD. However, it did not react with synthetic soluble In binding analysis, MAb 4396 recognized Aβ bound to GM1-containing liposomes. These results strongly suggest that GAβ adopts a conformation distinct from those of soluble and aggregated Aβs. Based on its unique molecular characteristics, including its extremely high aggregation potential and altered immunoreactivity, we concluded that Aβ binds to GM1 ganglioside at the early stage of AD and adopts an altered conformation, and subsequently accelerates the aggregation of soluble Aβ by acting as a seed. Following our initial report, several investigators performed *in vitro* experiments to confirm the potency of GAβ to serve as a seed. Surewicz and his colleagues prepared liposomes with or without GM1 ganglioside (Choo-Smith and Surewicz, 1997; Choo-Smith *et al.*, 1997). They incubated soluble Aβ with liposomes in a buffer with physiological ionic strength to avoid the non-specific binding of Aβ to liposomes, which is likely to occur under a condition of low ionic strength. They reported that soluble Aβ in the presence of GM1 ganglioside aggregates to form amyloid fibrils (Choo-Smith *et al.*, 1997). These results, taken together with our previous finding, strongly suggest that GAβ accelerates the aggregation of soluble Aβ by acting as a seed.

3.2.2 Cholesterol-dependent generation of GAβ

It remains to be determined how GAβ is generated *in vivo*; however, we found interesting phenomena in our previous studies, that suggested dependence of GAβ generation on cholesterol level (Mizuno *et al.*, 1998; Mizuno *et al.*, 1999). We investigated the molecular mechanism underlying the generation and secretion of Aβ using MDCK cells, which had been extensively studied in terms of cell polarity. We analyzed the Aβ species secreted into the apical and basolateral compartments from MDCK cells. Interestingly, in that experiment, we detected an Aβ species, only in the apical compartment, with unique molecular characteristics shared with GAβ, such as its appearance as a smear on a blot and its altered immunoreactivity. In regard to the mechanism underlying the generation of this unique Aβ species that was secreted into the apical compartment, apical Aβ we hypothesized that cholesterol and/or glycosphingolipids play an important role since these lipids are abundant in apical surface membranes as well as in apically transported vesicles (Simons and Ikonen, 1997). To examine this possibility, we treated cultured MDCK cells with compactin, an inhibitor of cholesterol synthesis. The cholesterol level in the cells decreased to 40% of the control level following treatment with compactin at concentrations ranging from 500 to 1000 nM. The treatment of the cells with compactin at concentrations higher than 5000 nM caused marked cell death, whereas cell death caused by treatment with compactin in the concentrations ranging from 500 to 1000 nM was negligible. Notably, the compactin treatment of the cells significantly decreased the extent of apical Aβ generation. This result suggests that apical Aβ generation is dependent on cholesterol level in membranes. The apical Aβ observed as a smear on immunoblots disappeared dramatically following treatment of cells with compactin. Since Aβ smearing suggested the potency of apical Aβ to act as a seed for the fibrillogenesis of soluble Aβ, we performed a thioflavin T (ThT) assay of the incubation mixture containing synthetic soluble Aβ and apical Aβ immunoprecipitated from the conditioned media of the apical compartment of the cell cultures. The fluorescent intensity of ThT increased without a lag phase and reached equilibrium hyperbolically in the presence of apical Aβ. In contrast, the ThT fluorescence intensity of the solution of synthetic Aβ in the presence of the Aβ immunoprecipitated from the conditioned media of the basolateral compartment did not increase. To further confirm the acceleration of the fibrillogenesis of soluble Aβ by apical Aβ, we performed electron microscopy analysis. Typical amyloid fibrils were observed in the incubation mixture containing synthetic and apical Aβs. Notably, apical Aβ lost its potency to act as a seed following the treatment of the cells with compactin. These results suggest that Aβ with a seeding activity is endogenously generated in a cholesterol-dependent manner by the cells.

Based on the finding that a seed Aβ is endogenously generated in a cholesterol-dependent manner, we then extended our study to investigate whether GAβ generation is also dependent on cholesterol level (Kakio *et al.*, 2001). In this study, we first prepared liposomes with lipids, *e.g.* cholesterol, phosphatidylcholine, phosphatidylserine, sphingomyelin and GM1 ganglioside of various concentrations. To avoid the nonspecific binding of Aβ to lipids under a low-ionic strength condition, we incubated liposomes with soluble Aβ in an incubation buffer containing 150 mM NaCl. We labeled synthetic Aβ with 7-diethylamonocoumarin-3-carbonyl group (DAC) at its amino terminus. The binding of DAC-labeled Aβ to liposomes was estimated by determining the fluorescence intensity from DAC. With this system, we found that soluble Aβ specifically binds to liposomes containing GM1 ganglioside. Other groups previously reported that soluble Aβ binds to phospholipid vesicles that lack ganglioside; however, they used an incubation buffer with a low ionic strength. Thus, as Surewicz and his colleagues clearly pointed out, it is highly likely that Aβ binding to ganglioside-free liposomes at a low ionic strength is due to nonspecific, purely electrostatic interaction between Aβ and lipids. We then examined how the lipid composition of liposomes modulates Aβ binding to GM1 ganglioside-containing liposomes. We found that Aβ binding is dependent not only on the concentration of GM1 ganglioside but also on cholesterol concentration in a given liposome. Interestingly, the effect of cholesterol concentration in the liposomes on the acceleration of Aβ binding was more marked in the case of a GM1 ganglioside-poor system (20% molar ratio) compared with that in a GM1 ganglioside-rich system (40% molar ratio). These suggest the following: first, GAβ generation *in vivo* likely occurs in

GM1 ganglioside- and cholesterol-rich environments; second, cholesterol plays a critical role in the regulation of Aβ binding to GM1 ganglioside *in vivo* since the concentration of GM1 ganglioside in the brain cannot be as high as 40 mol% in neuronal membranes. To elucidate the molecular mechanism underlying the cholesterol-dependent acceleration of Aβ binding to GM1 ganglioside, we labeled GM1 ganglioside with BODIPY, a well-known fluorophore, to form an excimer that emits redshifted fluorescence compared with monomers through collision of two dye molecules. The fluorescence emission spectra of BODIPY-GM1 ganglioside were recorded using various liposomes. We found that excimer fluorescence increases in a cholesterol-dependent manner in liposomes with a constant concentration of GM1 ganglioside. These results suggest that GM1 ganglioside forms a cluster in a cholesterol-rich environment (Figure 1).

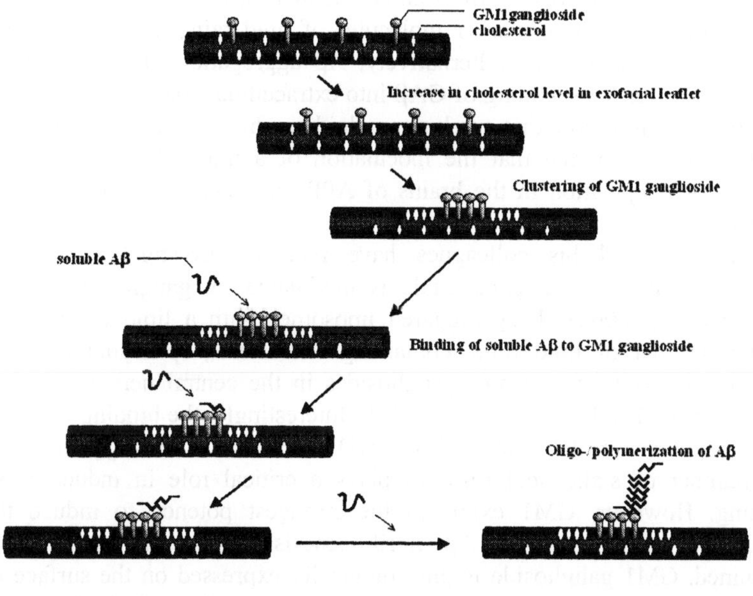

Figure 1. Hypothetical model of cholesterol-dependent oligo-/polymerization of Aβ through GAβ formation on GM1-ganglioside cluster in cholesterol-rich environment.

Recently, evidence has been growing to suggest that APP processing occurs in cholesterol- and GM1 ganglioside-rich microdomains, lipid rafts, in the cells (Ehehalt *et al.*, 2003). It has been suggested that β-cleavage, which is an initial step in Aβ generation from APP, is closely associated with lipid rafts (Fassbender *et al.*, 2001; Frears *et al.*, 1999). Moreover, γ-cleavage, which is the next step to β-cleavage, has also been recently suggested to be cholesterol-dependent (Wahrle *et al.*, 2002). Taken together with results of our previous studies, it is intriguing to speculate that Aβ generated in lipid rafts or raft-like microdomains start to aggregate following binding to GM1 ganglioside induced by as yet unknown alteration of the composition of lipids, including cholesterol and GM1 ganglioside, of the lipid rafts. This possibility is supported by the following studies: Firstly, Sawamura *et al.* (2000) previously reported that insoluble Aβ accumulates in the fraction of lipid rafts in the brains of the transgenic mouse model of AD. Secondly, Molamder-Melin and colleagues have recently analyzed lipids in the lipid rafts obtained from human brains and suggested that lipid composition is significantly altered in the lipid rafts obtained from AD brains (Molander-Melin *et al.*, 2003). Several investigators previously observed intracellular Aβ deposition prior to the extracellular formation of senile plaques (Gyure *et al.*, 2001; Gouras *et al.*, 2000). With this information, it is interesting to examine whether intracellularly accumulated Aβ is colocalized with marker molecules of lipid rafts, including flottilin and/or GM1 ganglioside. Alternatively, Aβ aggregation induced by GAβ may occur after the shedding of GAβ into extracellular space. In this regard, it is interesting to note that cerebral amyloid can be seeded *in vivo*. Kane et al. previously reported that the inoculation of a dilute AD brain extract induces Aβ deposition in the brains of APP-transgenic mice (Kane *et al.*, 2000).

Matsuzaki and his colleagues have recently investigated how the interaction of Aβ with ganglioside is modulated by ganglioside species (Kakio *et al.*, 2002). They prepared liposomes with a lipid composition similar to that of lipid rafts, containing cholesterol, sphingomyelin and ganglioside. They used major gangliosides in the central nervous system, including GM1, GD1a, GD1b and GT1b. Interestingly, the binding of Aβ to ganglioside was in the order of GM1<GD1b<GD1a=GT1b, suggesting that the number of sialic acid residues plays a critical role in inducing Aβ binding. However, GM1 exhibited the strongest potency to induce the generation of a seed for Aβ fibrillogenesis among the gangliosides examined. GM1 ganglioside is physiologically expressed on the surface of neurons, particularly on synaptic plasma membranes. It is likely that GM1 ganglioside plays a critical role in the modulation of neuronal function and viability. Thus, one could assume that Aβ binding to GM1 ganglioside exerts adverse effects on neurons. McLaurin and Chakrabartty previously reported that Aβ interaction with lipids, including ganglioside, induces membrane disruption (McLaurin and Chakrabartty, 1996). Furthermore, Matsuzaki and Horikiri investigated how Aβ binding to ganglioside affects the structure of the membranes and found that Aβ interaction with ganglioside-containing

such membranes induces the dehydration of lipid interfacial groups and the perturbation of the acyl chain orientation, suggesting that Aβ imposes a negative curvature strain on ganglioside-containing lipid bilayers (Matsuzaki and Horikiri, 1999). Taken together, all these results suggest that, aside from accelerating Aβ fibrillogenesis, Aβ binding to GM1 ganglioside directly suppresses neuronal viability through membrane damage.

3. CHOLESTEROL AND Aβ DEPOSITION *IN VIVO*

The deposition of Aβ in the brain as senile plaques is a fundamental process in AD. However, the precise mechanism underlying the formation of senile plaques remains to be determined. Probst et al. previously suggested that Aβ initially deposits on neuronal plasma membranes (Probst *et al.*, 1991). Yamaguchi et al. performed careful immunohistochemical analysis to clarify the ultrastructural localization of Aβ in diffuse plaques of brains with hereditary cerebral hemorrhage with amyloidosis and AD (Yamaguchi *et al.*, 2000). From their results, they suggested that Aβ starts to deposit in its membrane-bound form. These results suggest that the initial deposition of Aβ occurs on the neuronal surface although it remains to be determined how Aβ adsorbs onto lipids of neuronal membranes. Taken together with results of our studies, one of the possible scenario is that Aβ binds to GM1 ganglioside on the neuronal surface and accelerates the aggregation of soluble Aβ.

Alternatively, Mori *et al.* (2001) directly examined whether cholesterol is associated with senile plaques. They found the abnormal accumulation of cholesterol in cores of mature plaques but not in diffuse or immature plaques in AD brains. They also examined the transgenic mouse model of AD and obtained an essentially similar result. The results of Mori and collcagucs suggest that cholesterol plays a role in formation, progression and/or stabilization of senile plaques.

With regard to the initial site of Aβ deposition in the brain, Ihara and his colleagues performed immunochemical analysis of human prefrontal cortices and the brains of the transgenic mouse model of AD (Oshima *et al.*, 2001). They fractionated the cortices by sucrose density gradient centrifugation and then isolated detergent-insoluble low-density membrane (LDM) fractions. They determined Aβ level in those fractions and found that Aβ associated with LDM fractions reflected the extent of the level of extracellularly deposited Aβ. This result, together with the result of their previous study (Sawamura *et al*, 2000), suggests that Aβ initially deposits in LDM that is likely to be cholesterol- and GM1-ganglioside-rich microdomains.

4. CHOLESTEROL IN BRAIN AND NEURONAL MEMBRANES

Cholesterol is one of major lipids in cellular membranes and is essential for many cellular functions. This is also the case for cholesterol in the brain; however, little is known about cholesterol metabolism in the brain. The brain is highly enriched in cholesterol. It accounts for only 2% of the whole body mass but contains almost one-quarter of the unesterified cholesterol in the whole body (Dietschy and Turley, 2001). Although recent study provided the possibility of cholesterol transport across the blood-brain barrier (Refolo *et al.*, 2000), it has been generally accepted that cholesterol turnover is restricted only inside the brain, at a very low rate. Only 0.02% of the cholesterol pool in the brain undergoes a turnover each day (Dietschy and Turley, 2001). This is less than 1% as fast as the cholesterol turnover in the whole body (Dietschy and Turley, 2001). These lines of evidence imply that function and metabolism of cholesterol in the brain is unique. Again, although the results of recent study suggest that cholesterol input from plasma into the brain is not negligible (Refolo *et al.*, 2000), most cholesterol in the brain is likely to come from *in situ* synthesis. Interestingly, the rate of *in situ* cholesterol synthesis in many organs, including nervous system, depends on age (Cenedella and Shi, 1994; Stahlberg *et al.*, 1991; Goodrum, 1990; Popplewell and Azhar, 1987). In general, the rate of cholesterol synthesis is high in fetuses and newborn animals and decreases with age. Furthermore, Vance *et al.* (1994) previously reported that cholesterol is the only major lipid in the brain that cannot be synthesized at the end of neurites, implying that cholesterol level at the ends of neurites depends on two inputs, including axonal flow from the soma and uptake from the extracellular space through apolipoprotein receptors. Thus, it is reasonable to assume that the functions of neurons in the elderly depend on exogenous cholesterol level to a higher degree than those in young individuals. Aging is the strongest risk factor for the AD development, thus, it is interesting to speculate the pathological significance of aging for AD from the viewpoint of the aging-associated alteration of the synthesis and transport of cholesterol.

Neurons have a large plasma membrane compared with any other cell types. Moreover, part of the neuronal surface is specialized to integrate neuron-specific functions such as synaptic transmission. Cholesterol accounts for over 40 mol% of the total membrane lipids in synaptic plasma membranes (Schroeder *et al.*, 1991; Wood *et al.*, 1989). Thus, it is important to clarify how cholesterol in neuronal membranes, particularly in synaptic plasma membranes, is regulated. Wood and his colleagues and other groups have exerted much effort to clarify this issue and suggested the following (Wood *et al.*, 1999): First, cholesterol is not evenly distributed throughout neuronal membranes but is located in different pools, including cholesterol lateral domains and transbilayer cholesterol domains. In regard to cholesterol lateral domains, there seem to be two distinct pools: cholesterol in one pool is not stable but rather easily exchangeable following biochemical treatment, including incubation with small unilamellar vesicles, whereas that in the other pool is non-exchangeable. It was previously reported that the acetylcholine receptor is closely associated with the poorly exchangeable pool (Leibel *et al.*, 1987), thus, it is important in future studies to elucidate how the physiological function and pathological changes of neurons are associated with these two types of cholesterol lateral domain. Alternatively, regarding transbilayer cholesterol domains, previous studies have revealed an alteration in cholesterol distribution in the domain in association with the risk factors for AD development, as discussed below (Hayashi *et al.*, 2002; Igbavboa *et al.*, 1997; Igbavboa *et al.*, 1996).

It has been a controversial issue whether the net cholesterol content is altered in the AD brain. Mason *et al.* (1992) previously performed X-ray diffraction analysis of lipid membranes extracted from the cortical gray matter of AD brains. They found that the unesterified cholesterol:phospholipid molecular ratio significantly decreased by 30% in the AD brain compared to that in age-matched controls. Soderberg *et al.* (1992) performed a similar analysis using different brain regions from AD patients. In their experiments, the cortical area affected with AD showed an elevated phosphatidylinositol content; however, cholesterol level was highly variable but mostly unchanged. It seems likely that this controversy stems from the difference in techniques employed to determine lipid level. Another important point noted when we evaluated the data obtained from postmortem tissue is that autopsy samples may show modified the levels of lipids or proteins, which are different from those observed during the disease process.

Pathological changes, including the formation of senile plaques, can bc induced in the brains of elderly patients with Down's syndrome (DS). One possible explanation for the AD development in the DS brain is the high expression level of the APP gene by trisomy of chromosome 21. However, previous studies also provided evidence suggesting that lipid metabolism is altered in DS (Scott *et al.*, 1994; Naeim and Walford, 1980). Naeim and Walford reported that a membrane prepared from the peripheral mononuclear cells of DS patients showed an increased rigidity, suggesting the increased cholesterol level in the membrane (Naeim and Walford, 1980). Diomede et al. obtained additional results that supported the conclusion of the earlier study showing the increase in serum cholesterol level in DS. In order to clarify the mechanism underlying the altered lipid metabolism in DS, Diomede and co-workers analyzed liver cholesterol levels and the activities of the sterol regulatory element binding proteins, SREBP-1 and

SREBP-2, which are involved in the regulation of cholesterol synthesis (Diomede *et al.*, 1999). They found that trisomy 21 fetuses show an altered pattern of SREBP activation such as the sterol-independent maturation of SREBP-1. This result suggests that alteration in the regulation of cholesterol synthesis is induced by trisomy 21, resulting in high circulating and tissue cholesterol level.

Recently, novel information on cholesterol metabolism in AD has been obtained by determining 24S-hydroxycholesterol (24S-OH-Chol) level in cerebrospinal fluid. 24S-OH-Chol, an endogenously oxidized metabolite of cholesterol, is almost exclusively produced in the brain (Bjorkhem *et al.*, 1998), that is, the 24S-OH-Chol concentration in the brain is 30-1500-fold higher than those in any other organs. An important feature of 24S-OH-Chol is its potency to pass through the blood-brain barrier. It has been previously suggested that a continuous flux of 24S-OH-Chol across the blood-brain barrier is critical for cholesterol homeostasis in the brain. The plasma 24S-OH-Chol level is dependent on age; the level is high before age 20 and then falls after age 20. Based on the characteristics of 24S-OH-Chol, it is likely that the plasma 24S-OH-Chol level can be a peripheral marker of cholesterol metabolism in the brain. Lütjohann *et al.* (1996) performed determination of the plasma 24S-OH-Chol level in AD, non-AD demented and depressive patients and in healthy controls. The plasma 24S-OH-Chol levels in the AD and non-AD demented patients was elevated modestly but significantly compared with depressive patients and healthy controls. There was no significant difference in the 24S-OH-Chol levels between the AD and non-AD demented patients. Importantly, plasma 24S-OH-Chol level negatively correlated with the severity of dementia. This finding was supported by the study of Papassotiropoulos *et al.* (2000). The inheritance of the apoE4 allele is associated with the reduced plasma 24S-OH-Chol level in a manner independent of the severity of dementia in the patients examined. It has also been reported that the 24S-OH-Chol level was elevated in the cerebrospinal fluid in patients with mild cognitive impairment (MCI), an early stage of AD, and AD (Papassotiropoulos *et al.*, 2002; Schonknecht *et al.*, 2002), suggesting that cholesterol turnover is accelerated in the brain of the patients with AD. It remains to be determined how the plasma 24S-OH-Chol level is elevated in association with AD. In this regard, it is interesting to note that polymorphism in the cholesterol 24S-hydroxylase (CYP46) gene, which encodes enzyme involving in conversion of cholesterol to 24S-OH-Chol, is associated with AD (Papassotiropoulos *et al.*, 2003; Kolsch *et al.*, 2002). A genotype that is more prevalent in the AD group is associated with an increased 24S-OH-Chol/cholesterol ratio in CSF. Taken together, it is possible to assume that the accelerated conversion of cholesterol to 24S-OH-

Chol, which may or may not be with polymorphism of the CYP46 gene, predisposes to AD by altering cholesterol turnover in the brain.

Aging is the strongest risk factor for AD development. Thus, it is important to elucidate whether cholesterol distribution in neuronal membranes is altered with aging. As discussed above, there are two transbilayer cholesterol pools, including the cholesterol pool in exofacial and cytofacial lipid leaflets. Importantly, membrane fluidity and cholesterol distribution differ between these leaflets. A cytofacial leaflet contains seven times as much cholesterol as an exofacial leaflet (Igbavboa *et al.*, 1996). In accordance with this asymmetric distribution of cholesterol throughout the lipid bilayer, a cytofacial leaflet is much less fluid compared with an exofacial leaflet. Wood *et al.* (1999) previously carefully determined the age-dependent alteration of membrane fluidity and cholesterol distribution. They examined the synaptic plasma membranes (SPMs) obtained from mice in three different age groups, namely, 3-4, 14-15, and 24-25 months old. In that experiment, the exofacial leaflet of SPM from young mice was significantly more fluid than the cytofacial leaflet. In contrast, the difference in membrane fluidity between the exofacial and cytofacial leaflets was not significant in the SPM prepared from old mice. There was an approximately two-fold increase in the cholesterol level of the exofacial leaflet in the old mice compared with that in the young mice. Since the total amount of cholesterol in SPM did not change, they concluded that the asymmetric distribution of cholesterol throughout the lipid bilayer of SPM changes with age.

At this point, the mechanism underlying the generation of asymmetry in the cholesterol distribution throughout lipid bilayers remains to be determined. Wood and his colleagues suggested that apoE and the LDL receptor are involved in the regulation of the distribution of cholesterol in the SPM lipid bilayer from the results of their studies (Igbavboa *et al.*, 1997). Using apoE- and LDL-receptor-deficient mice, they performed experiments similar to those on aged mice. Interestingly, the cholesterol level in the SPM exofacial leaflets of the apoE- and LDL-receptor-deficient mice showed a twofold increase compared with that in wild-type mice. The possibility of apoE involvement in the regulation of the cholesterol distribution in the SPM lipid bilayer has been further supported by the results of our recent study. We analyzed the transbilayer distribution of cholesterol in the SPM of human apoE3- and apoE4-knock-in mice (Hayashi *et al.*, 2002). In these experiments, we prepared various subcellular fractions, including plasma membrane and endoplasmic fractions, and determined the levels of phospholipid and cholesterol (total and free) in the fractions. There were no significant differences in lipid concentration in the subcellular fraction among the wild-type, apoE3- and apoE4-knock-in mice. The asymmetric distribution of cholesterol in the SPM of the apoE3-knock-in mice did not differ from that of the wild-type mice; however, it was significantly altered in the apoE4-knock-in mice. The difference observed in the SPM prepared from the apoE4-knock-in mice was essentially the same as those observed in the aged and apoE-knock-out mice; that is, there was an approximately two-fold increase in the cholesterol concentration in the exofacial leaflet of SPM in the apoE4-knock-in mice compared with those in the other two groups of mice. Concomitantly, there was a significant decrease in the level of cytofacial leaflet cholesterol. The total amount of cholesterol in SPM did not significantly differ among the three mouse groups; thus, we concluded that apoE4 expression induces alteration in the cholesterol distribution throughout the lipid bilayer of SPM, thereby resulting in the exofacial leaflet having a higher cholesterol level than the cytofacial leaflet.

Aging and apoE4 expression are the major risk factors for AD development. Thus, it may be important that these two factors exert the same effect on the cholesterol distribution in SPM. It is also interesting to note that the extent of increase in the cholesterol level of the exofacial leaflet induced by these two factors are essentially the same. This result suggests that there are two pools of cholesterol in the lipid bilayer of SPM; cholesterol level in one pool changes in association with biological factors, including aging and apoE expression, whereas that in the other pool is rather stable.

The possible role of cholesterol in the acceleration of Aβ deposition in the brain has also been suggested by recent studies on Nieman-Pick type C disease (NPC). NPC is an autosomal recessive neurovisceral lipid storage disorder (Pentchev *et al.*, 1995). It is characterized by the abnormal accumulation of free cholesterol in lysosomes. The causative genes, including Niemann-Pick C1 (NPC1), have been identified and evidence is growing to indicate that the abnormal trafficking of exogenous cholesterol is closely associated with the progression of the disease. Previous pathological studies on NPC brains revealed the formation of neurofibrillary tangles, composed of hyperphosphorylated tau, which are indistinguishable from those formed in AD brains (Auer *et al.*, 1995; Suzuki *et al.*, 1995). Recently, it has been reported that Aβ deposition is induced in NPC brains. Murayama and his colleagues performed serial immunohistochemical examination of nine NPC brains (Saito *et al.*, 2002). They found Aβ deposition in the form of diffuse plaques in three of the nine brains. Notably, all the three NPC patients positive for Aβ deposition were below age 40 and were homozygous for apoE allele ε4. At age 40, it is is too early to observe Aβ deposition even in individuals homozygous for apoE allele ε4; thus, it is likely that the expression of the mutated *NPC* gene, combined with homozygosity for apoE allele ε4, is responsible for the induction of Aβ deposition in the brain. At this point, it remains to be elucidated how abnormal trafficking and/or cholesterol accumulation in lysosomes is associated with Aβ deposition in the brain. In this regard, a report by Sugimoto *et al.* (2001) should be to noted. They investigated the intracellular trafficking of lipids, including cholesterol and GM1 ganglioside, in NPC1-deficient cells. In their experiment, they found the accumulation of not only cholesterol but also GM1 ganglioside in NPC1-deficient cells. Although GM1 ganglioside and cholesterol accumulate in different vesicular compartments of NPC1-deficient cells, that is, cholesterol is accumulated in lysosomes whereas GM1 ganglioside in early endosomes, one can assume that the abnormal accumulation of these two lipids is responsible for Aβ deposition *in vivo.* Yamazaki *et al.* (2001) investigated whether Aβ aggregation is induced in NPC-deficient cells. They treated CHO cells with U18666A, which induces the NPC-mimicking accumulation of cholesterol in the cells. They, then, determined intracellular Aβ level and found that Aβ accumulates in its aggregated form in the late endosomes of the cells in association with intracellular cholesterol accumulation. The accumulated Aβ in the cells disappeared following U18666A withdrawal. Importantly, the level of Aβ secreted into the culture medium was not altered with U18666A treatment. These results suggest that intracellular Aβ accumulation is closely associated with the abnormal accumulation of cholesterol in the cells and that this abnormal Aβ accumulation is not due to accelerated Aβ generation. Recent studies suggest that an increase in cholesterol content potentially upregulates Aβ generation. However, evidence obtained from the NPC model cells suggests that an increase in intracellular cholesterol level facilitates Aβ aggregation through an as yet undetermined process, which is not dependent on accelerated Aβ generation induced by cholesterol.

5. CONCLUSIONS AND PERSPECTIVES

In the past few years, considerable attention has been focused on the possible link between cholesterol metabolism and AD development. However, we are still far from understanding the mechanism underlying the cholesterol-dependent acceleration of the pathological processes of AD. In regard to the putative acceleration of the early phase of amyloid cascade by cholesterol, evidence obtained to date suggests that both the generation and aggregation of Aβ is accelerated in a cholesterol-rich environment (Figure 2).

To further elucidate the pathological significance of cholesterol in Aβ metabolism, we need technological progress that will enable us to visualize cholesterol domains and to monitor cholesterol turnover inside neurons.

Cholesterol is not likely to be evenly distributed, but rather, concentrated in particular domains, including lipid rafts. Only recently, a probe for detecting cholesterol-rich microdomains has been developed. This probe will provide new approaches for the understanding of cholesterol homeostasis in neurons.

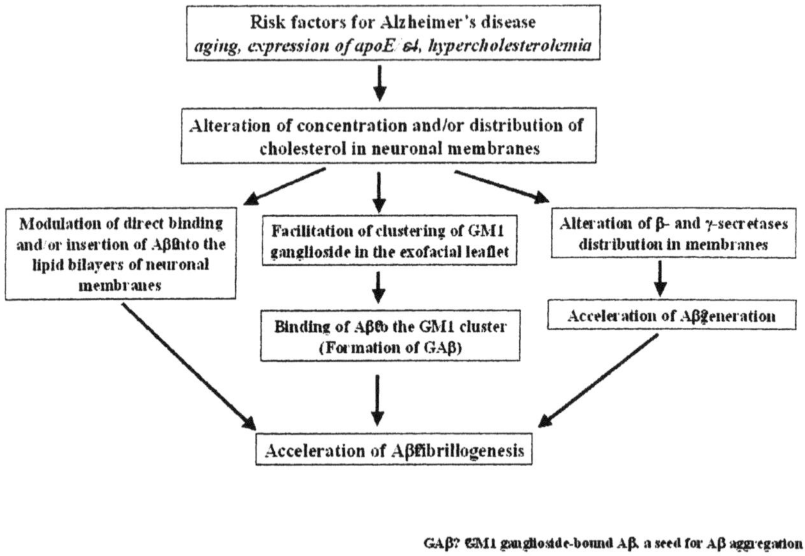

Figure 2. Possible mechanisms underlying cholesterol-dependent facilitation of Aβ fibrillogenesis.

Our studies also show that a decrease in cholesterol content in neurons causes pathological changes, including hyperphosphorylation (*see also* Chapter 20) and apoptosis, that are neuronal pathological features other than Aβ deposition in AD brains. At this point, we do not know whether too much or too little cholesterol is more crucial to AD development. AD is a disease whose pathological processes occur for more than a decade. Thus, it is likely that cholesterol is involved in the pathological process of AD in different manners during the long development of the disease. In conclusion, cholesterol is linked to AD development. A careful functional analysis of cholesterol from both physiological and pathological viewpoints will be helpful in future studies to further our understanding of AD.

REFERENCES

Auer, I.A., Schmidt, M.L., Lee, V.M., Curry, B., Suzuki, K., Shin, R.W., Pentchev, P.G., Carstea, E.D., and Trojanowski, J.Q., 1995, Paired helical filament tau (PHFtau) in Niemann-Pick type C disease is similar to PHFtau in Alzheimer's disease. *Acta. Neuropathol. (Berl)* **90:** 547-551.

Avdulov, N.A., Chochina, S.V., Igbavboa, U., Warden, C.S., Vassiliev, A.V., and Wood, W.G., 1997, Lipid binding to amyloid β-peptide aggregates: preferential binding of cholesterol as compared with phosphatidylcholine and fatty acids. *J. Neurochem.***69:** 1746-1752.

Bjorkhem, I., Lutjohann, D., Diczfalusy, U., Stahle, L., Ahlborg, G., and Wahren, J., 1998, Cholesterol homeostasis in human brain: turnover of 24S-hydroxycholesterol and evidence for a cerebral origin of most of this oxysterol in the circulation. *J. Lipid Res.* **39:** 1594-1600.

Cenedella, R.J., and Shi, H., 1994, Spatial distribution of 3-hydroxy-3-methylglutaryl coenzyme A reductase messenger RNA in the ocular lens: relationship to cholesterologenesis. *J. Lipid Res.* **35:** 2232-2240.

Choo-Smith, L.P., and Surewicz, W.K., 1997, The interaction between Alzheimer amyloid β(1-40) peptide and ganglioside GM1-containing membranes. *FEBS Lett.* **402:** 95-98.

Choo-Smith, L.P., Garzon-Rodriguez, W., Glabe, C.G., and Surewicz, W.K., 1997, Acceleration of amyloid fibril formation by specific binding of Aβ-(1-40) peptide to ganglioside-containing membrane vesicles. *J. Biol. Chem.***272:** 22987-22990.

Corder, E.H., Saunders, A.M., Strittmatter, W.J., Schmechel, D.E., Gaskell, P.C., Small, G.W., Roses, A.D., Haines, J.L., and Pericak-Vance, M.A., 1993, Gene dose of apolipoprotein E type 4 allele and the risk of Alzheimer's disease in late onset families. *Science* **261:** 921-923.

Dietschy, J.M., and Turley, S.D., 2001, Cholesterol metabolism in the brain. *Curr. Opin. Lipidol.* **12:** 105-112.

Diomede, L., Salmona, M., Albani, D., Bianchi, M., Bruno, A., and Salmona, S., Nicolini, U., 1999, Alteration of SREBP activation in liver of trisomy 21 fetuses. *Biochem. Biophys. Res. Commun.* **260:** 499-503.

Ehehalt, R., Keller, P., Haass, C., Thiele, C., and Simons, K., 2003, Amyloidogenic processing of the Alzheimer β-amyloid precursor protein depends on lipid rafts. *J. Cell Biol.* **160:** 113-123.

Fassbender, K., Simons, M., Bergmann, C., Stroick, M., Lutjohann, D., Keller, P., Runz, H., Kuhl, S., Bertsch, T., von Bergmann, K., Hennerici, M., Beyreuther, K., and Hartmann, T., 2001, Simvastatin strongly reduces levels of Alzheimer's disease β-amyloid peptides Aβ42 and Aβ40 in vitro and in vivo. *Proc. Natl. Acad. Sci. U S A* **98:** 5856-5861.

Frears, E.R., Stephens, D.J., Walters, C.E., Davies, H., and Austen, B.M., 1999, The role of cholesterol in the biosynthesis of β-amyloid. *Neuroreport* **10:** 1699-1705.

Gong, J.S., Kobayashi, M., Hayashi, H., Zou, K., Sawamura, N., Fujita, S.C., Yanagisawa, K., and Michikawa, M., 2002, Apolipoprotein E (ApoE) isoform-dependent lipid release from astrocytes prepared from human ApoE3 and ApoE4 knock-in mice. *J. Biol. Chem.* **277:** 29919-29926.

Goodrum, J.F., 1990, Cholesterol synthesis is down-regulated during regeneration of peripheral nerve. *J. Neurochem.***54:** 1709-1715.

Gouras, G.K., Tsai, J., Naslund, J., Vincent, B., Edgar, M., Checler, F., Greenfield, J.P., Haroutunian, V., Buxbaum, J.D., Xu, H., Greengard, P., and Relkin, N.R., 2000, Intraneuronal Aβ42 accumulation in human brain. *Am. J. Pathol.* **156:** 15-20.

Gyure, K.A., Durham, R., Stewart, W.F., Smialek, J.E., and Troncoso, J.C., 2001, Intraneuronal Aβ-amyloid precedes development of amyloid plaques in Down'ssyndrome. *Arch. Pathol. Lab. Med.***125:** 489-492.

Harris, J.R., 2002, In vitro fibrillogenesis of the amyloid β1-42 peptide: cholesterol potentiation and aspirin inhibition. *Micron* **33:** 609-626.

Hayashi, H., Igbavboa, U., Hamanaka, H., Kobayashi, M., Fujita, S.C., Wood, W.G., and Yanagisawa, K., 2002, Cholesterol is increased in the exofacial leaflet of synaptic plasma membranes of human apolipoprotein E4 knock-in mice. *Neuroreport* **13:** 383-386.

Igbavboa, U., Avdulov, N.A., Schroeder, F., and Wood, W.G., 1996, Increasing age alters transbilayer fluidity and cholesterol asymmetry in synaptic plasma membranes of mice. *J. Neurochem.* **66:** 1717-1725.

Igbavboa, U., Avdulov, N.A., Chochina, S.V., and Wood, W.G., 1997, Transbilayer distribution of cholesterol is modified in brain synaptic plasma membranes of knockout mice deficient in the low-density lipoprotein receptor, apolipoprotein E, or both proteins. *J. Neurochem.* **69:** 1661-1667.

Jarvik, G.P., Wijsman, E.M., Kukull, W.A., Schellenberg, G.D., Yu, C., and Larson, E.B., 1995, Interactions of apolipoprotein E genotype, total cholesterol level, age, and sex in prediction of Alzheimer's disease: a case-control study. *Neurology* **45:** 1092-1096.

Ji, S.R., Wu, Y., and Sui, S.F., 2002, Cholesterol is an important factor affecting the membrane insertion of β-amyloid peptide (Aβ1-40), which may potentially inhibit the fibril formation. *J. Biol. Chem* .**277:** 6273-6279.

Jick, H., Zornberg, G.L., Jick, S.S., Seshadri, S., and Drachman, D.A., 2000, Statins and the risk of dementia. *Lancet* **356:** 1627-1631.

Kakio, A., Nishimoto, S.I., Yanagisawa, K., Kozutsumi, Y., and Matsuzaki, K., 2001, Cholesterol-dependent formation of GM1 ganglioside-bound amyloid-protein, an endogenous seed for Alzheimer amyloid. *J. Biol. Chem* .**276:** 24985-24990.

Kakio, A., Nishimoto, S., Yanagisawa, K., Kozutsumi, Y., and Matsuzaki, K., 2002, Interactions of amyloid β-protein with various gangliosides in raft-like membranes: importance of GM1 ganglioside-bound form as an endogenous seed for Alzheimer amyloid. *Biochemistry* **41:** 7385-7390.

Kane, M.D., Lipinski, W.J., Callahan, M.J., Bian, F., Durham, R.A., Schwarz, R.D., Roher, A.E., and Walker, L.C., 2000, Evidence for seeding of β-amyloid by intracerebral infusion of Alzheimer brain extracts in β-amyloid precursor protein-transgenic mice. *J. Neurosci.* **20:** 3606-3611.

Kirsch, C., Eckert, G.P., and Mueller, W.E., 2002, Cholesterol attenuates the membrane perturbing properties of β-amyloid peptides. *Amyloid* **9:** 149-159.

Kivipelto, M., Helkala, E.L., Hanninen, T., Laakso, M.P., Hallikainen, M., Alhainen, K., Soininen, H., Tuomilehto, J., and Nissinen, A., 2001, Midlife vascular risk factors and late-life mild cognitive impairment: A population-based study. *Neurology* **56:** 1683-1689.

Kolsch, H., Lutjohann, D., Ludwig, M., Schulte, A., Ptok, U., Jessen, F., von Bergmann, K., Rao, M.L., Maier, W., and Heun, R., 2002, Polymorphism in the cholesterol 24S-hydroxylase gene is associated with Alzheimer's disease. *Mol. Psychiatry* **7:** 899-902.

Leibel, W.S., Firestone, L.L., Legler, D.C., Braswell, L.M., and Miller, K.W., 1987, Two pools of cholesterol in acetylcholine receptor-rich membranes from Torpedo. *Biochim. Biophys. Acta.* **897:** 249-260.

Lutjohann, D., Breuer, O., Ahlborg, G., Nennesmo, I., Siden, A., and Diczfalusy, U., Bjorkhem, I., 1996, Cholesterol homeostasis in human brain: evidence for an age-dependent flux of 24S-hydroxycholesterol from the brain into the circulation. *Proc. Natl. Acad. Sci. U S A* **93:** 9799-9804.

Mason, R.P., Shoemaker, W.J., Shajenko, L., Chambers, T.E., and Herbette, L.G., 1992, Evidence for changes in the Alzheimer's disease brain cortical membrane structure mediated by cholesterol. *Neurobiol. Aging* **13:** 413-419.

Matsuzaki, K., and Horikiri, C., 1999, Interactions of amyloid β-peptide (1-40) with ganglioside-containing membranes. *Biochemistry* **38:** 4137-4142.

McLaurin, J., and Chakrabartty, A., 1996, Membrane disruption by Alzheimer β-amyloid peptides mediated through specific binding to either phospholipids or gangliosides. Implications for neurotoxicity. *J. Biol. Chem* **.271:** 26482-26489.

McLaurin, J., Darabie, A. A., and Morrison, M.R., 2002, Cholesterol, a modulator of membrane-associated Aβ -fibrillogenesis. *Ann. N. Y. Acad. Sci.***977:** 376-383.

Michikawa, M., and Yanagisawa, K., 1998, Apolipoprotein E4 induces neuronal cell death under conditions of suppressed de novo cholesterol synthesis. *J. Neurosci. Res.***54:** 58-67.

Michikawa, M., Fan, Q.W., Isobe, I., and Yanagisawa, K., 2000, Apolipoprotein E exhibits isoform-specific promotion of lipid efflux from astrocytes and neurons in culture. *J. Neurochem.* **74:** 1008-1016.

Mizuno, T., Haass, C., Michikawa, M., and Yanagisawa, K., 1998, Cholesterol-dependent generation of a unique amyloid β-protein from apically missorted amyloid precursor protein in MDCK cells. *Biochim. Biophys. Acta.* **1373:** 119-130.

Mizuno, T., Nakata, M., Naiki, H., Michikawa, M., Wang, R., Haass, C., and Yanagisawa, K., 1999, Cholesterol-dependent generation of a seeding amyloid β-protein in cell culture. *J. Biol. Chem.***274:** 15110-15114.

Molander-Melin, M., Blennow, K., Bogdanovic, N., Dellheden, B., Mansson, J.E. and Fredman, P., 2003, Lipid analyses of human brain rafts. *J. Neurochemistry* **85:** 34.

Mori, T., Paris, D., Town, T., Rojiani, A.M., Sparks, D.L., Delledonne, A., Crawford, F., Abdullah, L.I., Humphrey, J.A., Dickson, D.W., and Mullan, M.J., 2001, Cholesterol accumulates in senile plaques of Alzheimer disease patients and in transgenic APP(SW) mice. *J. Neuropathol. Exp. Neurol.***60:** 778-785.

Naeim, F., and Walford, R.L., 1980, Disturbance of redistribution of surface membrane receptors on peripheral mononuclear cells of patients with Down's syndrome and of aged individuals. *J. Gerontol.* **35:** 650-655.

Notkola, I.L., Sulkava, R., Pekkanen, J., Erkinjuntti, T., Ehnholm, C., Kivinen, P., Tuomilehto, J., and Nissinen, A., 1998, Serum total cholesterol, apolipoprotein E epsilon 4 allele, and Alzheimer's disease. *Neuroepidemiology* **17:** 14-20.

Oshima, N., Morishima-Kawashima, M., Yamaguchi, H., Yoshimura, M., Sugihara, S., Khan, K., Games, D., Schenk, D., and Ihara, Y., 2001, Accumulation of amyloid β-protein in the low-density membrane domain accurately reflects the extent of β-amyloid deposition in the brain. *Am. J. Pathol.* **158:** 2209-2218.

Papassotiropoulos, A., Lutjohann, D., Bagli, M., Locatelli, S., Jessen, F., Rao, M.L., Maier, W., Bjorkhem, I., von Bergmann, K., and Heun, R., 2000, Plasma 24S-hydroxycholesterol: a peripheral indicator of neuronal degeneration and potential state marker for Alzheimer's disease, *Neuroreport* **11:** 1959-1962.

Papassotiropoulos, A., Lutjohann, D., Bagli, M., Locatelli, S., Jessen, F., Buschfort, R., Ptok, U., Bjorkhem, I., von Bergmann, K., and Heun, R., 2002, 24S-hydroxycholesterol in cerebrospinal fluid is elevated in early stages of dementia. *J. Psychiatr. Res.* **36:** 27-32.

Papassotiropoulos, A., Streffer, J.R., Tsolaki, M., Schmid, S., Thal, D., Nicosia, F., Iakovidou, V., Maddalena, A., Lutjohann, D., Ghebremedhin, E., Hegi, T., Pasch, T., Traxler, M., Bruhl, A., Benussi, L., Binetti, G., Braak, H., Nitsch, R.M., and Hock, C., 2003, Increased brain β-amyloid load, phosphorylated tau, and risk of Alzheimer disease associated with an intronic CYP46 polymorphism. *Arch. Neurol.* **60:** 29-35.

Pentchev, P.G., Vanier, M.T., Suzuki, K., and Patterson, M.C., 1995, Niemann-Pick disease, type C: a cellular cholesterol lipidosis, in:*The metabolic and Molecular Basis of Inherited Disease* (C. R Scriver, A. L. Beaudet, W.S. Sly, and D.Vall, eds), McGraw-Hill, New York. pp.2625-2640

Poirier, J., Davignon, J., Bouthillier, D., Kogan, S., Bertrand, P., Gauthier, S., 1993, Apolipoprotein E polymorphism and Alzheimer's disease. *Lancet* **342:** 697-699.

Popplewell, P.Y., and Azhar, S., 1987, Effects of aging on cholesterol content and cholesterol-metabolizing enzymes in the rat adrenal gland. *Endocrinology* **121:** 64-73.

Probst, A., Langui, D., Ipsen, S., Robakis, N., and Ulrich, J., 1991, Deposition of β/A4 protein along neuronal plasma membranes in diffuse senile plaques. *Acta. Neuropatho.l (Berl)* **83:** 21-29.

Puglielli, L., Tanzi, R.E., and Kovacs, D.M., 2003, Alzheimer's disease: the cholesterol connection. *Nat. Neurosci.***6:** 345-351.

Refolo, L.M., Malester, B., LaFrancois, J., Bryant-Thomas, T., Wang, R., Tint, G.S., Sambamurti, K., Duff, K., and Pappolla, M.A., 2000, Hypercholesterolemia accelerates the Alzheimer's amyloid pathology in a transgenic mouse model. *Neurobiol. Dis.***7:** 321-331.

Refolo, L.M., Pappolla, M.A., LaFrancois, J., Malester, B., Schmidt, S.D., Thomas-Bryant, T., Tint, G.S., Wang, R., Mercken, M., Petanceska, S.S., and Duff, K.E., 2001, A cholesterol-lowering drug reduces β-amyloid pathology in a transgenic mouse model of Alzheimer's disease. *Neurobiol. Dis.* **8:** 890-899.

Saito, Y., Suzuki, K., Nanba, E., Yamamoto, T., Ohno, K., and Murayama, S., 2002, Niemann-Pick type C disease: accelerated neurofibrillary tangle formation and amyloid β deposition associated with apolipoprotein E epsilon 4 homozygosity, *Ann. Neurol.* **52:** 351-355.

Saunders, A.M., Strittmatter, W.J., Schmechel, D., George-Hyslop, P.H., Pericak-Vance, M.A., Joo, S.H., Rosi, B.L., Gusella, J.F., Crapper-MacLachlan, D.R., Growden, J., Alberts, M.J., Hulette, C., Crain, B., Goldgaber, D., and Roses, A.D., 1993, Association of apolipoprotein E allele epsilon 4 with late-onset familial and sporadic Alzheimer's disease. *Neurology* **43:** 1467-1472.

Sawamura, N., Morishima-Kawashima, M., Waki, H., Kobayashi, K., Kuramochi, T., Frosch, M.P., Ding, K., Ito, M., Kim, T.W., Tanzi, R.E., Oyama, F., Tabira, T., Ando, S., and Ihara, Y., 2000, Mutant presenilin 2 transgenic mice. A large increase in the levels of Aβ 42 is presumably associated with the low density membrane domain that contains decreased levels of glycerophospholipids and sphingomyelin. *J. Biol. Chem.***275:** 27901-27908.

Schonknecht, P., Lutjohann, D., Pantel, J., Bardenheuer, H., Hartmann, T., von Bergmann, K., Beyreuther, K., and Schroder, J., 2002, Cerebrospinal fluid 24S-hydroxycholesterol is increased in patients with Alzheimer's disease compared to healthy controls. *Neurosci. Lett.* **324:** 83-85.

Schroeder, F., Nemecz, G., Wood, W.G., Joiner, C., Morrot, G., Ayraut-Jarrier, M., and Devaux, P.F., 1991, Transmembrane distribution of sterol in the human erythrocyte,.*Biochim. Biophys. Acta.* **1066:** 183-192.

Scott, R.B., Collins, J.M., and Hunt, P.A., 1994, Alzheimer's disease and Down'ssyndrome: leukocyte membrane fluidity alterations. *Mech. Ageing. Dev.* **75:** 1-10.

Simons, K., and Ikonen, E., 1997, Functional rafts in cell membranes. *Nature* **387:** 569-572.

Simons, M., Keller, P., De Strooper, B., Beyreuther, K., Dotti, C.G., and Simons, K., 1998, Cholesterol depletion inhibits the generation of β-amyloid in hippocampal neurons. *Proc. Natl. Acad. Sci. U S A* **95:** 6460-6464.

Simons, M., Keller, P., Dichgans, J., and Schulz, J.B., 2001, Cholesterol and Alzheimer's disease: is there a link? *Neurology* **57:** 1089-1093.

Soderberg, M., Edlund, C., Alafuzoff, I., Kristensson, K., and Dallner, G., 1992, Lipid composition in different regions of the brain in Alzheimer's disease/senile dementia of Alzheimer's type. *J. Neurochem* .**59:** 1646-1653.

Sparks, D.L., Scheff, S.W., Hunsaker, J.C., 3rd, Liu, H., Landers, T., and Gross, D.R., 1994, Induction of Alzheimer-like β-amyloid immunoreactivity in the brains of rabbits with dietary cholesterol. *Exp. Neurol.***126:** 88-94.

Stahlberg, D., Angelin, B., and Einarsson, K., 1991, Age-related changes in the metabolism of cholesterol in rat liver microsomes. *Lipids* **26:** 349-352.

Strittmatter, W.J., Weisgraber, K.H., Huang, D.Y., Dong, L.M., Salvesen, G.S., Pericak-Vance, M., Schmechel, D., Saunders, A.M., Goldgaber, D., and Roses, A.D., 1993, Binding of human apolipoprotein E to synthetic amyloid β peptide: isoform-specific effects and implications for late-onset Alzheimer disease. *Proc. Natl. Acad. Sci. U S A* **90:** 8098-8102.

Sugimoto, Y., Ninomiya, H., Ohsaki, Y., Higaki, K., Davies, J.P., Ioannou, Y.A., and Ohno, K., 2001, Accumulation of cholera toxin and GM1 ganglioside in the early endosome of Niemann-Pick C1-deficient cells. *Proc. Natl. Acad. Sci. U S A* **98:** 12391-12396.

Suzuki, K., Parker, C.C., Pentchev, P.G., Katz, D., Ghetti, B., D'Agostino, A.N., and Carstea, E.D., 1995, Neurofibrillary tangles in Niemann-Pick disease type C. *Acta. Neuropathol. (Berl)* **89:** 227-238.

Vance, J.E., Pan, D., Campenot, R.B., Bussiere, M., and Vance, D.E., 1994, Evidence that the major membrane lipids, except cholesterol, are made in axons of cultured rat sympathetic neurons. *J. Neurochem* .**62:** 329-337.

Wahrle, S., Das, P., Nyborg, A.C., McLendon, C., Shoji, M., Kawarabayashi, T., Younkin, L.H., Younkin, S.G., and Golde, T.E., 2002, Cholesterol-dependent β-secretase activity in buoyant cholesterol-rich membrane microdomains. *Neurobiol. Dis.* **9:** 11-23.

Wolozin, B., Kellman, W., Ruosseau, P., Celesia, G.G., and Siegel, G., 2000, Decreased prevalence of Alzheimer disease associated with 3-hydroxy-3-methyglutaryl coenzyme A reductase inhibitors. *Arch. Neurol.***57:** 1439-1443.

Wood, W.G., Cornwell, M., and Williamson, L.S., 1989, High performance thin-layer chromatography and densitometry of synaptic plasma membrane lipids. *J. Lipid Res.***30:** 775-779.

Wood, W.G., Schroeder, F., Avdulov, N.A., Chochina, S.V., and Igbavboa, U., 1999, Recent advances in brain cholesterol dynamics: transport, domains, and Alzheimer's disease. *Lipids* **34:** 225-234.

Yamaguchi, H., Maat-Schieman, M.L., van Duinen, S.G., Prins, F.A., Neeskens, P., Natte, R., and Roos, R.A., 2000, Amyloid β protein (Aβ) starts to deposit as plasma membrane-bound form in diffuse plaques of brains from hereditary cerebral hemorrhage with amyloidosis-Dutch type, Alzheimer disease and nondemented aged subjects. *J. Neuropathol. Exp. Neurol.***59:** 723-732.

Yamazaki, T., Chang, T.Y., Haass, C., and Ihara, Y., 2001, Accumulation and aggregation of amyloid β-protein in late endosomes of Niemann-pick type C cells. *J. Biol. Chem.***276:** 4454-4460.

Yanagisawa, K., Odaka, A., Suzuki, N., and Ihara, Y., 1995, GM1 ganglioside-bound amyloid β-protein (Aβ): a possible form of preamyloid in Alzheimer's disease. *Nat. Med.* **1:** 1062-1066.

Yanagisawa, K., McLaurin, J., Michikawa, M., Chakrabartty, A., and Ihara, Y., 1997, Amyloid β-protein (Aβ) associated with lipid molecules: immunoreactivity distinct from that of soluble Aβ, *FEBS Lett.* **420:** 43-46.

Yanagisawa, K., and Ihara, Y., 1998, GM1 ganglioside-bound amyloid β-protein in Alzheimer's disease brain, *Neurobiology of aging* **19:** 65-67.

Yip, C.M., Elton, E.A., Darabie, A.A., Morrison, M.R., and McLaurin, J., 2001, Cholesterol, a modulator of membrane-associated A β-fibrillogenesis and neurotoxicity, *J. Mol. Biol.* **311:** 723-734.

Chapter 10

Alzheimer's β-Amyloid: Insights into Fibril Formation and Structure from Congo Red Binding

Hideyo Inouye and Daniel A. Kirschner
Department of Biology, Boston College, Chestnut Hill, MA 02467-3811, USA

Abstract: We consider here the chemistry of Congo red (CR), its binding equilibrium to Alzheimer's β-amyloid, and the kinetics of β-amyloid formation. Spectroscopic UV/Vis measurements for the pH- and time-dependence binding of CR to Aβ analogues are analysed by Scatchard binding and the theory of nucleation-dependent fibril formation. CR likely binds electrostatically to the imidazolium sidechains of histidine residues that are exposed at the surface of amyloid fibrils. As revealed by atomic models of the Aβ protofilament, such as the nanotube β-helix and parallel β-sheet, the regular arrangement of histidines likely acts as a template for the end-to-end *J*-aggregation of CR molecules, which produces a red shift in UV/Vis absorption.

Key words: kinetics of amyloid aggregation, histidine, spectroscopy, β-helix, parallel β-sheet

1. INTRODUCTION

When histological sections of Alzheimer's brain tissue are viewed between crossed polarizers after treatment with Congo red (CR) in alkaline ethanol solution saturated with NaCl (Puchtler *et al.*, 1962), the amyloid plaques exhibit green birefringence. Saturated NaCl is needed to distinguish the amyloid from collagen and elastin fibrils, as the latter two give similar birefringence in dilute NaCl solution (Elghetany *et al.*, 1989). The birefringence of amyloid was initially thought to arise from CR binding to sugar; however, it was later shown to come from binding to proteins that have an extensive β-sheet conformation, which is one of the defining characteristics of amyloid fibrils in Alzheimer's and other amyloidoses (Sipe and Cohen, 2000; Glenner *et al.* 1972). Use of CR for diagnostic or therapeutic purposes has been proposed in the hope that this dye might detect early stages of amyloid formation or be used to disrupt or arrest amyloidogenesis (Linke, 2000; Kisilevsky *et al.*, 1995; Caughey and Race, 1992). CR effects are concentration dependent, *i.e.*, at lower concentration CR binds to the unfolded protein and can promote amyloidogenesis, whereas at higher concentration the dye acts as an inhibitor of amyloid fibril formation (Kim *et al.*, 2003). A similar aggregation effect of CR on β-lactoglobulin aggregation is evident from small-angle X-ray scattering (Khurana *et al.*, 2001).

Mechanisms that have been proposed for CR binding include: (1) hydrogen-bonding between the amino groups of CR and the hydroxyl group of the peptide or sugar (Puchtler *et al.*, 1962); (2) non-specific dye intercalation in amyloid (Cooper, 1974); (3) ionic interactions between the sulfonate group of CR and basic residues of the peptide (Klunk et al, 1999, 1994, 1989a,b); and (4) intercalation of CR between two anti-parallel hydrogen-bonded β-sheets (Carter and Chou, 1998; Turnell and Finch, 1992). Mechanisms (1), (2) and (4) are based on electron microscopy and/or X-ray diffraction, and (3) uses UV/Vis spectroscopy. The binding constant for CR to fibrils (including insulin, poly-lysine, -serine, and -glycine, and Aβ) has been estimated only with the UV/Vis method. CR binding has also been used to monitor amyloid fibril formation as a function of time (Inouye and Kirschner, 2000; Wood *et al.*, 1996) and pH (Inouye *et al.*, 2000), and in the presence or absence of the excluded volume effect due to dextran (Hatters *et al.*, 2002).

Molecules and macromolecular assemblies other than amyloid bind CR, *e.g.*, dehydrogenases (Edwards and Woody, 1979), RNA polymerase (Woody *et al.*, 1981), cellulose (Quenin and Henrissat, 1985), elastin (Tan *et al.*, 1991), chitin (Bartnicki-Garcia *et al.*, 1994), and HIV-1 protease (Ojala *et al.*, 1995), and CR inhibits the accumulation and propagation of prions (Caspi *et al.*, 1998). Polyanions that are more extensive than CR, such as DNA and RNA, can promote prion fibril formation, likely due to ionic interactions (Deleault *et al.*, 2003; Nandi *et al.*, 2002).

Two basic questions regarding CR binding are: (1) the strength of the binding, and (2) the number of binding sites as a function of time during fibril formation. With respect to the latter, a relation between the observed concentration of bound CR and the number of monomers in the fibril needs to be studied. In this paper we review the chemistry of CR, the binding equilibrium for CR (Scatchard, 1949), and the theory of nucleation-dependent aggregation as probed by the binding of CR (Inouye and

Kirschner, 2000; Ferrone, 1999; Oosawa and Asakura, 1975). We also consider whether current atomic models for Aβ fibrils, e.g., a nanotube β-helix (Perutz *et al.*, 2002) and a parallel β-sheet (Tycko, 2003), can account for the observed binding of this dye.

2. CHEMISTRY AND MOLECULAR STRUCTURE

Congo red (CR) is a sulfonated diazo compound which typically is used as a pH-dependent colour indicator. Thus, the colour change of CR in water from red at alkaline pH ≥4 to blue at acidic pH ≤4 arises from resonance effects due to protonation or to binding of cations to the unshared electron pairs of the nitrogen atoms of the naphthalene rings and azo groups (Mera and Davies, 1984). In non-aqueous media CR shows a different pH range, in which CR appears either blue or red, depending on the way that the local electrostatic field influences the proton titration (Inouye *et al.*, 2000). The sulfonate groups are protonated at pH ~2 and amino groups at ~4, as the proton dissociation constant for the aniline amino group is 4.6 in water (Gutbezahl and Grunwald, 1953). Thus, the net charge for CR is +2 at pH<~2, 0 at ~pH 2-4, and -2 at pH>~4 (Inouye *et al.*, 2000). The UV/Vis spectrum of CR at neutral pH shows a π-π* band at 498 nm, another π-π* band at 350 nm, and low intensity n-π* bands associated with the azo groups (Cooper and Stone, 1998; Mera and Davies, 1984). The absorption spectrum shows blue or red shifts when CR binds to peptides. For example, a red shift and *hyper*chromicity occur when CR binds to Alzheimer's Aβ1-40 (Klunk *et al.*, 1999), insulin fibrils (Klunk *et al.*, 1989a,b), the β form of poly-L-lysine (Klunk *et al.*, 1989a), and to $(AEAEAKAKAEAEAKAK)_9$ ("EAK9"), a nona-repeat peptide (Goeden-Wood *et al.*, 2003). By contrast, a blue shift and *hypo*chromicity occur when CR binds to the short peptide acetyl-YAAAKAAAAKAAAAKA-amide ("YAK123") (Cooper and Stone, 1998).

X-ray crystallographic analysis of CR shows that the molecule is extended and flat (Figure 1) (Ojala *et al.*, 1995). The central domain of biphenyl rings is hydrophobic, while the sulfonate and amino groups are hydrophilic. In the single crystal form the sulfonate groups are oriented *anti* with respect to the molecule's long axis (Ojala *et al.*, 1995); however, when complexed with insulin the sulfonates have a *syn* orientation (Turnell and Fintch, 1992), indicating torsional flexibility in the biphenyl region. CR molecules aggregate into a larger assembly which may be its form in binding to amyloid fibrils (Skowronek *et al.*, 1998), and which may underlie its birefringence and blue or red shift in UV/Vis absorption (Cooper and Stone, 1998). The hypsochromically-shifted *H*-aggregate (*H* for hypsochromic, or shift to a shorter wavelength in its absorption spectrum) and bathochromically-shifted *J*-aggregate (*J* for Jerry who studied this shift to a longer wavelength) of other dyes may be similar to that for the CR assembly. These two types of dye aggregates show characteristic absorption shifts from the absorption maximum of the monomer, i.e., a blue shift for *H*- and red shift for *J*-aggregation. X-ray and electron microscopy of dyes show highly ordered structures for dye aggregates, e.g., the rod-like helical structure of *H*-aggregates for merocyanine (Wurthner *et al.*, 2003), the columnar (*H*-) and sheet (*J*-) structures of polyphyrins (Shirakawa *et al.*, 2003), the *H*- and *J*-aggregates for the DNA-templated cyanine (Garoff *et al.*, 2002), and tubular cylinder of porphyrin (Gandini *et al.*, 2003). The helical peptide YAK123 causes CR to stack parallel, forming *H*-aggregates, while peptide EAK9 causes CR to stack end-to-end, forming *J*-aggregates (see above).

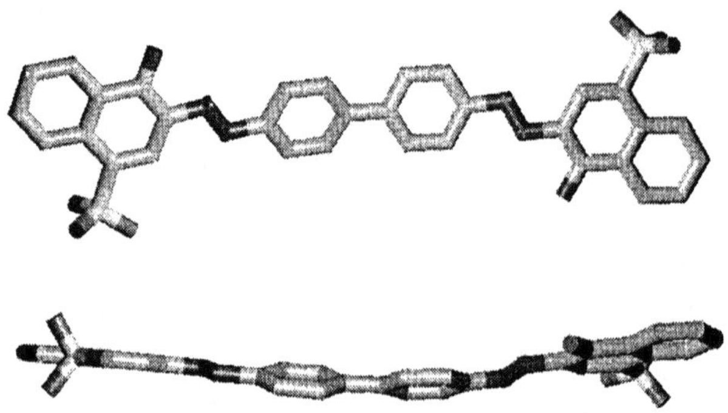

Figure 1. Crystallographic structure of Congo red (Ojala *et al.*, 1995). The sulfonate groups are oriented *anti* with respect to the long axis. The length of the molecule is ~24 Å, and the width across the planar face is ~6 Å. (The structure was rendered using RASTER3D [Merritt and Bacon, 1997] after being displayed with VMD software [Humphrey *et al.*, 1996]).

3. CR BINDING EQUILIBRIUM

3.1 Scatchard plot

According to the independent binding model (Inouye *et al.*, 2000; Scatchard, 1949), the binding equilibrium defines how strongly CR binds to a site and how many sites there are. As CR binding likely depends on the pH of the aqueous medium, the proton dissociation constant as well as the CR dissociation constant K_d must be explicitly written. Accordingly, the dissociation equilibrium is written as

$$R_iL \Leftrightarrow L_f + R_i$$

where L_f is free ligand, R_i is the binding site for the species, and R_iL is bound ligand. The dissociation constant $K_{d(i)}$ is given by

$$K_{d(i)} = [L_f][R_i]/[R_iL].$$

Given the total concentration of binding sites as $B_{\max(i)} = [R_i] + [R_iL]$, and assuming that the dissociation constants for different species are constant and that $K_{d(i)} = K_d$, then the concentration of bound CR for different sites

$\alpha = \sum_i [R_iL]$ is given by

$$\alpha = \{\beta - (\beta^2 - 4L_t \sum_i B_{\max(i)})^{1/2}\}/2,$$

where $\beta = \sum_i B_{\max(i)} + [L_t] + K_d$.

Assuming a single type of binding site, then $i = 1$ and $B_{\max(i)} = B_{\max}$, and the linearized form gives the classic Scatchard plot

$$[RL]/[L_f] = B_{\max}/K_d - [RL]/K_d.$$

3.2 Effect of pH on CR binding

If the CR binding site is titratable by protons, the concentrations of the protonated and unprotonated sites in the proton dissociation reaction

$$B_iH^+ \Leftrightarrow B_i + H^+$$

are given by the proton dissociation constant

$$K_i = [B_i][H^+]/[B_iH^+].$$

If CR binds only to positive-charged protonated sites, and one CR molecule (with two sulfonates) does not bind to more than two sites, then the concentration of the site is given by

$$B_{\max(i)} = [B_i H^+]/m,$$

where m is the binding stoichiometry 1 or 2. Using the total concentration of $B_{t(i)} = [B_i] + [B_i H^+]$, then

$$B_{\max(i)} = ([B_{t(i)}]/m)/(1 + 10^{pH - pK_i}).$$

4. UV/VIS SPECTROSCOPY OF CR BINDING

4.1 UV/Vis spectroscopy

Filtration and spectroscopy (using two wavelengths) are two methods for measuring the concentration of bound CR molecules (Klunk *et al.*, 1989a,b). For the spectroscopic method the extinction coefficients of both the free and bound forms of CR and peptide must be known to determine the concentration of bound CR. The concentration of CR, for example, can be calculated using the equation

$$C_b = A_{540}/25{,}295 - A_{480}/46{,}306,$$

where C_b is the concentration of bound CR, and A_{540} and A_{480} are the absorption at wavelengths 540 nm and 480 nm, respectively. This equation was derived from the binding of CR to insulin fibrils (Klunk *et al.*, 1989a,b) and has been applied to Aβ (Wood *et al.*, 1996). Corrected for light scattering and for the specific extinction coefficient of CR binding to Aβ fibrils (Goldsbury *et al.*, 2000; Klunk *et al.*, 1999), the equation becomes

$$C_b = A_{541}/47{,}800 - A_{403}/38{,}100.$$

The filtration method, on the other hand, needs only an extinction coefficient for free CR, e.g., $5 \times 10^4\ M^{-1}\ cm^{-1}$ (Inouye *et al.*, 2000; Edwards and Woody, 1979), because there is a physical separation of the bound and free forms. When used to study the binding of CR to Aβ analogues (Inouye *et al.*, 2000), this method employed 1K-size cut-off filters (MICROSEP Microconcentrators; Filtron Technology Corporation, Northborough, MA 01532) in which the pore size was much smaller than that used by Klunk *et al.* (1989a,b). For example, the concentration dependence of bound CR on total CR for Aβ11-28 is shown in Figures 2 and 3. For the experimental conditions illustrated, the number of binding sites n is 2.5 and the CR dissociation constant K_d is 10.8 μM.

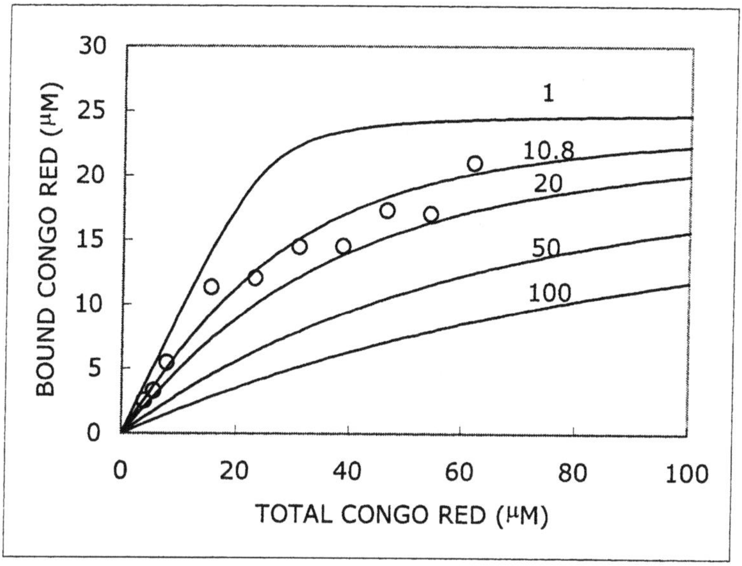

Figure 2. The concentration of bound CR vs. total CR for 10 μM Aβ11-28 at pH 4.0 (Inouye *et al.*, 2000) for different K_d values (from 1 to 100) and constant n=2.5. The continuous curves were derived from Scatchard plots defined by B_{max}=n[P], where n is the number of binding sites, *[P]* is the peptide concentration, and K_d is the dissociation constant (in μM). The best curve was for n=2.5 and K_d=10.8. Experimentally, peptide solution in a 1-ml Eppendorf microfuge was applied to the upper layer of the micro-concentrator and spun at 3000 x g for 3-4 h at 4 °C until all of the solution had passed through the filter. Here, 80% ethanol-solubilized CR but not aqueous CR at the same pH and ionic strength was completely filtered, likely due to self-aggregation of the aqueous dye (Edward and Woody, 1979). Lower concentration of ethanol (10%) also has been shown to prevent micelle formation (Klunk *et al.*, 1994).

4.2 pH titration shows specific CR binding site: histidine

Electrostatic effects on the binding of CR depend on the proton dissociation constant of the binding site for a given pH and ionic strength of the bulk medium. Analysis of the pH dependency of CR binding to Aβ amyloid analogues having different numbers of charged residues (Fraser *et al.*, 1992b), therefore, may help to identify where CR binds. Systematic pH titration experiments for a number of Aβ analogues (Table 1) including shorter peptides, peptides with amino acid substitutions, and full-length Aβ from other species, illustrate this approach (Inouye *et al.*, 2000).

The concentration of bound CR as a function of pH shows that for all amyloid analogues except 19-28 (which had nearly zero binding of CR over the examined pHs) the bound CR is at a much higher concentration at acidic pH ~4 than at alkaline pH (Table 2 and Figure 4). Above pH 6 all analogues showed almost no CR binding. This pH dependency is similar to the proton dissociation for the imidazolium sidechain of histidine, which has an intrinsic pK=6.3. At pH 4, all analogues showed their maximum binding, in the order Aβ1-40 > Aβ11-28 = Aβ13-28 > Aβ11-25 >> Aβ19-28. A similar pH dependency, i.e., stronger binding at acidic pH and weaker binding at alkaline pH, is observed for Aβ1-40 from different origins. The maximum binding at pH 4 decreased in the order Pr(40-1) > Pr(1-40) = Du(1-40) > Ro(1-40). Replacement of a single residue in the Aβ11-25 sequence showed different maximum binding concentration at pH 4, with the concentration of bound CR decreasing in the order D23K > 11-25 > H14D > H13D. This order correlates with the number of histidine residues.

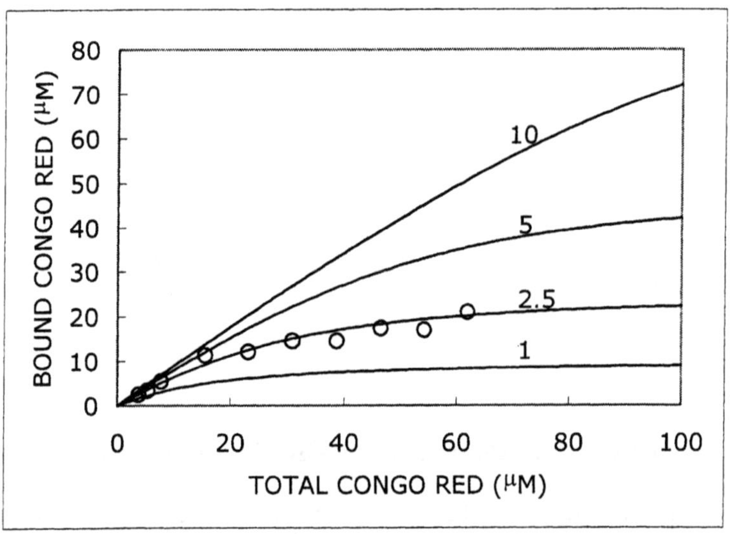

Figure 3. The dependence of the concentration of bound CR on total CR for 10 μM Aβ11-28 at pH 4.0 (Inouye *et al.*, 2000) for different *n* values (from 1 to 10) and constant K_d=10.8. See legend to Figure 2 for details.

The NH_2 group of CR becomes deprotonated at pH 4, and the two sulfonates are likely to be fully dissociated at pH>4. Thus, over the pH range 4.0-9.5 the net charge of CR is -2. Proton dissociation of the ionizable groups in protein are likely influenced by the local electrostatic field which shifts the intrinsic *pK* to apparent *pK* values (Inouye and Kirschner, 1988; Nozaki and Tanford, 1967).

Table 1. Sequences for Aβ1-40 and analogues

Analogue[a]	Sequence[b]
Pr1-40	DAEFRHDSGYEVHHQKLVFFAEDVGSNKGAIIGLMVGGVV
Ro1-40	DAEFGHDSGFEVRHQKLVFFAEDVGSNKGAIIGLMVGGVV
Du1-40	DAEFRHDSGYEVHHQKLVFFAQDVGSNKGAIIGLMVGGVV
Pr40-1	VVGGVMLGIIAGKNSGVDEAFFVLKQHHVEYGSDHRFEAD
11-28	EVHHQKLVFFAEDVGSNK
13-28	HHQKLVFFAEDVGSNK
19-28	FFAEDVGSNK
11-25	EVHHQKLVFFAEDVG
11-25 H13D	EV**D**HQKLVFFAEDVG
11-25 H14D	EVH**D**QKLVFFAEDVG
11-25 D23K	EVHHQKLVFFAE**K**VG

[a] *Pr*, primate; *Ro*, rodent; *Du*, Dutch type
[b] residue substitutions are **highlighted**

Table 2. Summary of characteristics related to CR binding

Analogue	*pI*[a]	*H*[b]	*b* (Å)	pK_H	K_d	pK_a
Pr1-40	5.4	3	11.7	5.6	5.9	5.3
Du1-40	6.0	3	11.7	5.6	5.9	5.3
Pr40-1	5.4	3	11.7	5.6	5.9	5.3
11-28	6.3	2	9.1	5.7	2.8	5.5
13-28	7.4	2	8.8	5.6	2.8	5.5
11-25	5.4	2	8.8	5.6	2.8	5.5
Ro1-40	4.9	2	11.7	5.9	2.8	5.5
D23K	7.9	2	8.6	5.6	2.8	5.5
H13D	4.2	1	8.6	5.9	5.5	5.0
H14D	4.2	1	8.6	5.9	5.5	5.0
19-28	4.2	0	7.4	-	-	-

[a] pI was obtained from pH at which net peptide charge is zero. Charge was calculated from intrinsic *pK* values for the ionizable residues including N and C termini. Effect of ionic strength on charge was not considered.

[b] *H*, number of histidine residues in the peptide; *b*, radius of peptide sphere used for calculating protein surface charge; pK_H, proton dissociation constant of His in medium with low dielectric constant (38.2 for 80% ethanol).

All peptides showed a lower pK_H than intrinsic value 6.3 of histidine. pK_a values estimated from concentration of bound CR were smaller than pK_H, indicating that the histidine residues are in a hydrophobic environment, and interact with the positive-charged residues.

Using the formula for CR binding as a function of pH, we calculated the concentration of bound dye by assuming CR only binds to histidines and with a 1:1 stoichiometry, and determined the best dye dissociation (K_d) and proton dissociation (pK_a) constants (Table 2 and Figure 4). Peptides having three histidines include Pr1-40, Du1-40 and Pr40-1, those having two histidines are 11-28, 13-28, 11-25, Ro1-40, and D23K, and those having a single histidine are H13D and H14D. The best values were pK_a=5.3, K_d=5.9 μM (relative error between observed and calculated bound CR, R=15.3%) for three histidines, pK_a=5.5, K_d=2.8 μM (R=16.1%) for two histidines, and pK_a=5.0, K_d=5.5 μM (R=26.1%) for one histidine (Inouye *et al.*, 2000). These K_d values are similar to those reported for CR binding to Aβ analogues (Klunk *et al.*, 1994).

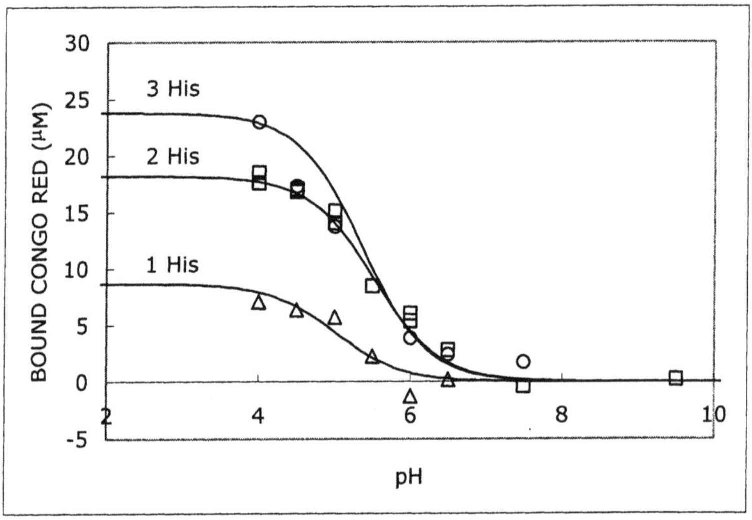

Figure 4. Concentration of bound CR vs. pH of incubation medium for amyloid analogues containing: O, three histidines (Pr1-40); □, two (Aβ11-28); and Δ, one (H13D) (Inouye *et al.*, 2000). Total concentrations were 10 μM peptide and 46.44 μM CR. The calculated pH titration curves for 1:1 dye binding were derived with (from the top) K_d=5.9 μM and pK=5.3 (three histidines), K_d=2.8 μM and pK=5.5 (two) and K_d=5.6 μM and pK=5.0 (one).

5. KINETIC STUDY OF AGGREGATION AND CR BINDING

The nucleation-dependent self-assembly model defines a critical aggregation number of nuclei, rate constant for production, and rate of monomer addition (Oosawa and Asakura, 1975). This model and a modified form have been used in explaining the time evolution of various protein assemblies (Inouye and Kirschner, 2000) including fibrils related to neurodegenerative diseases, *i.e.* Aβ amyloid, prion peptide fragments, yeast prion, Alzheimer's paired helical filaments, α-synuclein, and polyglutamine-containing huntingtin fragments. The experimental methods for studying fibril formation vary (Inouye and Kirschner, 2000), and include electron microscopy (Fraser *et al.*, 1991a,b), light scattering (Lomakin *et al.*, 1997; Tomski and Murphy, 1992), sedimentation (Burdick *et al.*, 1992), turbidity (Taylor *et al.*, 2003), thioflavinee T (LeVine, 1993; Naiki *et al.*, 1991), CR (Inouye and Kirschner, 2000; Wood *et al.* 1996), and atomic force microscopy (Khurana *et al.*, 2003). Equations relating the concentration of bound CR to the aggregation number (*i.e.* monomer concentration in fibrils) as a function of time and pH (Inouye and Kirschner, 2000) allow one to calculate the kinetic parameters and the size of the nucleus from experimental data on the concentration of bound CR (Wood *et al.*, 1996). This determination is based on the following:

- At pH ~3-7, there is a conformational change from the native monomeric structure to the fibrillar.
- Under these conditions Aβ sediments (Burdick *et al.*, 1992), forms β sheets (Barrow *et al.*, 1992), and binds CR (Wood *et al.*, 1996) and thioflavinee T (LeVine, 1993).
- Aspartate, glutamate, and histidine residues are likely involved in the pH dependency of β amyloid aggregation (Kirschner *et al.*, 1987), for the proton dissociation constants of their side chains are in this pH range (3-7).
- CR binds preferentially to the aggregated but not to the monomeric form of amyloid (Klunk *et al.*, 1989a,b).
- CR binding sites are likely to be exposed on the fibril surface while concealed in the native monomers.
- The histidine residues are exposed on the surface of the fibrils (Inouye *et al.*, 1993).
- Histidines are proposed to be the binding sites for CR, based on the pH dependency of CR binding to preformed fibrils (Inouye *et al.*, 2000; Fraser *et al.*, 1992a) and the likelihood that the imidazolium is positive-charged at pH ~3-7.

The concentration of the monomers within a fibril can be calculated from the total concentration of peptide and the proton dissociation constants of the carboxyl (pK_1) and imidazole (pK_2) groups. The reaction scheme for self-assembly of Aβ (Oosawa and Asakura, 1975) is

$$i_0 A_1 \underset{k_-^*}{\overset{k_+^*}{\Leftrightarrow}} A_{i_0};\ A_{i_0} + A_1 \underset{k_-}{\overset{k_+}{\Leftrightarrow}} A_{i_0+1};\ A_{i_0+1} + A_1 \underset{k_-}{\overset{k_+}{\Leftrightarrow}} A_{i_0+2};\bullet\bullet\bullet\bullet;$$

$$A_{i-1} + A_1 \underset{k_-}{\overset{k_+}{\Leftrightarrow}} A_i;\ A_i + A_1 \underset{k_-}{\overset{k_+}{\Leftrightarrow}} A_{i+1};\bullet\bullet\bullet\bullet;$$

where A denotes an Aβ protein molecule, A_1 is the monomer, A_{i_0} is the nucleus (considered to be the smallest assembly), A_i is the i-mer assembly, k_+^*, k_-^* are the rates of nuclei production and dissociation, and k_+ and k_- are the rates of polymer i production and dissociation (where i is larger than the number of monomers forming a nucleus i_0). The maximum i is assumed to be large and is not explicitly indicated. The concentration of Aβ nuclei is λ_{i_0} and that of i-mer assemblies is λ_i. At time zero $\lambda_1 = c_f$, where c_f is the total monomer concentration. The monomer concentration as a function of time is then

$$\ln\frac{1+(1-(\lambda_1/c_f)^{i_0})^{1/2}}{1-(1-(\lambda_1/c_f)^{i_0})^{1/2}} = (2i_0)^{1/2}(k_+^* k_+)^{1/2} c_f^{i_0/2} t.$$

If the polymer concentration m is constant as a function of time, the term $d\sum_i \lambda_i/dt$ becomes zero, and seeded assembly can be written as

$$\lambda_1(t) - \lambda_1(\infty) = \exp(-k_+ mt)[\lambda_1(0) - \lambda_1(\infty)].$$

Taking the logarithm of both sides,

$$\ln[\lambda_1(t) - \lambda_1(\infty)] = -k_+ mt + \ln[\lambda_1(0) - \lambda_1(\infty)],$$

where $\lambda_1(0) = c_f$ at $t = 0$ and $\lambda_1(\infty) = k_-/k_+$ since at equilibrium the term $d\lambda_1/dt = 0$. In this case k_+^* (the rate constant for nucleus formation) becomes 0, indicating that the seed is already present and no seeds are spontaneously created. This equation is the same as that for amyloid fibril elongation (Naiki *et al.*, 1991), and for the kinetics of fibril formation at low peptide concentration (Lomakin *et al.*, 1997). The parameters $[M]_e, k_{-1}, k_2$ in Naiki *et al.* (1991) correspond to $\lambda_1(\infty), k_-, k_+$ in the notation here (Inouye and Kirschner, 2000), and $k_e N_0$ in Lomakin *et al.* (1997) is identical to $k_+ m$ used here.

Using the concentration vs. time data for CR bound to Aβ1-40 at pH 5.8 (Wood *et al.*, 1996), nonlinear least squares (Inouye *et al.*, 1989) optimizes the kinetic parameters for two different assembly processes, i.e., (1) spontaneous and (2) seeded. Input constants include the total concentration of peptide (3.69 μM), total CR concentration (24 μM), the number of His (3) and the number of Asp and Glu (6) residues, proton dissociation constant pK_1 (3.0), and pH of the medium (5.8). The parameters include number of molecules per nucleus i_0, kinetic rate coefficient $k_+^* k_+$, dissociation constant K_d of bound CR, proton dissociation constant pK_2 for the former spontaneous model, and K_d, pK_2, and the rate coefficient $k_+ m$ for the latter (seeded) model. The calculated optimized values for spontaneous assembly are K_d=55.1, $k_+^* k_+ = 2.0$, pK_2=6.3, and i_0=3; and the ones for seeded assembly are K_d=53.7, $k_+ m$=5.0, and pK_2=6.22. The agreement between the observed and calculated values for both models is good; however, the seeded one fits slightly better than spontaneous assembly (Figure 5).

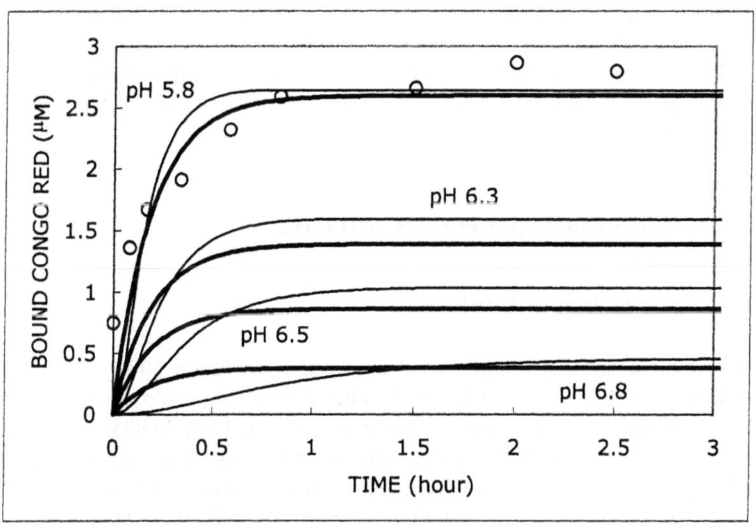

Figure 5. The concentration of bound CR as a function of time for Aβ1-40. The experimental data (O) are from Figure 2A of Wood *et al.* (1996). The thin and thick lines indicate spontaneous and seeded assembly, respectively, at pH 5.8, 6.3, 6.5 and 6.8. The total peptide concentration is 3.69 μM and the CR concentration is 24 μM. The kinetics parameters are cited in the text and in Inouye and Kirschner (2000).

6. CR BINDING SITE IN AMYLOID β FIBRILS

Electron microscopy and X-ray diffraction show that ~30 Å-diameter protofilaments constitute Aβ amyloid fibrils, both in vivo (Miyakawa *et al.*, 1986) as well as in different in vitro assemblies formed by Aβ analogues consisting of residues 6-25, 11-28, and 1-28 (Inouye *et al.*, 1993; Fraser *et al.*, 1991a,b; Kirschner *et al.*, 1987) as well as full sequence 1-40 (Malinchik *et al.*, 1998). The protofilaments are cylindrically arrayed, building a thicker fibril. Electron density profiles suggest that the subunit of the protofilament (or β-crystallite, as defined by X-ray diffraction) is mostly formed by the hydrophobic sequence Leu17-Val18-Phe19-Phe20-Ala21 within Aβ (Inouye and Kirschner, 1996; Inouye *et al.*, 1993). The neighbouring hydrophilic residues His13-His14 that are N-terminal to this domain are, therefore, likely to be on the surface of the protofilament. Previous studies (Fraser *et al.*, 1992a) show that fibrils formed by Aβ analogues having no histidine residues or at alkaline pH where histidines are uncharged do not show CR-induced lateral aggregation, suggesting that the protonated histidines on neighboring fibrils may be bridged by CR molecules to promote lateral packing. The histidine domain Val12-His13-His14-Gln15-Lys16-Leu17 in Aβ is similar to the known binding motif *XBBXB* (*B*=basic, *X*=hydrophobic) for heparan sulfate proteoglycan (HSGP) (Watson *et al.*, 1997; Brunden *et al.*, 1993; Kisilevsky, 1989). The negative-charged groups in CR and in HSGP (sulfonates and sulfates, respectively), therefore, must interact with the same positive-charged histidine residues that are exposed in Aβ fibrils (Ojala *et al.*, 1996).

7. AMYLOID β ATOMIC MODELS

7.1 Structure in solution

The atomic coordinates for Aβ determined by solution NMR (Protein Data Bank, 1AML; Sticht *et al.*, 1995) indicate an α-turn-α structure where the two helices are in the regions Gln15-Asp23 and Ile31-Met35, and the turn is at Gly25-Ser26-Asn27. The positions of the first helix and turn are nearly consistent with the ones predicted by the methods of Chou-Fasman and Garnier (Malinchik *et al.*, 1998). The binding of CR to the hydrophilic and flexible N-terminal domain (Figure 6) may facilitate amyloid formation (Kim *et al.*, 2003).

7.2 Nanotube β-helix

Based on analysis of X-ray diffraction and electron microscopy from a polyglutamine peptide assembly, $(Asp)_2$-$(Gln)_{15}$-$(Lys)_2$, Perutz and colleagues (2002) proposed an atomic model which they suggested might be valid also for other fibrillar amyloids, i.e., a water-filled nanotube. The authors interpreted the diffraction pattern as a fiber pattern, i.e., with the 4.8 Å reflection along the fibril axis. Their model depicts a β-helical array having 20 residues per turn; and the diameter of the helix as defined by the wrapping of the main chain is 20 Å. The 30 Å-outer diameter accounts for the observed low-angle reflection as well as for the size of the glutamine side chains of 5 Å, i.e., 30 Å - (2 × 5 Å) = 20 Å. Rather than attributing the prominent 8.3 Å reflection series to intersheet stacking of β sheets, the authors attributed it to a feature of unknown origin, and suggested that the absence of the typical 10 Å intersheet reflections meant that the β sheets are not stacked, but instead there is a single β sheet constituting the thin wall of the tube. The equatorial doublet at 3.2 Å and 3.6 Å was interpreted as arising from the repeat of amino acids in a parallel β-sheet (forming the tube), with the side chain amides in a circle facing the internal water-filled cavity (Perutz *et al.*, 2002). The structural consequence of this model is that two turns having 40 residues are held together by hydrogen bonds between

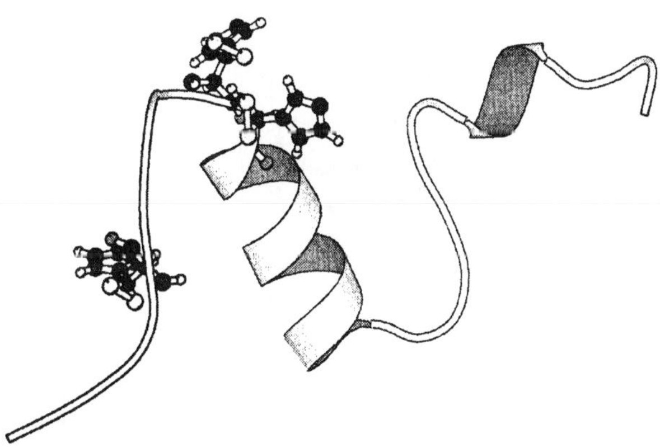

Figure. 6. NMR solution structure for Aβ1-40 (entry #1AML, Protein Data Bank; Sticht *et al.*, 1995). The three histidine residues (His6, His13 and His14, shown as balls and sticks) are located in the N-terminal flexible region, and Gly25-Ser26-Asn27 is located at the turn. Model drawn using MOLSCRIPT (Kraulis, 1991).

amides of successive turns, and this gives a stable nucleus for growth of a helical fibril. In this way the model cleverly accounts for the correlation between the neurodegenerative triplet-repeat CAG diseases that require a polyglutamine length greater than 40 residues.

Because the helical pitch is the 4.8 Å-hydrogen bonding distance, and the asymmetric unit on the discrete helix consists of two residues in the β-chain direction, then the helical selection rule is $1/c = n/P + m/h$ where c is the period along the fiber direction, P is pitch, and h is rise per unit. Because $c = P = 4.8$ Å, and $h = 4.8$Å$/10$, then $1 = n + 10m$, and the helical radius is 10 Å. This helical array predicts that the layer lines are at every (4.8 Å)$^{-1}$ along the fiber direction. The first layer line at 4.8 Å spacing corresponds to the J_1 Bessel function which gives an off-meridional intensity maximum in the radial direction at R=(35 Å)$^{-1}$. The nanotube model is consistent with the finding that Aβ with 40-42 residues gives a similar diameter protofilament and a 4.7 Å reflection (Perutz *et al.*, 2002; Malinchik *et al.*, 1998).

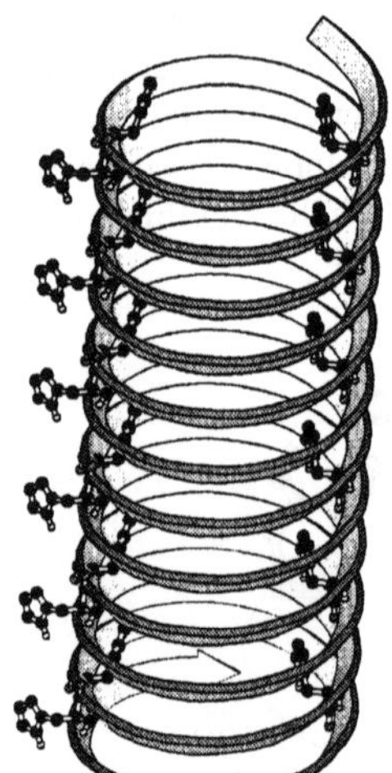

Figure 7. Model for Aβ1-40 as derived from nanotube model of polyglutamine (Perutz *et al.*, 2002). The model is viewed nearly normal to the fibril axis. We built the model given the assumed helical selection rule and radius of the helix (see text for details). Balls and sticks indicate the three histidines, where His6 and His14 face inward and His13 faces outward. Model drawn using MOLSCRIPT (Kraulis, 1991).

Using the β helical chain structure proposed for polyglutamine, we replaced the side chains with ones for Aβ1-40 (Figure 7). Within the putative domain for CR binding (Val12-His13-His14-Gln15-Lys16-Leu17) His13 faces outward, and His14 faces inward. Because His6 in the N-terminal domain also faces inward, then CR molecules that bind histidines should be localized both within and on the surface of the Aβ helical tube.

7.3 Parallel β sheet model

Solid state NMR indicates that the Aβ amyloid protofilament is built up from parallel β chains (Tycko, 2003). Aβ9-40 is modelled as two extended β chains with a turn at Ser26-Asn27-Lys28 (Figure 8). Two β chains interact across the intersheet space so that the stacking of the peptides along the hydrogen bonding direction gives parallel β sheets. Another similarly folded molecule placed along the intersheet direction accounts for the ~30 Å width of the protofilament, and there are four β chains in this direction (Malinchik *et al.*, 1998). The hydrophilic N-terminal domain is on the surface of the protofilament, while the hydrophobic C-terminal domain is located in the protofilament core. Both His13 and His14 residues are on the surface of the filament, and can be bound to CR.

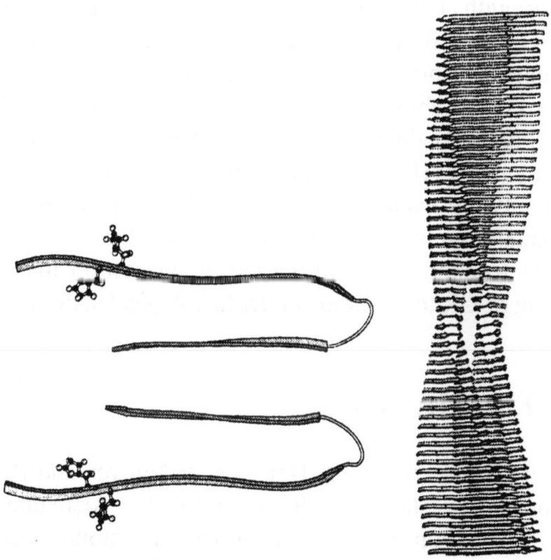

Figure 8. The parallel β chain model for Aβ (Tycko, 2003). (*Left*) Aβ9-40 has a turn at Ser26-Asn27-Lys28 which creates anti-parallel β chains in the inter-sheet direction (vertical axis). The hydrogen-bonding direction is normal to the page. His13 and His14 are indicated by the balls and sticks. (*Right*) The fibril grows in the hydrogen bonding direction (vertical direction axis). Models drawn using MOLSCRIPT (Kraulis, 1991).

8. CONCLUSIONS

Both ionic and non-ionic types of molecular interactions have been put forward to explain the binding of CR to Aβ fibrils. In the first type, negative-charged sulfonate groups of CR are proposed to interact with positive-charged histidine side chains of the peptide. Atomic structures for the β-helix nanotube and parallel β-sheet models both indicate a regular array of histidine residues on the surface of the fibrils, where ionic interactions between CR and the histidines (Inouye *et al.*, 2000) would facilitate lateral aggregation of protofilaments into fibrils (Fraser *et al.*, 1992a) either by reducing the electrostatic repulsion between edge- or surface-located histidines, or by the dye molecules chelating or bridging neighbouring protofilaments. A similar role for histidines in lateral packing of fibrils has been reported to underlie the Cu^{2+}-induced aggregation of betabellin (Lim *et al.*, 1999). Regular arrangement of histidines on the protofilament may act as a template for the end-to-end *J*-aggregate of CR molecules, which produces a red shift in UV/Vis absorption.

At neutral pH where histidine is deprotonated and no longer charged, or at high ionic strength, CR still binds to Aβ amyloid fibrils (Wood *et al.*, 1996). In fact, CR appears to bind universally to amyloid fibrils having the cross-β arrangement. When the length of the β-chain (~6 residues) (Inouye *et al.*, 1993) is comparable to the ~20 Å-length of the flat CR molecule which is parallel to the chain, van der Waals interactions between the peptide and the dye's π system (Hermel *et al.*, 1995) might be stronger, promoting dye binding to amyloid even in the absence of protonated histidines. Since the protofilaments from various amyloids are similar in size (Inouye and Kirschner, 1998; Cooper, 1974), this type of interaction may account for non-ionic binding of CR to amyloids under histological staining conditions.

ACKNOWLEDGMENTS

We thank our former colleagues Drs. Jack T. Nguyen and Paul E. Fraser for collaboration on the study of CR binding to the Alzheimer β-amyloid fibril, Dr. W. B. Gleason for the crystallographic atomic coordinates of Congo red, and Dr. R. Tycko for the solid-state NMR atomic coordinates of Aβ. Research at Boston College was supported by an Alzheimer's Association/T.L.L. Temple Discovery Award (to D.A.K.), and by institutional support from Boston College. D.A.K. also acknowledges a Fulbright Senior Research Scholar Award from the Binational US-Italian Fulbright Scholar Program during the preparation of this manuscript.

REFERENCES

Barrow, C.J., Yasuda, A., Kenny, P.T.M., and Zagorski, M.G., 1992, Solution conformations and aggregational properties of synthetic amyloid β-peptides of Alzheimer's disease. Analysis of circular dichroism spectra. *J. Mol. Biol.* **225:** 1075-1093.

Bartnicki-Garcia, S., Persson, J., and Chanzy, H., 1994, An electron microscope and electron diffraction study of the effect of calcofluor and congo red on the biosynthesis of chitin in vitro. *Arch. Biochem. Biophys.* **310:** 6-15.

Brunden, K.R., Richter-Cook, N.J., Chaturvedi, N., and Frederickson, R.C.A., 1993, pH-dependent binding of synthetic β-amyloid peptides to glycosaminoglycans. *J. Neurochem.* **61:** 2147-2154.

Burdick, D., Soreghan, B., Kwon, M., Kosmoski, J., Knauer, M., Henschen, A., Yates, J., Cotman, C., and Glabe, C., 1992, Assembly and aggregation properties of synthetic Alzheimer's A4/β amyloid peptide analogs. *J. Biol. Chem.* **267:** 546-554.

Carter, D.B., and Chou, K.-C., 1998, A model for structure-dependent binding of Congo red to Alzheimer β-amyloid fibrils. *Neurobiol. Aging.* **19:** 37-40.

Caspi, S., Halimi, M., Yanai, A., Sasson, S.B., Taraboulos, A., and Gabizon, R., 1998, The anti-prion activity of Congo red. Putative mechanism. *J. Biol. Chem.* **273:**3484-3489.

Caughey, B., and Race, R.E., 1992, Potent inhibition of scrapie-associated PrP accumulation by Congo red. *J. Neurochem.* **59:** 768-771.

Cooper, J.H., 1974, Selective amyloid staining as a function of amyloid composition and structure. Histochemical analysis of the alkaline Congo red, standardized toluidine blue, and iodine methods. *Lab. Invest.* **31:** 232-238.

Cooper, T.M., and Stone, M.O., 1998, Investigation of self-assembly upon formation of an electrostatic complex of Congo red and a helical peptide. *Lagmuir* **14:** 6662-6668.

Deleault, N.R., Lucassen, R.W., and Supattapone, S., 2003, RNA molecules stimulate prion protein conversion. *Nature* **425:** 717-720.

Edwards, R.A., and Woody, R.W., 1979, Spectroscopic studies of Cibacron Blue and Congo Red bound to dehydrogenases and kinases. Evaluation of dyes as probes of the dinucleotide fold. *Biochemistry* **18:** 5197-5204.

Elghetany, M.T., Saleem, A., and Barr, K., 1989, The congo red stain revisited. *Ann. Clin. Lab. Sci.* **19:** 190-195.

Ferrone, F., 1999, Analysis of protein aggregation kinetics. *Methods Enzymol.* **309:** 256-274

Fraser, P.E., Duffy, L.K., O'Malley, M.B., Nguyen, J., Inouye, H., and Kirschner, D.A., 1991a, Morphology and antibody recognition of synthetic β-amyloid peptides. *J. Neurosci. Res.* **28:** 474-485.

Fraser, P.E., Nguyen, J.T., Surewicz, W.K., and Kirschner, D.A., 1991b, pH-dependent structural transitions of Alzheimer amyloid peptides. *Biophys. J.* **60:** 1190-1201.

Fraser, P.E., Nguyen, J.T., Chin, D.T., and Kirschner, D.A., 1992a, Effects of sulfate ions on Alzheimer β/A4 peptide assemblies: Implications for amyloid fibril-proteoglycan interactions. *J. Neurochem.* **59:** 1531-1540.

Fraser, P.E., Nguyen, J.T., Inouye, H., Surewicz, W.K., Selkoe, D.J., Podlisny, M.B., and Kirschner, D.A., 1992b, Fibril formation by primate, rodent, and Dutch-hemorrhagic analogues of Alzheimer amyloid β-protein, *Biochemistry* **31:** 10716-10723.

Gandini, S.C.M., Gelamo, E.L., Itri, R., and Tabak, M., 2003, Small angle x-ray scattering study of meso-tetrakis(4-sulfonatophenyl) porphyrin in aqueous solution: A self-aggregation model. *Biophys. J.* **85:** 1259-1268.

Garoff, R.A., Litzinger, E.A., Connor, R.E., Fishman, I., and Armitage, B.A., 2002, Helical aggregation of cyanine dyes on DNA templates: Effect of dye structure on formation of homo- and heteroaggregates. *Langmuir* **18:** 6330-6337

Glenner, G.G., Eanes, E.D., and Page, D.L., 1972, The relation of the properties of Congo red-stained amyloid fibrils to the –conformation. *J. Histochem. Cytochem.* **20:** 821-826.

Goeden-Wood, N.L., Keasling, J.D., and Muller, S.J., 2003, Self-assembly of a designed protein polymer into β-sheet fibrils and responsive gels. *Macromolecules* **36:** 2932-2938.

Goldsbury, C.S., Wirtz, S., Muller, S.A., Sunderji, S., Wicki, P., Aebi, U., and Frey, P., 2000, Studies on the in vitro assembly of Aβ 1-40: Implications for the search for Aβ fibril formation inhibitors. *J. Struct. Biol.* **130:** 217-231.

Gutbezahl, B., and Grunwald, E., 1953, The effect of solvent on equilibrium and rate constants. II. The measurement and correlation of acid dissociation constants of anilinium and ammonium salts in the system ethanol-water. *J. Amer. Chem. Soc.* **75:** 559-574.

Hatters, D.M., Minton, A.P., and Howlett, G.J., 2002, Macromolecular crowding accelerates amyloid formation by human apolipoprotein C-II. *J. Biol. Chem.* **277:** 7824-7830.

Hermel, H., Holtje, H.D., Bergemann, S., De Rossi, U., and Kriwanek, J., 1995, Band-shifting through polypeptide β-sheet structures in the cyanine UV-Vis spectrum, *Biochim. Biophys. Acta* **1252:** 79-86.

Humphrey, W., Dalke, A., and Schulten, K., 1996, VMD: Visual molecular dynamics. *J. Mol. Graph.* **14:** 33-38.

Inouye, H., and Kirschner, D.A., 1988, Membrane interactions in nerve myelin: II. Determination of surface charge from biochemical data. *Biophys. J.* **53:** 247-260.

Inouye, H., and Kirschner, D.A., 1996, Refined fibril structures: The hydrophobic core in Alzheimer's amyloid β-protein and prion as revealed by X-ray diffraction. In: *The Nature and Origin of Amyloid Fibrils, Ciba Foundation Symposium* No. 199, John Wiley & Sons, New York, pp. 22-39.

Inouye, H., and Kirschner, D.A., 1998, Polypeptide chain folding in the hydrophobic core of hamster scrapie prion: Analysis by X-ray diffraction. *J. Struct. Biol.* **122:** 247-255.

Inouye, H., and Kirschner, D.A., 2000, Aβ fibrillogenesis: Kinetic parameters for fibril formation from Congo red binding. *J. Struct. Biol.* **130:** 123-129.

Inouye, H., Karthigasan, J., and Kirschner, D.A., 1989, Membrane structure in isolated and intact myelins. *Biophys. J.* **56:** 129-137.

Inouye, H., Fraser, P.E., and Kirschner, D.A., 1993, Structure of β-crystallite assemblies formed by Alzheimer β-amyloid protein analogues: analysis by x-ray diffraction. *Biophys. J.* **64:** 502-519.

Inouye, H., Nguyen, J.T., Fraser, P.F., Shinchuk, L.M., Packard, A.B., and Kirschner, D.A., 2000, Histidine residues underline Congo red binding to Aβ analogs. *Amyloid* **7:**179-188.

Khurana, R., Uversky, V.N., Nielsen, L., and Fink, A.L., 2001, Is Congo red an amyloid-specific dye?, *J. Biol. Chem.* **276:** 22715-22721.

Khurana, R., Ionescu-Zanetti, C., Pope, M., Li, J., Nielson, L., Ramírez-Alvarado, M., Regan, L., Fink, A.L., and Carter, S.A., 2003, A general model for amyloid fibril assembly based on morphological studies using atomic force microscopy. *Biophys. J.* **85:** 1135-1144.

Kim, Y.S., Randolph, T.W., Manning, M.C., Stevens, F.J., and Carpenter, J.F., 2003, Congo red populates partially unfolded states of an amyloidogenic protein to enhance aggregation and amyloid fibril formation. *J. Biol. Chem.* **278:** 10842-10850.

Kirschner, D.A., Inouye, H., Duffy, L.K., Sinclair, A., Lind, M., and Selkoe, D.J., 1987, Synthetic β-protein of Alzheimer disease forms amyloid-like fibrils *in vitro*. *Proc. Natl. Acad. Sci. USA*, **84:** 6953-6957.

Kisilevsky, R., 1989, Theme and variations on a string of amyloid. *Neurobiol. Aging.* **10:** 499-500.

Kisilevsky, R., Lemieux, L.J., Fraser, P.E., Kong, X., Hultin, P.G., and Szarek, W.A., 1995, Arresting amyloidosis in vivo using small-molecule anionic sulphonates or sulphates: implications for Alzheimer's disease. *Nat. Med.* **1:** 143-148.

Klunk, W. E., Pettegrew, J.W., and Abraham, D.J., 1989a, Quantitative evaluation of congo red binding to amyloid-like proteins with a β-pleated sheet conformation. *J. Histochem. Cytochem.* **37:** 1273-1281.

Klunk, W.E., Pettegrew, J.W., and Abraham, D.J., 1989b, Two simple methods for quantifying low-affinity dye-substrate binding. *J. Histochem. Cytochem.* **37:** 1293-1297.

Klunk, W.E., Debnath, M.L., and Pettegrew, J.W., 1994, Development of small molecule probes for the β-amyloid protein of Alzheimer's disease. *Neurobiol. Aging.* **15:** 691-698.

Klunk, W.E., Jacob, R.F., and Mason, R.P., 1999, Quantifying amyloid β-peptide (Aβ) aggregation using the Congo red-Aβ(CR-Aβ) spectrophotometric assay. *Anal. Biochem.* **266:** 66-76.

Kraulis, P.J., 1991, MOLSCRIPT: A program to produce both detailed and schematic plots of protein structures. *J. Appl. Crystallogr.* **24:** 946-950.

LeVine, H. III, 1993, Thioflavinee T interaction with synthetic Alzheimer's disease β-amyloid peptides: Detection of amyloid aggregation in solution, *Protein Sci.* **2:** 404-410.

Lim, A., Makhov, A.M., Saderholm, M.J., Griffith, J.D., and Erickson, B.W., 1999, Biophysical characterization of betabellin 16D: A β-sandwich protein that forms narrow fibrils which associate into broad ribbons. *Biochem. Biophys. Res. Commun.* **264:** 498-504.

Linke, R.P., 2000, Highly sensitive diagnosis of amyloid and various amyloid syndromes using Congo red fluorescence, *Virchows Arch.* **436:** 439-448.

Lomakin, A., Teplow, D.B., Kirschner, D.A., and Benedek, G.B., 1997, Kinetic theory of fibrillogenesis of amyloid β-protein. *Proc. Natl. Acad. Sci. USA* **94:** 7942-7947.

Malinchik, S.B., Inouye, H., Szumowski, K.E., and Kirschner, D.A., 1998, Structural analysis of Alzheimer's β(1-40) amyloid: Protofilament assembly of tubular fibrils. *Biophys. J.* **74:** 537-545.

Mera, S.L., and Davies, J.D., 1984, Differential Congo red staining: the effects of pH, non-aqueous solvents and the substrate. *Histochem. J.* **16:** 95-210.

Merritt, E.A., and Bacon, D.J., 1997, Raster3D: Photorealistic molecular graphics. *Methods Enzymol.* **277:** 505-524.

Miyakawa, T., Watanabe, K., and Katsuragi, S., 1986, Ultrastructure of amyloid fibrils in Alzheimer's disease and Down's syndrome. *Virchows Arch.[Cell Pathol.]* **52:** 99-106.

Naiki, H., Higuchi, K., Nakakuki, K., and Takeda, T., 1991, Kinetic analysis of amyloid fibril polymerization in vitro. *Lab. Invest.* **65:** 104-110.

Nandi, P.K., Leclerc, E., Nicole, J.-C., and Takahashi, M., 2002, DNA-induced partial unfolding of prion protein leads to its polymerisation to amyloid. *J. Mol. Biol.* **322:** 153-161.

Nozaki, Y., and Tanford, C., 1967, Examination of titration behavior. *Meth. Enzymol.* **11:** 715-734.

Ojala, W.H., Ojala, C.R., and Gleason, W.B., 1995, The X-ray crystal structure of the sulfonated azo dye Congo Red, a non-peptidic inhibitor of HIV-1 protease which also binds to reverse transcriptase and amyloid proteins. *Antiviral Chem. Chemother.* **6:** 25-33.

Ojala, W.H., Sudbeck, E.A., Lu, L.K., Richardson, T.I., Lovrien, R.E., and Gleason, W.B., 1996, Complexes of lysine, histidine, and arginine with sulfonated azo dyes: Model systems for understanding the biomolecular recognition of glycosaminoglycans by proteins. *J. Am. Chem. Soc.* **118:** 2131-2142.

Oosawa, F., and Asakura, S., 1975, *Thermodynamics of the Polymerization of Protein*, Academic Press, New York.

Perutz, M.F., Finch, J.T., Berriman, J., and Lesk, A., 2002, Amyloid fibers are water-filled nanotubes. *Proc. Natl. Acad. Sci. USA* **99:** 5591-5595.

Puchtler, H., Sweat, F., and Levine, M., 1962, On the binding of Congo red by amyloid. *J. Histochem. Cytochem.* **10:** 355-364.

Quenin, I., and Henrissat, B., 1985, Precipitation and crystallization of cellulose doped with dyes. *Makromol. Chem.* **6:** 737-741.

Scatchard, G., 1949, The attractions of proteins for small molecules and ions. *Ann. N.Y. Acad. Sci.* **51:** 660-672.

Shirakawa, M., Kawano, S., Fujita, N., Sada, K., and Shinkai, S., 2003, Hydrogen-bond-assisted control of H versus J aggregation model of porphyrins stacks in an organogel system. *J. Org. Chem.* **68:** 5037-5044.

Sipe, J.D., and Cohen, A.S., 2000, Review: History of the amyloid fibril. *J. Struct. Biol.* **130:** 88-98.

Skowronek, M., Stopa, B., Konieczny, L., Rybarska, J., Piekarska, B., Szneler, E., Bakalarski, G., and Roterman, I., 1998, Self-assembly of Congo red – A theoretical and experimental approach to identify its supramolecular organization in water and salt solutions,.*Biopolymers* **46:** 267-281.

Sticht, H., Bayer, P., Willbold, D., Dames, S., Hilbich, C., Beyreuther, K., Frank, R.W. and Rosch, P., 1995, Structure of amyloid A4-(1-40)-peptide of Alzheimer's disease. *Eur. J. Biochem.* **233:** 293-298.

Tan, O.T., Faris, B., and Franzblau, C., 1991, Aortic elastin fluorescence after in vivo labeling with Congo red. *J. Fluorescence* **1:** 147-151.

Taylor, B.M., Sarver, R.W., Fici, G., Poorman, R.A., Lutzke, B.S., Molinari, A., Kawabe, T., Kappenman, K., Buhl, A.E., and Epps, D.E., 2003, Spontaneous aggregation and cytotoxicity of the β-amyloid Aβ1-40: A kinetic model. *J. Protein Chem.* **22:** 31-40.

Tomski, S.J., and Murphy, R.M., 1992, Kinetics of aggregation of synthetic β-amyloid peptide. *Arch. Biochem. Biophys.* **294:** 630-638.

Turnell, W.G., and Finch, J.T., 1992, Binding of the dye Congo red to the amyloid protein pig insulin reveals a novel homology amongst amyloid-forming peptide sequences. *J. Mol. Biol.* **227:** 1205-1223.

Tycko, R., 2003, Insights into the amyloid folding problem from solid-state NMR. *Biochemistry* **42:** 3151-3159.

Watson, D.J., Lander, A.D., and Selkoe, D.J., 1997, Heparin-binding properties of the amyloidogenic peptides Aβ and amylin. Dependence on aggregation state and inhibition by Congo red. *J. Biol. Chem.* **272:** 31617-31624.

Wood, S.J., Maleeff, B., Hart, T., and Wetzel, R., 1996, Physical, morphological and functional differences between pH 5.8 and 7.4 aggregates of the Alzheimer's amyloid peptide Aβ. *J. Mol. Biol.* **256**: 870-877.

Woody, A.Y., Reisbig, R.R., and Woody, R.W., 1981, Spectroscopic studies of Congo red binding to RNA polymerase. *Biochim. Biophys. Acta* **655**: 82-88.

Wurthner, F., Yao, S., and Beginn, U., 2003, Highly ordered merocyanine dye assemblies by supramolecular polymerization and hierarchical self-organization. *Angew. Chem. Int. Ed. Engl.* **42**: 3247-3250.

Chapter 11

The Aluminium-Amyloid Cascade Hypothesis and Alzheimer's Disease

Aluminium and β-amyloid

Christopher Exley
Birchall Centre for Inorganic Chemistry and Materials Science, Keele University, Staffordshire, UK

Abstract: Aluminium (Al) is found associated with β-amyloid (Aβ) in the brain in Alzheimer's disease. Al precipitates Aβ *in vitro* as distinct fibrillar structures composed of β-pleated sheets of peptide. The aetiology of their association *in vivo* is not known. Al is known to increase the brain Aβ burden in experimental animals and this might be due to a direct influence upon Aβ anabolism or direct or indirect affects upon Aβ catabolism. It is difficult to rationalise from an evolutionary perspective the precipitation and persistence of Aβ *in vivo*. However, Al has not been subject to the same evolutionary pressures as Aβ, it is a recent addition to the biotic environment, and its precipitation of Aβ may have only been subjected to natural selection in the recent past. Whether AD is also part of this ongoing selection process remains to be elucidated

Key words: Aluminium, Aβ, APP metabolism, Aβ aggregation, Aβ catabolism, Aβ neurotoxicity;

1. INTRODUCTION

What is the connection between aluminium (Al) and the β-amyloid peptide (Aβ)? Was the first observation of their co-localisation in senile plaques serendipitous or erroneous (Duckett and Galle, 1980)? The case for serendipidity has remained open (Exley, 2001) whilst the question of whether Al is co-localised with Aβ in senile plaques has been addressed and Al has been shown to be one of a number of metals found associated with these structures (Beauchemin and Kisilevsky, 1998). The authors of this study were acutely aware of the purported problems associated with the contamination of reagents (Landsberg *et al.*, 1992) and took great care to account for any such possible sources of error. They showed unequivocally that Al was the second most abundant metal, behind iron, that was found associated with senile plaques isolated from Alzheimer's brain tissue. So why do some authors continue to contest this issue? Their objections seem to go beyond scientific reasoning.

Of course the role of Aβ in the aetiology of AD is itself contentious though this shall not be addressed herein. Regardless of the cause of AD the 'Amyloid Cascade' is a description of the probable metabolism of Aβ *in vivo* (Hardy and Higgins, 1992), and this established series of events will form the framework of this insight into the association of Aβ with Al. This review has been written so as to complement and not repeat the content of a recent publication in this field (Exley and Korchazhkina, 2001a).

2. ALUMINIUM AND THE ANABOLISM OF Aβ

It is equivocal as to whether Al influences the expression of the amyloid precursor protein (APP) (Exley and Korchazhkina, 2001a). However, the recent suggestion that APP metabolism might be linked to neuronal iron homeostasis (Rogers *et al.*, 2002) may also suggest a role for Al. The identification of an iron regulatory element (IRE) within the 5'-untranslated region of APP mRNA has implicated iron regulatory proteins (IRP1 and IRP2) in APP translation. IRE binding of an IRP signals a translation event and Al has been shown to both promote (*via* IRP2) (Yamanaka *et al.*, 1999) and inhibit (*via* IRP1) (Oshiro *et al.*, 1998) IRE binding of IRP's. In the case of IRP2 the effect was attributed to either an Al-induced stabilisation of the IRE/IRP2 complex or an Al-induced inhibition of the oxidative breakdown of IRP2. The authors speculated that IRP2 contained a binding site for iron which would also bind Al. The inhibition of IRE activity by Al has been attributed to IRP1 binding Al (Oshiro, 2002). IRP1 in the inactive state contains an iron-sulphur cluster (4Fe-4S). Activation of IRP1, such that it is bound by an IRE, is achieved by the oxidation of the 4Fe-4S cluster by the superoxide radical anion ($O_2^{\bullet-}$) and the subsequent release of Fe(II). The authors speculated that the substitution of the labile Fe in the Fe-S cluster by redox-inactive Al (thus, Al-3Fe-4S) would protect IRP1 against oxidation and so prevent it being bound by an IRE and signalling a translation event. Al substitution for Fe in an Fe-S cluster is not supported by the reported weak binding of Al by S ligands (Harris *et al.*, 1996). An alternative mechanism whereby Al might influence the oxidation of IRP1 without being bound by the cluster might involve it acting as a competitive substrate for $O_2^{\bullet-}$. The formation of the putative aluminium superoxide semi-reduced radical ion ($AlO_2^{\bullet 2+}$) (Exley, 2004) would limit the availability of $O_2^{\bullet -}$ and would reduce the rate of oxidation of the Fe-S cluster and so inhibit IRE binding of IRP1. The influence of Al on the activity of IRP's could provide a mechanism for an Al-induced increase or decrease in the translation of APP. These effects could be mediated *via* direct interactions with IRP's or through indirect actions upon neuronal iron homeostasis (Ward *et al.*, 2001).

It is not known if Al exerts any direct influence upon the secretases (α, β, γ) that have been charged with the proteolytic cleavage of APP. We do know that Al inhibits the activity of the serine protease, plasmin, Korchazhkina *et al.*, 2002) and that this enzyme is known to promote processing of APP *via* the non-amyloidogenic α-secretase pathway (Ledesma *et al.*, 2000). The latter may be achieved either by direct cleavage of APP, for example, plasmin has been shown to cleave Aβ between Lys_{28} and Gly_{29} (Korchazhkina *et al.*, 2002) and may also cleave at the α-secretase site (Lys_{16} – Leu_{17}) or by activation of MMP-9, a matrix metalloproteinase which is known to cleave Aβ at the α-secretase site (Backstrom *et al.*, 1996). In addition exposure of explanted human brain neural cells in primary culture to only 100 nM Al resulted in down-regulation of the gene for MDC9, a putative α-secretase (Koike *et al.*, 1999), and up-regulation of a gene for APP (Kuroda *et al.*, 2003). The action of Al in inhibiting the α-secretase pathway whilst concomitantly increasing APP expression would alter the physiological balance which exists between the α- and β-secretase pathways in favour of the formation of Aβ. Whether this balance is altered in the sporadic form of AD is equivocal whilst Aβ is produced in larger amounts in familial forms of the disease and in Down's syndrome. Interestingly, AD and Down's syndrome are linked in that in both conditions the gastrointestinal absorption of Al is significantly increased (Moore *et al.*, 2000; Moore *et al.*, 1997). Whether the increased production of Aβ in both of these diseases is as a consequence of a higher body burden of Al or *vice versa* is not known. The enzymatic processing of APP could be influenced directly by it binding Al in such a way as to alter the accessibility of the secretases to cleavage sites (Exley and Korchazhkina, 2001a) or indirectly by Al acting as a prooxidant (Exley, 2004) in altering the structure and composition of neuronal membranes or rafts where APP is localised. Until the requisite research is funded and carried out it will remain equivocal as to whether the burgeoning burden of brain Al is influencing the synthesis, transport, function and metabolism of APP.

3. ALUMINIUM AND THE AGGREGATION OF Aβ

A possible rationale for the co-localisation of Al and Aβ in senile plaques is that the peptide forms a complex with Al. Direct experimental evidence to support this contention was first presented over a decade ago when we demonstrated that Al disrupted the conformation of $A\beta_{40}$ in a membrane-mimicking solvent (Exley *et al.*, 1993). It was shown using circular dichroism spectroscopy that $A\beta_{40}$ adopted a partially helical conformation which was severely disrupted by a stoichiometric excess of Al. Abolition of α-helical structure was evident for a two-fold molar excess of Al and complete for a six-fold excess. The result demonstrated unequivocally that $A\beta_{40}$ complexed Al and that only a small molar excess of the metal was required to induce a change in its conformation in the presence of a membrane-mimicking solvent. This first observation of Aβ binding Al has been confirmed in a decade of further research (reviewed in Exley and Korchazhkina, 2001a) though it has not been supported by the determination of a formation constant for an Al-Aβ complex. The latter is not a trivial pursuit for either metals or metal-peptide complexes which are extremely insoluble at circumneutral pH and the lack of a thermodynamic constant does not, as has been suggested, preclude the existence of the complex.

There can be no question that Aβ forms β-pleated sheets in the presence of Al (Exley and Korchazhkina, 2001c). It does not form β-pleated sheets in the presence of copper (Zou *et al.*, 2001) and it only forms such in the presence of zinc after a considerable period of aging (House *et al.*, 2004). The classical indicator of β-pleated sheet formation is their binding of thioflavine T (ThT) to produce a fluor which upon excitation at *ca* 450 nm will emit a characteristic fluorescence at *ca* 482 nm (Levine, 1993). In solutions of Aβ which are aggregating Al has been shown to both promote and reduce the formation of ThT-reactive material. Generally, the aggregation of $A\beta_{40}$ is promoted whereas that for $A\beta_{42}$ is reduced (Exley and Korchazhkina, 2001c). However, it has recently come to our attention that there may be a simple explanation for such apparently opposite effects. By correlating ThT fluorescence (β-pleated sheet formation) with images obtained using transmission electron microscopy (TEM) we have found that Al promoted the self-aggregation of ThT-reactive material (β-pleated sheets) and that as a consequence it reduced the availability of binding sites for ThT. Thus Aβ in the presence of Al was precipitated in compact fibrillar structures which gave a lower ThT fluorescence than Aβ precipitated in the absence of additional Al. This property of Al was only observed when the metal was present to a molar excess such that crosslinking of peptide-bound Al *via* hydroxy-Al bridges would be favoured (House *et al.*, 2004). Aluminium's known activity as a cross-linking agent (Exley, 1998) may also explain how ligands which are competitive with hydroxide for binding Al, such as ATP (Exley, 1997) and inorganic phosphate (Bondy and Truong, 1999), increased ThT fluorescence of Aβ in the presence of Al by inhibiting the formation of hydroxy-Al bridges and thereby preventing the self-aggregation of ThT-reactive material. The high inherent propensity for $A\beta_{42}$ to aggregate would be expected to accentuate the influence of Al and might explain why this effect is less evident for $A\beta_{40}$ (Exley and Korchazhkina, 2001c).

Connected to the capacity for Al to induce self-aggregation of β-pleated sheets of $A\beta_{42}$ was the recent observation that *in vitro* preparations of $A\beta_{42}$ were contaminated with Al (House *et al.*, 2004). For example, we found that 1 μM concentrations of $A\beta_{42}$ which had been prepared in PIPES-buffered Krebs-Henseleit media contained *ca* 0.7 μM Al. This combination of $A\beta_{42}$ and Al resulted in the formation of ThT-reactive material and amyloid fibrils as viewed by TEM whereas the same combination in the presence of an Al chelator, DFO, neither resulted in ThT fluorescence nor amyloid fibrils. $A\beta_{42}$ did not form β-pleated amyloid fibrils when Al that had been present as a contaminant was chelated by DFO. The inability of 1 μM peptide to form fibrils was supported by the recent suggestion that the saturation constant for $A\beta_{42}$ was > 2 μM (Sengupta *et al.*, 2003). In our recent study Al was a prerequisite to amyloid fibril formation. Whether Al could act as a nidus for the precipitation of $A\beta_{42}$ as β-pleated sheets *in vivo* must now be considered a possibility.

4. ALUMINIUM AND THE CATABOLISM OF Aβ

A number of proteases which are able to degrade Aβ both *in vitro* and *in vivo* have been identified (Carson and Turner, 2002). All are known to cleave monomeric Aβ whilst only plasmin has additionally been implicated in the degradation of soluble oligomers (Selkoe *et al.*, 2003) and β-pleated sheet aggregates (Tucker *et al.*, 2000) of Aβ. Importantly, the degradation of Aβ by plasmin does not result in the formation of amyloidogenic Aβ fragments (Exley and Korchazhkina, 2001b) and consequently plasmin protects against the neurotoxicity of Aβ. Plasmin is also an α-secretase and its formation *via* the urokinase plasminogen activator system is genetically linked through chromosome 10 to late-onset AD (Finckh *et al.*, 2003). Plasmin is a serine protease and the activities of this class of enzymes are known to be influenced by Al (Clauberg and Joshi, 1993). We investigated the degradation of $A\beta_{25-35}$ by plasmin and showed that the peptide was cleaved between Lys_{28} and Gly_{29} to produce two non-amyloidogenic fragments. Al reduced the activity of plasmin by half and this interaction was attributed to a direct interaction with the enzyme as opposed to the $A\beta_{25-35}$

substrate (Korchazhkina *et al.*, 2002). It has not been determined if Al inhibits the activity of plasmin either in *in vitro* cell/tissue culture or *in vivo*. However, inhibition of plasmin activity could result in stimulation of the β-γ secretase pathway and reduced clearance of Aβ. In spite of the fact that the plasmin proteolytic cascade was induced by Aβ (Tucker *et al.*, 2000) brain tissue from AD has been shown to contain reduced levels of plasmin (Ledesma *et al.*, 2000). A physiologically significant concentration of Al (0.10 μM) down-regulated the gene for the urokinase plasminogen activator receptor in explanted human brain neural cells in primary culture (Kuroda *et al.*, 2003) and this would be expected to reduce the formation of plasmin *via* the urokinase plasminogen activator system. Whether the persistent down regulation of this system would be sufficient to influence the catabolism of Aβ over a lifetime remains to be determined.

5. ALUMINIUM AND THE NEUROTOXICITY OF Aβ

Aβ is clearly neurotoxic *in vitro*. However, we cannot be sure if neurotoxicity which follows either its injection into the brain in experimental animals or its endogenous expression in the brain of transgenic mice is due to a direct affect of Aβ (Caughey and Lansbury, 2003). However, what we do know is that the toxicity seen in each of these cases can be augmented by Al. For example, dissociated hippocampal and cortical neurones from embryonic rat brain were maintained in culture for up to six months and thereafter exposed to either freshly-prepared $A\beta_{40}$, aged $A\beta_{40}$ or aged $A\beta_{40}$ + Al. Both aged preparations of $A\beta_{40}$ resulted in degeneration though the extent of neuronal damage was far greater in the additional presence of Al (Kuroda *et al.* 1995). Similarly, Oka *et al.* combined $A\beta_{40}$ with Al to assure brain dysfunction following injection of the peptide into rat cerebral ventricle (Oka *et al.*, 1999). Most dramatically, in the only study thus far on the influence of Al on transgenic mice that over express human APP, dietary Al was shown to have exacerbated neurotoxicity (Pratico *et al.*, 2002). The latter was manifested as significant increases in soluble and insoluble Aβ ($A\beta_{40}$ and $A\beta_{42}$) in both the cerebral cortex and hippocampus, including the deposition of Aβ in plaques, and significant increases in isoprostane levels, a marker of *in vivo* oxidative stress, in both plasma and urine.

Whilst it is clear that Al can augment the neurotoxicity of Aβ there is additional evidence that exposure to Al accelerates the formation and deposition of endogenous Aβ (Miu *et al.*, 2003; Varner *et al.*, 1998; Zhang *et al.*, 2003). This has been observed both *in vitro* and *in vivo* and the link may well be the prooxidant activity of Al (Exley, 2004). For example, in the aforementioned transgenic study the exacerbation by Al of both oxidative stress and deposition of Aβ was abolished by the additional presence of vitamin E in the diet. The presence of antioxidant abolished neither the oxidative stress nor the amyloid burden which could be attributed to the over expression of APP but only the exacerbation of these effects by Al. Vitamin E has been shown to be similarly effective in abolishing Al-induced Aβ deposition in a non-transgenic animal study (Zhang *et al.*, 2003) and it is interesting to note that a number of antioxidants have been shown to abolish the prooxidant activity of Al without necessarily acting upon the background level of oxidative stress. Al is a prooxidant *in vivo* and a mechanism has been proposed to explain both this activity and the possible link with AD (Exley, 2004).

6. CONCLUSIONS

The recent literature purporting to review the interaction of Aβ with metals has ignored Al (Bush, 2003). Al has no known biological function but this lack of function hides both an ubiquitous presence and an extensive biochemistry. Unlike for other metals, both essential and non-essential, there is no specific or evolutionarily defined biological response to the presence of Al (Exley, 2003). The body copes with its interference in biochemical processes and the usual outcome to the many and varied (and unconscious!) compensatory mechanisms is no more than an additional drain upon energy stores. Aβ is produced *via* the metabolism of APP and its normal function is unknown. In AD it is found precipitated primarily as aggregates of β-pleated sheet conformers the persistence of which suggests that they are resistant to proteolytic degradation. The *in vivo* accumulation of Aβ would appear to be naïve from an evolutionary perspective and may simply be a consequence of its precipitation by the equally evolutionarily novel Al. Whether this coincidental combination leads to or from AD remains to be determined.

REFERENCES

Backstrom, J.R., Lim, G.P., Cullen, M. and Tokes, Z.A., 1996, Matrix metalloproteinase-9 (MMP-9) is synthesised in neurons of the human hippocampus and is capable of degrading the amyloid-β peptide (1-40). *J. Neurosci.* **16:** 7910-7919.

Beauchemin, D. and Kisilevsky, R., 1998, A method based on ICP-MS for the analysis of Alzheimer's amyloid plaques. *Anal. Chem.* **70:** 1026-1029.

Bondy, S. C. and Truong, A., 1999, Potentiation of beta-folding of beta-amyloid peptide 25-35 by aluminum salts. *Neurosci. Lett.* **267:** 25-28.

Bush, A.I., 2003, The metallobiology of Alzheimer's disease. *Trends in Neurosci.* **26:** 207-214.

Carson, J. A. and Turner, A. J., 2002, Beta-amyloid catabolism: roles for neprilysin (NEP) and other metallopeptidases? *J. Neurochem.* **81:** 1-8.

Caughey, B. and Lansbury, P. T., 2003, Protofibrils, pores, fibrils, and neurodegeneration: Separating the responsible protein aggregates from the innocent bystanders. *Ann. Rev. Neurosci.* **26:** 267-298.

Clauberg, M. and Joshi, J. G., 1993, Regulation of serine protease activity by aluminum - Implications for Alzheimer-disease. *Proc. Natl. Acad. Sci. USA* **90:** 1009-1012.

Duckett, S. and Galle, P., 1980, Electron microscope-microprobe studies of aluminium in the brains of cases of Alzheimer's disease and aging patients. *J. Neuropathol. Experiment. Neurol.* **39:** 350.

Exley, C., 1997, ATP-promoted amyloidosis of an amyloid beta peptide. *Neuroreport* **8:** 3411-3414.

Exley, C., 1998, The precipitation of mucin by aluminium. *J. Inorg. Biochem.* **70:** 195-206.

Exley, C. (ed.), 2001, *Aluminium and Alzheimer's Disease: The Science that Describes the Link.* Elsevier, Amsterdam.

Exley, C., 2003, A biogeochemical cycle for aluminium ? *J. Inorg. Biochem.* **97:** 1-7.

Exley, C., 2004, The prooxidant activity of aluminium. *Free Rad. Biol. Med.* **36:** 380-387.

Exley, C. and Korchazhkina, O., 2001a, The association of aluminium and βamyloid in Alzheimer's disease. In *Aluminium and Alzheimer's Disease: The Science that Describes the Link* (ed. C. Exley), pp. 421-434. Elsevier, Amsterdam.

Exley, C. and Korchazhkina, O. V., 2001b, Plasmin cleaves A beta 42 *in vitro* and prevents its aggregation into beta-pleated sheet structures. *Neuroreport* **12:** 2967-2970.

Exley, C. and Korchazhkina, O. V., 2001c, Promotion of formation of amyloid fibrils by aluminium adenosine triphosphate (AlATP). *J. Inorg. Biochem.* **84:** 215-224.

Exley, C., Price, N. C., Kelly, S. M. and Birchall, J. D., 1993, An interaction of beta-amyloid with aluminum *in vitro*. *Febs Lett.* **324:** 293-295.

Finckh, U., van Hadeln, K., Muller-Thomsen, T., Alberici, A., Binetti, G., Hock, C., Nitsch, R.M., Stoppe, G., Reiss, J. and Gal, A., 2003, Association of late-onset Alzheimer disease with a genotype of PLAU, the gene encoding urokinase-type plasminogen activator on chromosome 10q22.2. *Neurogenetics* **4:** 213-217.

Hardy, J. A. and Higgins, G. A., 1992, Alzheimers-disease - the amyloid cascade hypothesis. *Science* **256:** 184-185.

Harris, W. R., Berthon, G., Day, J. P., Exley, C., Flaten, T. P., Forbes, W. F., Kiss, T., Orvig, C. and Zatta, P. F., 1996, Speciation of aluminum in biological systems. *J. Toxicol. Environ. Health* **48:** 543-568.

House, E., Collingwood, J., Khan, A., Korchazhkina, O., Berthon, G and Exley, C., 2004, Aluminium, iron, zinc and copper influence the *in vitro* formation of amyloid fibrils of $A\beta_{42}$ in a manner which may have consequences for metal chelation therapy in Alzheimer's disease. *J. Alzh. Dis.* **6:** 291-301.

Koike, H., Tomioka, S., Sorimachi, H., Saido, T.C., Maruyama, K., Okuyama, A., Fujisawa-Sehara, A., Ohno, S., Suzuki, K and Ishiura, S., 1999, Membrane-anchored metalloprotease MDC9 has an α-secretase activity responsible for processing the amyloid precursor protein. *Biochem. J.* **343:** 371-375.

Korchazhkina, O. V., Ashcroft, A.E., Kiss, T., and Exley, C., 2002, The degradation of Aβ25-35 by the serine protease plasmin is inhibited by aluminium. *J. Alzh. Dis.* **4:** 357-367.

Kuroda, Y., Kobayashi, K., Ichikawa, M., Kawahara, M. and Muramoto, K., 1995, Application of long-term cultured neurons in aging and neurological research - aluminum neurotoxicity, synaptic degeneration and Alzheimers-disease. *Gerontology* **41:** 2-6.

Kuroda, H., Cui, J.G., Bazan, N.G. and Lukiw, W.J., 2003, Aluminum induces NF - kB signaling and inflammatory gene expression in human neural cells in primary culture:

implications for neurodegenerative disease. Program No. 669.9. *2003 Abstract Viewer/Itinerary Planner.* Washington, DC:Society for Neuroscience.

Landsberg, J. P., McDonald, B and Watt, F., 1992, Absence of aluminium in neuritic plaque cores in Alzheimer's disease. *Nature* **360:** 65-68.

Ledesma, M. D., Da Silva, J. S., Crassaerts, K., Delacourte, A., De Strooper, B. and Dotti, C. G., 2000, Brain plasmin enhances APP alpha-cleavage and A beta degradation and is reduced in Alzheimer's disease brains. *Embo Rep.* **1:** 530-535.

Levine, H., 1993, Thioflavine-T interaction with synthetic Alzheimers-disease beta-amyloid peptides - detection of amyloid aggregation in solution. *Prot. Sci.* **2:** 404-410.

Miu, A. C., Andreescu, C. E., Vasiu, R. and Olteanu, A. I., 2003, A behavioral and histological study of the effects of long-term exposure of adult rats to aluminum. *Int. J. Neurosci.* **113:** 1197-1211.

Moore, P. B., Day, J. P., Taylor, G. A., Ferrier, I. N., Fifield, L. K. and Edwardson, J. A., 2000, Absorption of aluminium-26 in Alzheimer's disease, measured using accelerator mass spectrometry. *Dement. Geriat. Cogn. Disord.* **11:** 66-69.

Moore, P. B., Edwardson, J. A., Ferrier, I. N., Taylor, G. A., Lett, D., Tyrer, S. P., Day, J. P., King, S. J. and Lilley, J. S., 1997, Gastrointestinal absorption of aluminum is increased in Down's syndrome. *Biol. Psychiat.* **41:** 488-492.

Oka, J. I., Suzuki, E., Goto, N. and Kameyama, T., 1999, Endogenous GLP-1 modulates hipocampal activity in beta-amyloid protein-treated rats. *Neuroreport* **10:** 2961-2964.

Oshiro, S., 2002, A new effect of aluminum on iron metabolism in mammalian cells. *Struct. Bond.* **104:** 59-77.

Oshiro, S., Kawahara, M., Mika, S., Muramoto, K., Kobayashi, K., Ishige, R., Nozawa, K., Hori, M., Yung, C., Kitajima, S. and Kuroda, Y., 1998, Aluminum taken up by transferrin-independent iron uptake affects the iron metabolism in rat cortical cells. *J. Biochem.* **123:** 42-46.

Pratico, D., Uryu, K., Sung, S., Tang, S., Trojanowski, J. Q. and Lee, V. M.-Y., 2002, Aluminum modulates brain amyloidosis through oxidative stress in APP transgenic mice. *FASEB J.* **16:** 1138-1140.

Rogers, J. T., Randall, J. D., Cahill, C. M., Eder, P. S., Huang, X. D., Gunshin, H., Leiter, L., McPhee, J., Sarang, S. S., Utsuki, T., Greig, N. H., Lahiri, D. K., Tanzi, R. E., Bush, A. I., Giordano, T. and Gullans, S. R., 2002, An iron-responsive element type II in the 5 '-untranslated region of the Alzheimer's amyloid precursor protein transcript. *J. Biol. Chem.* **277:** 45518-45528.

Selkoe, D.J., Shankar, G.M., Fadeeva, J.V. and Walsh, D.M., 2003, Modulation and further characterisation of cell-derived Aβ oligomers. Program No. 667.7. *2003 Abstract Viewer/Itinerary Planner.* Washington, DC: Society for Neuroscience.

Sengupta, P., Garai, K., Sahoo, B., Shi, Y., Callaway, D. J. E. and Maiti, S., 2003, The amyloid beta peptide (A beta(1-40)) is thermodynamically soluble at physiological concentrations. *Biochemistry* **42:** 10506-10513.

Tucker, H. M., Kihiko, M., Caldwell, J. N., Wright, S., Kawarabayashi, T., Price, D., Walker, D., Scheff, S., McGillis, J. P., Rydel, R. E. and Estus, S., 2000, The plasmin system is induced by and degrades amyloid-beta aggregates. *J. Neurosci.* **20:** 3937-3946.

Varner, J. A., Jensen, K. F., Horvath, W. and Isaacson, R. L., 1998, Chronic administration of aluminum-fluoride or sodium-fluoride to rats in drinking water: alterations in neuronal and cerebrovascular integrity. *Brain Res.* **784:** 284-298.

Ward, R. J., Zhang, Y. and Crichton, R. R., 2001, Aluminium toxicity and iron homeostasis. *J. Inorg. Biochem.* **87:** 9-14.

Yamanaka, K., Minato, N. and Iwai, K., 1999, Stabilization of iron regulatory protein 2, IRP2, by aluminum. *Febs Lett.* **462:** 216-220.

Zhang, Z. J., Qian, Y. H., Hu, H. T., Yang, H. and Yang, G. D., 2003, The herbal medicine Dipsacus asper Wall extract reduces the cognitive deficits and overexpression of beta-amyloid protein induced by aluminum exposure. *Life Sci.* **73:** 2443-2454.

Zou, J., Kajita, K. and Sugimoto, N., 2001, Cu2+ inhibits the aggregation of amyloid beta-peptide(1-42) *in vitro*. *Ange. Chem.-Int. Edit.* **40:** 2274-2277.

Chapter 12

Amyloid-β Metal Interaction and Metal Chelation

Math P. Cuajungco, [1]Christopher, J. Frederickson and [2]Ashley I. Bush
Harvard Medical School and Massachusetts Eye and Ear Infirmary, Boston MA 02114 USA..
[1]NeuroBioTex, Inc. and University of Texas Medical Branch, Galveston TX 77550 USA.
[2]Harvard Medical School and Massachusetts General Hospital, Boston MA 02114 USA

Abstract: Alzheimer's disease (AD) is associated with the abnormal aggregation of amyloid-beta (Aβ) protein. Aβ and its precursor protein (APP) interact with metal ions such as zinc, copper and iron. Evidence shows that these metals play a role in the precipitation and cytotoxicity of Aβ. Despite recent advances in AD research, there is a lack of therapeutic agents to hinder the apparent aggregation and toxicity of Aβ. Recent studies show that drugs with metal chelating properties could produce a significant reversal of amyloid-β plaque deposition *in vitro* and *in vivo*. Here we discuss the interaction of Aβ with metals, metal dyshomeostasis in the CNS of patients with AD, and the potential therapeutic effects of metal chelators.

Key words: Metal chelators, Zinc, Copper, Iron

1. INTRODUCTION

Alzheimer's disease (AD) is characterized by progressive loss of cholinergic neurons with concomitant deterioration of memory and cognition (Arendt *et al.*, 1984; Winblad *et al.*, 1985; Hyman *et al.*, 1986; Hu *et al.*, 2003; Luth *et al.*, 2003). In AD brain, there is marked accumulation of amyloid-β (Aβ) protein, the main constituent of senile plaques, particularly its longer and more amyloidogenic form Aβ1-42 (Glenner and Wong, 1984; Citron *et al.*, 1997). Evidence shows that formation of neurofibrillary tangles (NFT) and neuropil threads occur subsequent to amyloid deposition (Lombardo *et al.*, 2003). The Aβ protein is generated from the amyloid precursor protein (APP) by the proteolytic activities of β- and γ-secretases (Wolfe, 2003; Yang *et al.*, 2003). Majority of AD cases are classified as sporadic and approximately 5% of AD patients suffer from an early-onset familial AD (FAD) form. Mutations of the APP (chromosome 21), and presenilin (chromosomes 1 and 14) genes are linked to forms of FAD (Levy-Lahad *et al.*, 1995; Tanzi *et al.*, 1996; Levy-Lahad *et al.*, 1998). Several risk factors for late-onset AD have been identified such as apolipoprotein E (apo-E; ε4 allele) on chromosome 19 (Roses *et al.*, 1995) and α_2-macroglobulin (A2M) gene on chromosome 12 (Blacker *et al.*, 1998; Saunders *et al.*, 2003). In addition, a genetic locus on chromosome 10 was identified that may be also a risk factor for late-onset AD (Bertram *et al.*, 2000; Ertekin-Taner *et al.*, 2000).

The pathophysiologic effects of FAD gene mutations result in increased production of Aβ protein and its subsequent cerebral deposition, especially of Aβ1-42 isoform (Citron *et al.*, 1992; Citron *et al.*, 1997). Note, however, much is not known how amyloid plaques can form and accumulate in AD brain without any genetically determined elevation in Aβ synthesis. Furthermore, overproduction of Aβ1-42 associated with FAD or Down's syndrome (Citron *et al.*, 1992; Lemere *et al.*, 1996; Citron *et al.*, 1997) does not fully explain its pathologic deposition in the brain, suggesting that other neurochemical factors initiate Aβ deposition for FAD and sporadic or late-onset AD. Mitigating factors that may initiate amyloid deposition are the pathologic interactions of cerebral Aβ with transition metals such as zinc, copper or iron (Bush *et al.*, 1994b; Huang *et al.*, 1997; Atwood *et al.*, 1998). Zinc, copper, and iron have been implicated as possible pathogenic agents in AD due to high concentration gradients of these metals in the cortex, hippocampus and the cortical vasculature (Smith, 1983; Dwork *et al.*, 1988; Frederickson, 1989); brain regions that are severely affected by the pathological lesions of AD (Goedert *et al.*, 1991). Studies performed in our laboratory, as well as by other groups clearly demonstrated that these metals interact and induce Aβ protein precipitation (Lovell *et al.*, 1998; Garzon-Rodriguez *et al.*, 1999; Atwood *et al.*, 2000a; Cuajungco *et al.*, 2000b; Hirakura *et al.*, 2000; Miura *et al.*, 2000; Yang *et al.*, 2000; Kozin *et al.*, 2001; Suzuki *et al.*, 2001; Yoshiike *et al.*, 2001). The role of metals in AD pathogenesis remains unclear despite numerous reports and hypotheses that attempt to link them with AD. Notwithstanding, there is compelling evidence that chelation of transition metal, which prevents their promiscuous interactions with Aβ protein, has therapeutic potential for AD pathology.

2. CEREBRAL METAL ACCUMULATION IN ALZHEIMER'S DISEASE

Previous studies that attempt to quantify cerebral, cerebrospinal fluid (CSF), or peripheral levels of zinc, copper and iron in AD produce highly variable results (Hershey *et al.*, 1983; Sahu *et al.*, 1988; Basun *et al.*, 1991; Constantinidis, 1991a, b; Corrigan *et al.*, 1993; Samudralwar *et al.*, 1995; Tully *et al.*, 1995; Deibel *et al.*, 1996; Licastro *et al.*, 1996; Cornett *et al.*, 1998b; Cornett *et al.*, 1998a; Molina *et al.*, 1998; Gonzalez *et al.*, 1999). These inconsistencies are likely due to differences in methodology employed, technical difficulties encountered during tissue processing, and small sample size (Cuajungco and Faget, 2003). A consistent finding is that abnormal levels of these metals are associated with AD-affected brain areas that contain high amyloid burden (Lovell *et al.*, 1998; Suh *et al.*, 2000). It is worth noting that zinc also accumulate in the brains of APP transgenic mice, but is only present in mature, cored senile plaques (thioflavine-T staining positive), and not in diffuse or pre-amyloid plaques, suggesting that zinc may be responsible for plaque maturation (Lee *et al.*, 1999). A rather interesting observation is that when ZnT3, a neuronal vesicular zinc transporter, was knocked out ($ZnT3^{-}/^{-}$) and crossed with the APP Tg2576 mouse transgene, the amyloid burden in brains of resulting $APP2576^{+}$: $ZnT3^{-}/^{-}$ transgenes have significantly decreased when compared with controls (Lee *et al.*, 2002). Consequently, these mice have increased levels of soluble Aβ with no apparent adverse effects particularly on life span and at the cellular synapse (Lee *et al.*, 2002).

Like zinc levels, there are also inconsistent reports on cerebral copper levels in postmortem AD brain. Some observations show a significant decrease of copper concentrations in AD brain (Deibel *et al.*, 1996), while other studies report a 2-fold increase of copper levels in the CSF (Basun *et al.*, 1991), amyloid plaque rim (Lovell *et al.*, 1998), as well as an increase in brain and CSF ceruloplasmin levels, a known copper-binding/transporting protein (Loeffler *et al.*, 1994). There is also an indication that serum copper levels are significantly elevated among AD patients (Squitti *et al.*, 2002). It is interesting to note that *in situ*, copper ions found in plaques and tangles are redox active (Sayre *et al.*, 2000). Meanwhile, abnormal cerebral iron levels and its associated binding protein have been consistently observed in postmortem AD brain tissue (Connor *et al.*, 1992; Smith *et al.*, 1997; Lovell *et al.*, 1998; Smith *et al.*, 1998a). Iron ions present within amyloid plaques deposited in human brains (Smith *et al.*, 1997; Lovell *et al.*, 1998; Smith *et al.*, 1998b), as well as in amyloid-bearing APP transgenic mouse brains (Smith *et al.*, 1998b) are redox-active and likely to be neurotoxic (Schubert and Chevion, 1995). The implication of abnormal copper and iron levels in AD pathology is inherent to the redox properties of these two metals where they could likely contribute to oxidative stress in the central nervous system (Halliwell, 1992, 2001), a phenomenon typically seen in brain tissues of patients with AD and animal model of the disease (Smith *et al.*, 1998b; Huang *et al.*, 1999b; White *et al.*, 1999; Huang *et al.*, 2000; Perry *et al.*, 2003).

3. METAL-BINDING PROPERTIES AND METAL INTERACTIONS OF AMYLOID-B AND APP

3.1 APP is a metalloprotein

APP has specific and saturable binding sites for zinc (APP 181-200; K_A = 750 nM) (Bush *et al.*, 1993) and copper (APP 135-155; K_D = 10 nM) (Hesse *et al.*, 1994). These binding sites suggest that zinc and copper serve a physiological role for APP and that these sites have homology to all known members of the APP superfamily (Bush *et al.*, 1994a; Simons *et al.*, 2002), and the amyloid precursor-like proteins 1 and 2 (Wasco *et al.*, 1992; Wasco *et al.*, 1993) giving some weight on the importance of both metals in APP function and metabolism. In fact, overexpression of APP carboxyl-terminal fragment containing Aβ, results in significantly reduced copper and iron levels in the transgenic mouse brain (Maynard *et al.*, 2002). On the other hand, APP overexpression in Tg2576 mouse transgenes results in significantly reduced levels of copper, but not iron, even prior to the appearance of amyloid neuropathology and throughout the lifespan of the mouse (Maynard *et al.*, 2002). For reasons that are unclear, an associated increase in cerebral manganese levels was observed in both transgenic strains. Nevertheless, these observations confirm the roles for APP and Aβ in physiological metal regulation.

The role of zinc on APP is believed to be in a structural capacity. For example, zinc has been shown to enhance the binding of several proteoglycans and basement membrane proteins to APP (Bush *et al.*, 1994a; Multhaup, 1994; Multhaup *et al.*, 1994; Multhaup *et al.*, 1995). Zinc also inhibits cleavage of full-length APP from human platelets (Li *et al.*, 1995). An effect that could influence the generation of Aβ from APP and may increase the biological half-life of Aβ by protecting the peptide from proteolytic attack (Bush *et al.*, 1993; Bush *et al.*, 1994a). Nevertheless, there has been no report to date which shows that zinc binding to APP results in an increase production of Aβ from APP *in vivo*. In fact, zinc ($\leq$ 50 μM) was shown to specifically increase secreted APP, which consequently decreased the release of Aβ in the media of cultured Chinese hamster ovary (CHO) cells (Borchardt *et al.*, 2000). Like zinc, copper (10 μM) also produces a similar effect in CHO cells where it precludes generation of Aβ by enhancing synthesis of full-length, non-amyloidogenic APP (Borchardt *et al.*, 1999). One caveat, however, is that APP can reduce Cu(II) to Cu(I) which results in oxidation of Cys144 and Cys155 and the formation of a corresponding intramolecular disulfide bridge (Multhaup *et al.*, 1996; Ruiz *et al.*, 1999). The resulting APP-Cu(I) complex is prone to redox reactions resulting in site-specific and random APP fragmentation (Multhaup *et al.*, 1998) thus increasing the possibility of pathologic production of the amyloidogenic protein.

3.2 Aβ is a metalloprotein and interacts with transition metals: aggregation, oxidation and cytotoxicity

The aggregational state of Aβ is spatially- and temporally-dependent, and is influenced by its concentration, by pH, and by the ionic concentrations of zinc, copper, or iron (Bush *et al.*, 1994b; Huang *et al.*, 1997; Atwood *et al.*, 1998; Huang *et al.*, 1999b; Atwood *et al.*, 2000a). Aβ1-40 contains a high affinity binding site for zinc 1:1 (zinc:Aβ; K_D = 107 nM), and a 2:1 low affinity binding site (K_D = 5.2 μM) (Bush *et al.*, 1993; Bush *et al.*, 1994b). The zinc binding site spans AA positions 1-16 of the Aβ protein sequence (Kozin *et al.*, 2001). Zinc and copper binding sites are mediated by histidine residues at positions 13 and 14 (Liu *et al.*, 1999; Yang *et al.*, 2000). In addition, an intramolecular histidine bridge at position 6 of Aβ peptide may also coordinate the zinc or copper ion (Curtain *et al.*, 2001).

Zinc (≥ 10 μM) binds and precipitates Aβ protein over a wide pH range (pH 6.0-8.0) (Bush *et al.*, 1994b; Brown *et al.*, 1997; Huang *et al.*, 1997). Zinc preserves the α-helical conformation of Aβ1-40 and its complexation is completely reversed by chelation (Huang *et al.*, 1997). Indeed, this observation was recently substantiated and it was also observed that zinc-Aβ complexes (ie. zinc-induced aggregates) are soluble and stable for several months (Kozin *et al.*, 2001).

Copper ions [Cu(II)] are not much better than iron, aluminum, calcium, magnesium, manganese, mercury, lead, or nickel at precipitating Aβ protein at physiological pH (pH 7.0) (Atwood *et al.*, 1998). Note, however, that Cu(II) precipitates Aβ in a reaction that is potentiated by mildly acidic (pH 6.6) conditions (Atwood *et al.*, 1998; Miura *et al.*, 2000). Copper-Aβ complexes (ie. copper-induced aggregates) have been characterized as highly structured and soluble, similar to that of zinc-Aβ aggregates (Atwood *et al.*, 1998; Miura *et al.*, 2000; Curtain *et al.*, 2001; Suzuki *et al.*, 2001). Interestingly, although Aβ binds equimolar amounts of Cu(II) and Zn(II) at pH 7.4, Cu(II) displaces Zn(II) from Zn(II):Aβ complex under mildly acidic conditions (pH 6.6) (Atwood *et al.*, 1998; Atwood *et al.*, 2000a) by occupying the intramolecular histidine-bridge coordination sites (Curtain *et al.*, 2001). Copper seemingly competes for the zinc binding site when the Cu(II):Aβ ratio is 4, although a much lower copper concentration is needed to disrupt zinc-induced Aβ aggregation (Suzuki *et al.*, 2001). Note, however, that the stoichiometry of Cu(II):Aβ increases from zero (when Aβ is soluble) to 1.0-2.5 when Aβ is aggregated by Cu(II). Aβ1-40 has higher affinity (log K_{app} 10) and lower affinity (log K_{app} 7.0) binding sites for copper, but the affinity of copper for Aβ1-42 is much greater for both sites (log K_{app} 16.0 and log K_{app} 8.0, respectively) (Atwood *et al.*, 1998; Atwood *et al.*, 2000a). Under mildly acidic conditions (pH6.8), the affinity of Aβ1-40 and Aβ1-42 for Zn(II), but not Cu(II) decreases at the lower affinity binding sites (Atwood *et al.*, 2000a). Unlike Zn(II), both Cu(II) and Fe(III) induce greater Aβ aggregation under mildly acidic conditions (*e.g.* pH 6.6) (Atwood *et al.*, 1998). Recent studies indicate that copper's strong pH-dependence interaction with Aβ protein is due to its specific ability to bind either the N-τ of the histidine imidazole ring and main-chain amide nitrogens or N-π atom (Miura *et al.*, 2000; Suzuki *et al.*, 2001). The N-τ bridge results in copper-induced *insoluble* Aβ aggregation, while the N-π bridge produces *soluble* Aβ aggregates (Miura *et al.*, 2000; Suzuki *et al.*, 2001). Interestingly, displacement of Zn(II) by Cu(II) from Aβ's zinc binding sites supposedly inhibits zinc-induced Aβ aggregation via direct competition by Cu(II) at histidine sites. It is interesting to note that at physiological pH, zinc-induced Aβ aggregates have subtle differences from copper-induced Aβ aggregates in terms of its relative density and resolubilization state (*ie.* zinc-Aβ aggregates are denser, and harder to resolubilize than its copper-Aβ aggregates) (Moir *et al.*, 1999).

3.3 The neurotoxicity of Aβ is mediated by its oxidative state and redox metals

Many studies have confirmed that aged Aβ is neurotoxic *in vitro* (Pike *et al.*, 1991a, b) and *in vivo* (Emre *et al.*, 1992; Weldon *et al.*, 1998). While Aβ1-40 fragment is known to be neurotoxic (Yankner *et al.*, 1989; Pike *et al.*, 1991a, b; Kowall *et al.*, 1992), Aβ1-42 is more neurotoxic than Aβ1-40 (Huang *et al.*, 1999b), and is more likely to generate hydrogen peroxide (H_2O_2) than Aβ1-40 (Huang *et al.*, 1999a). In fact, the redox activity of Aβ species is greatest for Aβ1-42>Aβ1-40>>ratAβ1-40 (Huang *et al.*, 1999b). There is compelling evidence that H_2O_2 production is central to Aβ1-42's apparent cytotoxicity (Figure 1) (Behl *et al.*, 1994; Huang *et al.*, 1999b; Cuajungco *et al.*, 2000b). Binding of trace levels of redox active metals, Cu(II) or Fe(III), to Aβ1-42 engenders a cell-free catalytic production of H_2O_2 from O_2 via metal reduction (Huang *et al.*, 1999a; Huang *et al.*, 1999b)

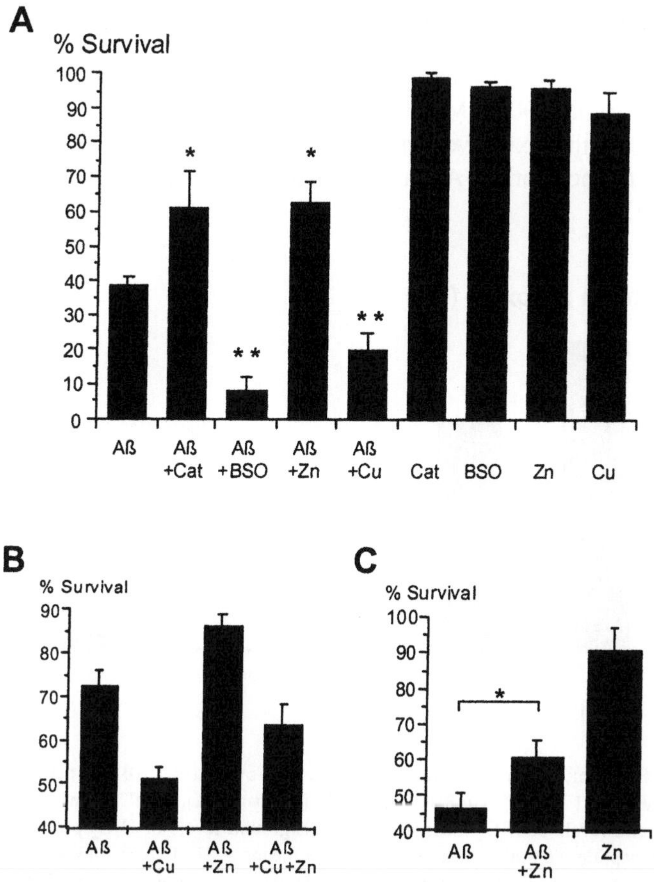

Figure 1. Effect of Zn^{2+} upon Aβ1-42 cytotoxicity. *A, B.* Primary rat neuronal cultures were incubated with Aβ1-42 (*A*, 20 μM; *B*, 10 μM) and/or other factors for 48 h, and cell survival measured by Live-Dead assay, compared to untreated cultures. *A*, Effects of co-incubation with catalase (Aβ+Cat), buthionine sulfoximine (Aβ+BSO), Zn^{2+} (20 μM, Aβ+Zn), Cu^{2+} (20 μM, Aβ+Cu), incubated and effects of these factors alone, upon neuronal survival are shown. *C*, human embryonic kidney 293 cells were with Aβ1-42 (10 μM) ± Zn^{2+} (10 μM) or Zn^{2+} alone, as shown. Surviving cells were assayed, compared to untreated control cultures. Data are represented as mean ± SEM, N = 5-6 experimental trials performed in triplicate wells (* $p < 0.01$, ** $p < 0.001$). Reprinted with permission, Cuajungco *et al.* 2000, *J. Biol. Chem.*, **275**:19439-19442. Copyright 2000, The American Society for Biochemistry and Molecular Biology, Inc.

and can be inhibited by high concentrations of zinc (Figure 2) (Cuajungco *et al.*, 2000b). On the other hand, copper competes with zinc, and is believed to augment the oxidation of Aβ which could affect its metabolism (Atwood *et al.*, 2000b) and neurotoxicity (Huang *et al.*, 1999b; Cuajungco *et al.*, 2000b).

Aβ's conformational state mediates its intrinsic toxicity and oxidative properties in the presence of redox active metals (Pike *et al.*, 1991a, b; Lorenzo and Yankner, 1994; Huang *et al.*, 1999b; Atwood *et al.*, 2000b; Cuajungco *et al.*, 2000b; Monji *et al.*, 2001, 2002). In fact, it was observed recently that the oxidative cytotoxicity of Aβ is caused by pre-fibrillar structure and not fibrillar Aβ, at least in a specific *C. elegans* strain (Drake *et al.*, 2003).

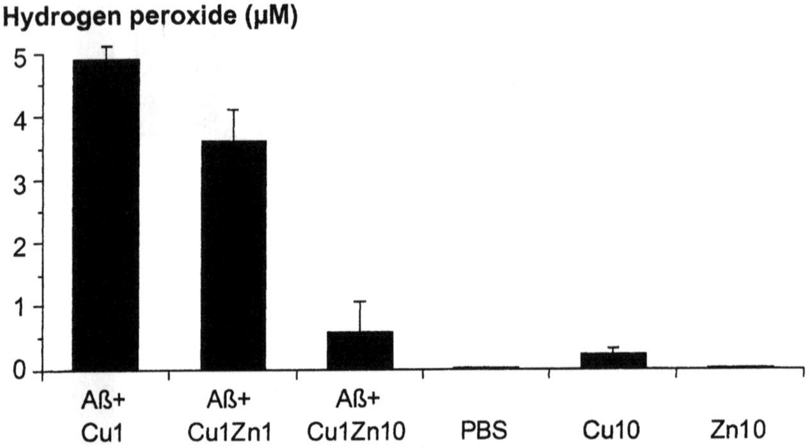

Figure 2. Effect of Zn^{2+} on cell-free H_2O_2 production by Aβ1-42. Aβ1-42 peptide (10 μM) was incubated for 1 h at 37°C in PBS with $CuCl_2$ (Cu1 = 1 μM), ± $ZnCl_2$ (Zn1 = 1 μM, Zn10 = 10 μM) and levels of H_2O_2 measured. The background levels of H_2O_2 production in the absence of peptide were also measured. Data are means ± SEM (N = 5 experimental trials performed in triplicate wells). Reprinted with permission, Cuajungco *et al.* 2000, *J. Biol. Chem.*, **275**:19439-19442. Copyright 2000, The American Society for Biochemistry and Molecular Biology, Inc.

Nevertheless, the biochemical relationship between Aβ's toxicity and oxidative stress in AD is rather complex and not well understood. Studies indicate that cerebral Aβ deposition and oxidative stress are considered closely related to the etiology of AD. Indeed, metabolic signs of oxidative stress such as oxygen radical-mediated damage of brain proteins, lipids and nucleic acids, as well as systemic signs of oxidative stress and the response of antioxidant systems have all been observed in AD tissue (Mecocci *et al.*, 1994; Smith *et al.*, 1997; Smith *et al.*, 1998b; Mecocci *et al.*, 2002). Likewise, Aβ induces lipid peroxidation (Butterfield *et al.*, 1999a; Butterfield *et al.*, 2002a; Butterfield *et al.*, 2002b; Butterfield and Lauderback, 2002) and exerts toxicity through mechanisms involving the generation of cellular H_2O_2 through redox metal interactions (Behl *et al.*,

1994; Huang *et al.*, 1999b; Cuajungco *et al.*, 2000b). This toxic effect is abolished by superoxide/H_2O_2 scavengers, antioxidants, and metal chelators (Schubert and Chevion, 1995; Bruce *et al.*, 1996; Butterfield *et al.*, 1999b; Sayre *et al.*, 2000; Rottkamp *et al.*, 2001). In a clinical setting, there is evidence that treatment of patients suffering from AD with an antioxidant (vitamin E) delays their cognitive decline (Sano *et al.*, 1997). Taken together, it seems that the oxidative stress in AD occurs via elevated H_2O_2 levels and promote cytotoxicity by incessantly taxing cellular antioxidant system, and by interacting with transition metals that could produce the highly toxic hydroxyl radicals (Halliwell, 1992, 2001). These reports reinforce the belief that the pathologic basis of AD is through oxidative stress mitigated by abnormal metabolisms of Aβ and its interaction with redox active metals.

4. METAL CHELATORS: THERAPEUTIC POTENTIAL FOR ALZHEIMER'S DISEASE

4.1 Basic science studies: chelator-induced neutralization of Aβ-mediated oxidative activity and resolubilization of amyloid plaques

Aβ reduces Cu(II) and Fe(III) to Cu(I) and Fe(II), respectively, and that Aβ can generate H_2O_2 through a metal-dependent reaction (Huang *et al.*, 1999a; Huang *et al.*, 1999b; Cuajungco *et al.*, 2000b). Henceforth, using metal chelating agents to disrupt this interaction with redox active metals and oxidative activity seems to be the most promising and appropriate therapy for AD. Indeed, Huang *et al.* previously reported that the metal chelators, bathophenantroline disulfonic acid (BP) and diethylenetetraamine pentaacetic acid (DTPA), significantly hindered with the redox activity of Aβ by quenching the H_2O_2 produced through Aβ-metal interactions (Huang *et al.*, 1999b). Furthermore, desferrioxamine (DFO; Fe[III]-, and Cu[II]-selective), and DTPA (Fe[III]-, Fe[II], Cu[II]- and Zn[II]-selective) have been found to neutralize Aβ-metal-generated H_2O_2 *in situ* (Sayre *et al.*, 2000). Note that when Aβ is pretreated with DFO, its neurotoxicity *in vitro* is abolished (Rottkamp *et al.*, 2001).

To assess the potential of metal chelating agents to abstract metals from Aβ and solubilize the protein from senile plaques, Cherny *et al.* used N,N,N',N'-tetrakis-(2-pyridylmethyl)-ethylenediamine (TPEN), bathocuproine disulfonic acid (BC), and ethylene glycol-bis-(β-aminoethyl ether)-N,N,N',N'-tetraacetic acid (EGTA) and successfully showed an enhancement in resolubilization of Aβ deposits from post-mortem AD and non-AD brain samples (Figure 3) (Cherny *et al.*, 1999). Likewise, using either triethylenetetraamine (TETA; a high affinity Cu[II] chelator) or bicinchoninic acid (BCA; a Cu[I]-selective chelator), marked increase in resolubilization of Aβ deposits from brain samples of APP Tg2576 was also observed (Gray *et al.*, 1998). This led Cherny *et al.* to perform a pilot blind study using clioquinol (iodochlorhydroxyquin, CQ; a hydrophobic drug with metal chelating property that crosses the blood-brain barrier) as an oral therapy for 21-month old APP Tg2576 transgenic mice (Cherny *et al.*, 2001). The group found that after 9 weeks, a 49% decrease in brain Aβ deposition (-375 μg/g wet weight, $p = 0.0001$) was observed. In addition to the lowering of cerebral amyloid burden, there was a modest increase in soluble Aβ (1.45% of total cerebral Aβ), while the levels of APP, synaptophysin, and glial fibrillary acidic protein (GFAP) were not significantly affected. It appeared that the animals showed an overall increase in weight, general health, motor activity, and alertness relative to untreated controls; however, these improvements did not affect the lifespan of these animals.

While the above results show some promise on chelation therapy, one caveat for using metal chelators in the brain is that it has been reported that certain chelating agents, regardless of their affinity or specificity for a particular brain metal, must be used with caution since distinct compounds can produce severe side effects that could lead to seizure and significant neuronal death when administered directly into the brain (Cuajungco and Lees, 1996, 1998; Lees *et al.*, 1998; Armstrong *et al.*, 2001).

4.2 Clinical studies: past and present

The principle of a therapeutic drug abstracting a metal on a protein target is well-developed in pharmacology. A number of well-known antibiotic, anticonvulsive, antitumor, and antiinflammatory drugs exert their pharmacological effect by interacting with the Cu-, Zn- or Fe-active sites of their target protein (Cuajungco *et al.*, 2000a). Historically, the use of the iron chelator, DFO, in a clinical setting was the first published attempt as therapeutics for AD. DFO proved to have some positive effects on the rate of cognitive decline among AD patients (Crapper McLachlan *et al.*, 1991).

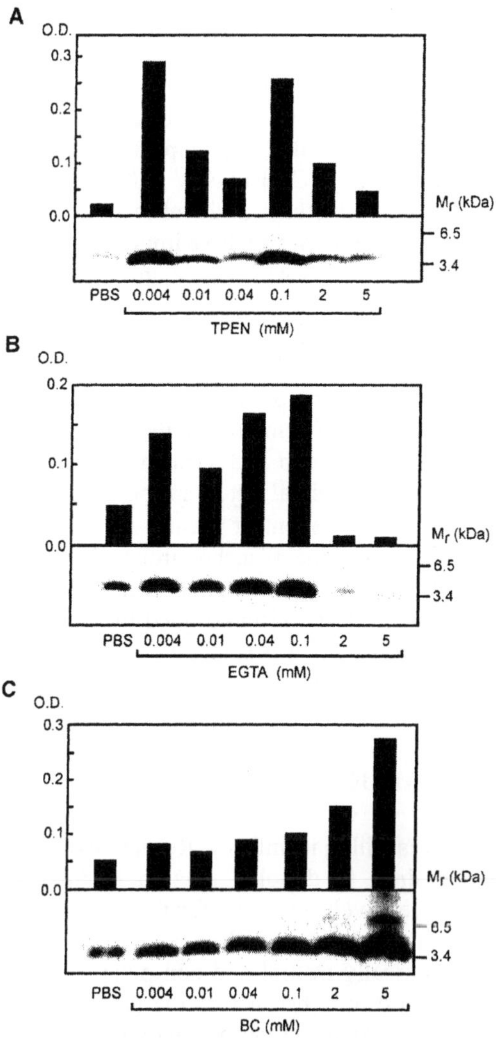

Figure 3. Release of Aβ from sedimentable deposits by chelators. Frontal cortex from an AD brain was homogenized in PBS, pH 7.4, with or without increasing concentrations of TPEN (*A*), EGTA (*B*), or BC (*C*). Following centrifugation, Aβ in the supernatants was visualized by Western blot using anti-Aβ monoclonal antibody WO2 (*lower panels*), and quantified by densitometry (*graphs* above corresponding blots). Although there is considerable variation in the optimum chelator concentration for the maximal recovery of Aβ from case to case, these data are representative of 17 AD cases. Reprinted with permission, Cherny *et al.* 1999, *J. Biol. Chem.*, **274**:23223-23228. Copyright 1999, The American Society for Biochemistry and Molecular Biology, Inc.

There was, however, no follow-up study done to verify this result. DFO, while it binds iron with high avidity, also chelates zinc and copper (Cuajungco *et al.*, 2000a; Cuajungco and Faget, 2003). CQ, a USP antibiotic with metal chelating properties, shows some promise for clinical use in AD as current reports showed that the drug appears to be well-tolerated in mice when administered orally (Cherny *et al.*, 2001; Nitzan *et al.*, 2003). Recently, a short-term Swedish study on the efficacy of CQ on 10 AD patients (80 mg/day) resulted in modest cognitive improvement after 21 days (Regland *et al.*, 2001). Because this was a very short study with small patient number, it was hard to conclude if CQ was really effective. More recently, though, Ritchie *et al.* reported a more promising result on the use of CQ for AD (Ritchie *et al.*, 2003). The effect of CQ treatment was significant in the more severely affected group with a baseline cognitive subscale score of greater than or equal to 25 using the Alzheimer's Disease Assessment Scale (ADAS). The placebo group had substantial worsening on ADAS scores when compared with minimal deterioration for the CQ-treated group. It is noteworthy that measurement of plasma Aβ1-42 levels declined in the CQ-treated group while the placebo group showed an increase. While the drug was well tolerated, plasma zinc levels rose in the CQ-treated group with no apparent ill effects. It was argued that while the sample size in this pilot Phase 2 clinical study was small, the results warrant further investigation on this novel treatment strategy for AD.

5. CONCLUDING REMARKS

Metal chelation therapy for AD is still in its infancy. Recent observations indicate that an ideal chelating drug to dissolve Aβ would involve a molecule that avidly binds and is relatively selective for Cu(I), Zn(II) and possibly Fe(III), but does not sequester other abundant, yet crucial cations like calcium and magnesium. Electrically neutral and non-polar molecules are ideal chelators, since they are best absorbed across the gastrointestinal tract and achieve a broad distribution throughout various tissues. Abstraction of metals from protein still has many obstacles due to the potential difficulties in: (a) route of administration; (b) reaching the specific target brain tissue due to protection by the blood-brain barrier; (c) possible associated non-specific problems of systemic metal ion depletion; and (d) potential severe side effects by the drug itself. Further investigations on the molecular and physiological effects of metal chelating agents are necessary to circumvent these potentially severe complications. Notwithstanding, can we ask the question therefore whether metal chelators will be pharmacologically useful for Alzheimer's disease? The answer to this question remains uncertain, although very encouraging on the bases of several pilot Phase 2 clinical trials of clioquinol. More clinical investigations to establish unequivocal drug efficacy of clioquinol and other candidate metal chelators for patients with AD are clearly warranted.

REFERENCES

Arendt, T., Bigl, V., Tennstedt, A., and Arendt, A., 1984, Correlation between cortical plaque count and neuronal loss in the nucleus basalis in Alzheimer's disease. *Neurosci. Lett.* **48:** 81-85.

Armstrong, C., Leong, W., and Lees, G.J., 2001, Comparative effects of metal chelating agents on the neuronal cytotoxicity induced by copper (cu+2), iron (fe+3) and zinc in the hippocampus. *Brain Res.* **892:** 51-62.

Atwood, C.S., Scarpa, R.C., Huang, X., Moir, R.D., Jones, W.D., Fairlie, D.P., Tanzi, R.E., and Bush, A.I., 2000a, Characterization of copper interactions with Alzheimer amyloid beta peptides: Identification of an attomolar-affinity copper binding site on amyloid beta1-42. *J. Neurochem.* **75:** 1219-1233.

Atwood, C.S., Moir, R.D., Huang, X., Scarpa, R.C., Bacarra, N.M, Romano, D.M., Hartshorn, M.A., Tanzi, R.E., and Bush, A.I., 1998, Dramatic aggregation of Alzheimer abeta by cu(ii) is induced by conditions representing physiological acidosis. *J. Biol. Chem.* **273:** 12817-12826.

Atwood, C.S., Huang, X., Khatri, A., Scarpa, R.C., Kim, Y.S., Moir, R.D., Tanzi, R.E., Roher, A.E., and Bush, A.I., 2000b, Copper catalyzed oxidation of Alzheimer abeta. *Cell. Mol. Biol.* **46:** 777-783.

Basun, H., Forssell, L,G, Wetterberg, L, and Winblad, B., 1991, Metals and trace elements in plasma and cerebrospinal fluid in normal aging and Alzheimer's disease. *J. Neural Transm. Park. Dis. Dement. Sect.* **3:** 231-258.

Behl, C., Davis, J.B., Lesley, R., and Schubert, D., 1994, Hydrogen peroxide mediates amyloid beta protein toxicity. *Cell* **77:** 817-827.

Bertram, L., Blacker, D., Mullin, K., Keeney, D., Jones, J., Basu, S., Yhu, S., McInnis, M.G., Go, R.C., Vekrellis, K., Selkoe, D.J., Saunders, A.J., and Tanzi, R.E., 2000, Evidence for genetic linkage of Alzheimer's disease to chromosome 10q. *Science* **290:** 2302-2303.

Blacker, D., Wilcox, M.A., Laird, N.M., Rodes, L., Horvath, S.M., Go, R.C., Perry, R., Watson, B., Jr., Bassett, S.S., McInnis, M.G., Albert, M.S., Hyman, B.T., and Tanzi, R.E., 1998, Alpha-2 macroglobulin is genetically associated with Alzheimer disease. *Nature Genet.* **19:** 357-360.

Borchardt, T., Camakaris, J., Cappai, R., Masters, C.L., Beyreuther, K., and Multhaup, G., 1999, Copper inhibits beta-amyloid production and stimulates the non-amyloidogenic pathway of amyloid-precursor-protein secretion. *Biochem. J.* **344** : 461-467.

Borchardt, T., Schmidt, C., Camarkis, J., Cappai, R., Masters, C.L., Beyreuther, K., and Multhaup, G., 2000, Differential effects of zinc on amyloid precursor protein (app) processing in copper-resistant variants of cultured Chinese hamster ovary cells. *Cell. Mol. Biol.* **46:** 785-795.

Brown, A.M., Tummolo, D.M., Rhodes, K.J., Hofmann, J.R., Jacobsen, J.S., Sonnenberg-and Reines, J., 1997, Selective aggregation of endogenous beta-amyloid peptide and soluble amyloid precursor protein in cerebrospinal fluid by zinc. *J. Neurochem.* **69:** 1204-1212.

Bruce, A.J., Malfroy, B., Baudry, M., 1996, Beta-amyloid toxicity in organotypic hippocampal cultures: Protection by euk-8, a synthetic catalytic free radical scavenger. *Proc. Natl. Acad. Sci. U.S.A.* **93:** 2312-2316.

Bush, A.I., Pettingell, W.H., Jr., de Paradis, M., Tanzi, R.E., and Wasco, W., 1994a, The amyloid beta-protein precursor and its mammalian homologues. Evidence for a zinc-modulated heparin-binding superfamily. *J. Biol. Chem.* **269:** 26618-26621.

Bush, A.I., Multhaup, G., Moir, R.D., Williamson, T.G., Small, D.H., Rumble, B., Pollwein, P., Beyreuther, K., and Masters, C.L, 1993, A novel zinc(ii) binding site modulates the function of the beta a4 amyloid protein precursor of alzheimer's disease. *J. Biol .Chem.* **268:** 16109-16112.

Bush AI, Pettingell, WH, Multhaup, G, d Paradis, M, Vonsattel, JP, Gusella, JF, Beyreuther, K, Masters, CL, Tanzi, RE (1994b) Rapid induction of Alzheimer A beta amyloid formation by zinc. Science 265:1464-1467.

Butterfield, D.A., and Lauderback, C.M., 2002, Lipid peroxidation and protein oxidation in Alzheimer's disease brain: Potential causes and consequences involving amyloid beta-peptide-associated free radical oxidative stress. *Free Radic. Biol. Med.* **32:** 1050-1060.

Butterfield, D.A., Yatin, S.M., and Link, C.D., 1999a, In vitro and in vivo protein oxidation induced by Alzheimer's disease amyloid beta-peptide (1-42). Ann. N.Y. Acad. Sci. **893:** 265-268.

Butterfield, D.A., Koppal, T., Subramaniam, R., and Yatin, S., 1999b, Vitamin e as an antioxidant/free radical scavenger against amyloid beta-peptide-induced oxidative stress in neocortical synaptosomal membranes and hippocampal neurons in culture: Insights into Alzheimer's disease. *Rev. Neurosci.* **10:** 141-149.

Butterfield, D.A., Castegna, A., Lauderback, C.M., and Drake, J., 2002a, Evidence that amyloid beta-peptide-induced lipid peroxidation and its sequelae in Alzheimer's disease brain contribute to neuronal death. *Neurobiol. Aging* **23:** 655-664.

Butterfield, D.A., Griffin, S., Munch, G., and Pasinetti, G.M., 2002b, Amyloid beta-peptide and amyloid pathology are central to the oxidative stress and inflammatory cascades under which alzheimer's disease brain exists. *J. Alzheimer's Dis.* **4:** 193-201.

Cherny, R.A., Legg, J.T., McLean, C.A., Fairlie, D.P, Huang, X., Atwood, C.S., Beyreuther, K, Tanzi, R.E., Masters, C.L., and Bush, A.I., 1999, Aqueous dissolution of Alzheimer's disease a.beta amyloid deposits by biometal depletion. *J. Biol. Chem.* **274:** 23223-23228.

Cherny, R.A., Atwood, C.S., Xilinas, M.E., Gray, D.N, Jones, W.D., McLean, C.A., Barnham, K.J., Volitakis, I., Fraser, F.W., Kim, Y., Huang, X., Goldstein, L.E., Moir, R.D., Lim, J.T., Beyreuther, K., Zheng, H., Tanzi, R.E., Masters, C.L., and Bush, A.I., 2001, Treatment with a copper-zinc chelator markedly and rapidly inhibits beta-amyloid accumulation in alzheimer's disease transgenic mice. *Neuron* **30:** 665-676.

Citron, M., Oltersdorf, T., Haass, C., McConlogue, L., Hung, A,Y., Seubert, P., Vigo-Pelfrey, C., Lieberburg, I., and Selkoe, D.J., 1992, Mutation of the beta-amyloid precursor protein in familial Alzheimer's disease increases beta-protein production. *Nature* **360:** 672-674.

Citron, M., Westaway, D., Xia, W., Carlson, G., Diehl, T., Levesque, G., Johnson-Wood, K., Lee, M., Seubert, P., Davis, A., Kholodenko, D., Motter, R., Sherrington, R., Perry, B., Yao, H, Strome, R, Lieberburg, I, Rommens, J, Kim, S, Schenk, D, Fraser, P, St George Hyslop, P., and Selkoe, D.J., 1997, Mutant presenilins of Alzheimer's disease increase production of 42-residue amyloid beta-protein in both transfected cells and transgenic mice. *Nature Med.* **3:** 67-72.

Connor, J.R., Menzies, S.L., St Martin, S.M., and Mufson, E.J., 1992, A histochemical study of iron, transferrin, and ferritin in Alzheimer's diseased brains. *J. Neurosci. Res.* **31:** 75-83.

Constantinidis, J., 1991a, The hypothesis of zinc deficiency in the pathogenesis of neurofibrillary tangles. *Med. Hypoth.* **35:** 319-323.

Constantinidis, J., 1991b, Hypothesis regarding amyloid and zinc in the pathogenesis of Alzheimer disease: Potential for preventive intervention. *Alzheimer Di.s Assoc. Disord.* **5:** 31-35.

Cornett, C.R., Markesbery, W.R., and Ehmann, W.D., 1998a, Imbalances of trace elements related to oxidative damage in Alzheimer's disease brain. *Neurotoxicol.* **19:** 339-345.

Cornett, C.R., Ehmann, W.D., Wekstein, D.R, and Markesbery, W.R, 1998b, Trace elements in Alzheimer's disease pituitary glands. *Biol. Trace Elem. Res.* **62:** 107-114.

Corrigan, F.M., Reynolds, G.P., Ward, N.I., 1993, Hippocampal tin, aluminum and zinc in Alzheimer's disease. *Biometals* **6:** 149-154.

Crapper McLachlan, D.R., Dalton, A.J., Kruck, T.P., Bell, M.Y., Smith, W.L., Kalow, W., and Andrews, D.F., 1991, Intramuscular desferrioxamine in patients with Alzheimer's disease. *Lancet* **337:** 1304-1308.

Cuajungco, M.P., and Lees, G.J., 1996, Prevention of zinc neurotoxicity in vivo by n,n,n',n'-tetrakis (2-pyridylmethyl) ethylene-diamine (tpen). *Neuroreport* **7:** 1301-1304.

Cuajungco, M.P., and Lees, G.J., 1998, Diverse effects of metal chelating agents on the neuronal cytotoxicity of zinc in the hippocampus. *Brain Res.* **799:** 97-107.

Cuajungco, M.P., and Faget, K.Y., 2003, Zinc takes the center stage: Its paradoxical role in Alzheimer's disease. *Brain Res. Brain Res. Rev.* **41:** 44-56.

Cuajungco, M.P., Faget, K.Y., Huang, X., Tanzi, R.E., and Bush, A.I., 2000a, Metal chelation as a potential therapy for Alzheimer's disease. *Ann. N. Y. Acad. Sci.* **920:** 292-304.

Cuajungco, M.P., Goldstein, L.E., Nunomura, A., Smith, M.A., Lim, J.T., Atwood, C.S., Huang, X., Farrag, Y.W., Perry, G., and Bush, A.I., 2000b, Evidence that the beta-amyloid plaques of Alzheimer's disease represent the redox-silencing and entombment of abeta by zinc. *J. Biol. Chem.* **275:**.19439-19442.

Curtain, C.C., Ali, F., Volitakis, I., Cherny, R.A., Norton, R.S., Beyreuther, K., Barrow, C.J., Masters, C.L., Bush, A.I., and Barnham, K.J., 2001, Alzheimer's disease amyloid-beta binds copper and zinc to generate an allosterically ordered membrane-penetrating structure containing superoxide dismutase-like subunits. *J. Biol. Chem.* **276:** 20466-20473.

Deibel, M.A., Ehmann, W.D., and Markesbery, W.R., 1996, Copper, iron, and zinc imbalances in severely degenerated brain regions in Alzheimer's disease: Possible relation to oxidative stress. *J. Neurol. Sci.* **143:** 137-142.

Drake, J., Link, C.D., and Butterfield, D.A., 2003, Oxidative stress precedes fibrillar deposition of Alzheimer's disease amyloid beta-peptide (1-42) in a transgenic caenorhabditis elegans model. *Neurobiol. Aging* **24:** 415-420.

Dwork, A.J., Schon, E.A, and Herbert, J., 1988, Nonidentical distribution of transferrin and ferric iron in human brain. *Neuroscience* **27:** 333-345.

Emre, M., Geula, C., Ransil, B.J., and Mesulam, M.M., 1992, The acute neurotoxicity and effects upon cholinergic axons of intracerebrally injected beta-amyloid in the rat brain. *Neurobiol. Aging* **13:** 553-559.

Ertekin-Taner, N., Graff-Radford, N., Younkin, L.H., Eckman, C., Baker, M., Adamson, J., Ronald, J., Blangero, J., Hutton, M., and Younkin, S.G., 2000, Linkage of plasma abeta42 to a quantitative locus on chromosome 10 in late-onset Alzheimer's disease pedigrees. *Science* **290:** 2303-2304.

Frederickson, C.J., 1989, Neurobiology of zinc and zinc-containing neurons. *Int. Rev. Neurobiol.* **31:** 145-238.

Garzon-Rodriguez, W., Yatsimirsky, A.K., and Glabe, C.G., 1999, Binding of zn(ii), cu(ii), and fe(ii) ions to Alzheimer's A beta peptide studied by fluorescence. *Bioorg. Med. Chem. Lett.* **9:** 2243-2248.

Glenner, G.G., and Wong, C.W., 1984, Alzheimer's disease: Initial report of the purification and characterization of a novel cerebrovascular amyloid protein. *Biochem. Biophys. Res. Commun.* **120:** 885-890.

Goedert, M., Sisodia, S.S., and Price, D.L., 1991, Neurofibrillary tangles and beta-amyloid deposits in Alzheimer's disease. *Curr. Opin. Neurobiol.* **1:** 441-447.

Gonzalez, C., Martin, T., Cacho, J., Brenas, M.T., Arroyo, T., Garcia-Berrocal, B., Navajo, J.A., and Gonzalez-Buitrago, J.M., 1999, Serum zinc, copper, insulin and lipids in Alzheimer's disease epsilon 4 apolipoprotein e allele carriers. *Eur. J. Clin. Invest.* **29:** 637-642.

Gray, D.N., Cherny, R., Masters, C.L., Tanzi, R.E., and Bush, A.I., 1998, Resolubilization of Alzheimer and app transgenic beta amyloid plaque by copper chelators. *Soc. Neurosci. Abstr.* **24:** 722.

Halliwell, B., 1992, Reactive oxygen species and the central nervous system. *J. Neurochem.* **59:** 1609-1623.

Halliwell, B., 2001, Role of free radicals in the neurodegenerative diseases: Therapeutic implications for antioxidant treatment. *Drugs Aging* **18:** 685-716.

Hershey, C.O., Hershey, L.A., Varnes, A., Vibhakar, S.D., Lavin, P., and Strain, W.H., 1983, Cerebrospinal fluid trace element content in dementia: Clinical, radiologic, and pathologic correlations. *Neurology* **33:** 1350-1353.

Hesse, L., Beher, D., Masters, C.L.,and Multhaup, G., 1994, The beta a4 amyloid precursor protein binding to copper. *FEBS Lett.* **349:** 109-116.

Hirakura, Y., Yiu, W.W., Yamamoto, A., and Kagan, B.L., 2000, Amyloid peptide channels: Blockade by zinc and inhibition by congo red (amyloid channel block). *Amyloid* **7:** 194-199.

Hu, L., Wong, T.P., Cote, S.L., Bell, K.F., and Cuello, A.C., 2003, The impact of abeta-plaques on cortical cholinergic and non-cholinergic presynaptic boutons in Alzheimer's disease-like transgenic mice. *Neuroscience* **121:** 421-432.

Huang, X., Cuajungco, M.P., Atwood, C.S., Moir, R.D., Tanzi, R.E., and Bush, A.I., 2000, Alzheimer's disease, beta-amyloid protein and zinc. *J. Nutr.* **130:** 1488S-1492S.

Huang, X., Atwood, C.S., Moir, R.D., Hartshorn, M.A., Vonsattel, J.P., Tanzi, R.E., Bush, A.I., 1997, Zinc-induced Alzheimer's abeta1-40 aggregation is mediated by conformational factors. *J. Biol. Chem.* **272:** 26464-26470.

Huang, X., Atwood, C.S., Hartshorn, M.A., Multhaup, G., Goldstein, L.E., Scarpa, R.C., Cuajungco, M.P., Gray, D.N., Lim, J., Moir, R.D., Tanzi, R.E., and Bush, A.I., 1999a, The a beta peptide of Alzheimer's disease directly produces hydrogen peroxide through metal ion reduction. *Biochemistry* **38:** 7609-7616.

Huang, X., Cuajungco, M.P., Atwood, C.S., Hartshorn, M.A., Tyndall, J.D., Hanson, G.R., Stokes, K.C., Leopold, M., Multhaup, G., Goldstein, L.E., Scarpa, R.C., Saunders, A.J., Lim, J., Moir, R.D., Glabe, C., Bowden, E.F., Masters, C.L., Fairlie, D.P., Tanzi, R.E., and Bush, A.I., 1999b, Cu(ii) potentiation of Alzheimer abeta neurotoxicity. Correlation with cell-free hydrogen peroxide production and metal reduction. *J. Biol. Chem.* **274:** 37111-37116.

Hyman, B.T., Van Hoesen, G.W., Kromer, L.J., and Damasio, A.R., 1986, Perforant pathway changes and the memory impairment of Alzheimer's disease. *Ann. Neurol.* **20:** 472-481.

Kowall, N.W., McKee, A.C., Yankner, B.A., and Beal, M.F., 1992, In vivo neurotoxicity of beta-amyloid [beta(1-40)] and the beta(25-35) fragment. *Neurobiol. Aging* **13:** 537-542.

Kozin, S.A., Zirah, S., Rebuffat, S., Hoa, G.H., and Debey, P., 2001, Zinc binding to Alzheimer's abeta(1-16) peptide results in stable soluble complex. *Biochem. Biophys. Res. Commun.* **285:** 959-964.

Lee, J.Y., Mook-Jung, I., and Koh, J.Y., 1999, Histochemically reactive zinc in plaques of the swedish mutant beta-amyloid precursor protein transgenic mice. *J. Neurosci.* **19:** RC10.

Lee, J.Y., Cole, T.B., Palmiter, R.D., Suh, S.W., and Koh, J.Y., 2002, Contribution by synaptic zinc to the gender-disparate plaque formation in human swedish mutant app transgenic mice. *Proc. Natl. Acad. Sci. USA* **99:** 7705-7710.

Lees, G.J., Cuajungco, M.P., and Leong, W., 1998, Effect of metal chelating agents on the direct and seizure-related neuronal death induced by zinc and kainic acid. *Brain Res.* **799:** 108-117.

Lemere, C.A., Blusztajn, J.K., Yamaguchi, H., Wisniewski, T., Saido, T.C., and Selkoe, D.J. 1996, Sequence of deposition of heterogeneous amyloid beta-peptides and apo e in Down syndrome: Implications for initial events in amyloid plaque formation. *Neurobiol. Dis.* **3:** 16-32.

Levy-Lahad, E., Tsuang, D., Bird, T.D., 1998, Recent advances in the genetics of Alzheimer's disease. *J. Ger. Psych. Neurol.* **11:** 42-54.

Levy-Lahad, E., Wasco, W., Poorkaj, P., Romano, D.M., Oshima, J., Pettingell, W.H., Yu, C.E., Jondro, P.D., Schmidt, S.D., Wang, K., et al., 1995, Candidate gene for the chromosome 1 familial Alzheimer's disease locus. *Science* **269:** 973-977.

Li, Q.X., Evin, G., Small, D.H., Multhaup, G., Beyreuther, K., and Masters, C.L., 1995, Proteolytic processing of Alzheimer's disease beta a4 amyloid precursor protein in human platelets. *J. Biol. Chem.* **270:** 14140-14147.

Licastro, F., Davis, L.J., Mocchegiani, E., and Fabris, N., 1996, Impaired peripheral zinc metabolism in patients with senile dementia of probable Alzheimer's type as shown by low plasma concentrations of thymulin. *Biol. Trace Elem. Res.* **51:** 55-62.

Liu, S.T., Howlett, G., and Barrow, C.J., 1999, Histidine-13 is a crucial residue in the zinc ion-induced aggregation of the a beta peptide of Alzheimer's disease. *Biochemistry* **38:** 9373-9378.

Loeffler, D.A., DeMaggio, A.J., Juneau, P.L., Brickman, C.M., Mashour, G.A., Finkelman, J.H., Pomara, N., and LeWitt, P.A., 1994, Ceruloplasmin is increased in cerebrospinal fluid in Alzheimer's disease but not Parkinson's disease. *Alz. Dis. Assoc. Disord.* **8:** 190-197.

Lombardo, J.A., Stern, E.A., McLellan, M.E., Kajdasz, S.T., Hickey, G.A., Bacskai, B.J., and Hyman, B.T., 2003, Amyloid-beta antibody treatment leads to rapid normalization of plaque-induced neuritic alterations. *J. Neurosci.* **23:** 10879-10883.

Lorenzo, A., and Yankner, B.A., 1994, Beta-amyloid neurotoxicity requires fibril formation and is inhibited by congo red. *Proc. Natl. Acad. Sci. USA* **91:** 12243-12247.

Lovell, M.A., Robertson, J.D., Teesdale, W.J., Campbell, J.L., and Markesbery, W.R., 1998, Copper, iron and zinc in Alzheimer's disease senile plaques. *J. Neurol. Sci.* **158:** 47-52.

Luth, H.J., Apelt, J., Ihunwo, A.O., Arendt, T., and Schliebs, R., 2003, Degeneration of beta-amyloid-associated cholinergic structures in transgenic app sw mice. *Brain Res.* **977:** 16-22.

Maynard, C.J., Cappai, R., Volitakis, I., Cherny, R.A., White, A.R., Beyreuther, K., Masters, C.L., Bush, A.I., and Li, Q.X., 2002, Overexpression of Alzheimer's disease amyloid-beta opposes the age-dependent elevations of brain copper and iron. *J. Biol. Chem.* **277:** 44670-44676.

Mecocci P, MacGarvey, U, Beal, MF (1994) Oxidative damage to mitochondrial DNA is increased in Alzheimer's disease. Ann Neurol 36:747-751.

Mecocci, P., Polidori, M.C., Cherubini, A., Ingegni, T., Mattioli, P., Catani, M., Rinaldi, P., Cecchetti, R., Stahl, W., Senin, U., and Beal, M.F., 2002, Lymphocyte oxidative DNA damage and plasma antioxidants in Alzheimer disease. *Arch. Neurol.* **59:** 794-798.

Miura, T., Suzuki, K., Kohata, N., and Takeuchi, H., 2000, Metal binding modes of Azheimer's amyloid beta-peptide in insoluble aggregates and soluble complexes. *Biochemistry* **39:** 7024-7031.

Moir, R.D., Atwood, C.S., Romano, D.M., Laurans, M.H., Huang, X., Bush, A.I., Smith, J.D., and Tanzi, R.E., 1999, Differential effects of apolipoprotein e isoforms on metal-induced aggregation of a beta using physiological concentrations. *Biochemistry* **38:** 4595-4603.

Molina, J.A., Jimenez-Jimenez, F.J., Aguilar, M.V., Meseguer, I., Mateos-Vega, C.J., Gonzalez-Munoz, M.J., de Bustos, F., Porta, J., Orti-Pareja, M., Zurdo, M., Barrios, E., and Martinez-Para, M.C., 1998, Cerebrospinal fluid levels of transition metals in patients with Alzheimer's disease. *J. Neural Transm.* **105:** 479-488.

Monji, A., Utsumi, H., Ueda, T., Imoto, T., Yoshida, I., Hashioka, S., Tashiro, K., and Tashiro, N., 2001, The relationship between the aggregational state of the amyloid-beta peptides and free radical generation by the peptides. *J. Neurochem.* **77:** 1425-1432.

Monji, A., Utsumi, H., Ueda, T., Imoto, T., Yoshida, I., Hashioka, S., Tashiro, K., and Tashiro, N., 2002, Amyloid-beta-protein (a beta) (25-35)-associated free radical generation is strongly influenced by the aggregational state of the peptides. *Life Sci.* **70:** 833-841.

Multhaup, G., 1994, Identification and regulation of the high affinity binding site of the Alzheimer's disease amyloid protein precursor (app) to glycosaminoglycans. *Biochimie* **76:** 304-311.

Multhaup, G., Mechler, H., and Masters, C.L., 1995, Characterization of the high affinity heparin binding site of the Alzheimer's disease beta a4 amyloid precursor protein (app) and its enhancement by zinc(ii). *J. Mol. Recognit.* **8:** 247-257.

Multhaup, G., Bush, A.I., Pollwein, P., and Masters, C.L., 1994, Interaction between the zinc (ii) and the heparin binding site of the Alzheimer's disease beta a4 amyloid precursor protein (app). *FEBS Lett.* **355:** 151-154.

Multhaup, G., Schlicksupp, A., Hesse, L., Beher, D., Ruppert, T., Masters, C.L., and Beyreuther, K., (1996, The amyloid precursor protein of Alzheimer's disease in the reduction of copper(ii) to copper(i). *Science* **271:** 1406-1409.

Multhaup, G., Ruppert, T., Schlicksupp, A., Hesse, L., Bill, E., Pipkorn, R., Masters, C.L., and Beyreuther, K., 1998, Copper-binding amyloid precursor protein undergoes a site-specific fragmentation in the reduction of hydrogen peroxide. *Biochemistry* **37:** 7224-7230.

Nitzan, Y.B., Sekler, I., Frederickson, C.J., Coulter, D.A., Balaji, R.V., Liang, S.L., Margulis, A., Hershfinkel, M., and Silverman, W.F., 2003, Clioquinol effects on tissue chelatable zinc in mice. *J. Mol. Med.* **81:** 637-644.

Perry, G., Taddeo, M.A., Petersen, R.B., Castellani, R.J., Harris, P.L., Siedlak, S.L., Cash, A.D., Liu, Q., Nunomura, A., Atwood, C.S., and Smith, M.A., 2003, Adventiously-bound redox active iron and copper are at the center of oxidative damage in alzheimer disease. *Biometals* **16:** 77-81.

Pike, C.J., Walencewicz, A.J., Glabe, C.G., and Cotman, C.W., 1991a, Aggregation-related toxicity of synthetic beta-amyloid protein in hippocampal cultures. *Eur. J. Pharmacol.* **207:** 367-368.

Pike, C.J., Walencewicz, A.J., Glabe, C.G., and Cotman, C.W., 1991b, In vitro aging of beta-amyloid protein causes peptide aggregation and neurotoxicity. *Brain Res.* **563:** 311-314.

Regland, B., Lehmann, W., Abedini, I., Blennow, K., Jonsson, M., Karlsson, I., Sjogren, M., Wallin, A., Xilinas, M., and Gottfries, C.G., 2001, Treatment of Alzheimer's disease with clioquinol. *Dement. Geriatr. Cogn. Disord.* **12**: 408-414.

Ritchie, C.W., Bush, A.I., Mackinnon, A., Macfarlane, S., Mastwyk, M., MacGregor, L., Kiers, L., Cherny, R., Li, Q.X., Tammer, A., Carrington, D., Mavros, C., Volitakis, I., Xilinas, M., Ames, D., Davis, S., Beyreuther, K., Tanzi, R.E., and Masters, C.L., 2003,

Metal-protein attenuation with iodochlorhydroxyquin (clioquinol) targeting abeta amyloid deposition and toxicity in Alzheimer disease: A pilot phase 2 clinical trial. *Arch. Neuro.l* **60:** 1685-1691.

Roses, A.D., Saunders, A.M., Alberts, M.A., Strittmatter, W.J., Schmechel, D., Gorder, E., and Pericak-Vance, M.A., 1995, Apolipoprotein e e4 allele and risk of dementia. *Jama* **273:** 374-375; author reply 375-376.

Rottkamp, C.A., Raina, A.K., Zhu, X., Gaier, E., Bush, A.I., Atwood, C.S., Chevion, M., Perry, G., and Smith, M.A., 2001, Redox-active iron mediates amyloid-beta toxicity. *Free Radic. Biol. Med.* **30:** 447-450.

Ruiz, F.H., Gonzalez, M., Bodini, M., Opazo, C., and Inestrosa, N.C., 1999, Cysteine 144 is a key residue in the copper reduction by the beta-amyloid precursor protein. *J. Neurochem.* **73:** 1288-1292.

Sahu, R.N., Pandey, R.S., Subhash, M.N., Arya, B.Y., Padmashree, T.S., and Srinivas, K.N., 1988, Csf zinc in Alzheimer's type dementia. Biol Psychiatry 24:480-482.

Samudralwar, D.L., Diprete, C.C., Ni, B.F., Ehmann, W.D., and Markesbery, W.R., 1995, Elemental imbalances in the olfactory pathway in Alzheimer's disease. *J. Neurol. Sci.* **130:** 139-145.

Sano, M., Ernesto, C., Thomas, R.G., Klauber, M.R., Schafer, K., Grundman, M., Woodbury, P., Growdon, ., Cotman, C.W., Pfeiffer, E., Schneider, L.S., and Thal, L.J., 1997, A controlled trial of selegiline, alpha-tocopherol, or both as treatment for Alzheimer's disease. The Alzheimer's disease cooperative study. *N. Engl. J. Med.* **336:** 1216-1222.

Saunders, A.J., Bertram, L., Mullin, K., Sampson, A.J., Latifzai, K., Basu, S., Jones, J., Kinney, D., MacKenzie-Ingano, L., Yu, S., Albert, M.S., Moscarillo, T.J., Go, R.C., Bassett, S.S., Daly, M.J., Laird, N.M., Wang, X., Velicelebi, G., Wagner, S.L., Becker, D.K., Tanzi, R.E., and Blacker, D., 2003, Genetic association of Alzheimer's disease with multiple polymorphisms in alpha-2-macroglobulin. *Hum. Mol. Genet.* **12:** 2765-2776.

Sayre, L.M., Perry, G., Harris, P.L., Liu, Y., Schubert, K.A., and Smith, M.A., 2000, In situ oxidative catalysis by neurofibrillary tangles and senile plaques in Alzheimer's disease: A central role for bound transition metals. *J. Neurochem* **74:** 270-279.

Schubert, D., and Chevion, M., 1995, The role of iron in beta amyloid toxicity. *Biochem. Biophys. Res. Commun.* **216:** 702-707.

Simons, A., Ruppert, T., Schmidt, C., Schlicksupp, A., Pipkorn, R., Reed, J., Masters, C.L., White, A.R., Cappai, R., Beyreuther, K., Bayer, T.A., and Multhaup, G., 2002, Evidence for a copper-binding superfamily of the amyloid precursor protein. *Biochemistry* **41:** 9310-9320.

Smith, M.A., Harris, P.L., Sayre, L.M., and Perry, G., 1997, Iron accumulation in Alzheimer disease is a source of redox-generated free radicals. *Proc. Natl. Acad. Sci. USA* **94:** 9866-9868.

Smith, M.A., Wehr, K., Harris, P.L., Siedlak, S.L., Connor, J.R., and Perry, G., 1998a, Abnormal localization of iron regulatory protein in Alzheimer's disease. *Brain Res.* **788:** 232-236.

Smith, M.A., Hirai, K., Hsiao, K., Pappolla, M.A., Harris, P.L., Siedlak, S.L., Tabaton, M., and Perry, G., 1998b, Amyloid-beta deposition in Alzheimer transgenic mice is associated with oxidative stress. *J. Neurochem.* **70:** 2212-2215.

Smith, R., ed., 1983, Copper in the developing brain. Human Press, New Jersey.

Squitti, R., Lupoi, D., Pasqualetti, P., Dal Forno, G., Vernieri, F., Chiovenda, P., Rossi, L., Cortesi, M., Cassetta, E., and Rossini, P.M., 2002, Elevation of serum copper levels in Alzheimer's disease. *Neurology* **59:** 1153-1161.

Suh, S.W., Jensen, K.B., Jensen, M.S., Silva, D.S., Kesslak, P.J., Danscher, G., and Frederickson, C.J., 2000, Histochemically-reactive zinc in amyloid plaques, angiopathy, and degenerating neurons of Alzheimer's diseased brains. *Brain Res.* **852:** 274-278.

Suzuki, K., Miura, T., and Takeuchi, H., 2001, Inhibitory effect of copper(ii) on zinc(ii)-induced aggregation of amyloid beta-peptide. *Biochem. Biophys. Res. Commun.* **285:** 991-996.

Tanzi, R.E., Kovacs, D.M., Kim, T.W., Moir, R.D., Guenette, S.Y., and Wasco, W., 1996, The gene defects responsible for familial Alzheimer's disease. *Neurobiol. Dis.* **3:** 159-168.

Tully, C.L., Snowdon, D.A., and Markesbery, W.R., 1995, Serum zinc, senile plaques, and neurofibrillary tangles: Findings from the nun study. *Neuroreport* **6:** 2105-2108.

Wasco, W., Bupp, K., Magendantz, M., Gusella, J.F., Tanzi, R.E., and Solomon, F., 1992, Identification of a mouse brain cdna that encodes a protein related to the Alzheimer disease-associated amyloid beta protein precursor. *Proc. Natl. Acad. Sci. USA* **89:** 10758-10762.

Wasco, W., Gurubhagavatula, S., Paradis, M.D., Romano, D.M., Sisodia, S.S., Hyman, B.T., Neve, R.L., and Tanzi, R.E., 1993, Isolation and characterization of aplp2 encoding a homologue of the Alzheimer's associated amyloid beta protein precursor. *Nature Genet.* **5:** 95-100.

Weldon, D.T., Rogers, S.D., Ghilardi, J.R., Finke, M.P., Cleary, J.P., O'Hare, E., Esler, W.P., Maggio, J.E., and Mantyh, P.W., 1998, Fibrillar beta-amyloid induces microglial phagocytosis, expression of inducible nitric oxide synthase, and loss of a select population of neurons in the rat cns in vivo. *J. Neurosci.* **18:** 2161-2173.

White, A.R., Multhaup, G., Maher, F., Bellingham, S., Camakaris, J., Zheng, H., Bush, A.I., Beyreuther, K., Masters, C.L., and Cappai, R., 1999, The Alzheimer's disease amyloid precursor protein modulates copper-induced toxicity and oxidative stress in primary neuronal cultures. *J. Neurosci.* **19:** 9170-9179.

Winblad, B., Hardy, J., Backman, L., and Nilsson, L.G., 1985, Memory function and brain biochemistry in normal aging and in senile dementia. *Ann. N. Y. Acad. Sci.* **444:** 255-268.

Wolfe, M.S., 2003, The secretases of Alzheimer's disease. *Curr. Top. Dev. Biol.* **54:** 233-261.

Yang, D.S., McLaurin, J., Qin, K., Westaway, D., and Fraser, P.E., 2000, Examining the zinc binding site of the amyloid-beta peptide. *Eur. J. Biochem.* **267:** 6692-6698.

Yang, L.B., Lindholm, K., Yan, R., Citron, M., Xia, W., Yang, X.L., Beach, T., Sue, L., Wong, ., Price, D., Li, R., and Shen, Y., 2003, Elevated beta-secretase expression and enzymatic activity detected in sporadic Alzheimer disease. *Nature Med.* **9:** 3-4.

Yankner, B.A., Dawes, L.R., Fisher, S., Villa-Komaroff, L., Oster-Granite, M.L., and Neve, R.L., 1989, Neurotoxicity of a fragment of the amyloid precursor associated with Alzheimer's disease. *Science* **245:** 417-420.

Yoshiike, Y., Tanemura, K., Murayama, O., Akagi, T., Murayama, M., Sato, S., Sun, X., Tanaka, N., andTakashima, A., 2001, New insights on how metals disrupt amyloid beta-aggregation and their effects on amyloid-beta cytotoxicity. *J. Biol. Chem.* **276:** 32293-32299.

Chapter 13

The Interaction of Amyloid-β with ApoE

The form of beta amyloid is a partial determinant of the interaction with Apolipoprotein E

Donald B. Carter

CNS Therapeutics, Pfizer Corporation, Kalamazoo, MI, USA

Abstract: Brain plaque deposition in the form of amyloid-beta (Aβ) peptide is a pathological hallmark of Alzheimer's disease (AD). Apolipoprotein E (ApoE) is thought to be involved in plaque formation. Individuals afflicted with AD carrying the ApoE4 isoform have shown a greater number of Aβ plaques when compared to ApoE3 carriers, and inheritance of an ApoE4 allele increases the risk of AD when compared to ApoE2 and ApoE3 carriers. The role of ApoE in the pathogenesis of AD is not well understood but a hypothesis gaining widespread support is that ApoE is involved in deposition or clearance of Aβ by direct protein-to-protein interaction. We have established that various human Aβ conformations apparent during spontaneous aggregation confer differing degrees of binding to the three ApoE isoforms. When associated with lipid, ApoE4 bound preferentially to an intermediate aggregated form of Aβ and had higher avidity than did ApoE2 or ApoE3.

Key words: Apolipoprotein E (ApoE), amyloid-beta (Aβ), protofibrils, cerebral spinal fluid (CSF)

1. INTRODUCTION

Alzheimer's disease is characterized by progressive neurodegeneration associated with the deposition of extracellular proteinaceous amyloid-beta peptide (Aβ is a collection of peptides derived from amyloid precursor protein including peptides with lengths from positions 1-12 to positions 38-43) in the form of senile plaques (Selkoe, 2002). The ApoE4 allele of apolipoprotein E is a well established risk factor for late onset Alzheimer's disease (LOAD) (Saunders *et al.*, 1993). The mechanisms underlying the increased risk of AD in ApoE4 carriers are currently under investigation in numerous labs but there is accumulating evidence that the capacity of ApoE to interact with Aβ and mediate its clearance or conformation is of prime importance. Genetic epidemiologic studies have shown that ApoE4 allele is also associated with cerebral amyloid angiopathy (CAA), another neurodegenerative disease in which excessive deposition of Aβ occurs in cerebral blood vessels.

ApoE is a 34kDa (299 amino acids) lipid transport protein which is primarily synthesized in liver and brain and is found associated with compartment specific lipid-vesicles in plasma and brain (Mahley, 1988). ApoE exists as three primary isoforms in the human population, E2 (Cys^{112}, Cys^{158}), E3 (Cys^{112}, Arg^{158}), and E4 (Arg^{112}, Arg^{158}), with E3 being the most common allele at approximately 80% frequency in the Caucasian population (Corbo *et al.*, 1999). In the central nervous system ApoE is the most abundant apolipoprotein in the CSF and is secreted predominantly by glia in a high-density lipoprotein (HDL) like particle (Pitas *et al.*, 1987a,b; LaDu *et al.*, 1998; Fagan *et al.*, 1999). Within the brain ApoE is derived from inside the blood-brain-barrier where it is found in the CSF at approximately 5 mg/ml (Linton *et al.*, 1991). Strittmatter et al. (1993a) first showed that the E4 allele of ApoE was a genetic risk factor for LOAD. Subsequently other researchers confirmed this finding (Corder *et al.*, 1993; Mayeux *et al.*, 1993; Saunders *et al.*, 1993; Corder *et al.*, 1994; Strittmatter and Roses, 1996). A potential interaction between ApoE and amyloid proteins was first suggested when ApoE immunoreactivity was detected in extracellular amyloid deposits in the brains of patients with AD, Down's syndrome, cerebral and systemic amyloid (Namba *et al.*, 1991; Wisniewski and Frangione, 1992). More recently data was published which suggests that C-terminal residues of ApoE bind to Aβ plaques and that apoE may help aid in the progression of small Aβ deposits to larger deposits (Cho *et al.*, 2001). These results are consistent with early *in vitro* data which implicated the C-terminal residues of ApoE in Aβ binding. Early *in vitro* studies have also shown that either lipid free or lipid associated ApoE can bind Aβ with isoform specificity (Strittmatter *et al.*, 1993a,b; LaDu *et al.*, 1993; LaDu *et al.*, 1995).

2. *IN VITRO APOE-Aβ* INTERACTIONS

2.1. ApoE isoform dependence of Aβ interaction (delipidated ApoE)

Strittmatter and colleagues (1993a,b) first described the interaction between de-lipidated forms of ApoE with membrane bound synthetic Aβ(1-40), Aβ(12-28) or Aβ(1-28). They found that formation of an SDS-stable complex of ApoE with Aβ was dependent on oxidation of the ApoE and the pH of the medium. SDS-stable binding of ApoE with Aβ was maximal at pH 7.5 and decreased rapidly as pH was lowered. However it is possible that other physiologically relevant complexes were formed which were not SDS-stable and this possibility was not addressed in these studies. Also the forms of Aβ actually attached to the filters were not defined thus making interpretation of these experiments difficult since it was subsequently shown that the binding of ApoE and Aβ is dependent on the form of Aβ (Shuvaev and Siest, 1996). Strittmatter *et al.* (1993a,b) also demonstrated that peptide fragments from the C-terminal portion of ApoE were capable of forming SDS-stable complexes with Aβ(1-28) but N-terminal fragments from ApoE were not competent to bind the peptide Aβ(1-28). This work was extended to examination of the effects of ApoE on the conformation of Aβ (Sanan *et al.*, 1994) which showed that ApoE/Aβ coincubates yielded monofibrils 7 nm in diameter. The precise forms of the Aβ conformations were dependent on the ApoE isoform in the incubation and both ApoE3 and ApoE4 were seen to be uniformly distributed along the Aβ monofibrils with ApoE4 forming a more dense matrix of monfibrils. Several other studies have suggested that there is an interaction between soluble forms of Aβ and delipidated ApoE which influences fibrillogenesis of Aβ (Ma *et al.*, 1994; Wisniewski *et al.*, 1994; Castano *et al.*, 1995; Soto *et al.*, 1996). Ma *et al.* (1994). They observed the interaction of Aβ(1-42) with lipid-free plasma derived ApoE2, 3 and 4 using electron microscopy starting from fibril free preparations of Aβ(1-42). ApoE4 was concluded to be the most active at promoting filament formation, ApoE2 the least active and ApoE3 having an intermediate level of activity when the peptide:ApoE molar ratio was 200:1. This molar ratio of Aβ:ApoE reflects the ratio found in localized amyloid plaques in diseased brain (Castano *et al.*, 1995). In addition, Ma *et al.*, (1994) discovered that ApoE2 was capable of competing against ApoE4 in the reaction. These results were confirmed and extended to Aβ(1-40) using recombinant human ApoE at a ratio Aβ(1-40):ApoE of 100:1 (Wisniewski *et al.*, 1994). In this study the Aβ(1-40) peptide was denatured in 1% trifluroacetic-50% acetonitrile before stock solution preparation in order to disaggregate the peptide. Another report showed that the interaction between delipidated ApoE and Aβ is dependent on the Aβ peptide conformation (Golabek *et al.*, 1996). These workers found no ApoE isoform dependence on the Aβ interaction strength and characterized the ApoE-Aβ(1-40) interaction by saturation binding analysis of soluble ApoE incubated with Aβ(1-40) immobilized in microtiter plastic wells. A K_D of 20 nM was computed from the saturation binding curves and competition studies with other forms of Aβ indicated that the K_i for competitors is inveresely proportional to the amount of β-sheet conformation contained in the peptide. In a similar solid-state binding paradigm Shuvaev and Siest (1996) showed by surface plasmon resonance that for immobilized Aβ(1-40) and soluble delipidated ApoE3 the binding constant for soluble delipidated ApoE3 was 10 nM, in good agreement with the saturation studies of Golabek *et al.* (1995). Shuvaev and Siest (1996) also showed that only a small fraction of the immobilized Aβ was capable of binding the ApoE and confirms the importance of Aβ conformation on binding. Levine (2000) used I^{125} radiolabelled ApoE purified from pooled human VLDL in a competitive binding filtration assay to measure the IC_{50} for displacement of I^{125}-ApoE from fibrillar Aβ(1-40) by various lipoproteins. The I^{125}-ApoE was presented to the fibrillar Aβ(1-40) in solution phase and the bound aggregate captured on filters. The IC_{50} values for displacement of I^{125}-ApoE from fibrillar Aβ(1-40) for delipidated VLDL ApoE and baculovirus expressed ApoE2, ApoE3, ApoE4 were 43nM, 9.5 nM, 50 nM, and 13 nM respectively which compares well with the K_D values of Golabek *et al.* (1996). In the Levine (2000) study, reconstituted lipidated ApoE was also used as competitor but no controls were shown to verify that the preparation contained intact lipoprotein particles (it is known that lipoprotein particle integrity is compromised by freeze-thaw). The question of whether the N- or C- terminal thrombin fragments of delipidated ApoE are responsible for binding Aβ was originally investigated by Strittmatter *et al.* (1993) but the peptide used was Aβ(1-28). This question was also pursued by Pillot *et al.* (1999) using tryptophan fluorescence quenching with synthetic ApoE peptides but the Aβ peptides used were Aβ(29-40), Aβ(29-42), Aβ(22-42) and Aβ(12-42). The results of this study showed that the C-terminal peptides of ApoE2 and ApoE3 were primarily responsible for the binding of ApoE-Aβ(1-40 and 29-40) peptides and not N-terminal ApoE peptides. Golabek *et al.* (2000) returned to this question but they utilized the more physiologically relevant Aβ(1-40) in their binding studies. Also using tryptophan fluorescence quenching similar to studies carried out by Pillot *et al.* (1999), Golabek *et al.* (2000) determined that the 22 Kd N-terminal thrombin fragment of ApoE3 binds Ab(1-40) with a K_D of ~ 11 nM and the C-terminal 10 kD thrombin fragment binds Aβ(1-40) with a K_D ~ 45 nM in excellent agreement with their previous results with whole delipidated ApoE (Golabek *et al.*, 1996) and with those of Shuvaev and Siest (1996). Three other studies showing the importance of the N-terminal of delipidated ApoE in Aβ(1-40) interactions have been published (Evans *et al.*, 1995; Wood *et al.*, 1996; Chan *et al.*, 1996). Evans *et al.* (1995) showed that the N-terminal domain of thrombin cleaved ApoE3 or 4 was an inhibitor of Aβ(1-40) nucleation at concentrations as low as 40 nM with an ApoE/Aβ(1-40) ratio of 1/1000. The lag time for fibrillogenesis is extended in a dose-dependent manner for all three isoforms of lipid free ApoE. Their studies also revealed that the ApoE3 dimer was a more potent inhibitor than either ApoE3 or ApoE4 monomer. Chan *et al.* (1996) also showed that the N-terminal of ApoE3 (amino acids 1-191) was capable of interacting with Aβ(1-40) under native conditions of binding. However, the experiments with N- and C-terminal thrombin-cleaved ApoE fragments may not extrapolate to the intact protein. It is well known that there are N- and C-terminal domain interactions in ApoE which determine what forms of lipoprotein particles each ApoE-isoform associates with as well as effects on receptor-binding activity (Weisgraber, 1994). This sequence of studies with various forms of Aβ teaches the importance of the primary structure as well as the secondary structure of Aβ interaction with various lipid free-ApoE protein fragments.

2.2. ApoE isoform dependence of Aβ interaction (lipidated ApoE)

In vivo ApoE is always found associated with lipoprotein particles and in the CNS these particles resemble HDL-like lipoproteins (Pitas et al., 1987a,b; LaDu et al., 1998). One of the first studies (LaDu et al., 1994) carried out on the interaction of lipid associated ApoE with Aβ(1-40) used the technique of assessing SDS-stable complexes of ApoE-Aβ on PAGE introduced by Strittmatter et al. (1993a). The results of these experiments suggested that lipidated ApoE3 and ApoE4 were capable of forming an ApoE-Aβ(1-40) complex and that ApoE3 was more competent to form the complex than ApoE4. In addition LaDu et al. (1994) showed that similar phenomena occurred with the truncated peptide Aβ(1-28) and that formation of both the ApoE3 and ApoE4 Aβ(1-40) interactions are sensitive to the presence of mercaptoethanol. As noted above the physiological significance of SDS-stable ApoE-Aβ(1-40) is unknown and the observation that the SDS-stable complex is an undetermined fraction of the total ApoE-Aβ(1-40) complex that could be formed if the reaction were complete presents problems with interpretation. These results are in direct contrast to the observations of Strittmatter *et al.* (1993) who concluded that ApoE4 formed more stable-SDS complexes with Aβ(1-40) than ApoE3 using delipidated ApoE. The source of LaDu's ApoE was from HEK cells transfected with human ApoE3 and ApoE4 cDNAs and human plasma VLDL fractions from individuals homozygous for ApoE3 or ApoE4. Other than different sources of ApoE, these two studies differed by how the Aβ(1-40) was presented to ApoE; in the case of LaDu *et al.,* (1994; 1995) the peptide was in solution at 250 mM and probably consisted of a mixture of monomeric and polymeric forms of Ab(1-40) and in Strittmatter's *et al.* (1993) study the Aβ(1-40) peptide was immobilized in a undefined conformation on filters. Thus the differences in isoform specific binding of Aβ(1-40) to ApoE3 or 4 could have been due to the conformation of Aβ(1-40) presented to ApoE as suggested by the studies of Golabek *et al.* (1996) and Shuvaev and Siest (1996) or the manner of presentation of the Aβ peptides (immobilized on filters or in solution). Similar results for formation of SDS-stable ApoE-Aβ(1-40) complexes were obtained for native ApoE3 and ApoE4 expressed by Raw-624 macrophages (Zhou *et al.*, 1996), from CHO cells (Yang *et al.,* 1997) and from BHK-21 cells (Aleshkov *et al.*, 1997), where the format of Aβ presentation to ApoE was in solution. The ApoE structural requirements for the formation of SDS-stable complexes with Aβ(1-40) were investigated by Bentley *et al.* (2002). It was found that changing residue cysteine 112 in ApoE to alanine or lysine abolished complex formation whereas serine substitution at position 112 retained the ability to form the complex. Bentley *et al.* (2002) concluded that the nature of the cysteine residue at position 112 and interactions between the N- and C-terminal domains of ApoE are important for formation of stable SDS-ApoE complexes. More recently Tokuda *et al.* (2000) used a solid-phase assay to report the binding of lipid particle-associated and purified delipidated ApoE3 and ApoE4 to both immobilized Aβ(1-40) and Aβ(1-42). In an attempt to control the conformation of the immobilized Aβ, Tokuda *et al.* (2002) attached two varieties of Aβ peptide to polystyrene microtitre plates: freshly prepared and presumably non-aggregated species and 72 hour aggregated species. Detection of bound ApoE was done with monoclonal anti-ApoE 3D12 antibody. Saturation binding experiments with native ApoE derived from RAW-264 and HEK cell lines as well as delipidated ApoE from these cell lines and Sf9 baculovirus expressed ApoE were carried out in an ELISA format with the Aβ peptides immobilized on plastic and dissociation constants (K_d) determined by analysis of the binding isotherms. The results indicated that regardless of the source of the native ApoE, there was a consistently higher K_d for the ApoE4 isoform than the ApoE3 isoform by a factor of ~2. For the lipid free ApoE the binding was shown to be 5-10 fold lower than with lipid associated ApoE but the K_d values did not depend on the isoform of ApoE used. These observations were independent of the species of Aβ used in the assays and the results are in reasonable agreement with those from the earlier studies with lipid free ApoE (Golabek *et al.*, 1996). Thus it appears that Aβ has a high affinity for ApoE (~5-20 nM) as determined by different labs using Ab immobilized on several different surfaces independently of the lipidation state of the ApoE protein. Whether this is the presentation format of Aβ to ApoE *in vivo* is an important question. Others have postulated that ApoE is a pathological chaperone which binds Aβ and induces beta-pleated sheets and in these experiments soluble forms of Aβ have been mixed with soluble lipid-free ApoE to simulate a solution phase interaction of the two molecules (Wisniewski and Frangione, 1992; Wisniewski *et al.*, 1994; Sanan *et al.*, 1994; Ma *et al.*, 1994). One could envision that a physical ApoE-Ab solution interaction would reasonably occur in cerebral spinal fluid (CSF). A series of informative experiments were done by Zhou *et al.* (1996) with CSF from ApoE3 and ApoE4 homozygotes in which two forms of ApoE-associated Ab were distinguished: one dissociable from ApoE in the presence of SDS and the other, resistant to SDS. Zhou *et al.* (1996) also used conditioned media from mouse macrophages stably transfected with human ApoE3 and ApoE4 cDNA as a source of lipid-associated ApoE. The ApoE concentration in the conditioned media was between 0.4 and 0.8 mM and Aβ(1-40) was added at a concentration of 100 mM (these concentrations are similar to those used by LaDu *et al.* (1994)). For the SDS dissociable complexes most of the Aβ was monomeric with smaller amounts of dimeric and oligomcric species detectable on the SDS polyacrylamide gels. The SDS resistant ApoE-Aβ complex showed a ~ 9-fold increase in association with Aβ(1-40) for ApoE3 over that for ApoE4 and a ~2-fold increase in the SDS-releasable association with ApoE3 over ApoE4. Greater than 50% of the of the ApoE3-associated Aβ(1-40) was SDS-resistant and the bulk of the ApoE4-associated Aβ(1-40) was SDS-releasable. When CSF was used as the source of the ApoE, the same pattern of preference of ApoE3-Aβ(1-40) was seen in the SDS-resistant fraction but the ability of the CSF ApoE3 to form a complex was dramatically lower. In addition the capacity of the ApoE3 CSF to form SDS-resistant complexes was only 3-fold higher than the ability of the ApoE4 CSF. One interpretation is that CSF might contain molecules which inhibit complex formation and reduce the isoform dependence of SDS-resistance complex formation. Another interpretation could be that the conditioned media recombinant ApoE from the cultured cells was different from the CSF ApoE-bearing lipoprotein particles. Since no controls for the integrity of the lipoprotein particles present in the conditioned media or the CSF were reported in the references contained in this review, firm conclusions regarding the impact of ApoE isoform on Aβ interactions are difficult to make. An effort to examine the interaction of lipidated ApoE isoforms with spontaneously aggregating Aβ(1-40) species in solution was carried out by Stratman *et al.* (2003). The integrity of the ApoE reconsituted lipoprotein particles (made the same day of the experiments) was checked by ultracentrifugation prior to use in the experiments and the Aβ(1-40) aggregation reactions were carried out in a standardized format which gave highly reproducible results (Taylor *et al.*, 2003). The Aβ(1-40) aggregation reaction proceeded in a three step process consistent with that reported by Taylor *et al.* (2003); a lag phase of 40 minutes, a phase in which intermediate sized species were formed and a third phase in which the formation of fibrils occurred. Based on kinetic anaylsis the intermediate species of Aβ reached its highest concentration at 60 minutes which coincided with the time that the lipidated ApoE demonstrated an isoform specific binding to the Aβ(1-40) aggregate. These binding interactions revealed the formation of ApoE4 complexes with 60 minute Aβ(1-40) aggregates at levels that were 2-3 fold greater than either ApoE2 or ApoE3-Aβ(1-40) complex levels. Nonlipidated apoE isoforms did not show isoform-specific binding differences to any of the Aβ(1-40) aggregated species. The binding interactions could only be detected by antibodies specific for the N-terminal of ApoE and not the C-terminal, implying that the ApoE-Aβ(1-40) interaction domain was in the C-terminal region of ApoE. The *in vitro* approach to understanding possible *in vivo* ApoE-Aβ interactions has not produced a unified picture of how the ApoE-Aβ compex is formed but it has pointed to the possible variables that must be accounted for in an attempt to deduce information from *in vivo* models of ApoE and Aβ interaction. Some of those variables are: the state of the Aβ peptide, the isoform of ApoE, where the ApoE is expressed and how the ApoE and Aβ are presented to each other. In addition there may be other molecules *in vivo* which are not present *in vitro* which can influence the binding of ApoE lipoprotein particles to various forms of Aβ.

3. *IN VIVO* MODELS OF ApoE AND Aβ INTERACTION

3.1. *In vivo* Models of ApoE and Aβ Interaction (human brain)

In humans the association between ApoE, AD and Aβ was first suggested by Namba *et al.* (1991 and 1992) and Wisniewski and Frangione (1992) when they observed ApoE- immunoreactivity localized with extracellular amyloid deposits in the brains of subjects with AD, Down's syndrome, and Cruetzfeld-Jakob disease. Furthermore Rebeck *et al.* (1993), Schmechel *et al.* (1993) and Hyman *et al.* (1995) observed that the density of Aβ deposits in ApoE4 individuals was increased significantly relative to the ApoE3 homozygotes and that plaque density tended to be related to the dose of the ApoE4 allele. When fibrillar Aβ was isolated from individuals with AD, ApoE was found to be stably bound with the Aβ (Naslund *et al.*, 1995; Wisniewski *et al.*, 1995). ApoE has also been found associated with soluble forms of Aβ derived from normal human brain (Russo *et al.*, 1998). A large portion of water soluble brain Aβ derived from normal brain tissue was shown to be complexed with ApoE in a SDS-resistant species which migrated at a molecular weight of ~40 kD in SDS PAGE (Russo *et al.* 1998). Presumably this complex could be composed of a monomer or dimer of Aβ(1-40 or 42) and one molecule of ApoE. However Russo *et al.* (1998) found that the level of the SDS-resistant ApoE-Aβ complex was markedly decreased in soluble extracts from AD brains with the greater decrease apparent in the ApoE4 homozygotes as compared to the ApoE3 homozygotes. These results are complemented by those of Naslund *et al.* (1995) which showed that the the majority of SDS-stable ApoE-Aβ complexes from AD brains was found in 70% formic acid extracted fractions which presumably solubilizes most of brain Aβ (Gravino *et al.*, 1995). In addition Naslund *et al.* (1995) found that the complexed structure of ApoE was the C-terminal fragment originating in the region after amino acid residue 158 which would not include any of the common ApoE polymorphic sites. In a recent study in which ApoE domains were quantitated in Aβ deposits in AD and cerebral amyloid angiopathy (CAA) brains using ApoE

N- and C-terminal specific antibodies, it was determined that plaque cores contained the entire ApoE molecule whereas C-terminal fragments of ApoE were found only in the outer regions of the Aβ plaque deposits (Cho *et al.*, 2001). The numbers of plaques labelled by N- and C-terminal ApoE antibodies were equal but accounted for only 60% of the total Aβ plaque number and most of the Aβ positive and ApoE negative plaques were of the smallest variety (< 150 mm^2). These data were interpreted to suggest that C-terminal residues of ApoE bind to Aβ and aid in the progression of small Aβ deposits to larger deposits (Cho *et al.*, 2001). The observations by Cho *et al.* (2001) are consistent with the observations of Wisniewski *et al.* (1995) who biochemically isolated amyloid beta from senile plaques and found that a carboxyl-terminal fragment (residues 216-299) of apolipoprotein E co-purified. Thus it appears that in human brains with Alzheimer's disease, ApoE is most likely to be associated with amyloid deposits in which the C-terminal portion of ApoE has been consistently identified as the moiety associated with Aβ in various laboratories. Since the C-terminal portion of ApoE (amino acids 193-299) does not contain the common polymorphic regions of ApoE, there is no straightforward interpretation to explain the rank order of AD risk conferred by ApoE2, 3 or 4. It is known that the N- and C-terminals of ApoE can interact intramolecularly to cause a preference of ApoE3 for HDL and ApoE4 for VLDL in plasma (Dong *et al.*, 1994; Dong and Weisgraber, 1996). One could conjecture that the ApoE first binds either the soluble Aβ before deposition or the insoluble Aβ after deposition (at which point the ApoE-isoform preferences for Aβ are realized), then is cleaved by unknown proteases which release the N-terminal fragment of ApoE. It is known that ApoE does bind the diffuse plaques in young Down's syndrome brains (Arai *et al.*, 1995; Lemere *et al.*, 1996). Whether ApoE binds to nascent diffuse plaques and converts the bound Aβ to β-sheet structures capable of binding thioflavine-S or Congo Red is not known but that is one possible interpretation of the data presented above.

In human brain tissue from AD patients, increased Aβ senile plaque frequency in E4 genotypes was largely attributed to a significant increase in Aβ(1-40) immunoreactive plaques, both in number and ratio of Aβ(1-40) to Aβ(1-42) plaques in contrast to lack of differences detected in Aβ(1-42) immunoreactive plaques with either E3 or E4 genotypes (Gearing *et al.*, 1996). This study was done with 68 AD brain samples 10 of which were E4 homozygotes and 32 E3/E4. Another smaller study with 28 AD brains showed that in humans, the E4 genotype is associated with increases in both Aβ(1-40) and Aβ(1-42) using immunohistochemistry (McNamara *et al.*, 1998). Mann *et al.* (1997) also observed that preferential deposition of Aβ(1-40) was associated with a gene dosage effct of ApoE4 in a sample of 54 patients with AD. Using formic acid extraction from 36 AD brains, Aβ(1-

40) levels were found by ELISA to be significantly higher in E3/E4 and E4/E4 cases than in E3/E3 cases (Ishii *et al.*, 1997). There is strong agreement between the Gearing *et al.* (1995), Mann *et al.* (1997) and Ishii *et al.* (1997) studies showing that the presence of ApoE4 significantly increased the levels of Aβ(1-40) relative to the levels of Aβ(1-42) in AD brain tissue. One interpretation of these results posits that the ApoE4 allele caused the selective deposition of Aβ(1-40) relative to Aβ(1-42) or that clearance of Aβ(1-40) was selectively retarded relative to Aβ(1-42). Morishima-Kawashima *et al.* (2000) observed that ApoE4 tended to foster the deposition of Aβ(1-42) in younger brains (< 50 years of age) and at ages of >60 years Aβ(1-40) showed increased deposition relative to Aβ(1-42). These data from human brain tissue have not been explained adequately by the data from *in vitro* binding experiments with various forms of either Aϑ or ApoE. However there is a correlation of this body of data with the observations from *in vivo* mouse models of ApoE-Aβ interactions to be discussed in the following section.

3.2. *In vivo* models of ApoE and Aβ interaction (transgenic mouse brain)

Studies using genetically modified mice have demonstrated that ApoE-Aβ interactions are important in the simulation of aspects of human AD and CAA pathology. The genetically modified mice overexpressed amyloid precursor protein (APP) with the AD familial mutations APP(V717F) or APPsw (Games *et al.*, 1995; Hsiao *et al.*, 1996). In both these models APP is overexpressed primarily in neurons and develop age-dependent Aβ deposition which was either diffuse (thioflavine-S and Congo red negative) or dense neuritic plaques which stain positively with thioflavine-S and Congo red. The APPsw mice also develop CAA-like Aβ deposition at 12 months of age. Aβ deposition is present in the cortex and hippocampus and is also prominent within leptomeningeal and cortical blood vessels of all APPsw*ApoE (+/+) mice. Deposition of Aβ in cerebral blood vessels is not detectable in the APP(V717F) mouse model of AD; whether this is because of the use of the PDGF promoter in the APP(V717F) mice versus the prion promoter in the APPsw model is unknown. Bales *et al.* (1997) published a landmark paper showing the impact of ApoE on Aβ deposition in the APP(V717F) mouse. Using mice homozygous for the APP(V717F) transgene bred into an ApoE knockout (-/-) background Bales *et al.* (1997) showed that Aβ deposition still occurred but that when compared to 6 month APP(V717F) mice with endogenous ApoE (+/+), there were two important differences in the pattern of Aβ deposition. First, the level of Aβ deposition in the mice on the ApoE (-/-) background was markedly lower than the amount of Aβ deposition in the mice on the wild type ApoE background.

The second was the lack of formation of thioflavine-S or Congo red positive plaques in the brains of the mice on the ApoE (-/-) background at 6 months of age. In addition there was a dose effect for ApoE for both attributes discussed above. In ApoE (+/-) APP(V717F) mice the neuropatholgy was intermediate to the ApoE wild type and the ApoE (-/-) mice. Furthermore an extension of this study to mice 22 months of age revealed that APP(V717F) ApoE (+/+) mice had large brain deposits of thioflavine-S and Congo Red positive Aβ plaques in cortex and hippocampus (Bales *et al.,* 1999). In contrast, 22 month APP(V717F) ApoE (-/-) mice had developed thioflavine-S and Congo red negative Aϑ deposition only in the hippocampus. In addition Bales *et al.,* (1999) observed that > 99% of the thioflavine-S fluorescent Aβ deposits were also ApoE immunoreactive. Moreover the absence or presence of ApoE at time points before deposition did not impact the levels of 5.5 M guanidine extractable Aβ in brain tissue homogenates. Similar results were seen in the APPsw*ApoE(+/+), ApoE(+/-) ApoE(-/-) mice (Holtzman *et al.,* 2000b). This observation goes to the question raised in the previous section of this review; namely whether ApoE interacts with the extracellular Aβ before or after deposition. One can envision several scenarios whereby ApoE could influence the quantity and quality of deposited Aβ from this study. In one scenario ApoE could bind a soluble monomer or oligomer of Aβ and influence the secondary structure of Aβ to assume a β-pleated sheet which then deposits on cellular surfaces. Another scenario would be the deposition of the oligomeric Aβ first, followed by ApoE binding and conversion of the deposited Aβ to the beta sheet structures capable of binding thioflavine-S or Congo red. It is also possible that ApoE could influence aspects of Aβ deposition by both mechanisms.

ApoE isoform-dependent Aβ deposition was observed in the APP(V717F) mouse model by replacement of the endogenous mouse ApoE with human ApoE 3 or 4 under the control of the glial fibrillary acid protein promoter (Holtzman *et al.,* 2000a). The results of this study confirmed those of Bales *et al.*(1997) and showed that expression of either human ApoE3 or ApoE4 restored the deposition of thioflavine-S positive Aβ to levels seen with the endogenous ApoE(+/+) mice and substantially more fibrillar deposits were seen with ApoE4 than ApoE3 in mice at 15 months of age. Fagan *et al.* (2002) extended the study of the APP(V717)*hApoE mouse model to include hApoE2 replacement of the endogenous mouse ApoE gene. Analysis of APP(V717) (+/-) mice at 18-24 months of age showed that Aϑ deposition followed the rank order ApoE4(+/-)> ApoE3(+/-) ~ApoE2(+/-) ~ ApoE(-/-). Fagan *et al.* (2002) also observed that in APP(V717F)*ApoE(-/) mice greater than 20 months of age, Aβ deposits could be detected which were thioflavine-S positive. Thus Fagan and colleagues (2002) concluded that ApoE facilitates but is not required for formation of Aβ thioflavine-S positive deposits. In a study of Aβ deposition in the APPsw mouse in which human ApoE isoforms are expressed on the ApoE wild type background under the control of the transferrin promoter Carter *et al.* (2001) found that Aβ deposition was accelerated in APPsw(+/-)*ApoE4(+/-) mice relative to APPsw(+/-)*ApoE3(+/-) mice at 8-10 months of age. In this case the excess Aβ deposition could be accounted for entirely by an increase in the amount of Aβ(1-40) versus Aβ(1-42). This phenomenom appears to recapitulate the observations made with regard to ApoE isoform-dependence of Aβ deposition in humans (Gearing *et al.*, 1996; Mann *et al.*, 1997 and Ishii *et al.*, 1997). In the APP(V717F)*ApoE models there is no isoform-specific effect on the length of the Aβ peptide deposited since no alteration of the Aβ(1-42)/ Aβ total ratio was observed at either young, middle or older ages (Holtzman *et al.*, 2000a; Fagan *et al.*, 2002). The differences between the APP(V717F) and APPsw*ApoE models could be the due to the different promoters used for transgene expression, the genetic backgrounds of the mice or most likely to the APP mutation used for transgene expression. It is known that the APP(V717F) mutation confers greater ability for expression of the Aβ(1-42) versus total Aβ *in vitro* whereas the Swedish mutation APPsw constructs increases both Aβ(1-40) and Aβ(1-42) synthesis proportionately in neurons (Suzuki *et al.*, 1994; Citron *et al.*, 1992). The discovery by Stratman *et al.* (2003) that lipidated ApoE4 has greater affinity for oligomeric species of Aβ than either ApoE3 or ApoE2 *in vitro* is consistent with the hypothesis that ApoE4 can interact with Aβ(1-40) preferentially.

4. CONCLUSIONS

Increasing evidence suggests that interactions between CNS derived ApoE-associated lipoproteins and various forms of Aβ are critical in the etiology of Alzheimer's disease. *In vitro* and *in vivo* data suggest that ApoE interactions with Aβ can affect mechanisms of both clearance and deposition of Aβ. Experimental evidence also suggests that ApoE interactions with Aβ can also influence the secondary structure of Aβ to cause the maturation of diffuse plaques into neuritic plaques. Mouse models of the human Aβ-ApoE interaction appear to faithfully reproduce key aspects of Alzheimer neuropathology in human brain tissue. With a more complete understanding of how parenchymal Aβ in the brain interacts with ApoE, new therapeutic approaches may become evident for treating Alzheimer's disease and perhaps CAA.

REFERENCES

Aleshkov, S., Abraham, C.R., and Zannis, V.I., 1997. Interaction of nascent ApoE2, ApoE3, and ApoE4 isoforms expressed in mammalian cells with amyloid peptide beta (1-40). Relevance to Alzheimer's disease. *Biochemistry* **36:** 10571-10580.

Aleshkov, S.B., Li, X., Lavrentiadou, S.N., and Zannis, V.I., 1999, Contribution of cysteine 158, the glycosylation site threonine 194, the amino- and carboxy-terminal domains of apolipoprotein E in the binding to amyloid peptide beta (1-40). *Biochemistry* **38:** 8918-8925.

Arai, Y., Mizuguchi, M., Ikeda, K., and Takashima, 1995, Developmental changes of apolipoprotein E immunoreactivity in Down syndrome brains. *Brain Res. Dev.* **87:** 228-232.

Bales, K.R., Verina, T., Dodel, R.C., Du, Y., Altstiel, L., Bender, M., Hyslop, P., Johnstone E.M., Little, S.P., Cummins, D.J., Piccardo, P., Ghetti, B., and Paul, S.M., 1997, Lack of apolipoprotein E dramatically reduces amyloid beta-peptide deposition. *Nature Genet.* **17:** 263-264.

Bales, K.R., Verina, T., Cummins, D.J., Du, Y., Dodel, R.C., Saura, J., Fishman, C.E., DeLong, C.A., Piccardo, P., Petegnief, V., Ghetti, B., and Paul, S.M., 1999, Apolipoprotein E is essential for amyloid deposition in the APP(V717F) transgenic mouse model of Alzheimer's disease. *Proc. Natl. Acad. Sci. U.S.A.* **96:** 15233-15238.

Bentley, N.M., Ladu, M.J., Rajan, C., Getz, G.S., and Reardon, C.A., 2002, Apolipoprotein E structural requirements for the formation of SDS-stable complexes with beta-amyloid-(1-40): the role of salt bridges. *Biochem. J.* **366:** 273-279.

Carter, D.B., Dunn, E., McKinley, D.D., Stratman, N.C., Boyle, T.P., Kuiper, S.L., Oostveen, J.A., Weaver, R.J., Boller, J.A., and Gurney, M.E., 2001, Human apolipoprotein E4 accelerates beta-amyloid deposition in APPsw transgenic mouse brain. *Ann. Neurol.* **50:** 468-475.

Castano, E.M., Prelli, F., Wisniewski ,T., Golabek, A., Kumar, R.A., Soto, C., and Frangione, B., 1992, Fibrillogenesis in Alzheimer's disease of amyloid beta peptides and apolipoprotein E. *Biochem. J.* **306:** 599-604.

Chan, W,, Fornwald, J., Brawner, M., and Wetzel, R., 1996, Native complex formation between apolipoprotein E isoforms and the Alzheimer's disease peptide A beta. *Biochemistry* **35:** 7123-7130.

Cho, H.S., Hyman, B.T., Greenberg, S.M., and Rebeck, G,W., 2001, Quantitation of apoE domains in Alzheimer disease brain suggests a role for apoE in Abeta aggregation. *J. Neuropathol. Exp. Neurol.* **60:** 342-349.

Citron, M., Oltersdorf, T., Haass, C., McConlogue, L., Hung, A.Y., Seubert, P., Vigo-Pelfrey, C., Lieberburg, I., and Selkoe, D.J., 1992, Mutation of the beta-amyloid precursor protein in familial Alzheimer's disease increases beta-protein production. *Nature.* **360:** 672-674.

Corbo, R.M., and Scacchi, R., 1999, Apolipoprotein E (APOE) allele distribution in the world. Is APOE*4 a 'thrifty' allele? *Ann.Hum .Genet.* **63:** 301-310.

Corder, E.H., Saunders, A.M., Strittmatter, W.J., Schmechel, D.E., Gaskell, P.C., Small, G.W., Roses, A.D., Haines, J.L., and Pericak-Vance, M.A., 1993, Gene dose of apolipoprotein E type 4 allele and the risk of Alzheimer's disease in late onset families. *Science* **261:** 921-923.

Corder, E.H., Saunders, A.M., Risch, N..J., Strittmatter, W.J., Schmechel, D.E., Gaskell, P.C. Jr, Rimmler, J.B., Locke, P.A., Conneally, P.M., Schmader, K.E., and Roses, A.D., 1994, Protective effect of apolipoprotein E type 2 allele for late onset Alzheimer disease. *Nature Genet.* **7:** 180-184.

Dong, L.M., Wilson, C., Wardell, M.R., Simmons, T., Mahley, R.W., Weisgraber, K.H., and Agard, D.A., 1994, Human apolipoprotein E. Role of arginine 61 in mediating the lipoprotein preferences of the E3 and E4 isoforms. *J. Biol. Chem.* **269:** 22358-22365.

Dong, L.M., and Weisgraber, K.H., 1996, Human apolipoprotein E4 domain interaction. Arginine 61 and glutamic acid 255 interact to direct the preference for very low density lipoproteins. *J. Biol. Chem.* **271:** 19053-19057.

Fagan, A.M., Holtzman, D.M., Munson, G., Mathur, T., Schneider, D., Chang, L.K., Getz, G.S., Reardon, C.A., Lukens, J., Shah, J.A., and LaDu, M,J., 1999, Unique lipoproteins secreted by primary astrocytes from wild type, apoE (-/-), and human apoE transgenic mice. *J. Biol.Chem.* 1999 **274:** 30001-30007.

Fagan, A.M., Watson, M., Parsadanian, M., Bales, K.R., Paul, S.M., and Holtzman, D.M., 2002, Human and murine ApoE markedly alters A beta metabolism before and after plaque formation in a mouse model of Alzheimer's disease. *Neurobiol. Dis.* **9:** 305-318.

Games, D., Adams, D., Alessandrini, R., Barbour, R., Berthelette, P., Blackwell, C., Carr, T., Clemens, J., Donaldson, T., and Gillespie F, 1995, Alzheimer-type neuropathology in transgenic mice overexpressing V717F beta-amyloid precursor protein. *Nature* **373:** 523-527.

Gearing, M., Mori, H., and Mirra, S.S., 1996, Aβ- peptide length and apolipoprotein E genotype in Alzheimer's disease. *Ann. Neurol.* **39:** 395-399.

Golabek, A.A., Soto, C., Vogel, T., and Wisniewski, T., 1996, The interaction between apolipoprotein E and Alzheimer's amyloid beta-peptide is dependent on beta-peptide conformation. *J. Biol. Chem.* **271:** 10602-10606.

Golabek, A.A., Kida, E., Walus, M., Perez, C., Wisniewski, T., and Soto, C., 2000, Sodium dodecyl sulfate-resistant complexes of Alzheimer's amyloid beta-peptide with the N-terminal, receptor binding domain of apolipoprotein E. *Biophys. J.* **79:** 1008-1015.

Gravina, S.A., Ho, L., Eckman, C.B., Long, K.E., Otvos, L.,Jr, Younkin, L.H., Suzuki, N., and Younkin, S.G., 1995, Amyloid beta protein (A beta) in Alzheimer's disease brain. Biochemical and immunocytochemical analysis with antibodies specific for forms ending at A beta 40 or A beta 42(43). *J. Biol. Chem.* **270:** 7013-7016.

Holtzman, D.M., Bales, K.R., Wu, S., Bhat, P., Parsadanian, M., Fagan, A.M., Chang, L.K., Sun, Y., Paul, S.M., 1999, Expression of human apolipoprotein E reduces amyloid-beta deposition in a mouse model of Alzheimer's disease. *J. Clin. Invest.* **103:** R15-R21.

Holtzman, D..M., Bales, K.R., Tenkova, T., Fagan, A.M., Parsadanian, M., Sartorius, L.J., Mackey, B., Olney, J., McKeel, D., Wozniak, D., and Paul, S.M., 2000a, Apolipoprotein E isoform-dependent amyloid deposition and neuritic degeneration in a mouse model of Alzheimer's disease. *Proc. Natl. Acad. Sci. U.S.A.* **97:** 2892-2897.

Holtzman, D.M., Fagan, A.M., Mackey, B., Tenkova, T., Sartorius, L., Paul, S.M., Bales, K., Ashe, K.H., Irizarry, M.C., and Hyman, B,T., 2000b, Apolipoprotein E facilitates neuritic and cerebrovascular plaque formation in an Alzheimer's disease model. *Ann. Neurol.* **47:** 739-747.

Hsiao, K., Chapman, P., Nilsen, S., Eckman, C., Harigaya, Y., Younkin, S., Yang, F., and Cole, G., 1996, Correlative memory deficits, Abeta elevation, and amyloid plaques in transgenic mice. *Science* **274:** 99-102.

Ishii, K., Tamaoka, A., Mizusawa, H., Shoji, S., Ohtake, T., Fraser, P.E., Takahashi, H., Tsuji, S., Gearing, M., Mizutani, T., Yamada, S., Kato, M., St George-Hyslop, P.H., Mirra, S.S., and Mori, H., 1997, Abeta1-40 but not Abeta1-42 levels in cortex correlate with apolipoprotein E epsilon4 allele dosage in sporadic Alzheimer's disease. *Brain Res.* **748:** 250-252.

LaDu, M.J., Falduto, M.T., Manelli, A.M., Reardon, C.A., Getz, G.S., and Frail, D.E., 1994, Isoform-specific binding of apolipoprotein E to beta-amyloid. *J. Biol. Chem.* **269:** 23403-23406.

LaDu, M.J., Pederson, T.M., Frail, D.E., Reardon, C.A., Getz, G.S., and Falduto, M.T., 1995, Purification of apolipoprotein E attenuates isoform-specific binding to beta-amyloid. *J. Biol. Chem.* **270:** 9039-9042.

LaDu, M.J., Gilligan, S.M., Lukens, J.R., Cabana, V.G., Reardon, C.A., Van Eldik, L.J., and Holtzman, D.M., 1998, Nascent astrocyte particles differ from lipoproteins in CSF. *J. Neurochem.* **70:** 2070-2081.

Lemere, C.A., Blusztajn, J.K., Yamaguchi, H., Wisniewski, T., Saido, T.C., and Selkoe, D.J., 1996, Sequence of deposition of heterogeneous amyloid beta-peptides and APO E in Down syndrome: implications for initial events in amyloid plaque formation. *Neurobiol. Dis.* **3:** 16-32.

LeVine, H., 2000, 125I-labeled ApoE binds competitively to beta(1-40) fibrils with pathological chaperone proteins. *Amyloid* **7:** 83-89.

Linton, M.F., Gish, R., Hubl, S.T., Butler, E., Esquivel, C., Bry, W.I., Boyles, J.K., Wardell, M.R., and Young, S.G., 1991, Phenotypes of apolipoprotein B and apolipoprotein E after liver transplantation. *J. Clin. Invest.* **88:** 270-281.

Ma, J., Yee, A., Brewer, H.B. Jr, Das, S., and Potter, H., 1994, Amyloid-associated proteins alpha 1-antichymotrypsin and apolipoprotein E promote assembly of Alzheimer beta-protein into filaments. *Nature* **372:** 92-94.

Mann, D.M., Iwatsubo, T., Pickering-Brown, S.M., Owen, F., Saido, T.C., and Perry, R.H., 1997, Preferential deposition of amyloid beta protein (Abeta) in the form Abeta40 in Alzheimer's disease is associated with a gene dosage effect of the apolipoprotein E E4 allele. *Neurosci. Lett.* **221:** 81-84.

McNamara, M.J., Gomez-Isla, T., and Hyman, B.T., 1998, Apolipoprotein E genotype and deposits of Abeta40 and Abeta42 in Alzheimer disease. *Arch. Neurol.* **55:** 1001-1004.

Meyer, M.R., Tschanz, J.T., Norton, M.C., Welsh-Bohmer, K.A., Steffens, D.C., Wyse, B.W., and Breitner, J.C., 1998, APOE genotype predicts when--not whether--one is predisposed to develop Alzheimer disease. *Nature Genet.* **19:** 321-322.

Morishima-Kawashima, M., Oshima, N., Ogata, H., Yamaguchi, H., Yoshimura, M., Sugihara, S., and Ihara, Y., 2000, Effect of apolipoprotein E allele epsilon4 on the initial phase of amyloid beta-protein accumulation in the human brain. *Am. J. Pathol.* **157:** 2093-2099.

Namba, Y., Tomonaga, M., Kawasaki, H., Otomo, E., and Ikeda, K., 1991, Apolipoprotein E immunoreactivity in cerebral amyloid deposits and neurofibrillary tangles in Alzheimer's disease and kuru plaque amyloid in Creutzfeldt-Jakob disease. *Brain Res.* **541:** 163-166.

Namba, Y., Tsuchiya, H., and Ikeda, K. , 1992, Apolipoprotein B immunoreactivity in senile plaque and vascular amyloids and neurofibrillary tangles in the brains of patients with Alzheimer's disease. *Neurosci. Lett.* **134:** 264-266.

Naslund, J., Thyberg, J., Tjernberg, L.O., Wernstedt, C., Karlstrom, A.R., Bogdanovic, N., Gandy, S.E., Lannfelt, L., Terenius, L., and Nordstedt, C., 1995, Characterization of stable complexes involving apolipoprotein E and the amyloid beta peptide in Alzheimer's disease brain. *Neuron* **15:** 219-228.

Pillot, T., Goethals, M., Najib, J., Labeur, C., Lins, L., Chambaz, J., Brasseur, R., Vandekerckhove, J., and Rosseneu, M., 1999, Beta-amyloid peptide interacts specifically with the carboxy-terminal domain of human apolipoprotein E: relevance to Alzheimer's disease. *J. Neurochem.* **72:** 230-237.

Pitas, R.E., Boyles, J.K., Lee, S.H., Foss, D., and Mahley, R.W., 1987a, Astrocytes synthesize apolipoprotein E and metabolize apolipoprotein E-containing lipoproteins. *Biochim. Biophys. Acta* **917:** 148-161.

Pitas, R.E., Boyles, J.K., Lee, S.H, Hui, D., and Weisgraber, K,H., 1987b, Lipoproteins and their receptors in the central nervous system. Characterization of the lipoproteins in cerebrospinal fluid and identification of apolipoprotein B,E(LDL) receptors in the brain. *J. Biol. Chem.* **262:** 14352-14360.

Rebeck, G.W., Reiter, J.S., Strickland, D.K., and Hyman, B.T., 1993, Apolipoprotein E in sporadic Alzheimer's disease: allelic variation and receptor interactions. *Neuron* **11:** 575-580.

Sanan, D.A., Weisgraber, K.H., Russell, S.J., Mahley, R.W., Huang, D., Saunders, A., Schmechel, D., Wisniewski, T., Frangione, B., Roses, A.D., and Strittmatter, W.J., 1994), Apolipoprotein E associates with beta amyloid peptide of Alzheimer's disease to form novel monofibrils. Isoform apoE4 associates more efficiently than apoE3. *J. Clin. Invest.* **94:** 860-869.

Saunders, A.M., Strittmatter, W.J., Schmechel, D., St.George-Hyslop, P., et al., 1993), Association of apolipoprotein E allele epsilon 4 with late-onset familial and sporadic Alzheimer's disease. *Neurology* **43:** 1467-1472.

Schmechel, D.E., Saunders, A.M., Strittmatter, W.J., Crain, B.J., Pericak-Vance, S.H., et al., 1993, Increased amyloid beta-peptide deposition in cerebral cortex as a consequence of apolipoprotein E genotype in late-onset Alzheimer disease. *Proc Natl Acad Sci .U.S.A.* **90:** 9649-9653.

Selkoe, D.J., 2002, Alzheimer's disease is a synaptic failure. *Science* **298:** 789-791.

Shuvaev, V.V., and Siest, G., 1996, Interaction between human amphipathic apolipoproteins and amyloid beta-peptide: surface plasmon resonance studies. *FEBS Lett.* **383:** 9-12.

Soto, C., Golabek, A., Wisniewski, T., and Castano, E.M., 1996, Alzheimer's beta-amyloid peptide is conformationally modified by apolipoprotein E *in vitro*. *Neuroreport* **7:** 721-725.

Stratman, N.C., Castle, C., Taylor, B.M., Epps, D.E., Melchior, G.W., and Carter, D.B., 2003, Isoform specific interactions of human apolipoprotein E to and intermediate conformation of human alzheimer amyloid-beta peptide. *Neurobiol. Dis.* in review.

Strittmatter, W.J., Saunders, A.M., Schmechel, D., Pericak-Vance, M., Enghild, J., Salvesen, G.S., and Roses, A.D., 1993a, Apolipoprotein E: high-avidity binding to beta-amyloid and increased frequency of type 4 allele in late-onset familial Alzheimer disease. *Proc. Natl. Acad. Sci. U.S.A.* **90:** 1977-1981.

Strittmatter, W.J., Weisgraber, K.H., Huang , D.Y., Dong, L.M., Salvesen, G.S., Pericak-Vance, M., Schmechel, D., Saunders, A.M., Goldgaber, D., and Roses, A.D., 1993b, Binding of human apolipoprotein E to synthetic amyloid beta peptide: isoform-specific effects and implications for late-onset Alzheimer disease. *Proc. Natl. Acad. Sci. U.S.A.* **90:** 8098-8102.

Strittmatter, W.J., and Roses, A.D., 1996, Apolipoprotein E and Alzheimer's disease. *Ann. Rev. Neurosci.* **19:** 53-77.

Suzuki, N., Cheung, T.T., Cai, X.D., Odaka, A., Otvos, L. Jr,, Eckman, C., Golde, T.E., and Younkin, S.G., 1994, An increased percentage of long amyloid beta protein secreted by familial amyloid beta protein precursor (beta APP717) mutants. *Science* **264:** 1336-1340.

Taylor, B.M., Sarver, R.W., Fici, G., Poorman, R.A., Lutzke, B.S., Molinari, A., Kawabe, T., Kappenman, K., Buhl, A.E., and Epps, D.E., 2003, Spontaneous aggregation and cytotoxicity of the beta-amyloid Abeta1-40: a kinetic model. *J. Protein Chem.* **22:** 31-40.

Tokuda, T., Calero, M., Matsubara, E., Vidal, R., Kumar, A., Permanne, B., Zlokovic, B., Smith, J.D., Ladu, M.J., Rostagno, A., Frangione, B., and Ghiso, J., 2000, Lipidation of apolipoprotein E influences its isoform-specific interaction with Alzheimer's amyloid beta peptides. *Biochem. J.* **348:** 359-365.

Weisgraber, K.H., 1994, Apolipoprotein E: Structure-Function Relationships. In: *Advances in Protein Chemistry* **45**, pp. 249-295.

Wisniewski, T., and Frangione, B., 1992, Apolipoprotein E: a pathological chaperone protein in patients with cerebral and systemic amyloid. *Neurosci. Lett.* **135:** 235-238.

Wisniewski, T., Castano, E.M., Golabek, A., Vogel, T., and Frangione, B., 1994, Acceleration of Alzheimer's fibril formation by apolipoprotein E *in vitro. Am. J. Pathol.* **145:** 1030-1035.

Wisniewski, T., Lalowski, M., Golabek, A., Vogel, T., and Frangione, B., 1995, Is Alzheimer's disease an apolipoprotein E amyloidosis? *Lancet* **345:** 956-958.

Zhou, Z., Smith J.D., Greengard, P., and Gandy, S., 1996, Alzheimer amyloid-beta peptide forms denaturant-resistant complex with type epsilon 3 but not type epsilon 4 isoform of native apolipoprotein E. *Mol. Med.* **2:** 175-180.

Yang, D.S., Smith, J.D., Zhou, Z., Gandy, S.E., Martins, R.N., 1997, Characterization of the binding of amyloid-beta peptide to cell culture-derived native apolipoprotein E2, E3, and E4 isoforms and to isoforms from human plasma. *J. Neurochem.* **68:** 721-725.

Chapter 14

Clusterin and Alzheimer's Disease

Miguel Calero [1,2], Agueda Rostagno[2], Blas Frangione[2], Jorge Ghiso[2]
[1]*Centro Nacional de Microbiología, Instituto de Salud Carlos III, 28220 Majadahonda, Madrid, Spain and* [2]*Department of Pathology, New York University School of Medicine, New York, NY 10016, USA*

Abstract: Clusterin (apolipoprotein J) is a ubiquitous multifunctional glycoprotein with the capability to interact with a broad spectrum of molecules, among them the Alzheimer's Aβ peptide. Due to its co-localization with fibrillar deposits in systemic and cerebral amyloid disorders, clusterin is also considered an amyloid-associated protein. Although no genuine function has been attributed to this protein so far, it has been implicated in a wide variety of physiological and pathological processes, a role that may vary according to the protein maturation, sub-cellular localization, and the presence of certain tissue- or cell-specific factors. This review focuses on the importance of clusterin in health and disease conditions, with particular emphasis in its role in Aβ amyloidosis and other disorders of protein folding.

Key words: apolipoprotein J; amyloid β; cerebral amyloidosis; neuroprotection; neurotoxicity

1. INTRODUCTION

Clusterin[1] (also known as apolipoprotein J, apoJ) is a fascinating highly conserved glycoprotein with a nearly ubiquitous tissue distribution, that is remarkably regulated under certain conditions among them apoptosis, cancer, inflammation, as well as in nearly all neurodegenerative disorders

[1] Alternative names commonly used in the literature: sulfated glycoprotein 2, SGP2; complement-associated protein, SP-40,40; complement lysis inhibitor, CLI; testosterone-repressed prostate message 2, TRPM2.

described to date. Clusterin is a unique molecule both from a structural and a functional perspective and some of its most interesting aspects are often displayed as a dichotomy of opposite effects or a multiplicity of attributes.

The impressive number of ligands reported and the many, and often opposite, functions attributed to clusterin clearly indicate the need for a better understanding of the protein structure-function relationship, as well as the determination of its physiologically relevant functions. Fortunately, recent studies are starting to shed light on its role in certain physiological processes and the actual significance of many of its attributed ligand-binding interactions. The development of clusterin-deficient mouse models, the understanding of the disparity of roles played by secreted and nuclear isoforms, and the appreciation of the involvement of unstructured regions of the molecule in binding to hydrophobic ligands have played a key role for gaining new insights into the genuine function of this molecule.

2. CLUSTERIN STRUCTURE AND EXPRESSION

2.1 Secreted and intracellular forms of clusterin

Clusterin, a 70-80 kDa secreted protein (Kirszbaum *et al.*, 1989), was first described in 1983 as one of the major proteins present in ram rete testis fluid and was named after its ability to elicit clustering of different cell types (Blaschuk *et al.*, 1983; Fritz *et al.*, 1983). In humans, clusterin is coded by a single gene located on chromosome 8p21-p12 (Slawin *et al.*, 1990; Purrello *et al.*, 1991; Tobe *et al.*, 1991) that is organized into nine exons spanning a region of 16,580 bp (Wong *et al.*, 1994). Heat shock and apoptotic signals in rat and human cells are able to induce alternative splicing of the clusterin gene resulting in the generation of a variant mRNA lacking exon V and coding for a protein isoform missing the C-terminal two-thirds of the molecule (Kimura and Yamamoto, 1996; Kimura *et al.*, 1997). Recently, another alternative splicing has been described that omits exon II, which contains the first translation start site (AUG) and codes for the endoplasmic reticulum-targeting peptide. In this variant, a downstream AUG in exon III becomes the first available translation start site originating a shorter mRNA and resulting in a 49-kDa intracellular protein (Leskov *et al.*, 2003) normally present at low levels and which expression presumably increases under cellular stress conditions.

Clusterin gene promoter is highly conserved among vertebrates and contains several potential regulatory elements such as AP-1, AP-2, Spl and

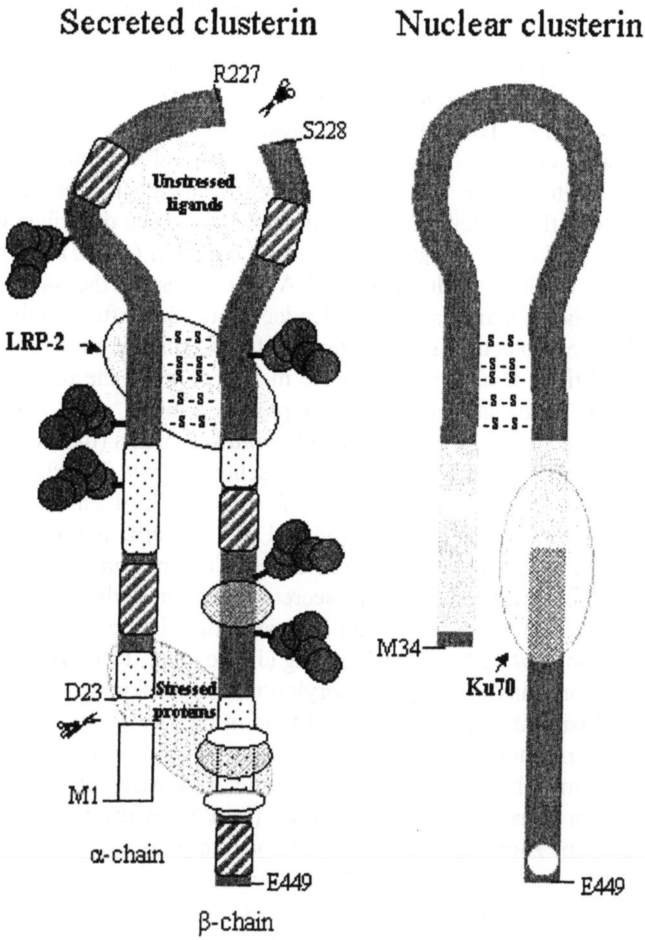

Figure 1: Schematic representation of secreted and nuclear clusterin. The secreted form contains a leader sequence (amino acids 1-22, open box), six N-glycosylation sites (positions 86, 103, 145, 291, 354 and 374), five putative amphipatic helices (hatched boxes), four predicted disordered regions (dotted boxes), two heparin-binding sites (open ovals), and two segments with cell adhesion activity (shadowed ovals). The regions involved in the binding of stressed proteins, LRP-2 and unstressed ligands are indicated. The nuclear form contains two coiled-coil segments (light grey areas) and two nuclear localization signals (open circles). The Ku70 binding site is indicated. Both, secreted and nuclear forms contain five disulphide bonds (-S-S-).

NF (Herault *et al.*, 1992; Wong *et al.*, 1994; Michel *et al.*, 1995; Michel *et al.*, 1997) that result in differential expression in response to a number of growth factors, cytokines, and stress and apoptosis inducing agents in a cell-type or tissue specific manner (Jin and Howe, 1997; Michel *et al.*, 1997; Gutacker *et al.*, 1999; Trougakos and Gonos, 2002).

In humans, clusterin is mainly a secreted glycoprotein, although intracellular -cytoplasmic and nuclear- isoforms originated by the above described alternative splicing mechanisms have also been described (Jin and Howe, 1997; Leskov *et al.*, 2003). Secreted clusterin (sCLU) results from the translation of the full clusterin mRNA that codes for the 449 residues-long protein and contains a 22-mer leader sequence. After synthesis, the structure of the molecule is stabilized by the formation of five disulfide bonds (Choi-Miura *et al.*, 1992) and the mature form of the protein finally generated by posttranslational cleavage at peptide bond R^{227}-S^{228} in the trans- or post-Golgi, rendering two ~40 kDa antiparallel subunits (named α- and β-chains) (Murphy *et al.*, 1988; Jenne and Tschopp, 1989; Kirszbaum *et al.*, 1989; de Silva *et al.*, 1990a; de Silva *et al.*, 1990b; Kirszbaum *et al.*, 1992) (Figure 1). Clusterin is heavily glycosylated and the carbohydrate molecules attached to the six existing sites (three on each chain) constitute about 30-40% of the molecular mass of the secreted protein (Urban *et al.*, 1987; Kapron *et al.*, 1997). Other reported post-translational modifications include sulfation (Urban *et al.*, 1987), lipid binding (Jenne *et al.*, 1991; Burkey *et al.*, 1992), iodination, mannose-6-phosphorylation (Lemansky *et al.*, 1999), and glycation (Mitsuhashi *et al.*, 1997). In addition, similarity searches to identify protein motifs and potential post-translational modifications performed against the Swiss-Prot and Prosite data bases at the ExPASY (Expert Protein Analysis System) server (http//www.expasy.org) of the Swiss Institute of Bioinformatics indicate putative phosphorylation sites for protein kinase C, casein kinase II and tyrosine kinase.

Cytoplasmic clusterin (cCLU) is normally expressed at very low levels as the translation product of a shorter mRNA produced by alternative splicing, as described above. The translated protein, lacking any signal peptide, is unable to be secreted and, probably as a result of retrograde transport from the Golgi to the ER, accumulates as a 49-kDa cytoplasmic, disulfide-linked, non-glycosylated, and uncleaved isoform. Under cellular stress, cCLU is assumed to be post-translationally modified, presumably phosphorylated, and subsequently translocated into the nucleus most likely via two functional nuclear localization signals. In the nucleus it is found as a 50-53 kDa nuclear clusterin (nCLU) (Leskov *et al.*, 2003; O'Sullivan *et al.*, 2003) that remains uncleaved at the peptide bond R^{227}-S^{228} (Park *et al.*, 1997) (Figure 1). Recently, this nuclear form has been shown to bind Ku70, a DNA damage sensor and key double-strand break repair protein that mediates cell death responses, through its C-terminal domain (Leskov *et al.*, 2003).

2.2 Tissue expression: an ubiquitous protein

Clusterin mRNA is present in almost all mammalian tissues (Collard and Griswold, 1987; de Silva *et al.*, 1990a) and protein expression in nearly all body fluids (de Silva *et al.*, 1990c; de Silva *et al.*, 1990b; Sylvester *et al.*, 1991; Aronow *et al.*, 1993). The normal concentration of clusterin in plasma, where it is primarily distributed within the high-density lipoproteins (de Silva *et al.*, 1990a; de Silva *et al.*, 1990c; de Silva *et al.*, 1990b; Jenne *et al.*, 1991; Stuart *et al.*, 1992), ranges between 35 and 105 μg/ml (0.44 - 1.35 μM) (Murphy *et al.*, 1988). About 30% of the protein seems to be sequestered by the alpha granules of platelets and can be released upon platelet activation (Witte *et al.*, 1993). Clusterin is about four-times more concentrated in seminal fluid than in plasma (O'Bryan *et al.*, 1990), while the CSF values vary between 1.2 and 3.6 μg/ml (15-45 nM) (Choi-Miura *et al.*, 1992; Harr *et al.*, 1996) with clusterin also co-localizing with high-density lipoproteins.

Within a given tissue, clusterin is expressed predominantly in particular cell types. Secreted clusterin is constitutively and developmentally expressed in epithelial cells, mainly at the fluid-tissue interface in the case of biologically active fluids such as gastric, pancreatic, and biliary secretions, urine and CSF (Aronow *et al.*, 1993; French *et al.*, 1993). In this sense it is present in the epithelial lining of the esophagus, biliary ducts, gallbladder, urinary bladder, ureter, kidney distal convoluted tubules, gastric glands, Brunner's glands, choroid plexus, ependyma, ocular ciliary body, endometrium, cervix, vagina, testis, epididymus, and visceral yolk sac (Aronow *et al.*, 1993). Clusterin is also expressed at high levels in several non-epithelial secretory cell types that line fluid compartments, such as synovial lining cells and ovarian granulosa cells (Aronow *et al.*, 1993).

The anatomical and cellular distribution of clusterin within the CNS has been studied in particular detail. In the adult rat brain the mRNA shows regional differences with the highest levels in the ependymal lining of the ventricles followed by various gray matter areas, mainly the hypothalamic and brainstem nuclei, the habenular complex, and the motorneurons of the ventral horn of the spinal cord (Danik *et al.*, 1993). In situ hybridization combined with immunocytochemistry studies (Pasinetti *et al.*, 1994) demonstrated clusterin mRNA in astrocytes throughout the normal adult brain as well as in specific neurons, but not in microglial cells. Among neuronal areas, clusterin was prevalent in pontine nuclei and in the red nucleus of the midbrain tegmentum whereas it was expressed in only a subset of substantia nigra dopaminergic neurons or locus ceruleus noradrenergic neurons (tyrosine hydroxylase immunopositive). Clusterin immunopositive cells were also observed in the Purkinje cell layer of the cerebellum, medial and interposed cerebellar nuclei, trigeminal motor and red nucleus, but not in the striatum (Pasinetti *et al.*, 1994). Interestingly, clusterin expression in the adult rat CNS changes dramatically following brain injury from a constitutive restricted neuronal distribution to a widespread inducible astrocytic expression (Danik *et al.*, 1993; Pasinetti *et al.*, 1994).

3. A CONSERVED STRUCTURE FOR A MYRIAD OF LIGANDS

Clusterin is a highly conserved protein amongst mammals, showing 70-80% sequence identity at the amino acid level. Secondary structure analysis by circular dichroism shows a majority of α-helical domains, predicted to be forming three amphipathic α -helices and two coiled-coil regions (de Silva *et al.*, 1990a; de Silva *et al.*, 1990b) (Figure 1). It has recently been proposed that clusterin likely includes three extended regions of natively disordered or molten globule-like structures, flexible and thereby highly sensitive to trypsin digestion, that contain the putative amphipathic α-helices. It has been suggested that these natively disordered regions may confer to clusterin the ability to bind a variety of molecules (Bailey *et al.*, 2001).

A short review of the literature reveals that clusterin binds to an impressive array of different ligands including apolipoproteins, lipids, immune system related molecules, cellular receptors, amyloid- forming proteins and peptides, enzymes, bacterial proteins and heparin. More recently, it has been demonstrated that clusterin is also able to bind to a number of stressed, partially folded proteins (Poon *et al.*, 2000; Lakins *et al.*, 2002; Trougakos and Gonos, 2002). Except for the amyloid-forming peptides that will be further analyzed below, all described ligands of clusterin are summarized in Table 1 and will not be discussed in detail.

According to Lankins and collaborators (2002), secreted clusterin has at least three distinct sites by which it binds three different types of ligands, namely: i) unstressed ligands such as the $A\beta_{1\text{-}40}$ peptide, Complement Components C7, C8, C9, and IgG, ii) stressed proteins, and iii) cell-surface receptors. Experimental data and sequence analysis predictions indicate that these sites are located at different parts of the molecule (Figure 1). The binding site for unstressed ligands may involve the amphipathic helices located at the C-terminal of the β-chain and the N-terminal of the α-chain. The binding to stressed proteins – seemingly of high affinity and low specificity- is apparently mediated by natively disordered or molten globule-

Table 1. Clusterin ligands

Ligands	References
Apolipoproteins & Lipids	
ApoAI	Tschopp J, 1993; Jenne DE, 1991; Ehnholm C, 1991
ApoCII	Hatters DM, 2002
Paraoxonase	Kelso GJ, 1994
Lipids	Jenne DE, 1991; Yokoyama M, 1988; Burkey BF, 1992
Bacterial proteins	
Streptococcus pyogenes (Prot SIC)	Akesson P, 1996
Streptococcus sanguis	Hogg SD and Embery G, 1979
Staphylococcus aureus	Partridge SR, 1996
Staphylococcus epidermidis	Li DQ and Ljungh A, 2001
Bacteriorhodopsin	Bailey RW, 2001
Immune system related molecules	
Complement components	Murphy BF, 1988; Kirszbaum L, 1989; Tschopp J, 1993; Tschopp J and French LE, 1994
Immunoglobulins	Wilson MR, 1991; Wilson MR and Easterbrook-Smith SB, 1992
Intracellular ligands	
Ku proteins (Ku70, Ku86)	Leskov KS, 2001; Yang CR, 1999; Yang CR, 2000; Trougakos IP and Gonos ES, 2002
Glycoproteins H, J, K	Palmer DJ and Christie DL, 1992
Cellular receptors	
TGFβ receptors	Golabek A, 1995
Megalin/LRP-2/gp330	Zlokovic BV, 1996; Kounas MZ, 1995
Amyloid forming proteins	
Aβ peptide	McGeer PL, 1992; Oda T, 1995; Matsubara E., 1996; Calero M, 2000, De Mattos RB, 2002
PrP	Sasaki K, 2002; McHattie S and Edington N, 1999; Calero M, 2000
Corneal amyloid GDCD and LCD-I	Nishida K, 1999
α-synuclein	Sasaki K, 2002
ABri	Ghiso J, 1995; Ghiso J, 2002
ADan	Ghiso J, 2002
Stressed ligands	
Catalase, β-lactalbumin, BSA	Humphreys DT, 1999
Denatured IgG, ovotransferrin, lysozyme	Hochgrebe T, 2000
Other molecules	
Glutatione-S-transferase	Humphreys D, 1997
α1-microglubulin	Calero M, unpublished observation
Heparin	de Silva HU, 1990; Pankhurst GS, 1998
Leptin	Bajari TM, 2003
MT-6 MMP	Matsuda A., 2003

GDCD: gelatinous drop-like corneal dystrophy

LCD-I: lattice corneal dystrophy type I

TGFβ: transforming growth factor beta

MT-6 MMP: membrane-type 6 matrix metalloproteinase

like regions at the N-terminal of the β-chain and the C-terminal of the α-chain containing the putative amphipathic α-helices. On the contrary, the highly conserved and well structured disulphide- linked central region of the molecule may be involved in the highly specific, and probably fundamental for its function, binding to the LRP-2 cell receptor (Bailey *et al.*, 2001;

Lakins *et al.*, 2002). More recently, studies with the intracellular form of clusterin have indicated a fourth type of binding site, probably located within the C-terminal coiled-coil region of the protein, involved in cell death responses through the interaction with Ku70 (Leskov *et al.*, 2003) (see Figure 1 and Table 1).

Interestingly, at mildly acidic pH clusterin shows increased binding for some of its ligands -heparin (Pankhurst *et al.*, 1998), IgG, C9, apoAI and GST (Hochgrebe *et al.*, 2000)- most likely due to modifications in its aggregation state than to conformational changes resulting from the change in pH. In this sense it should be noted that at physiological pH clusterin exists in an array of multimeric forms, whereas mildly acidic pH favors the formation of clusterin monomers more easily available for binding interactions. This enhancement has potential physiological relevance as clusterin is locally produced at sites of tissue damage and inflammation, where local acidosis also occurs (Hochgrebe *et al.*, 2000).

4. CLUSTERIN AS A MULTIFUNCTIONAL PROTEIN

4.1 The many professions of clusterin

Mostly as a result of the interaction with the many ligands described above, clusterin has been implicated in a number of diverse biological processes including cell-cell interactions (Fritz *et al.*, 1983; Silkensen *et al.*, 1995), sperm maturation (Collard and Griswold, 1987; Sylvester *et al.*, 1991), apoptosis (Buttyan *et al.*, 1989), complement inhibition (Murphy *et al.*, 1988), lipid transport (de Silva *et al.*, 1990c; Jenne *et al.*, 1991), tissue remodeling, membrane recycling (Danik *et al.*, 1991; Palmer and Christie, 1992), and clearance of cellular debris (Bartl *et al.*, 2001), as well as degradation of the extracellular matrix through the interaction with the membrane-type 6 matrix metalloproteinase (Matsuda *et al.*, 2003).

The structural nature of the different binding sites described above may explain, at least in part, many of clusterin's putative functions that result from its competence for binding such a diverse array of molecules. For instance, the presence of the amphipathic α-helices, a type of secondary structure that mediates interactions with hydrophobic molecules (de Silva *et al.*, 1990a; de Silva *et al.*, 1990b) may originate broad and non-specific interactions with some of the ligands shown in Table 1 that, in turn, may not be physiologically significant. In this sense, recent data suggest that clusterin may not be a relevant regulator of complement activation at physiological concentrations (Hochgrebe *et al.*, 1999) in spite of its ability to bind components of the membrane attack complex (Blaschuk *et al.*, 1983). In contrast, the interaction with LRP-2, the only known cellular receptor for clusterin, is highly specific and it most likely plays a crucial role in clusterin function by either triggering signal transduction pathways or by directing internalization and lysosomal degradation of different ligands through receptor-mediated endocytosis. Other relevant biological function of clusterin may relate to its capacity to participate in signal transduction pathways such as regulation of the transcription factor NF-κB signaling (Santilli *et al.*, 2003).

Although some results are controversial, a number of evidences support the idea that clusterin is directly involved in the response to cellular stress and mediates in apoptotic signaling. Clusterin mRNA level is increased in regressing tissues (Viard *et al.*, 1999) and endothelial cells undergoing apoptosis show transcriptome changes that include not only down regulation of survival signals, and alteration of cell cycle elements and apoptosis regulators, but a significant increment in clusterin expression (Johnson *et al.*, 2004). Controversially, a number of findings point out to a cytoprotective/anti-apoptotic role. Clusterin over-expression is able to protect cells in culture from the cytotoxic effect of TNF-α (Humphreys *et al.*, 1997), and blocking clusterin biosynthesis by anti-sense oligonucleotides or by silencing gene expression with small interfering RNAs induces apoptosis, reduces growth ability and enhances susceptibility for oxidative stress (Viard *et al.*, 1999; Trougakos *et al.*, 2004). Mouse models also provide contradictory information evidencing once more that the role of clusterin may vary according to particular circumstances, the final maturation and localization of the protein, as well as tissue or cell specific factors. Whereas studies performed on clusterin-deficient mice have demonstrated that, under certain conditions, the protein may have a protective effect against apoptosis (Bailey *et al.*, 2002), mice suffering hypoxic-ischemic brain injuries, a model of cerebral palsy, show clusterin accumulation in dying neurons contributing to caspase-3-independent brain injury (Han *et al.*, 2001)

In order to fully comprehend clusterin's diverse and many times contradictory functions, the importance of protein maturation and sub-cellular localization, as well as the potentially differential roles of the secreted and intracellular isoforms should be taken into consideration. Although secreted clusterin seems to act as an inherent pro-survival and anti-apoptotic protein under stress, intracellular clusterin becomes highly toxic and strongly accumulates in ubiquitinated form in juxtanuclear aggregates leading to profound alterations of the mitochondrial network (Debure *et al.*, 2003) presumably triggering cell death signals through Ku70/Ku80 binding (Yang *et al.*, 2000). Additionally, different stress conditions including ionizing radiation or apoptosis provoked by treatment with TNF-α or antiestrogens, induce nuclear translocation of intracellular clusterin.

4.2 Clusterin as an extracellular chaperone

Clusterin has been described as the first identified extracellular mammalian chaperone which binds to a wide variety of partly unfolded stressed proteins, inhibiting their stress-induced precipitation, via an ATP-independent mechanism (Poon *et al.*, 2000; Wilson and Easterbrook-Smith, 2000). Although the ultimate biological function of clusterin is still not well defined, its role as an extracellular chaperone is consistent with its involvement in tissue remodeling and protective responses against cellular stress. Chaperone activity may arise from the presence of the amphipathic helices in combination with natively disordered stretches in the protein structure that allows the formation of a putative, dynamic, molten globule-like binding site with ability to transiently bind to a variety of molecules (Bailey *et al.*, 2001), preferentially partly folded protein intermediates that are slowly aggregating as a result of stress. Through this interaction, clusterin contributes to solubilize high molecular weight complexes and inhibits rapid and irreversible protein precipitation producing a pool of inactive but stabilized molecules from which in turn, other intracellular refolding chaperones, such as heat shock protein 70 (HSP70) can subsequently rescue functional proteins (Poon *et al.*, 2000; Poon *et al.*, 2002a). Interestingly, clusterin is also the first chaperone molecule shown to be activated by mildly acidic pH and not by increased temperature. This unique mode of activation appears to result from an increased solvent-exposed hydrophobicity of the molecule with acidic pH, independent of any major changes in the secondary or tertiary structures (Hochgrebe *et al.*, 2000; Poon *et al.*, 2002b).

5. AMYKLOID β AMYLOIDOSIS: A DISORDER OF PROTEIN FOLDING

Protein misfolding and aggregation associate with a variety of human disorders. Among them, particular interest has been devoted to a large group of chronic and progressive neurodegenerative conditions characterized by the selective loss of neurons associated either with cognitive, motor or sensory systems as well as the intra- or extra-cellular deposition of protein aggregates and/or fibrils in different regions of the CNS (Table 2).

Table 2. Disorders of protein folding, characteristic neuropathology and deposited proteins

Disease	Neuropathology	Protein
Alzheimer's disease, Down's syndrome, normal aging, Sporadic cerebral amyloid angiopathy, Familial AD	Neuritic plaques, pre-amyloid lesions neurofibrillary tangles, cerebrovascular amyloid deposits.	Aβ
Familial British dementia	Cerebrovascular amyloid deposits, neurofibrillary tangles, neuritic plaques, pre-amyloid lesions.	ABri
Familial Danish dementia	Cerebrovascular amyloid deposits, pre-amyloid lesions, neurofibrillary tangles	ADan
Creutzfeldt-Jakob disease, Gerstmann-Straussler-Schinker disease, fatal familial insomnia, Kuru	Spongiform degeneration , gliosis, neuronal loss	PrP^{Sc}
Fronto-temporal dementia, Pick's disease	Loss and athophy of neurons, Pick's inclusion bodies	tau
Diffuse Lewy body disease	Diffuse Lewy bodies	α-synuclein
Parkinson's disease	Neuronal loss, Lewy bodies in pigmented neurons	α-synuclein, parkin
Huntington's disease	Neuronal loss	Huntingtin
Spinocerebellar ataxia Type 1	Neuronal loss	Ataxin 1
Spinocerebellar ataxia Type 2	Neuronal loss	Ataxin 2
Spinocerebellar ataxia Type 3	Neuronal loss	Ataxin 3
Spinocerebellar ataxia Type 6	Neuronal loss	α_{1A}V-Ca C
Spinocerebellar ataxia Type 7	Neuronal loss	Ataxin 7
Spinocerebellar ataxia Type 8	Neuronal loss	untranslated
Multiple-system atrophy	Neuronal and glial inclusions	α-synuclein
Friedereich's ataxia	Neuronal loss in sensory and motor systems	Frataxin
Dentatorubropallidolusyan atrophy	Neuronal loss	Atrophin I
Amyotropic lateral sclerosis	Upper and lower motor neurons loss	SOD1
Familial Spastic paraparesis	Upper motor neurons loss	Paraplegin
Spinal muscular atrophy	Lower motor neurons loss	SMN
Spinal and bulbar muscular atrophy	Lower motor neurons loss	Androgen receptor

A subset of these diseases, collectively known as cerebral amyloidosis, is characterized by the deposition of poorly soluble, long, twisted amyloid filaments composed of low molecular weight proteins that are normally soluble under physiologic conditions. In cases in which the amyloid lesions are primarily restricted to the cerebral vessel walls, the common clinical manifestation is stroke; contrastingly, a widespread distribution throughout selected parenchymal areas, particularly the limbic structures, is associated with dementia. Alzheimer's disease (AD) is the most common form of cerebral amyloidosis in humans and the major cause of dementia. In hereditary as well as in sporadic AD, extracellular amyloid deposits in the form of amyloid plaques and cerebral amyloid angiopathy co-exist with intraneuronal neurofibrillary tangles in the brain parenchyma. The major component of the amyloid lesions, Aβ, is a 40-42 residues internal proteolytic fragment of a larger type I transmembrane precursor molecule APP codified by a single multi-exonic gene located on chromosome 21. Although a soluble form of Aβ (sAβ, predominantly 40 residues in length) has been identified in biological fluids, systemic deposits of Aβ cannot be demonstrated in AD patients (Ghiso and Frangione, 2002). Whether the circulating soluble forms of Aβ represent immediate precursors of the deposited amyloid, reflect brain clearance, or both, are not solved issues. Certainly, the blood-brain barrier has the capability to modulate this equilibrium.

The brain uptake of free sAβ entails the participation of two different receptors: RAGE (the receptor for advanced glycation end-products) (Yan *et al.*, 1996) and SR-A (scavenger receptor type A) (Christie *et al.*, 1996; El Khoury *et al.*, 1996). RAGE is present in different cell types, including vascular endothelial and smooth-muscle cells, choroid plexus epithelium and phagocytes (Brett *et al.*, 1993) whereas SR is expressed on macrophages, microglia and vascular endothelial cells (Lucarelli *et al.*, 1997). RAGE mediates patho-physiological responses in the vasculature when occupied with glycated ligands or Aβ (Brett *et al.*, 1993; Yan *et al.*, 1996; Mackic *et al.*, 1998), whereas SR-A promotes endocytosis and degradation of oxidized LDL and glycated ligands (Christie *et al.*, 1996). Both RAGE and SR-A modulate brain endothelial endocytosis and transcytosis of Aβ from the luminal side of the blood-brain barrier (Mackic *et al.*, 1998; Deane *et al.*, 2003). *In vitro*, Aβ binding to brain microvascular endothelial cell monolayers is time-dependent, polarized to the apical side, saturable, and susceptible to inhibition by anti-RAGE antibodies (~63%) and by acetylated LDL (~33%) (Mackic *et al.*, 1998). Consistent with these data, both RAGE-transfected cells and macrophage SR-A displayed binding and internalization of radiolabeled Aβ (Mackic *et al.*, 1998) and systemic administration of soluble RAGE reduced accumulation of Aβ in brain parenchyma of transgenic APP mice (Deane *et al.*, 2003).

The brain clearance mechanism of the Aβ peptide appears to be in part regulated by the LDL receptor-related protein 1 (LRP-1), a promiscuous receptor highly expressed in the CNS. More than twenty ligands have been identified for LRP-1 (Hussain *et al.*, 1999), including apoE-containing lipoproteins, chylomicron remnants, activated α-2 macroglobulin and Receptor Associated Protein (RAP). When injected into the brain, radiolabeled Aβ is rapidly removed across the blood-brain barrier with a $t_{1/2}$ ~25 minutes being its clearance significantly inhibited by RAP and by antibodies against LRP-1 and α2 –macroglobulin. In *in vitro* experiments,

LRP-1 has also been implicated in Aβ clearance by smooth-muscle cells, neurons and fibroblasts (Narita *et al.*, 1997; Urmoneit *et al.*, 1997; Jordan *et al.*, 1998; Qiu *et al.*, 1999). Interestingly, LRP-1 is abundant in brain microvessels of young mice but is down-regulated in older animals (Shibata *et al.*, 2000), suggesting that a failure in the clearance mechanism leads to the accumulation of Aβ in the brain with the consequent amyloid formation.

6. CLUSTERIN AND ITS RELATION TO Aβ

6.1 Clusterin as an Aβ-carrier molecule

Soluble Aβ co-localizes with high-density lipoprotein (HDL) particles in serum (Koudinov *et al.*, 1994) and CSF (Koudinov *et al.*, 1996). Under normal circumstances, about ninety percent of the circulating sAβ is associated with plasma HDL fractions (Matsubara *et al.*, 1999), largely complexed to the carrier clusterin (Ghiso *et al.*, 1993).

The Aβ-clusterin complex is transported from the blood to the brain across the blood-brain barrier through the only known clusterin receptor, the low-density lipoprotein receptor related protein 2 (LRP-2; also known as megalin or gp330) (Zlokovic *et al.*, 1996), broadly expressed in vascular CNS tissues including the choroid plexus, the blood-brain barrier endothelium and the ependyma (reviewed in (Calero *et al.*, 2000). LRP-2 interacts with clusterin with high affinity, mediates clusterin endocytosis and its subsequent lysosomal degradation (Kounnas *et al.*, 1995; Hammad *et al.*, 1997) or transcytosis (Zlokovic *et al.*, 1996). In the case of sAβ40-clusterin complexes, the transport mechanism is identical to that of free clusterin, being specifically abolished by native clusterin as well as by anti-LRP-2 and RAP. Free sAβ peptides which, as described above, are internalized by a completely different mechanism do not exert any inhibitory effect.

In *in vitro* binding experiments, clusterin is specifically retrieved from plasma and CSF by immobilized Aβ species. Clusterin-Aβ interaction is saturable and specific, with Kd values in the low nanomolar range (2-4 nM), and is not affected by the degree of lipidation of the apolipoprotein (Matsubara *et al.*, 1995; Matsubara *et al.*, 1996; Calero *et al.*, 1999). As a result of the complex formation, *in vitro* peptide aggregation, polymerization and amyloid fibril assembly is significantly prevented (Oda *et al.*, 1995; Matsubara, *et al.*, 1996, Calero *et al.*, 2000), as it is the Aβ neurotoxicity (Oda *et al.*, 1995; Boggs *et al.*, 1996; Lambert *et al.*, 1998). These protective properties of clusterin, consistent with its extracellular chaperone activity described above, are not restricted to the Aβ peptide. Clusterin was shown to prevent *in vitro*, in a dose-dependent manner, the spontaneous aggregation of a synthetic peptide homologous to residues 106-126 of human prion protein (McHattie and Edington, 1999). Sub-stoichiometric levels of clusterin inhibited apoC-II amyloid formation, although they did not promote fibril dissociation (Hatters *et al.*, 2002). Clusterin binds to slowly aggregating proteins, such as some forms of stressed α-lactalbumin, α-crystallin and lysozyme preventing their aggregation (Poon *et al.*, 2002a). Moreover, as supported by data obtained from a clusterin knock-out mouse model, the expression of clusterin participate in the prevention of progressive glomerulopathy in aging mice, which is characterized by the deposition of immune complexes consisting of IgG, IgM and IgA together with complement components (Rosenberg *et al.*, 2002).

6.2 Clusterin as an amyloid associated protein

Paradoxically to its chaperone activity, clusterin has been found co-deposited with all amyloid molecules tested so far, not only in cerebral forms but also in systemic types of amyloidosis. There is extensive immunohistochemical data indicating that besides clusterin, a variety of unrelated proteins -serum amyloid P-component, α_1-antichymotrypsin, apolipoprotein E, complement components, vitronectin, cytokines, glycosaminoglycans and extracellular matrix proteins, among many others- co-deposit in the lesions composed of different amyloid molecules, including Aβ in senile plaques and vascular deposits. These components, collectively known as amyloid associated proteins, co-localize with the lesions but are not a structural part of the final fibril. It is still not clear whether these elements are innocent bystanders or if their presence is related to the mechanism of amyloidogenesis / fibrillogenesis. Several lines of investigation favor the latter notion, at least for some of these proteins, since various studies suggest that they can modulate the formation of amyloid-like fibrils *in vitro* (Ghiso and Frangione, 2002).

As illustrated in Figure 2, anti-clusterin antibodies effectively highlight parenchymal and vascular lesions in Aβ-related disorders [AD (panel A), Down syndrome (panel B), FAD cases associated to the Iowa and Dutch mutations (panels C and D)] as well as in non-Aβ cerebral amyloidosis [CJD panel E), HCHWA-I (panel F), familial British dementia (panel G) and familial Danish dementia (panel H)]. In all these cases, clusterin is present (in both, fibrillar (amyloid) and non-fibrillar (pre-amyloid) deposits.

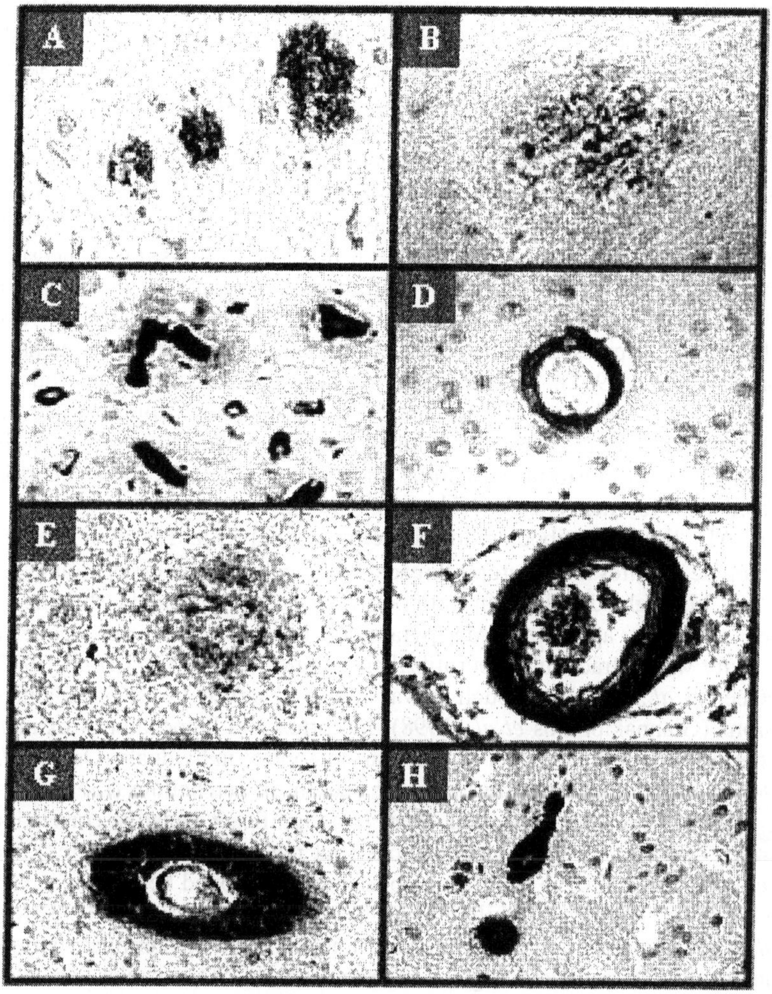

Figure 2: Clusterin co-localizes with parenchymal and cerebrovascular amyloid deposits. Clusterin in **A:** AD, Aβ plaques; **B:** Down syndrome, primitive Aβ plaques; **C:** familial AD (Iowa kindred), cerebrovascular AβN23 deposits; **D:** familial AD (Dutch kindred), cerebrovascular AβQ22 deposits; **E:** Creutzfeldt-Jakob disease, PrP plaques; **F:** hereditary cerebral hemorrhage with amyloidosis - Icelandic type, leptomeningeal Cystatin C Q68 deposits; **G:** familial British dementia, perivascular ABri plaques; **H:** familial Danish dementia, cerebrovascular ADan deposits. Magnification X880.

In addition, clusterin co-localizes with plaque-type deposits but not with punctuate-type prion protein (PrP) lesions in human and animal TSEs (Manuelidis *et al.*, 1997; McHattie *et al.*, 1999; Sasaki *et al.*, 2002b; Sasaki *et al.*, 2002c). It is associated with α-synuclein aggregates in cortical Lewy bodies and glial cytoplasmic inclusions in cases with multiple system atrophy, Parkinson's disease, and dementia with Lewy bodies (Sasaki *et al.*, 2002a). Furthermore, clusterin also co-localizes with amyloid in gelatinous drop-like and lattice type I corneal dystrophies (Nishida *et al.*, 1999). Abnormal staining for clusterin has been described in dystrophic neurites, some neurofibrillary tangles in the Parkinson's dementia of Guam and in ischemic Purkinje cells (Yasuhara *et al.*, 1994). Clusterin is also differentially expressed in the retinas of patients with retinitis pigmentosa (Jones *et al.*, 1992), and elevated levels were described in human cerebrospinal fluid in a number of acute neuropathies (Polihronis *et al.*, 1993; Schreiber *et al.*, 1993). Complete cerebral ischemia also leads to the accumulation of clusterin in neurons and in multiple extracellular deposits located close to microvessels (Kida *et al.*, 1995).

In AD tissue, antibodies to clusterin strongly stain senile plaques, dystrophic neurites and neuropil threads, while cellular processes around Aβ are not highlighted. Clusterin antibody also shows punctuate staining of some normally appearing AD pyramidal neurons, and very scattered reactivity with intracellular neurofibrillary tangles (Choi-Miura *et al.*, 1992; McGeer *et al.*, 1992), although it was rarely observed in NFT-containing neurons (Giannakipoulos *et al.*, 1998). In cortex and hippocampus extracts clusterin is about 40% higher in AD than in control individuals (Oda *et al.*, 1994), while normal clusterin levels are present in cerebellum (Lidstrom *et al.*, 1998). Within the AD group, there is a significant negative correlation between clusterin levels in hippocampus (but not in the frontal cortex) and severity of dementia. No significant correlations are found between clusterin levels and the number of senile plaques or neurofibrillary tangles (Lidstrom *et al.*, 1998). Interestingly, there are no changes in CSF-clusterin levels from patients with different neurological disorders such as AD, vascular dementia, Parkinson's disease or acute stroke, when compared to controls (Harr *et al.*, 1996; Lidstrom *et al.*, 2001).

6.3 The role of clusterin tested in a transgenic mouse model of AD

The protective properties of clusterin, preventing aggregation and polymerization of Aβ *in vitro*, drastically contrast with its co-localization with amyloid and pre-amyloid deposits. An interesting model to test the role of clusterin in amyloid formation emerged from the crossing of the PDAPP mice, homozygous for the APPV717F transgene (Games *et al.*, 1995; Bales *et al.*, 1997) with a clusterin knock-out mouse (McLaughlin *et al.*, 2000; Han *et al.*, 2001). If clusterin prevented fibril formation as indicated by *in vitro* experiments, an increase in the amyloid load in the off-springs would be expected. However, at 12 months of age, PDAPP-clusterin $^{-/-}$ animals had a similar degree of Aβ deposition compared with PDAPP-clusterin $^{+/+}$ counterparts although the deposits were largely thioflavin negative, indicating that fewer fibrillar Aβ (amyloid) deposits were formed (DeMattos *et al.*, 2002). Although electronmicroscopic evaluation of the lesions was not performed, these results suggest that, opposite to what was observed *in vitro*, clusterin may have a role in facilitating the conversion of soluble Aβ into forms with high β-sheet content *in vivo*. Interestingly, these results were similar to those previously obtained for PDAPP/apoE$^{-/-}$ mice (Bales *et al.*, 1997; Bales *et al.*, 1999).

Surprisingly, breeding PDAPP-apoE$^{-/-}$ with PDAPP-clusterin$^{-/-}$ mice resulted in an unexpected phenotype. Opposite to both single knock-out models, the double knock-out PDAPP-apoE$^{-/-}$-clusterin$^{-/-}$ mice revealed earlier onset and greater levels of Aβ deposition with a parallel increase in both Aβ40 and Aβ42 in interstitial fluid and CSF but not in plasma (DeMattos *et al.*, 2004). Contrary to previous conclusions achieved through the single knockout models, these results suggest that apoE and clusterin cooperatively suppress Aβ deposition. Whether the discrepancies observed between the single and the double knock-out animals are the consequence of significant interactions between both molecules that modulate their responses or a phenomenon self-related to the particular transgenic used in the experiments remains to be elucidated. Certainly, a comparison of the biochemical properties of the Aβ deposits in all these models would help with the interpretation.

7. CONCLUDING REMARKS

Current data indicate divergent and controversial functions for clusterin, ranging from cytoprotection to cytotoxicity. As a result, it is still difficult to assign a definitive role for clusterin in the molecular mechanism(s) of Aβ amyloidogenesis. The double knockout data suggest that clusterin, likely in cooperation with apoE, may normally function as a neuroprotective and anti-amyloidogenic molecule. Consistent with its extracellular chaperone function, clusterin transports the Aβ peptide in biological fluids, maintains its solubility, modulates its uptake by the brain across the blood-brain barrier and contributes to its clearance. In addition, clusterin has the capability to inhibit the formation of the membrane-attack complex during the complement activation cascade and to bind to partly unfolded stressed structures, both contributing elements to the inflammatory conditions seen in

AD. As an amyloid associated protein, however, it co-localizes with Aβ amyloid and pre-amyloid lesions without being a structural part of the final fibrils, possibly allowing the conversion of soluble Aβ into forms of high β-sheet content. Under these circumstances, clusterin may also promote neurotoxicity by triggering uncontrolled cell death signals. Whether these opposite effects reflect structural differences (i.e. full-length vs. truncated forms), specific sub-cellular distribution (i.e. secreted vs. nuclear) or still undefined regulatory interactions with some of its multiple ligands and/or unknown partners remains unknown. Certainly the use of DNA microarrays and comprehensive functional proteomic analysis will provide a large-scale gene expression assessment and a panoramic view of the relevant pathways and binding molecules associated with clusterin regulation in AD. These pathways / physiologic ligands may, in turn, become targets for therapeutic interventions.

ACKNOWLEDGEMENTS

The authors acknowledge the support from the National Institute of Health grants NS38777, AG05891, AG08721, and from the Alzheimer Association.

REFERENCES

Aronow, B.J., Lund, S.D., Brown, T.L., Harmony, J.A., and Witte, D.P., 1993, Apolipoprotein J expression at fluid-tissue interfaces: potential role in barrier cytoprotection. *Proc. Natl. Acad. Sci. USA* **90:** 725-729.

Bailey, R.W., Aronow, B., Harmony, J.A., and Griswold, M.D., 2002, Heat shock-initiated apoptosis is accelerated and removal of damaged cells is delayed in the testis of clusterin/ApoJ knock-out mice. *Biol. Reprod.* **66:** 1042-1053.

Bailey, R.W., Dunker, A.K., Brown, C.J., Garner, E.C., and Griswold, M.D., 2001, Clusterin, a binding protein with a molten globule-like region. *Biochemistry* **40:** 11828-11840.

Bales, K.R., Verina, T., Cummins, D.J., Du, Y., Dodel, R.C., Saura, J., Fishman, C.E., DeLong, C.A., Piccardo, P., Petegnief, V., Ghetti, B., and Paul, S.M., 1999, Apolipoprotein E is essential for amyloid deposition in the APP(V717F) transgenic mouse model of Alzheimer's disease. *Proc. Natl. Acad. Sci. USA* **96:** 15233-15238.

Bales, K.R., Verina, T., Dodel, R.C., Du, Y., Altstiel, L., Bender, M., Hyslop, P., Johnstone, E.M., Little, S.P., Cummins, D.J., Piccardo, P., Ghetti, B., and Paul, S.M., 1997, Lack of apolipoprotein E dramatically reduces amyloid beta-peptide deposition. *Nature Genet.* **17:** 263-264.

Bartl, M.M., Luckenbach, T., Bergner, O., Ullrich, O., and Koch-Brandt, C., 2001, Multiple receptors mediate apoJ-dependent clearance of cellular debris into nonprofessional phagocytes. *Ex.p. Cel.l. Res.* **271:** 130-141.

Blaschuk, O., Burdzy, K., and Fritz, I.B., 1983, Purification and characterization of a cell-aggregating factor (clusterin), the major glycoprotein in ram rete testis fluid. *J. Bio. Chem.* **258:** 7714-7720.

Boggs, L.N., Fuson, K.S., Baez, M., Churgay, L., McClure, D., Becker, G., and May, P.C., 1996, Clusterin (Apo J) protects against in vitro amyloid-beta (1-40) neurotoxicity. *J. Neurochem.* **67:** 1324-1327.

Brett, J., Schmidt, A.M., Yan, S.D., Zou, Y.S., Weidman, E., Pinsky, D., Nowygrod, R., Neeper, M., Przysiecki, C., Shaw, A., Migheli, A., and Stern, D., 1993, Survery of the distribution of a newly characterized receptor for advanced glycation end products in tissues. *Am. J. Pathol.* **143:** 1699-1712.

Burkey, B.F., Stuart, W.D., and Harmony, J.A., 1992, Hepatic apolipoprotein J is secreted as a lipoprotein. *J. Lipid Re.s* **33:** 1517-1526.

Buttyan, R., Olsson, C.A., Pintar, J., Chang, C., Bandyk, M., Ng, P.Y., and Sawczuk, I.S., 1989, Induction of the TRPM-2 gene in cells undergoing programmed death. *Mol. Cell Biol.* **9:** 3473-3481.

Calero, M., Rostagno, A., Matsubara, E., Zlokovic, B., Frangione, B., and Ghiso, J., 2000, Apolipoprotein J (clusterin) and Alzheimer's disease. *Microsc. Res. Tech.* **50:** 305-315.

Calero, M., Tokuda, T., Rostagno, A., Kumar, A., Zlokovic, B., Frangione, B., and Ghiso, J., 1999, Functional and structural properties of lipid-associated apolipoprotein J (clusterin). *Biochem.. J.* **344:** 375-383.

Choi-Miura, N.H., Ihara, Y., Fukuchi, K., Takeda, M., Nakano, Y., Tobe, T., and Tomita, M., 1992, SP-40,40 is a constituent of Alzheimer's amyloid. *Acta Neuropatho.l (Berl.)* **83:** 260-264.

Christie, D.L., Freeman, M., and Hyman, B.T., 1996, Expression of the macrophage scavenger receptor, a multifunctional lipoprotein receptor, in microglia associated with senile plaques in Alzheimer's disease. *Am. J. Pathol.* **148:** 399-403.

Collard, M.W., and Griswold, M.D., 1987, Biosynthesis and molecular cloning of sulfated glycoprotein 2 secreted by rat Sertoli cells. *Biochemistry* **26:** 3297-3303.

Danik, M., Chabot, J.G., Hassan-Gonzalez, D., Suh, M., and Quirion, R., 1993, Localization of sulfated glycoprotein-2/clusterin mRNA in the rat brain by in situ hybridization. *J. Comp. Neurol.* **334:** 209-227.

Danik, M., Chabot, J.G., Mercier, C., Benabid, A.L., Chauvin, C., Quirion, R., and Suh, M., 1991, Human gliomas and epileptic foci express high levels of a mRNA related to rat testicular sulfated glycoprotein 2, a purported marker of cell death. *Proc. Natl. Acad. Sci. USA* **88:** 8577-8581.

de Silva, H.V., Harmony, J.A., Stuart, W.D., Gil, C.M., and Robbins, J., 1990a, Apolipoprotein J: structure and tissue distribution. *Biochemistry* **29:** 5380-5389.

de Silva, H.V., Stuart, W.D., Park, Y.B., Mao, S.J., Gil, C.M., Wetterau, J.R., Busch, S.J., and Harmony, J.A., 1990b, Purification and characterization of apolipoprotein J. *J. Biol. Chem.* **265:** 14292-14297.

de Silva, H.V., Stuart, W.D., Duvic, C.R., Wetterau, J.R., Ray, M.J., Ferguson, D.G., Albers, H.W., Smith, W.R., and Harmony, J.A., 1990c, A 70-kDa apolipoprotein designated ApoJ is a marker for subclasses of human plasma high density lipoproteins. *J. Biol. Chem.* **265:** 13240-13247.

Deane, R., Du Yan, S., Submamaryan, R.K., LaRue, B., Jovanovic, S., Hogg, E., Welch, D., Manness, L., Lin, C., Yu, J., Zhu, H., Ghiso, J., Frangione, B., Stern, A., Schmidt, A.M., Armstrong, D.L., Arnold, B., Liliensiek, B., Nawroth, P., Hofman, F., Kindy, M., Stern, D., and Zlokovic, B., 2003, RAGE mediates amyloid-beta peptide transport across the blood-brain barrier and accumulation in brain. *Nature Med.* **9:** 907-913.

Debure, L., Vayssiere, J.L., Rincheval, V., Loison, F., Le Drean, Y., and Michel, D., 2003, Intracellular clusterin causes juxtanuclear aggregate formation and mitochondrial alteration. *J. Cell Sci.* **116:** 3109-3121.

DeMattos, R.B., O'Dell M, A., Parsadanian, M., Taylor, J.W., Harmony, J.A., Bales, K.R., Paul, S.M., Aronow, B.J., and Holtzman, D.M., 2002, Clusterin promotes amyloid plaque formation and is critical for neuritic toxicity in a mouse model of Alzheimer's disease. *Proc. Nat.l Acad. Sci. USA* **99:** 10843-10848.

DeMattos, R.B., Cirrito, J., Parsadanian, M., May, P.C., O'Dell, M.A., Taylor, J.W., Harmony, J.A., Aronow, B., Bales, K.R., Paul, S.M., and Holtzman, D., 2004, ApoE and clusterin cooperatively suppress Ab levels and deposition: Evidence that apoE regulates extracellular Ab metabolism in vivo. *Neuron* **41:** 193-202.

El Khoury, J., Hickman, S.E., Thomas, C.A., Cao, L., Silverstein, S.C., and Loike, J.D., 1996, Scavenger receptor-mediated adhesion of microglia to beta-amyloid fibrils. *Nature* **382:** 716-719.

French, L.E., Chonn, A., Ducrest, D., Baumann, B., Belin, D., Wohlwend, A., Kiss, J.Z., Sappino, A.P., Tschopp, J., and Schifferli, J.A., 1993, Murine clusterin: molecular cloning and mRNA localization of a gene associated with epithelial differentiation processes during embryogenesis. *J. Cell Biol.* **122:** 1119-1130.

Fritz, I.B., Burdzy, K., Setchell, B., Blaschuk, O., Lyon, M.F., and Setchell, B.P., 1983, Ram rete testis fluid contains a protein (clusterin) which influences cell-cell interactions in vitro. Evidence for a defective seminiferous tubule barrier in testes of Tfm and Sxr mice. *Biol. Reprod.* **28:** 1173-1188.

Games, D., Adams, D., Alessandrini, R., Barbour, R., Berthelette, P., Blackwell, C., Carr, T., Clemens, J., Donaldson, T., Gillespie, F., and *et al.*, 1995, Alzheimer-type neuropathology in transgenic mice overexpressing V717F beta-amyloid precursor protein. *Nature* **373:** 523-527.

Ghiso, J., and Frangione, B., 2002, Amyloidosis and Alzheimer's disease. *Adv. Drug Deliv. Rev.* **54:** 1539-1551.

Ghiso, J., Matsubara, E., Koudinov, A., ChoiMiura, N.H., Tomita, M., Wisniewski, T., and Frangione, B., 1993, The cerebrospinalfluid soluble formof Alzheimer's amyloid beta is complexed to SP40,40 (apolipoprotein J), an inhibitor of the complement membrane attack complex. *Biochem. J.* **293:** 27-30.

Giannakipoulos, P., Kovari, E., French, L., Viard, I., Hof, P.R., and Bouras, C., 1998, Possible neuroprotective role of clusterin in Alzheimer's disease: a quantitative immunocytochemical study. *Acta Neuropathol. (Berl.)* **95:** 387-394.

Gutacker, C., Klock, G., Diel, P., and Koch-Brandt, C., 1999, Nerve growth factor and epidermal growth factor stimulate clusterin gene expression in PC12 cells. *Biochem. J.* **339:** 759-766.

Hammad, S.M., Ranganathan, S., Loukinova, E.B., Twal, W.O., and Argraves, W.S., 1997, Interaction of apolipoprotein J-amyloid beta-peptide complex with low density lipoprotein receptor-related protein-2/megalin. A mechanism to prevent pathological accumulation of amyloid beta-peptide. *J. Biol. Chem.* **272:** 18644-18649.

Han, B.H., DeMattos, R.B., Dugan, L.L., Kim-Han, J.S., Brendza, R.P., Fryer, J.D., Kierson, M., Cirrito, J., Quick, K., Harmony, J.A., Aronow, B., and Holtzman, D., 2001, Clusterin contributes to caspase-3 independent brain injury following neonatal hypoxia-ischemia. *Nature Med.* **7:** 338-343.

Harr, S.D., Uint, L., Hollister, R., Hyman, B.T., and Mendez, A.J., 1996, Brain expression of apolipoproteins E, J, and A-I in Alzheimer's disease. *J. Neurochem.* **66:** 2429-2435.

Hatters, D.M., Wilson, M.R., Easterbrook-Smith, S.B., and Howlett, G.J., 2002, Suppression of apolipoprotein C-II amyloid formation by the extracellular chaperone, clusterin. *Eur. J. Biochem.* **269:** 2789-2794.

Herault, Y., Chatelain, G., Brun, G., and Michel, D., 1992, V-src-induced-transcription of the avian clusterin gene. *Nucleic Acids Res.* **20:** 6377-6383.

Hochgrebe, T., Pankhurst, G.J., Wilce, J., and Easterbrook-Smith, S.B., 2000, pH-dependent changes in the in vitro ligand-binding properties and structure of human clusterin. *Biochemistry* **39:** 1411-1419.

Hochgrebe, T.T., Humphreys, D., Wilson, M.R., and Easterbrook-Smith, S.B., 1999, A reexamination of the role of clusterin as a complement regulator. *Exp. Cell Res.* **249:** 13-21.

Humphreys, D., Hochgrebe, T.T., Easterbrook-Smith, S.B., Tenniswood, M.P., and Wilson, M.R., 1997, Effects of clusterin overexpression on TNF alpha- and TGF beta-mediated death of L929 cells. *Biochemistry* **36:** 15233-15243.

Hussain, M.M., Strickland, D.K., and Bakillah, A., 1999, The mammalian low-density lipoprotein receptor family. *Ann.. Rev.. Nutr.* **19:** 141-172.

Jenne, D.E., and Tschopp, J., 1989, Molecular structure and functional characterization of a human complement cytolysis inhibitor found in blood and seminal plasma: identity to sulfated glycoprotein 2, a constituent of rat testis fluid. *Proc Natl Acad Sci USA* **86:** 7123-7127.

Jenne, D.E., Lowin, B., Peitsch, M.C., Bottcher, A., Schmitz, G., and Tschopp, J., 1991, Clusterin (complement lysis inhibitor) forms a high density lipoprotein complex with apolipoprotein A-I in human plasma. *J. Biol. Chem.* **266:** 11030-11036.

Jin, G., and Howe, P.H., 1997, Regulation of clusterin gene expression by transforming growth factor beta. *J. Biol. Chem.* **272:** 26620-26626.

Johnson, N.A., Sengupta, S., Saidi, S.A., Lessan, K., Charnock-Jones, S.D., Scott, L., Stephens, R., Freeman, T.C., Tom, B.D.M., Harris, M., Denyer, G., Sundaram, M., Sasisekharan, R., Smith, S.K., and Print, C.G., 2004, Endothelial cells preparing to die by apoptosis initiate a program of transcriptome and glycome regulation. *Faseb J.* **18:** 188-190.

Jones, S.E., Meerabux, J.M., Yeats, D.A., and Neal, M.J., 1992, Analysis of differentially expressed genes in retinitis pigmentosas retinas. Altered expression of clusterin mRNA. *FEBS Lett.* **300:** 279-282.

Jordan, J., Galindo, M.F., Miller, R.J., Reardon, C.A., Getz, G.S., and LaDu, M.J., 1998, Isoform-specific effect of apolipoprotein E on cell survival and beta-amyloid-induced toxicity in rat hippocampal pyramidal neuronal cultures. *J. Neurosci.* **18:** 195-204.

Kapron, J.T., Hilliard, G.M., Lakins, J.N., Tenniswood, M.P., West, K.A., Carr, S.A., and Crabb, J.W., 1997, Identification and characterization of glycosylation sites in human serum clusterin. *Protein Sci.* **6:** 2120-2133.

Kida, E., Pluta, R., Lossinsky, A.S., Golabek, A., Choi-Miura, N.H., Wisniewski, H.E., and Mossakowski, M.J., 1995, Complete cerebral ischemia with short-term survival in rat induced by cardiac arrest. II. Extracellular and intracellular accumulation of apolipoproteins E and J in the brain. *Brain Res.* **674:** 341-346.

Kimura, K., and Yamamoto, M., 1996, Modification of the alternative splicing process of testosterone-repressed prostate message-2 (TRPM-2) gene by protein synthesis inhibitors and heat shock treatment. *Biochim. Biophys. Acta* **1307:** 83-88.

Kimura, K., Asami, K., and Yamamoto, M., 1997, Effect of heat shock treatment on the production of variant testosterone-repressed prostate message-2 (TRPM-2) mRNA in culture cells. *Cell Biochem. Funct.* **15:** 251-257.

Kirszbaum, L., Bozas, S.E., and Walker, I.D., 1992, SP-40,40, a protein involved in the control of the complement pathway, possesses a unique array of disulphide bridges. *FEBS Lett.* **297**: 70-76.

Kirszbaum, L., Sharpe, J.A., Murphy, B., d'Apice, A.J., Classon, B., Hudson, P., Walker, I.D., Leskov, K.S., Klokov, D.Y., Li, J., Kinsella, T.J., Boothman, D.A., Fritz, I.B., Burdzy, K., Setchell, B., Blaschuk, O., Slawin, K., Sawczuk, I.S., Olsson, C.A., Buttyan, R., Tobe, T., Minoshima, S., Yamase, S., Choi, N.H., Tomita, M., Shimizu, N., Purrello, M., Bettuzzi, S., Di Pietro, C., Mirabile, E., Di Blasi, M., Rimini, R., Grzeschik, K.H., Ingletti, C., Corti, A., Sichel, G., Wong, P., Taillefer, D., Lakins, J., Pineault, J., Chader, G., and Tenniswood, M., 1989, Molecular cloning and characterization of the novel, human complement-associated protein, SP-40,40: a link between the complement and reproductive systems. *EMBO J.* **8**: 711-718.

Koudinov, A., Matsubara, E., Frangione, B., and Ghiso, J., 1994, The soluble form of Alzheimer's amyloid beta protein is complexed to high density lipoprotein 3 andvery high density lipoprotein in normal human plasma. *Biochem. Biophys. Res. Commun.* **205**: 1164-1171.

Koudinov, A., Koudinova, N., Kumar, A., Beavis, R.C., and Ghiso, J., 1996, Biochemical characterization of Alzheimer's soluble amyloid beta protein in human cerebrospinal fluid: Association with high density lipoproteins. *Biochem. Biophys. Res. Commun.* **223**: 592-597.

Kounnas, M.Z., Loukinova, E.B., Stefansson, S., Harmony, J.A., Brewer, B.H., Strickland, D.K., and Argraves, W.S., 1995, Identification of glycoprotein 330 as an endocytic receptor for apolipoprotein J/clusterin. *J. Biol. Chem.***270**: 13070-13075.

Lakins, J.N., Poon, S., Easterbrook-Smith, S.B., Carver, J.A., Tenniswood, M.P., and Wilson, M.R., 2002, Evidence that clusterin has discrete chaperone and ligand binding sites. *Biochemistry* **41**: 282-291.

Lambert, M.P., Barlow, A.K., Chromy, B.A., Edwards, C., Freed, R., Liosatos, M., Morgan, T.E., Rozovsky, I., Trommer, B., Viola, K.L., Wals, P., Zhang, C., Finch, C.E., Krafft, G.A., and Klein, W.L., 1998, Diffusible, nonfibrillar ligands derived from Abeta1-42 are potent central nervous system neurotoxins. *Proc. Natl. Acad. Sci. USA* **95**: 6448-6453.

Lemansky, P., Brix, K., and Herzog, V., 1999, Subcellular distribution, secretion, and posttranslational modifications of clusterin in thyrocytes. *Exp. Cell Re.s* **251**: 147-155.

Leskov, K.S., Klokov, D.Y., Li, J., Kinsella, T.J., and Boothman, D.A., 2003, Synthesis and functional analyses of nuclear clusterin, a cell death protein. *J. Biol. Chem.* **278**: 11590-11600.

Lidstrom, A.M., Bogdanovic, N., Hesse, C., Volkman, I., Davidsson, P., and Blennow, K., 1998, Clusterin (apolipoprotein J) protein levels are increased in hippocampus and in frontal cortex in Alzheimer's disease. *Exp. Neurol.* **154**: 511-521.

Lidstrom, A.M., Hesse, C., Rosengren, L., Fredman, P., Davidsson, P., and Blennow, K., 2001, Normal levels of clusterin in cerebrospinal fluid in Alzheimer's disease, and no change after acute ischemic stroke. *J. Alzheimers Dis.* **3**: 435-442.

Lucarelli, M., Gennarelli, M., Cardelli, P., Novelli, G., Scarpa, S., Dallapiccola, B., and Strom, R., 1997, Expression of receptors for native and chemically modified low-density lipoproteins in brain microvessels. *FEBS Lett.* **401**: 53-58.

Mackic, J.B., Stins, M., McComb, J.G., Calero, M., Ghiso, J., Kim, K.S., Yan, S.D., Stern, D., Schmidt, A.M., Frangione, B., and Zlokovic, B.V., 1998, Human blood-brain barrier receptors for Alzheimer's amyloid-beta 1- 40. Asymmetrical binding, endocytosis, and transcytosis at the apical side of brain microvascular endothelial cell monolayer. *J. Clin. Invest.* **102**: 734-743.

Manuelidis, L., Fritch, W., and Xi, Y.G., 1997, Evolution of a strain of CJD that induces BSE-like plaques. *Science* **277:** 94-98.

Matsubara, E., Frangione, B., and Ghiso, J., 1995, Characterization of lipoprotein J Alzheimer's A-beta interaction. *J. Biol. Chem.* **270:** 7563-7567.

Matsubara, E., Soto, C., Governale, S., Frangione, B., and Ghiso, J., 1996, Apolipoprotein J and Alzheimer's amyloid beta solubility. *Biochem. J.* **316:** 671-679.

Matsubara, E., Ghiso, J., Frangione, B., Amari, M., Tomidokoro, Y., Ikeda, Y., Harigaya, Y., Okamoto, K., and Shoji, M., 1999, Lipoprotein-free amyloidogenic peptides in plasma are elevated in patients with sporadic Alzheimer's disease and Down's syndrome. *Ann. Neurol.* **45:** 537-541.

Matsuda, A., Itoh, Y., Koshikawa, N., Akizawa, T., Yana, I., and Seiki, M., 2003, Clusterin, an abundant serum factor, is a possible negative regulator of MT6-MMP/MMP-25 produced by neutrophils. *J. Biol. Chem.* **278:** 36350-36357.

McGeer, P.L., Kawamata, T., and Walker, D.G., 1992, Distribution of clusterin in Alzheimer brain tissue. *Brain Res.* **579:** 337-341.

McHattie, S., and Edington, N., 1999, Clusterin prevents aggregation of neuropeptide 106-126 in vitro. *Biochem. Biophys. Re.s Commun.* **259:** 336-340.

McHattie, S., Wells, G.A., Bee, J., and Edington, N., 1999, Clusterin in bovine spongiform encephalopathy (BSE). *J. Comp. Pathol.* **121:** 159-171.

McLaughlin, L., Zhu, G., Mistry, M., Ley-Ebert, C., Stuart, W.D., Florio, C.J., Groen, P.A., Witt, S.A., Kimball, T.R., Witte, D.P., Harmony, J.A., and Aronow, B.J., 2000, Apolipoprotein J/clusterin limits the severity of murine autoimmune myocarditis. *J. Clin. Invest.* **106:** 1105-1113.

Michel, D., Chatelain, G., Herault, Y., and Brun, G., 1995, The expression of the avian clusterin gene can be driven by two alternative promoters with distinct regulatory elements. *Eur. J. Biochem.* **229:** 215-223.

Michel, D., Chatelain, G., North, S., and Brun, G., 1997, Stress-induced transcription of the clusterin/apoJ gene. *Biochem. J.* **328:** 45-50.

Mitsuhashi, T., Li, Y.M., Fishbane, S., and Vlassara, H., 1997, Depletion of reactive advanced glycation endproducts from diabetic uremic sera using a lysozyme-linked matrix. *J. Clin. Invest.* **100:** 847-854.

Murphy, B.F., Kirszbaum, L., Walker, I.D., and d'Apice, A.J., 1988, SP-40,40, a newly identified normal human serum protein found in the SC5b-9 complex of complement and in the immune deposits in glomerulonephritis. *J. Clin. Invest.* **81:** 1858-1864.

Narita, M., Holtzman, D.M., Schwartz, A.L., and Bu, G., 1997, Alpha2-macroglobulin complexes with and mediates the endocytosis of beta-amyloid peptide via cell surface low-density lipoprotein receptor-related protein. *J. Neurochem.* **69:** 1904-1911.

Nishida, K., Quantock, A.J., Dota, A., Choi-Miura, N.H., and Kinoshita, S., 1999, Apolipoproteins J and E co-localise with amyloid in gelatinous drop-like and lattice type I corneal dystrophies. *Br. J. Ophthalmol.* **83:** 1178-1182.

O'Bryan, M.K., Baker, H.W., Saunders, J.R., Kirszbaum, L., Walker, I.D., Hudson, P., Liu, D.Y., Glew, M.D., d'Apice, A.J., and Murphy, B.F., 1990, Human seminal clusterin (SP-40,40). Isolation and characterization. *J. Clin. Invest.* **85:** 1477-1486.

Oda, T., Pasinetti, G.M., Osterburg, H.H., Anderson, C., Johnson, S.A., and Finch, C.E., 1994, Purification and characterization of brain clusterin. *Biochem. Biophys. Res. Commun.* **204:** 1131-1136.

Oda, T., Wals, P., Osterburg, H.H., Johnson, S.A., Pasinetti, G.M., Morgan, T.E., Rozovsky, I., Stine, W.B., Snyder, S.W., Holzman, T.F., and *et al.*, 1995, Clusterin (apoJ) alters the aggregation of amyloid beta-peptide (Abeta 1-42) and forms slowly sedimenting A beta complexes that cause oxidative stress. *Exp. Neurol.* **136:** 22-31.

O'Sullivan, J., Whyte, L., Drake, J., and Tenniswood, M., 2003, Alterations in the post-translational modification and intracellular trafficking of clusterin in MCF-7 cells during apoptosis. *Cell Death Differ.* **10:** 914-927.

Palmer, D.J., and Christie, D.L., 1992, Identification of molecular aggregates containing glycoproteins III, J, K (carboxypeptidase H), and H (Kex2-related proteases) in the soluble and membrane fractions of adrenal medullary chromaffin granules. *J. Biol. Chem.* **267:** 19806-19812.

Pankhurst, G.J., Bennett, C.A., and Easterbrook-Smith, S.B., 1998, Characterization of the heparin-binding properties of human clusterin. *Biochemistry* **37:** 4823-4830.

Pasinetti, G.M., Johnson, S.A., Oda, T., Rozovsky, I., and Finch, C.E., 1994, Clusterin (SGP-2): a multifunctional glycoprotein with regional expression in astrocytes and neurons of the adult rat brain. *J. Comp. Neurol* **.339:** 387-400.

Polihronis, M., Paizis, K., Carter, G., Sedal, L., and Murphy, B., 1993, Elevation of human cerebrospinal fluid clusterin concentration is associated with acute neuropathology. *J. Neurol. Sci.* **115:** 230-233.

Poon, S., Easterbrook-Smith, S.B., Rybchyn, M.S., Carver, J.A., and Wilson, M.R., 2000, Clusterin is an ATP-independent chaperone with very broad substrate specificity that stabilizes stressed proteins in a folding-competent state. *Biochemistry* **39:** 15953-15960.

Poon, S., Treweek, T.M., Wilson, M.R., Easterbrook-Smith, S.B., and Carver, J.A., 2002a, Clusterin is an extracellular chaperone that specifically interacts with slowly aggregating proteins on their off-folding pathway. *FEBS Lett.* **513:** 259-266.

Poon, S., Rybchyn, M.S., Easterbrook-Smith, S.B., Carver, J.A., Pankhurst, G.J., Wilson, M.R., Treweek, T.M., Lakins, J.N., and Tenniswood, M.P., 2002b, Mildly acidic pH activates the extracellular molecular chaperone clusterin. *J.Biol. Chem.* **277:** 39532-39540.

Purrello, M., Bettuzzi, S., Di Pietro, C., Mirabile, E., Di Blasi, M., Rimini, R., Grzeschik, K.H., Ingletti, C., Corti, A., Sichel, G., Wong, P., Taillefer, D., Lakins, J., Pineault, J., Chader, G., and Tenniswood, M., 1991, The gene for SP-40,40, human homolog of rat sulfated glycoprotein 2, rat clusterin, and rat testosterone-repressed prostate message 2, maps to chromosome 8. *Genomics* **10:** 151-156.

Qiu, Z., Strickland, D.K., Hyman, B.T., and Rebeck, G.W., 1999, Alpha2-macroglobulin enhances the clearance of endogenous soluble beta-amyloid peptide via low-density lipoprotein receptor-related protein in cortical neurons. *J. Neurochem.* **73:** 1393-1398.

Rosenberg, M.E., Girton, R., Finkel, D., Chmielewski, D.H., Barie, A., III, Witte, D.P., Zhu, G., Bissler, J.J., Harmony, J.A., and Aronow, B., 2002, Apolipoprotein J/Clusterin prevents a progressive glomerulopathy of aging. *Mol. Cell Biol.* **22:** 1893-1902.

Santilli, G., Aronow, B.J., and Sala, A., 2003, Essential requirement of apolipoprotein J (clusterin) signaling for IkappaB expression and regulation of NF-kappaB activity. *J. Biol. Chem.* **278:** 38214-38219.

Sasaki, A., Shoji, M., Harigaya, Y., Kawarabayashi, T., Ikeda, M., Naito, M., Matsubara, E., Abe, K., and Nakazato, Y., 2002a, Amyloid cored plaques in Tg2576 transgenic mice are characterized by giant plaques, slightly activated microglia, and the lack of paired helical filament-typed, dystrophic neurites. *Virchows Arch.* **441:** 358-367.

Sasaki, K., Doh-ura, K., Ironside, J.W., and Iwaki, T., 2002b, Increased clusterin (apolipoprotein J) expression in human and mouse brains infected with transmissible spongiform encephalopathies. *Acta Neuropathol. (Berl.)* **103:** 199-208.

Sasaki, N., Takeuchi, M., Chowei, H., Kikuchi, S., Hayashi, Y., Nakano, N., Ikeda, H., Yamagishi, S., Kitamoto, T., Saito, T., and Makita, Z., 2002c, Advanced glycation end products (AGE) and their receptor (RAGE) in the brain of patients with Creutzfeldt-Jakob disease with prion plaques. *Neurosci. Lett.* **326:** 117-120.

Schreiber, S.S., Tocco, G., Najm, I., and Baudry, M., 1993, Seizure activity causes a rapid increase in sulfated glycoprotein-2 messenger RNA in the adult but not in the neonatal rat brain. *Neurosci. Lett.* **153:** 17-20.

Shibata, M., Yamada, S., Kumar, S.R., Calero, M., Bading, J., Frangione, B., Holtzman, D.M., Miller, C.A., Strickland, D.K., Ghiso, J., and Zlokovic, B.V., 2000, Clearance of Alzheimer's amyloid-ss(1-40) peptide from brain by LDL receptor-related protein-1 at the blood-brain barrier. *J. Clin. Invest.* **106:** 1489-1499.

Silkensen, J.R., Skubitz, K.M., Skubitz, A.P., Chmielewski, D.H., Manivel, J.C., Dvergsten, J.A., and Rosenberg, M.E., 1995, Clusterin promotes the aggregation and adhesion of renal porcine epithelial cells. *J. Clin. Invest.* **96:** 2646-2653.

Slawin, K., Sawczuk, I.S., Olsson, C.A., Buttyan, R., Blaschuk, O., Burdzy, K., Fritz, I.B., Tobe, T., Minoshima, S., Yamase, S., Choi, N.H., Tomita, M., Shimizu, N., Purrello, M., Bettuzzi, S., Di Pietro, C., Mirabile, E., Di Blasi, M., Rimini, R., Grzeschik, K.H., Ingletti, C., Corti, A., Sichel, G., Wong, P., Taillefer, D., Lakins, J., Pineault, J., Chader, G., and Tenniswood, M., 1990, Chromosomal assignment of the human homologue encoding SGP-2. *Biochem. Biophys. Res. Commun.* **172:** 160-164.

Stuart, W.D., Krol, B., Jenkins, S.H., and Harmony, J.A., 1992, Structure and stability of apolipoprotein J-containing high-density lipoproteins. *Biochemistry* **31:** 8552-8559.

Sylvester, S.R., Morales, C., Oko, R., and Griswold, M.D., 1991, Localization of sulfated glycoprotein-2 (clusterin) on spermatozoa and in the reproductive tract of the male rat. *Biol. Reprod* **.45:** 195-207.

Tobe, T., Minoshima, S., Yamase, S., Choi, N.H., Tomita, M., Shimizu, N., Purrello, M., Bettuzzi, S., Di Pietro, C., Mirabile, E., Di Blasi, M., Rimini, R., Grzeschik, K.H., Ingletti, C., Corti, A., Sichel, G., Wong, P., Taillefer, D., Lakins, J., Pineault, J., Chader, G., and Tenniswood, M., 1991, Assignment of a human serum glycoprotein SP-40,40 gene (CLI) to chromosome 8. *Cytogenet. Cell Genet.* **57:** 193-195.

Trougakos, I.P., and Gonos, E.S., 2002, Clusterin/apolipoprotein J in human aging and cancer. *Int. J. Biochem. Cell Biol.* **34:** 1430-1448.

Trougakos, I.P., So, A., Jansen, B., Gleave, M.E., and Gonos, E.S., 2004, Silencing expression of the clusterin/apolipoprotein J gene in human cancer cells using small interfering RNA induces spontaneous apoptosis, reduced growth ability, and cell sensitization to genotoxic and oxidative stress. *Cancer Res.* **64:** 1834-1842.

Urban, J., Parczyk, K., Leutz, A., Kayne, M., and Kondor-Koch, C., 1987, Constitutive apical secretion of an 80-kD sulfated glycoprotein complex in the polarized epithelial Madin-Darby canine kidney cell line. *J. Cell. Biol.***105:** 2735-2743.

Urmoneit, B., Prikulis, I., Wihl, G., D'Urso, D., Frank, R., Heeren, J., Beisiegel, U., and Prior, R., 1997, Cerebrovascular smooth muscle cells internalize Alzheimer amyloid beta protein via a lipoprotein pathway: implications for cerebral amyloid angiopathy. *Lab. Invest.* **77:** 157-166.

Viard, I., Wehrli, P., Jornot, L., Bullani, R., Vechietti, J.L., Schifferli, J.A., Tschopp, J., and French, L.E., 1999, Clusterin gene expression mediates resistance to apoptotic cell death induced by heat shock and oxidative stress. *J. Invest. Dermatol.* **112:** 290-296.

Wilson, M.R., and Easterbrook-Smith, S.B., 2000, Clusterin is a secreted mammalian chaperone. *Trends Biochem. Sci.* **25:** 95-98.

Witte, D.P., Aronow, B.J., Stauderman, M.L., Stuart, W.D., Clay, M.A., Gruppo, R.A., Jenkins, S.H., and Harmony, J.A., 1993, Platelet activation releases megakaryocyte-synthesized apolipoprotein J, a highly abundant protein in atheromatous lesions. *Am. J. Pathol.***143:** 763-773.

Wong, P., Taillefer, D., Lakins, J., Pineault, J., Chader, G., and Tenniswood, M., 1994, Molecular characterization of human TRPM-2/clusterin, a gene associated with sperm maturation, apoptosis and neurodegeneration. *Eur. J. Biochem.* **221:** 917-925.

Yan, S.D., Chen, X., Fu, J., Chen, M., Zhu, H., Roher, A., Slattery, T., Zhao, L., Nagashima, M., Morser, J., Migheli, A., Nawroth, P., Stern, D., and Schmidt, A.M., 1996, RAGE and amyloid-beta peptide neurotoxicity in Alzheimer's disease. *Nature* **382:** 685-691.

Yang, C.R., Leskov, K., Hosley-Eberlein, K., Criswell, T., Pink, J.J., Kinsella, T.J., and Boothman, D.A., 2000, Nuclear clusterin/XIP8, an x-ray-induced Ku70-binding protein that signals cell death. *Proc. Natl. Acad. Sc.i USA* **97:** 5907-5912.

Yasuhara, O., Aimi, Y., and McGeer, P.L., 1994, Expression of the complement membrane attack complex and its inhibitors in Pick disease brain. *Brain Res.* **652:** 346-349.

Zlokovic, B.V., Martel, C.L., Matsubara, E., McComb, J.G., Zheng, G., McCluskey, R.T., Frangione, B., and Ghiso, J., 1996, Glycoprotein 330/megalin: probable role in receptor-mediated transport of apolipoprotein J alone and in a complex with Alzheimer disease amyloid beta at the blood-brain and blood-cerebrospinal fluid barriers. *Proc. Natl. Acad. Sci. USA* **93:** 4229-4234.

Chapter 15

Acetylcholinesterase Interaction with Alzheimer Amyloid β

Nibaldo C. Inestrosa, Juan Paulo Sagal and Marcela Colombres
FONDA-Biomedical Center, P. Catholic University of Chile, Santiago, Chile

Abstract: Acetylcholinesterase (AChE) is an enzyme involved in cholinergic and non-cholinergic functions in both the central and peripheral nervous system, most of the AChE is found as a tetrameric form bound to neuronal membranes. Early cytochemical studies have demonstrated that the AChE associated with senile plaques differs enzymatically from the AChE associated with neurons in several respects. Biochemical studies indicated that AChE induces amyloid fibril formation and form highly toxic AChE-Aβ complexes. A 3.5 kDa peptide containing a tryptophan of the enzyme peripheral binding site (PAS) mimics the effect of the whole enzyme on amyloid formation. The neurotoxicity induced by AChE-Aβ complexes indicated that they trigger more neurodegeneration than those of the Aβ peptide alone, both *in vitro* (hippocampal neurons) and *in vivo* (rats injected in the dorsal hippocampus as a model of Alzheimer). The fact that AChE is able to accelerate amyloid formation and that such effect is sensitive to drugs that block PAS of the enzyme, suggests that specific and new AChE inhibitors may well provide an attractive possibility for treating Alzheimer's disease.

Key words: Amyloid β peptide (Aβ), Acetylcholinesterase (AChE), Alzheimer's Disease (AD), Peripheral Anionic Site (PAS).

1. INTRODUCTION

AChE has been described in cholinergic and non-cholinergic processes in both the central and peripheral nervous system (Massoulié *et al.*, 1993; Grisaru *et al.*, 1999). The enzyme is secreted and becomes associated with extracellular structures, namely the synaptic basal lamina at the neuromuscular junction and the amyloid plaques of Alzheimer's disease (AD) brain (Inestrosa and Perelman, 1989; Inestrosa *et al.*, 1997). Most of the AChE is found as a tetrameric form bound to neuronal membranes (Inestrosa *et al.*, 1987). However, in AD brains, AChE appears to be mainly associated with insoluble material found between cells -i.e. an extracellular matrix of amyloid fibrils, called senile plaques, whose main component is the amyloid-β-peptide (Aβ) (Soto *et al.*, 1994; Selkoe, 2001). Histochemical studies have demonstrated that the AChE associated with senile plaques differs enzymatically from the enzyme associated with normal nerve fibrils and neurons in several respects (Geula and Mesulam, 1994). We have also shown that AChE promotes the assembly of Aβ into amyloid fibrils (Inestrosa *et al.*, 1996), and that a monoclonal antibody directed against the peripheral binding site of AChE inhibits the effect of the enzyme upon amyloid formation (Reyes *et al.*, 1997).

AD is a progressive dementia paralleled by selective neuronal cell death, which is probably caused by Aβ fibrils or oligomers (Selkoe, 2001). According to the "amyloid cascade" hypothesis, neurodegeneration in AD begins with abnormal processing of the amyloid precursor protein (APP) and results in the production and aggregation of Aβ peptide forming oligomers and amyloid fibrils, which form the senile plaques. It is thought that either Aβ species trigger the formation of neurofibrillary tangles and eventually neuronal cell death. In fact, the "amyloid cascade" hypothesis has received considerable support from genetic studies into the early-onset familial forms of AD (FAD), for which mutations in APP or presenilin genes lead to an increase in Aβ production. Nevertheless, while the study of FAD individuals has proved to be significant in the understanding of the onset of the disease in its familial form (~1 % of AD cases), the onset as well as the reasons for the selective neuronal vulnerability of AD sporadic cases (~99 %) remains open. AD is normally diagnosed only after death when the examination of the brain of an AD patient reveals the presence of amyloid plaques which are the hallmark of the disease. These deposits have been shown to be composed not only of the Aβ peptide but other proteins have been detected and identified. Here we will discuss our studies concerning the interaction of AChE with the Aβ peptide, the role of the enzyme on amyloid formation and brain deposition and the effect of AChE-Aβ complexes on neuronal toxicity.

2. MOLECULAR CHANGES IN AChE IN AD BRAIN

Several studies on brains displaying AD lesions have shown changes in the expression and distribution of AChE, the enzyme responsible for the hydrolysis of the neurotransmitter acetylcholine (Kasa *et al.*, 1997; Talesa, 2001). Firstly, AChE activity is lost in specific regions of the AD brain, such as the cortex, hippocampus, amygdala and nucleus basal of Meynert (Hyman *et al.*, 1984; Geula and Mesulam, 1994). Secondly, the relative proportions of the different forms of AChE change, with a decrease in the tetrameric globular G_4 AChE form (Atack *et al.*, 1983; Fishman *et al.*, 1986) and a parallel increase in the monomeric G_1 and the asymmetric AChE forms (Younkin *et al.* 1986), although the latter normally corresponds to a small fraction of the total enzyme present in the human brain (Figure 1).

Finally, AChE has also been found to co-localize with Aβ deposits, such as those present in pre-amyloid diffuse deposits, mature senile plaques and cerebral blood vessels (Geula and Mesulam, 1989; Ulrich *et al.*, 1990; Kalaria *et al..*, 1992). Aβ deposition is an important step in the neurodegenerative processes associated with AD (Masters *et al.*, 1985; Roher *et al.,* 1986). However, there is no explanation for the selective death of specific populations of neurons in AD brain. It is possible that senile plaque components may be involved in this selective neuronal cell death and AChE is one plausible candidate, suggesting an important role in AD pathogenesis. In fact, among the neurons that degenerate in AD, there are populations that do not contain choline acetyltransferase yet still express AChE, suggesting that a major characteristic of degenerating neurons could be the expression of AChE. Furthermore, most of the cortical AChE activity present in AD brain is probably promoting the aggregation of Aβ into amyloid fibrils, thus increasing its toxicity (Inestrosa *et al.*, 1997; Alvarez *et al.*, 1998).

3. AMYLOID DEPOSITION AND THE ROLE OF AChE

Although AD is a complex disorder with mutations in many genes accounting for 1% of AD patients, the large majority of AD cases occur sporadically (Inestrosa *et al.*, 1997). There is therefore a need to search for the mechanisms responsible for the progressive cognitive decline observed in these cases. It is presently thought that the amyloidogenic process that converts soluble Aβ from its relatively inert form to its putative pathogenic state into amyloid fibrils is a nucleation-dependent process (Jarret and

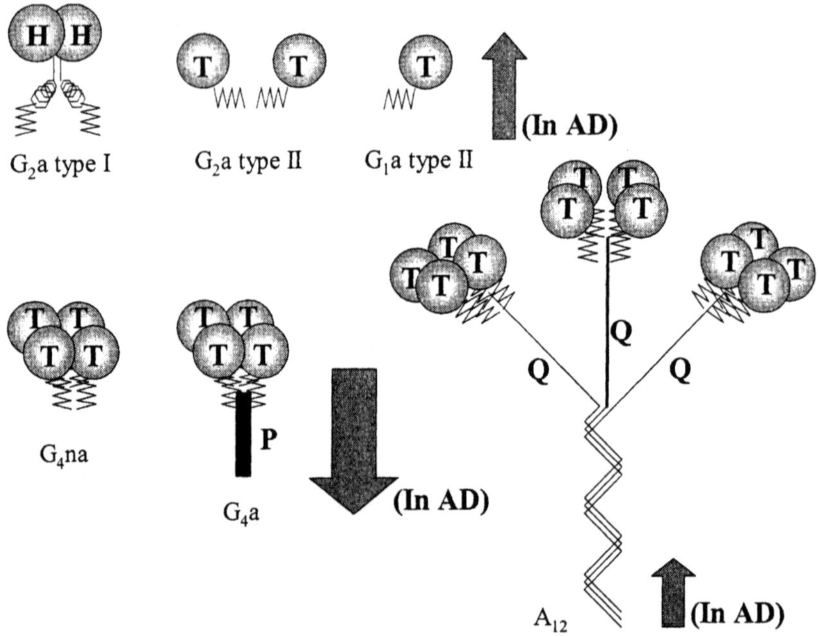

Figure 1. AChE Isoforms. AChE can have a variety of quaternary structures. The catalytic monomer can associate with other monomers to form complexes that are different in their hydrodynamic parameters and ionic or hydrophobic interactions. Then, they can be separated in the laboratory because they show different sedimentation coefficients. Molecular forms of the enzyme are of two types: Globular (G) and Asymmetric (A), and according to the number of catalytic monomers in them they are named G_1, G_2, G_4 and A_4, A_8 and A_{12}. The globular forms can have different amphiphilicity depending on if they are attached to the plasma membrane or not (G_2a type I form is attached to the membrane through a GPI anchor, and the G_4a form is anchored through its 20 kDa. hydrophobic P subunit). The other forms are fully soluble (G_4na) or are weakly associated to the membrane, being easily solubilized without detergents (G_1a type II and G_2a type II). The asymmetric forms are formed by three tetramers of catalytic subunits attached to the basal lamina through a collagen-like tail (ColQ). The catalytic monomer H in humans is unique to cells of the haematopoietic lineage. In the central nervous system, the G_1a type II, G_2a type II, G_4 and G_4na forms are usually found, while the asymmetric forms are found in the postsynaptic space of the neuromuscular junction. In AD there has been shown that variations in the levels of different molecular forms occurs with an increase in the monomeric and asymmetric forms and a decrease in the tetrameric globular forms. The fact that AChE is found associated to senile plaques, accelerates the Aβ peptide aggregation and induced selective neuronal cell death suggest that this enzyme plays a critical role on the pathogenesis of AD.

Lansbury, 1992) that probably requires structural transitions of Aβ (Soto *et al.*, 1995). Although the molecular factors underlying this transition *in vivo* remain unknown, the participation of additional senile plaque constituents known as "companion" or "chaperone" molecules has been proposed. AChE is an enzyme that in AD brains is associated predominantly with the amyloid core of mature senile plaques, pre-amyloid diffuse deposits, and cerebral blood vessels (Geula and Mesulam, 1994), and those brain areas where senile plaques are present are strongly AChE positive (Ulrich *et al.*, 1990).

Interestingly when AChE became associated with the amyloid fibrils, the enzyme changes some of its characteristics including, sensitivity to low pH and high substrate concentration, under these conditions the AChE-Aβ complexes are more toxic than the amyloid fibrils alone (Table 1 and Figure 2B). Considering that AChE has been designated as a marker for the cholinergic deficit observed in AD brains, this data suggests that "companion" molecules detected in senile plaques may indeed represent the link between amyloid deposition and the selective death of neuronal populations in AD brains.

Table 1. Effect of different inhibitors on bovine AChE in its free state and complexed with the Aβ peptide.

Inhibitors	AChE	$A\beta_{1-40}$+AChE
Active Site (IC_{50})		
Tacrine (10^{-9} M)	445.0 ± 17.6	1074.8 ± 73.3
Edrophonium (10^{-6} M)	5.36 ± 0.48	17.6 ± 4.6
BW284c51 (10^{-9} M)	57.1 ± 1.3	116.9 ± 2.7
Peripheric Site (IC_{50})		
Propidium (10^{-6} M)	34.6 ± 1.2	72.0 ± 4.8
Gallamine (10^{-3} M)	8.76 ± 0.46	13.3 ± 0.45
Fasciculin (10^{-11} M)	24.9 ± 1.5	274.6 ± 28.0

For determination of IC_{50} values, samples were incubated in the presence of appropriate concentrations of inhibitors 15-30 minutes at 37° C and the enzymatic activity was determined. The IC_{50} values were obtained from the inhibition curves. Values represent the mean ± s.d. (Adapted from Alarcón R. 1999.)

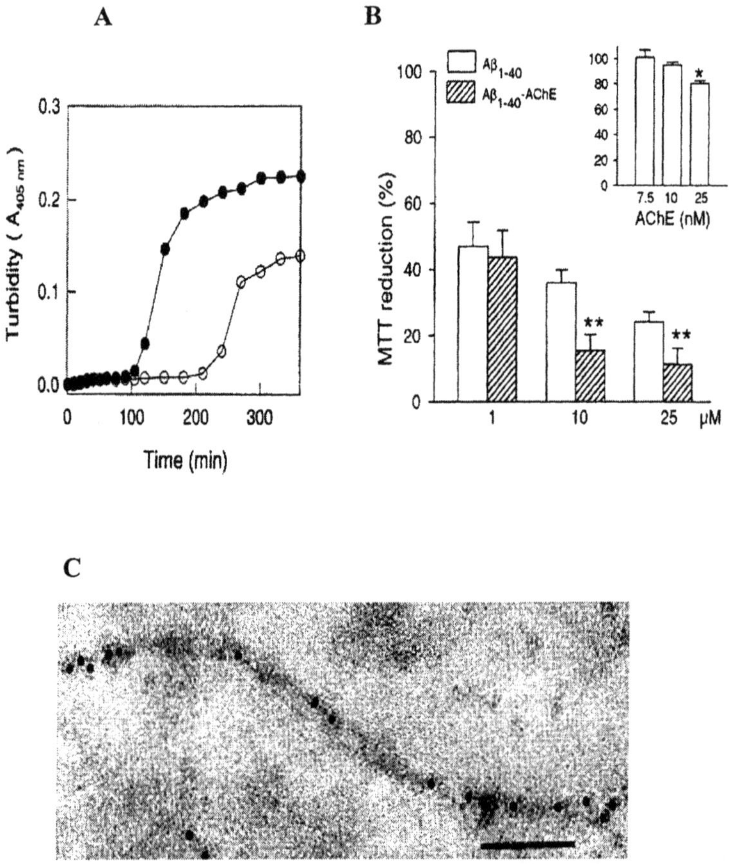

Figure 2. AChE accelerates the aggregation of the Aβ peptide, forms a stable complex with the fibers and through this tight interaction confers strong neurotoxicity to Aβ fibrils. **(A)** Aggregation assay of 100 μM Aβ peptide alone (empty circles) or with 0.1 μM of AChE (filled circles). Note that the lag period is decreased in the presence of the enzyme. (Adapted from Alvarez *et al.*, 1998). **(B)** Neurotoxic effect of different concentrations of the complex AChE-Aβ on PC12 cells followed through the MTT reduction assay, an indicator of cell redox activity (from Muñoz and Inestrosa, 1999). **(C)** Amyloid fibril labeled with anti-AChE conjugated to gold particles. Scale bar: 80 nm. (Adapted from Alvarez *et al.*, 1998).

A few years ago we found that AChE was able to promote the formation of amyloid fibrils (Inestrosa *et al.*, 1996). The enzyme seems to affect the kinetics of the Aβ aggregation process, as well as the lag period of fibrillogenesis, although the final amount of amyloid formed is similar in both cases (Figure 2A). Moreover, at concentrations sufficient to increase the nucleation of $A\beta_{1-40}$, AChE did not affect the thermodynamic solubility of the peptide, indicating that AChE acts as a kinetic and not a thermodynamic activator (Inestrosa and Alarcón, 1998; Alarcón, 1999).

It was also found that stable AChE-Aβ complexes are formed during fibrillogenesis *in vitro* (Figure 2C) (Alvarez *et al.*, 1997). AChE associated to amyloid was found to be more resistant to low pH conditions than free enzyme when AChE-Aβ complexes were incubated under varying pH conditions (Alvarez *et al.*, 1998). Further analysis of the catalytic activity of the AChE incorporated into these complexes shows an anomalous behaviour reminiscent of the AChE associated with senile plaques (Alvarez *et al.*, 1998). In a broad context, our results support the notion that the AChE-Aβ fibril association determines changes in the enzymatic properties of AChE, as occurs in the senile plaques of AD brain.

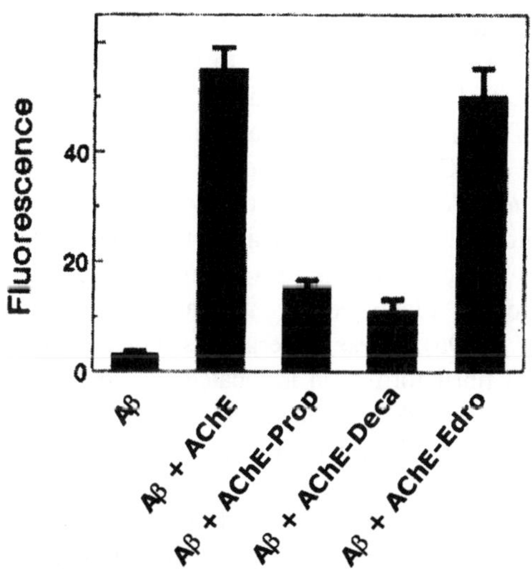

Figure 3. Effect of different inhibitors of AChE over amyloid formation. Amyloid formation quantitated by the Thioflavine-T assay, a specific probe for amyloid fibrils. Note that the use of Propidium (a specific PAS inhibitor) and Decamethonium (a dual active site and PAS inhibitor) block the effect of AChE on amyloid formation. Edrophonium, an active site inhibitor, did not have any effect. (Adapted from Inestrosa *et al.*, 1996).

Previous studies have suggested that AChE interacts with Aβ in a hydrophobic environment close to the peripheral anionic binding site (PAS) of the enzyme, and thus promotes amyloid fibril formation (Inestrosa *et al.*,

1996). In addition, AChE is incorporated into senile plaques *in vitro*, by forming macromolecular complexes with the growing Aβ fibrils (Alvarez *et al.*, 1997) which are resistant to the action of detergent and high ionic strength conditions (Alvarez *et al.*, 1998). Moreover under some conditions, it is possible to show cross-linking between the Aβ peptide and the AChE (Opazo and Inestrosa, 1998).

Several anti-cholinesterase drugs are able to decrease the effect of AChE on amyloid formation (Inestrosa and Alarcón, 1998). No effect is observed when active site inhibitors, like tacrine or edrophonium are used, on the contrary, propidium and fasciculin, anti-cholinesterase agents that inhibit the PAS of the enzyme, are able to block amyloid formation (Figure 3). These findings are entirely consistent with studies carried out using a monoclonal antibody directed against the PAS site of AChE (Reyes *et al.*, 1997).

3.1 Structural motifs of AChE involved in amyloid formation

To identify the AChE motif that promotes Aβ fibril formation, molecular dynamic techniques have been used in order to model the docking of Aβ onto the catalytic subunit of AChE. Using this approach four potential sites were identified, one of them (site I) spans a major hydrophobic sequence exposed on the surface of AChE, a polypeptide of 3.4 kDa called H peptide (aa. 274-308 in *Torpedo californica* AChE) that was able to accelerate Aβ fibril formation, as shown by turbidity measurements (De Ferrari *et al.*, 2001) (Figure 4). This site corresponds to a hydrophobic AChE sequence (L_{281} - M_{315}), previously identified by its capacity to interact with membranes (Shin *et al.*, 1996). These experiments indicate that the AChE motif that promotes Aβ fibril formation is located in a small hydrophobic peptide that contains a conserved tryptophan (W279), an amino acid residue which belongs to the PAS site of the catalytic subunit of AChE (Inestrosa *et al.*, 2004a).

Efficient AChE therapeutic inhibitors should be able to reach the active site at the base of the gorge as well as to the PAS. This latter site contributes to the catalytic efficiency by the transient binding of ACh until they reach the acylation site, disrupting interactions between AChE and Aβ. Ligands, which are able to bind to the PAS, can act by inhibiting or accelerating reactions towards the acylation site. In fact, caproctamine, a polyamine, was developed as a non-covalent inhibitor of AChE, which was able to bind at PAS (Tumiatti *et al.*, 2003).

On the other hand, binding of ligands to the acylation site can alter the conformation of the PAS showing an intimate structural communication between both sites. Structural data involved in the regulation of catalysis by

PAS ligands like propidium, decidium and gallamine, offer information on the residues which interact with other molecules and which participate in the nucleation process of Aβ fibrils. Novel inhibitors, which could bind both, the catalytic and the PAS sites of AChE, could offer an effective therapeutic approach for the treatment of AD. Such mechanism would include the blockade of the acceleration effect of AChE on the Aβ aggregation and at the same time keep the cognitive abilities of the AD patients (Colombres *et al.*, 2004).

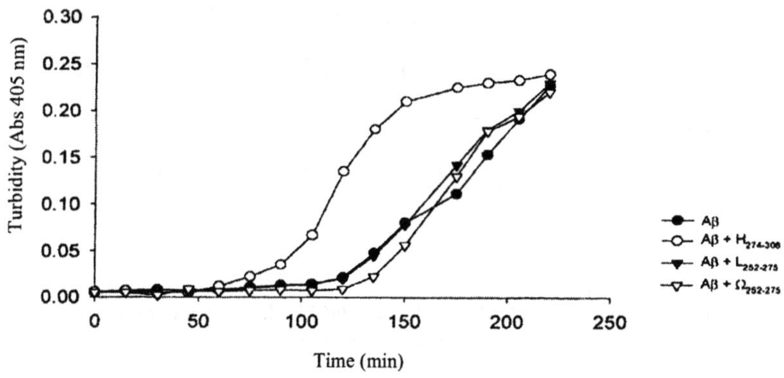

Figure 4. A hydrophobic peptide close to the PAS induces amyloid formation. Aggregation assay of 100 μM Aβ alone or with 1 μM of each peptide used. The aggregation assay shows that the H peptide derived from *Tc*AChE was able to accelerate the aggregation of the Aβ peptide as also occurs with the native AChE. The Ω peptide ($\Omega_{252-275}$), also derived from the *Tc*AChE, was chosen to perform the same experiment due to the fact that it is an exposed region of the enzyme that blocks the PAS when the enzyme is in its dimeric form. Ω has a disulfide bridge and L is its linear counterpart. Note that these two peptides have no effect over the aggregation process. (Adapted from De Ferrari *et al.*, 2001)

A short sequence in human AChE has been found to form amyloid fibrils *in vitro* (Cottingham *et al.*, 2002). It is a hypothetical tryptic fragment derived from the C-terminal tail of the synaptic T isoform of the enzyme ($AChE_{586-599}$), which presents certain sequence identity with the N-terminal sequence of Aβ peptide (Greenfield and Vaux, 2002) and is also conserved in the sequence of AChE from other mammalian sources (Figure 5).

The C-terminal tail of AChE corresponds to the oligomerization domain in the dimeric and tetrameric molecular forms of the enzyme. The peptide $AChE_{586-599}$ shows to be amyloidogenic, and considering that butyrylcholinesterase (BuChE) is also present in senile plaques but does not contain the PAS which interacts with Aβ peptide, it was proposed as an additional motif of interaction between AChE and the Aβ peptide. Given the amyloidogenic properties of the C-terminal peptide, it also remains to be determined whether the C-terminal tail interacts with the Aβ peptide and whether it accelerates its deposition.

Human APP	$_{674}$D**AEF**RHD**S**G**Y**E**VH**H**Q**KLV$_{691}$
Human AChE	$_{585}$K**AEF**HRW**S**S**Y**M**VH**WKNQF$_{602}$
Rat AChE	$_{585}$K**AEF**HRW**S**S**Y**M**VH**WKNQF$_{602}$
Mouse AChE	$_{585}$K**AEF**HRW**S**S**Y**M**VH**WKNQF$_{602}$
Human BuChE	$_{572}$K**A**G**F**HRWNN**Y**MMDWKNQF$_{590}$
Mouse BuChE	$_{572}$K**A**G**F**HRW**S**N**Y**MMDW**Q**NQF$_{590}$

Figure 5. Protein alignment of APP, AChE and BuChE. In bold are shown underlined residues that match with the human APP region corresponding to Aβ.

We performed some experiments to evaluate this possibility. Results suggest that this AChE peptide could play some influence in the process of aggregation of Aβ peptide as well as in the effect of the enzyme over the process (Figure 6). However, we were not able to see any amyloid formation from the AChE derived peptide when aggregated alone at high concentrations, as reported previously (Cottingham, *et al.*, 2002).

Along with its amyloidogenic properties, the $AChE_{586\text{-}599}$ peptide is cytotoxic. Greenfield and colleagues have proposed that this neurotoxicity is mediated by the activation of voltage-gated calcium channels in the membrane leading to an excessive influx of Ca^{+2} ions triggering apoptotic signals in the cell (Day and Greenfield, 2003).

On the other hand, AChE has been shown to have non-cholinergic properties independent of its enzymatic activity, such as effects on neuromuscular junction development, neurite outgrowth (Layer *et al.*, 1993; Bigbee *et al.*, 2000), hematopoiesis and osteogenesis (Grisaru *et al.*, 1999). In fact, it presents neurotrophic (Muñoz *et al.*, 1999) and adhesive (Johnson and Moore, 1999) properties mediated through the PAS region.

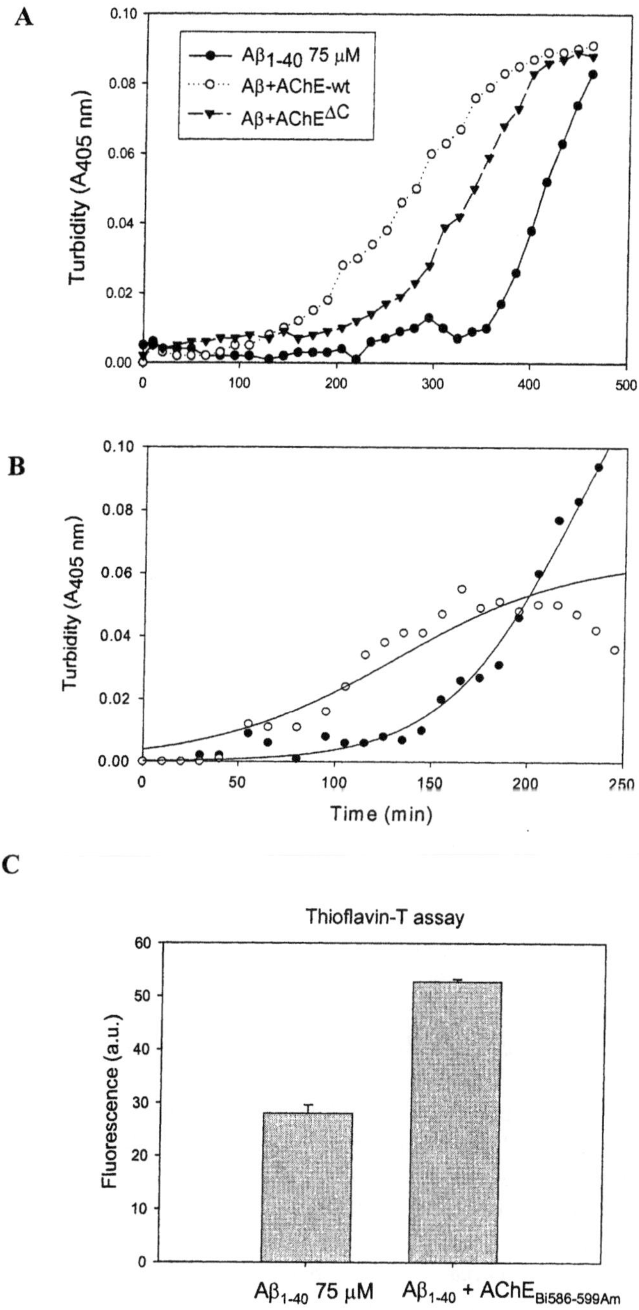
A
Aβ1-40 75 μM
Aβ+AChE-wt
Aβ+AChE^ΔC
Turbidity (A405 nm)
0.00
0.02
0.04
0.06
0.08
0.10
0
100
200
300
400
500
B
Turbidity (A405 nm)
0.00
0.02
0.04
0.06
0.08
0.10
0
50
100
150
200
250
Time (min)
C
Thioflavin-T assay
Fluorescence (a.u.)
0
10
20
30
40
50
60
Aβ1-40 75 μM
Aβ1-40 + AChE Bi586-599Am

Figure 6. (A) Aggregation assay of Aβ peptide in the presence of AChE. In Cottingham's work, he shows that an $AChE_{586\text{-}599}$ peptide is able to form amyloid fibrils, suggesting that the location of this peptide on the wild-type enzyme could be an additional site of interaction with senile plaques (Cottingham *et al.*, 2002). We evaluated this possibility using a truncated form of the recombinant human AChE which lacks the last 40 aa. of the C-terminal tail (T_{40} peptide) (Kronman, *et al.* 1992; Cohen, *et al.*, 2001). This region includes the $AChE_{586\text{-}599}$ peptide and it has been shown to be the oligomerization domain of the enzyme that is responsible for the contact with other catalytic monomers to form the different molecular forms present in any organism (G_2 and G_4). The kinetic stirred assay was carried out using 19 μg of both enzymes. This preliminary experiment shows that the wild-type enzyme reduces the lag phase of the aggregation as we have shown previously in our work using a bovine source for the enzyme (see figure 2 for comparison). The effect of the truncated form of the enzyme ($AChE^{\Delta C}$) over the Aβ peptide aggregation seems to be a partial effect since it does not reach the effect obtained with the wild-type enzyme, suggesting that the region corresponding to the AChE peptide might have some influence in the aggregation process of the Aβ peptide. (unpublished results).

(B) Aggregation assay of Aβ peptide in the presence of the $AChE_{Bi586\text{-}599Am}$ peptide. As for the other experiments, a kinetic stirred assay was carried out to evaluate the effect of the $AChE_{Bi586\text{-}599Am}$ peptide over the aggregation of the Aβ peptide. This assay measures solution turbidity, which is due to the formation of particles that are able to cause an interference with the light beam that crosses the solution of the peptide. $AChE_{Bi586\text{-}599Am}$ peptide is a chemically modified variant of the original $AChE_{586\text{-}599}$ peptide, which has a biotinylated N-terminus, and an amidated C-terminus to exclude the effect of charged terminals of the peptide (Cottingham *et al.*, 2002). The assay shows that the AChE peptide accelerates the aggregation but probably forming smaller particles.

(C) To evaluate the formation of amyloid fibrils we used the Thioflavine-T assay (Th-T assay). Th-T binds specifically to amyloid fibrils inducing a shift in its emission spectra. The Th-T assay shows that there is much amyloid formed in the presence of the AChE peptide. The mass ratio used in this experiment was 1:50 for Aβ and AChE peptide respectively. In previous experiments (not shown), we tested the ability of $AChE_{586\text{-}599}$ (the unmodified peptide with its charged terminals) to form amyloid fibrils using concentrations up to 300 μM, however the peptide showed low turbidity and the Th-T assay indicated that the nature of these aggregates was not amyloid. This is in contrast to previous studies (Cottingham *et al.* 2002). In spite of our observation, results suggest that this peptide might be exerting some influence in the aggregation process of Aβ peptide. Further experiments are necessary to evaluate this possibility. (Unpublished results).

3.2 AChE induces Aβ deposition in mammalian brain

In vivo studies also indicate that AChE has the ability to enhance Aβ aggregation and amyloid fibril formation. In fact, when AChE is infused stereotaxically into the CA_1 region of the rat hippocampus, novel plaque-like structures are formed (Chacón *et al.*, 2003). These amyloid plaques formed *de novo* share features with the lesions actually found in Alzheimer's brain, including amyloid deposits reactive to antibodies anti-Aβ and to thioflavine-S. The fact that rats never form amyloid plaques under normal conditions (its Aβ has 3 amino acids substitutions in relation to the human sequence), indicate that it is highly remarkably that the presence of AChE in the rat hippocampus was able to trigger the deposition of endogenous rat Aβ.

AChE has been previously shown to be neurotoxic for neuronal and glial-like cells *in vitro* (Calderón *et al.*, 1998) at low concentrations (nM range). Toxicity depends on the enzyme concentration, time of incubation, and cell culture density. The effect of AChE over cells involves alterations of the actin filament skeleton and induction of DNA fragmentation, along with cytoplasmic shrinkage and nuclear condensation, morphological characteristics of an apoptotic process. These effects are independent of the enzymatic activity of AChE, since the use of specific active site inhibitors do not change the neurotoxicity of the enzyme, and the same effects are observed after heat inactivation of the enzyme, a fact that suggests that the three-dimensional conformation of the enzyme is required to excert such effects.

It has been also observed that APP processing and AChE expression are modulated through the same environmental stimuli, *i.e.* cell differentiation and cell-cell or cell-matrix interactions. Since both molecules are membrane-associated proteins and they are thought to be delivered to their functional space via the secretory pathway, common signaling mechanisms may underlie basic aspects of their normal metabolism (Bronfman *et al.*, 1996).

3.3 AChE-Aβ complexes increase the neurotoxicity of Alzheimer's fibrils *in vitro* and *in vivo*

Since it has been demonstrated that amyloid fibrils are toxic to neuronal cells in culture, we hypothesised that the fibrils of the AChE-Aβ complexes would also be neurotoxic. To evaluate this, experiments using AChE-Aβ complexes and Aβ aggregates (aged for 5 days) were carried out in cell cultures and *in vivo* using hippocampal infusion. Firstly, cells exposed to the AChE-Aβ complexes presented dramatic changes in comparison to those incubated with the Aβ aggregates (Alvarez *et al.*, 1998). Similarly, when neuronal cells were processed using TdT mediated dUTP nick end labeling (TUNEL) to detect cell death *in situ*, cells treated with AChE displayed patterns of DNA fragmentation, indicating that apoptotic cell death is at least one of the cytotoxic mechanisms of AChE (Calderón *et al.*, 1998).

In vivo studies carried out using the injection of AChE into the hippocampus of rats as a model of AD (Chacón *et al.*, 2003) show neurological changes into the site of injection seen as neuronal cell loss and hypertrophic astrocytes, (detected by the glial fibrillary acidic protein immunoreactivity, GFAP), an indication of inflammatory processes in response to brain injury. Behavioural studies show that injection of AChE produces cognitive impairment demonstrated by an altered Morris water maze performance (Chacón *et al.*, 2003) Moreover, the toxicity of the AChE-Aβ complex is dependent on AChE concentration in the complexes (Muñoz and Inestrosa, 1999; Muñoz *et al.*, 2002; Chacón *et al.*, 2003). Furthermore, the neurotoxicity seen in our model confirmed previous studies in a transgenic mouse model that over-expresses the human AChE (Beeri *et al.*, 1995). In this model, besides learning and memory impairments (Cohen *et al.*, 2002), there are hypertrophic hippocampal astrocytes just as those seen in the rats injected with the enzyme and in AD brains, suggesting that activation of astrocytes is triggered as a response against a detrimental factor, i.e. an excess of AChE. A remarkable fact using the rat model is that these animals never form amyloid plaques under normal conditions. Neuropathological changes generated by human Aβ fibrils and AChE-Aβ complexes have been recently compared in rat hippocampus *in vivo.* Results showed that AChE-Aβ complexes trigger a more dramatic response *in situ* than Aβ fibrils alone as characterized by the following features observed 8 weeks after treatment: (1) amyloid deposits were larger than those produced in the absence of AChE. In fact, AChE strongly stimulates rat Aβ aggregation *in vitro* as shown by turbidity measurements, Congo red binding as well as electron microscopy, suggesting that AChE-Aβ deposits observed *in vivo* probably recruited endogenous Aβ peptide; (2) the appearance of laminin expressing neurons surrounding AChE-Aβ deposits. Such deposits are resistant to disaggregation by laminin *in vitro*; (3) an extensive astrocytosis revealed by both GFAP immunoreactivity and number counting of reactive hypertrophic astrocytes and; (4) a stronger neuronal cell loss in comparison with Aβ injected animals (Inestrosa and Reyes, 1998; Reyes *et al.*, 2004). We conclude that the hippocampal injection of AChE-Aβ complexes results in the appearance of some features reminiscent of Alzheimer-like lesions in rat brain. Our studies are consistent with the notion that AChE-Aβ complexes are more toxic than Aβ fibrils and that AChE triggered some of the neurodegenerative changes observed in AD brains (Reyes *et al.*, 2004).

Recent studies support our initial observation that AChE accelerates Aβ deposition (Inestrosa *et al.*, 1998). In fact, a double transgenic mice over expressing both the human APP containing the Swedish mutation and the human AChE has been developed, such double transgenic starts to form amyloid plaques around 3 months earlier than mice expressing only the APP transgene, moreover the double AChE-APP transgenic presents more plaques and bigger ones than control animals, as well as some behavioural deterioration, as shown by a working memory test (Rees *et al.*, 2003).

In conclusion, the above data indicate that AChE, being an enzyme associated with amyloid deposits in the brain of AD patients, may indeed play a critical role in contributing to the toxic effects that may trigger neurodegeneration in AD brain.

4. THE Wnt SIGNALLING PATHWAY RESCUES NEURONS FROM Aβ-INDUCED DAMAGE

The neurotoxicity produced by Aβ and induced by AChE-Aβ complexes can be prevented by the activation of a cellular process, which has been described as the Wnt signalling pathway (De Ferrari and Inestrosa, 2000; Inestrosa *et al.*, 2002). The Wnt ligand binds to a receptor of the Frizzled family and transduces its signal to the intracellular inactivating glycogen synthase kinase-3β (GSK-3β). As a consequence of GSK-3β inactivation, intracellular levels of β-catenin increase, allowing its binding to components of the high mobility group family of transcription factors (T-cell factor/lymphoid enhancer-binding factor; Tcf/LEF), which regulate the expression of multiple target genes. Conversely, in the absence of Wnt ligand (or in this case, the prescence of Aβ), the GSK-3β activity is switched on phosphorylating β-catenin for ubiquitin-proteasome-mediated degradation. As a net result, β-catenin levels are diminished within the cytosol and therefore the expression of Wnt target genes is repressed. Previous studies have suggested a relationship between Aβ and AChE-Aβ complexes induced neurotoxicity and lower cytoplasmatic levels of β-catenin. On the other hand, inhibition of GSK-3β by lithium protects rat neurons from Aβ-induced damage (De Ferrari *et al.*, 2003). These evidence led to the idea that a sustained loss of Wnt signaling function may be involved in the Aβ dependent neurodegeneration as observed in AD (Inestrosa *et al.*, 2004b; Alvarez *et al.*, 2004).

5. CONCLUDING REMARKS

The above data indicate that AChE is able to accelerate amyloid formation, and that such effect is sensitive to drugs that block the peripheral site of the enzyme; these studies suggest that specific and new AChE inhibitors may well provide an attractive therapeutic possibility for treating AD. *In vivo* studies confirm that the neurotoxicity and the behavioral impairments generated by the AChE-Aβ complexes is dependent on the enzyme concentration in such complexes and that this effect can be prevented by the activation of the Wnt signalling pathway.

ACKNOWLEDGEMENTS

We would like to thank to Dr. A. Shafferman (Israel Institute for Biological Research, Ness-Ziona, Israel) for the mutant AChE used to study the role of the C-terminus of AChE on the Aβ aggregation process.

This work was supported by grants from FONDAP - Biomedicine N° 13980001, Millennium Institute for Fundamental and Applied and Biology (MIFAB) and a predoctoral fellowship from Conicyt to MC.

REFERENCES

Alarcón, R., 1999, Studies of the association and enzymatic behaviour of the Acetylcholinesterase present in the amyloid β-peptide fibers. M.Sc. Thesis, Dept. Biochemistry, University of Chile.

Alvarez, A., Opazo, C., Alarcón, R., Garrido, J., and Inestrosa, N.C., 1997, Acetylcholinesterase promotes the aggregation of amyloid-β-peptide fragments by forming a complex with the growing fibrils. *J. Mol. Biol.* **272:** 348-361.

Alvarez, A., Alarcón, R., Opazo, C., Campos, E.O., Muñoz, F.J, Calderón F.H., Dajas, F., Gentry, M.K., Bhupendra, P. Doctor, De Mello, F.G., and Inestrosa, N.C., 1998, Stable complexes involving acetylcholinesterase and amyloid-β peptide change the biochemical properties of the enzyme and increase the neurotoxicity of Alzheimer's fibrils. *J. Neurosci.* **18:** 3213-3223.

Alvarez, A., Godoy, J.A., Mullendorff, K., Olivares, G.H., Bronfman, M., and Inestrosa, N.C., 2004, Wnt-3a overcomes β-amyloid toxicity in rat hippocampal neurons. *Exp. Cell Res.* (In press).

Atack, J.R., Perry, E.K., Bonham, J.R., Perry, R.H., Tomlinson, B.E., Candy, J., Blessed, G. ,and Fairbairn, A., 1983, Molecular forms of acetylcholinesterase in senile dementia of Alzheimer type: selective loss of the intermediate (10S) form. *Neurosci. Lett.* **40:** 199-204.

Beeri, R., Andres, C., Lev-Lehman, E., Timberg, R., Huberman, T., Shani, M., and Soreq, H., 1995, Transgenic expression of human acetylcholinesterase induces progressive cognitive deterioration in mice. *Curr. Biol.* **9:** 1063-1071.

Bigbee, J.W., Sharma, K.V., Chan, E.L., and Bogler, O., 2000, Evidence for the direct role of acetylcholinesterase in neurite outgrowth in primary dorsal root ganglion neurons. *Brain Res.* **861:** 354-362.

Bronfman, F.C., Fernández, H.L., and Inestrosa, N.C., 1996, Amyloid precursor protein fragment and acetylcholinesterase increase with cell confluence and differentiation in a neuronal cell line. *Exp. Cell Res.* **229:** 93-99.

Calderón, F.H., von Bernhardi, R., De Ferrari, G., Luza, S., Aldunate, R., and Inestrosa, N.C., 1998, Toxic effects of acetylcholinesterase on neuronal and glial-like cells in vitro. *Mol. Psychiatry* **3:** 247-255.

Chacón, M.A., Reyes, A.E., and Inestrosa, N.C., 2003, Acetylcholinesterase induces neuronal cell loss, astrocyte hypertrophy and behavioral deficits in mammalian hippocampus. *J. Neurochem.* **87:** 195-204.

Cohen, O., Erb, C., Ginzberg, D., Pollak, Y., Seidman, S., Shoham, S., Yirmiya, R., and Soreq, H., 2002, Neuronal overexpression of readthrough acetylcholinesterase is associated with antisense-supressible behavioral impairments. *Mol. Psychiatry* **7:** 874-885.

Cohen, O., Kronman, C., Chitlaru, T., Ordentlich, A., Velan, B., and Shafferman A., 2001, Effect of chemical modification of recombinant human acetylcholinesterase by polyethylene glycol on its circulatory longevity. *Biochem. J.* **357:** 795-802.

Colombres, M., Sagal, J.P., and Inestrosa, N.C., 2004, An Overview of the Current and Novel Drugs for Alzheimer's Disease with particular reference to anti-cholinesterase compounds. *Curr. Pharmacol. Design*, In Press.

Cottingham, M.G., Hollinshead, M.S., and Vaux, D.J., 2002, Amyloid fibril formation by a synthetic peptide from a region of human acetylcholinesterase that is homologous to the Alzheimer's amyloid-β peptide. *Biochemistry* **41:** 13539-13547.

Day, T., and Greenfield, S.A., 2003, A peptide derived from acetylcholinesterase induces neuronal cell death: characterisation of possible mechanisms. *Exp. Brain Res.* **153:** 334-342.

De Ferrari, G.V., and Inestrosa, N.C., 2000, Wnt signaling function in Alzheimer's disease. *Brain Res. Rev.***33:** 1-12.

De Ferrari, G.V., Canales, M.A., Shin, I., Weiner, L.M., Silman, I., and Inestrosa, N.C., 2001, A structural motif of acetylcholinesterase that promotes amyloid β-peptide fibril formation. *Biochemistry* **40:** 10447-10457.

De Ferrari, G.V., Chacón, M.A., Barría, M.I., Garrido, J.L., Godoy, J.A., Olivares, G., Reyes, A.E., Alvarez, A., Bronfman, M., and Inestrosa, N.C., 2003, Activation of Wnt signaling rescues neurodegeneration and behavioral impairments induced by β-amyloid fibrils. *Mol. Psychiatry* **8:** 195-208.

Fishman, E.B., Siek, G.C., MacCallum, R.D., Bird, E.D., Volicer, L., and Marquis, J.K., 1986, Distribution of the molecular forms of acetylcholinesterase in human brain: alterations in dementia of the Alzheimer type. *Ann. Neurol.* **19:** 246-252.

Geula, C., and Mesulam, M.M., 1989, Special properties of cholinesterase in the cerebral cortex of Alzheimer's disease. *Brain Res.* **498:** 185-189.

Geula, C., and Mesulam, M.M., 1994, Cholinergic systems and related neuropathological predilection patterns in Alzheimer disease. In: *Alzheimer Disease* (Terry, R.D., Katzman, R. and Bick, K.L., eds.), pp. 263-291, Raven Press, New York.

Greenfield, S., and Vaux, D.J., 2002, Parkinson's disease, Alzheimer's disease and motor neurone disease: identifying a common mechanism. *Neurosci.* **113:** 485-492.

Grisaru, D., Sternfeld, M., Eldor, A., Glick, D., and Soreq, H., 1999, Structural roles of acetylcholinesterase variants in biology and pathology. *Eur. J. Biochem.* **264:** 672-686.

Hyman, B.T., Van Horsen, G.W., Damasio, A.R., and Barnes, C.L., 1984, Alzheimer's disease: cell-specific pathology isolates the hippocampal formation. *Science* **225:** 1168-1170.

Inestrosa, N.C., Roberts, W.L., Marshall, T.D., and Rosenberry, T.L., 1987), Acetylcholinesterase from bovine caudate nucleus is attached to membranes by a novel subunit distinct from those of acetylcholinesterases in other tissues. *J. Biol. Chem.* **262:** 4441-4444.

Inestrosa, N.C., and Perelman, A., 1989, Distribution and anchoring of the molecular forms of acetylcholinesterase. *Trends Pharmacol. Sci.* **10:** 325-329.

Inestrosa, N.C., Alvarez, A., Perez, C.A., Moreno, R.D., Vicente, M., Linker, C., Casanueva, O.I., Soto, C., and Garrido, J., 1996, Acetylcholinesterase accelerates assembly of amyloid-β-peptides into Alzheimer's fibrils: possible role of the peripheral site of the enzyme. *Neuron* **16:** 881-891.

Inestrosa, N.C., Alvarez, A., Garrido, J., Calderón, F., Bronfman, F.C., Dajas, F., Gentry, M.K., and Doctor, B.P., 1997, Acetylcholinesterase promotes Alzheimer β-amyloid Fibril Formation. In: *Alzheimer's Disease: Biology, Diagnosis and Therapeutics,*

(K. Iqbal, B. Winblad, T. Nishimura, M. Takeda, H.M. Wisniewski, Eds.), pp. 499-508, J. Wiley & Sons Ltd., London, U.K.

Inestrosa, N.C., and Alarcón, R., 1998, Molecular interactions of acetylcholinesterase with senile plaques. *J. Physiol. (Paris)* **92:** 341-344.

Inestrosa, N.C., and Reyes, A.E., 1998, Acetylcholinesterase induces amyloid formation and increases neurotoxicity of Alzheimer's fibrils. *Neurobiol Aging* **19:** S44 (Abst).

Inestrosa, N.C., De Ferrari, G.V., Garrido, J.L., Alvarez, A., Olivares, G.H., Barría, M.I., Bronfman, M., and Chacón, M.A., 2002, Wnt signaling involvement in β-amyloid-dependent neurodegeneration. *Neurochem. Int.* **41:** 341-344.

Inestrosa, N.C., De Ferrari G.V., Opazo, C., and Alvarez A., 2004[a], Neurodegenerative processes in Alzheimer's disease: role of Aβ-AChE complexes and Wnt signaling. In: *XIth International Symposium on Cholinergic Mechanisms*, (Silman, I., Soreq, H. Fisher, A., and Anglister L, eds.), St. Moritz, Switzerland, Ch. 51 (In Press).

Inestrosa, N.C., Urra, M.S., and Colombres, C., 2004b, Acetylcholinesterase (AChE) - amyloid-β-peptide complexes in Alzheimer's Disease. The Wnt signaling pathway connection. *Curr. Alzheimer Res.* (In Press).

Jarrett, J.T., and Lansbury, P.T. Jr., 1992, Amyloid fibril formation requires a chemically discriminating nucleation event: studies of an amyloidogenic sequence from the bacterial protein OsmB. *Biochemistry* **31:** 12345-12352.

Johnson, G., and Moore, S.W., 1999, The adhesion function on acetylcholinesterase is located at the peripheral anionic site. *Biochem. Biophys. Res. Commun.* **258:** 758-762.

Kalaria, R.N., Kroon, S.N., Grahovac, I., and Perry, G., 1992, Acetylcholinesterase and its association with heparan sulphate proteoglycans in cortical amyloid deposits of Alzheimer's disease. *Neurosci.* **51:** 177-184.

Kasa, P., Rakonczay, Z., and Gulya, K., 1997, The cholinergic system in Alzheimer's disease. *Prog. Neurobiol.* **52:** 511-535.

Kronman, C., Velan, B., Gozes, Y., Leitner, M., Flashner, Y., Lazar, A., Marcus, D., Sery, T., Papier, Y., Grosfeld, H., Cohen, S., and Shafferman, A., 1992, Production and secretion of high levels of recombinant human acetylcholinesterase in cultured cell lines: microheterogeneity of the catalytic subunit. *Gene* **121:** 295-304.

Layer, P.G., Weikert, T., and Alber, R., 1993, Cholinesterases regulate neurite growth of chick nerve cells in vitro by means of a non-enzymatic mechanism. *Cell Tissue Res.* **273:** 219-226.

Massoulie, J., Pezzementi, L., Bon, S, Krejci, E., and Vallette, F.M., 1993, Molecular and cellular biology of cholinesterases. *Prog. Neurobiol.* **41:** 31-91.

Masters, C.L., Simms, G., Weinman, N.A., Multhaup, G., McDonald, B.L., and Beyreuther, K., 1985, Amyloid plaque protein in Alzheimer disease and Down's syndrome. *Proc. Natl. Acad. Sci. USA* **82:** 4245-4249.

Muñoz, F.J., and Inestrosa, N.C.,1999, Neurotoxicity of acetylcholinesterase-amyloid β-peptide aggregates is dependent on the type of Aβ-peptide and the AChE concentration present in the complexes. *FEBS Lett.* **450**: 205-209.

Muñoz, F.J., Aldunate, R., and Inestrosa, N.C., 1999, Peripheral binding site is involved in the neurotrophic activity of acetylcholinesterase. *NeuroReport* **10:** 3621-3625.

Muñoz, F.J., Opazo, C., Gil-Gómez, G., Tapia, G., Fernández, V., Valverde, M.A., and Inestrosa, N.C., 2002, Vitamin E but not 17β-Estradiol protects against vascular toxicity induced by β-amyloid wild-type and the Dutch amyloid variant. *J. Neurosci.* **22:** 3081-3089.

Opazo, C., and Inestrosa, N.C., 1998, Crosslinking of amyloid-β peptide to brain acetylcholinesterase. *Mol. Chem. Neuropathol.* **33:** 39-49.

Rees, T., Hammond, P.I., Soreq, H., Younkin, S., and Brimijoin, S., 2003, Acetylcholinesterase promotes β-amyloid plaques in cerebral cortex. *Neurobiol. Aging* **24:** 777-787.

Reyes, A.E., Pérez, D.R., Alvarez, A., Garrido, J., Gentry, M.K., Doctor, B.P., and Inestrosa, N.C., 1997, A monoclonal antibody against acetylcholinesterase inhibits the formation of amyloid fibrils induced by the enzyme. *Biochem. Biophys. Res. Commun.* **232:** 652-655.

Reyes, A.E. Chacón, M.A., Dinamarca, M.C., Cerpa, W., Morgan, C., and Inestrosa, N.C., 2004, Acetylcholinesterase-Aβ complexes are more toxic than Aβ fibrils in rat hippocampus: Effect on rat β-amyloid aggregation, laminin expression, reactive astrocytosis and neuronal cell loss. *Am. J. Pathol.* (In Press).

Roher, A. E., Wolfe, D., Palutke, M., and Kukurga, D., 1986, Purification, ultrastructure, and chemical analysis of Alzheimer disease amyloid plaque core protein. *Proc. Natl. Acad. Sci. USA* **83:** 2662-2666.

Selkoe, D.J., 2001, Alzheimer's disease: genes, proteins, and therapy. *Physiol. Rev.* **8:** 741-766.

Shin, I., Silman, I., and Weiner, L.M., 1996, Interaction of partially unfolded forms of Torpedo acetylcholinesterase with liposomes. *Protein Sci.* **5:** 42-51.

Soto, C., Brañes, M.C., Alvarez, J., and Inestrosa, N.C., 1994, Structural determinants of the Alzheimer's amyloid β-peptide. *J. Neurochem.* **63:** 1191-1198.

Soto, C., Castaño, E., Frangione, B., and Inestrosa, N.C., 1995, The α-helical to β-strand transition in the amino-terminal fragment of the amyloid β-peptide modulates amyloid formation. *J. Biol. Chem.* **270:** 3063-3067.

Talesa, V.N., 2001, Acetylcholinesterase in Alzheimer's disease. *Mech. Ageing Devel.* **122:** 1961-1969.

Tumiatti, V., Rosini, M., Bartolini, M., Cavalli, A., Marucci, G., Andrisano, V., Angeli, P., Banzi, R., Minarini, A., Recanatini, M., and Melchiorre, C., 2003, Structure-Activity Relationships of Acetylcholinesterase Noncovalent Inhibitors Based on a Polyamine Backbone. 2. Role of the Substituents on the Phenyl Ring and Nitrogen Atoms of Caproctamine. *J. Med. Chem.* 46: 954-966.

Ulrich, J., Meier-Ruge, W., Probst, A., Meier, E., and Ipsen, S., 1990, Senile plaques: staining for acetylcholinesterase and A4 protein. A comparative study in the hippocampus and entorhinal cortex. *Acta Neuropathol.* **80:** 624-628.

Younkin, S.G., Goodridge, B., Katz, J., Lockett, G., Nafziger, D., Usiak, M.F., and Younkin, L.H., 1986, Molecular forms of acetylcholinesterase in Alzheimer's disease. *Fed. Proc.* **45:** 2982-2988.

Chapter 16

Membrane Disordering Effects of β-Amyloid Peptides

[1]Gunter P. Eckert, [2]W.G. Wood and [1]W.E. Müller

[1]Department of Pharmacology, ZAFES, Biocenter University of Frankfurt, Germany. [2]Department of Pharmacology, University of Minnesota School of Medicine, Minneapolis, USA

Abstract: The interaction of Aβ with synaptosomal plasma membranes decreases membrane fluidity. Using model membrane/liposome systems the interaction of Aβ with specific lipids (*e.g.* phospholipids, gangliosides, cholesterol) has been defined. The formation of the β-sheet structure of Aβ when undergoing peptide aggregation is important for Aβ's membrane perturbing properties. This effect can be correlated with the peptide length of Aβ, the longer $A\beta_{1-42}$ having the greatest effect on membrane fluidity and on neurotoxicity.

Key words: β-amyloid (Aβ), membrane fluidity, brain membranes, cholesterol, phospholipid, ganglioside

1. INTRODUCTION

Beta-amyloid (Aβ) protein is the major constituent of senile plaques and cerebrovascular deposits characteristic of Alzheimer's disease (AD). The causal relationship between β-amyloid deposition and AD-specific neuropathological lesions, like neurodegeneration and cortical atrophy is still not known. Observations that Aβ has neurotoxic properties (Pike *et al.*,1991; Yankner *et al.*,1990) represented the first link between Aβ formation and the

AD specific neuropathological lesions. Subsequent studies have confirmed the neurotoxic effect of Aβ (Kawahara and Kuroda, 2000; Small and McLean, 1999) and have further suggested that its mechanism is probably related to the amplifying effect of Aβ on cellular calcium signaling and/or the induction of oxidative stress (Butterfield *et al.*, 1999; Hartmann *et al.*, 1993; Kawahara and Kuroda, 2000; Keller *et al.*, 1997; Mattson and Horitiri,1993; Small and McLean, 1999). The molecular mechanism of the Ca^{2+} amplifying and the oxidative stress inducing effects are not known. The possibility that Aβ alters membrane properties was originally supported by preliminary findings indicating that Aβ may disrupt membranes (McLaurin and Chakrabartty, 1996) and directly decrease the fluidity of artificial unilamellar liposomes (Chauhan *et al.*, 1993) or of mouse brain membranes (Mattson *et al.*, 1993).

Membrane fluidity is a general term used to describe the movement of lipids in membranes. Actually, fluidity consists of different components including, for example, rate of probe motion, a dynamic component, and extent of probe motion, a static component (Lakowicz, 1983; Yeagle,1989). A variety of physical techniques have been utilized to assess membrane fluidity. These include among others different fluorescence techniques. When added to cellular or even artificial membranes, the fluorescent dye DPH (1,6-diphenylhexa-1,2,3-triene) is localized preferentially in the hydrocarbon core, while its cationic trimethylamino derivative TMA-DPH remains at the level of the polar heads of the phospholipid bilayer (Figure 1). DPH intercalates preferentially axial between the acyl chains of membrane fatty acids (Kaiser and London,1998). Accordingly, its mobility predominantly depends on acyl chain flexibility that is restricted by the physico-chemical environment of the membrane. Pyrene diffuses laterally in the hydrocarbon core of the membrane and mainly labels the interphase between both membrane leaflets (Figure 1). Excimer formation of the fluorescent probe pyrene represents predominantly the lateral motion of the probe within the lipid bilayer (bulk fluidity), while annular fluidity as measured by energy transfer from tryptophan residues to pyrene gives information about the fluidity of lipids close to membrane proteins(Igbavboa *et al.*, 1996). Herein we use the term fluidity in the broadest sense to describe an average lateral motion in the membrane lipid environment without making a distinction between dynamic and static sates of fluorescent probe motion.

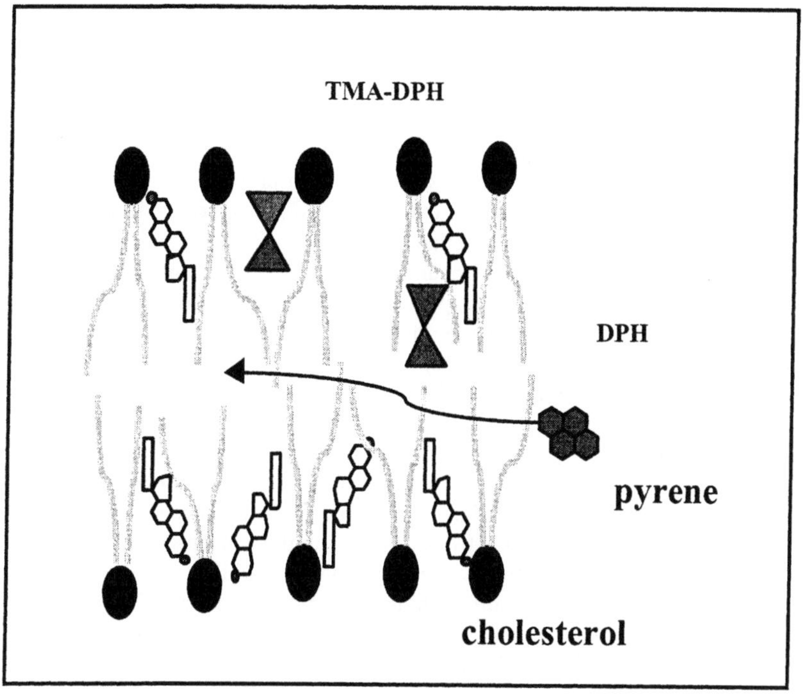

Figure 1. Location of fluorescence probes within the membrane. The physico-chemical properties of phospholipid bilayers can be examined using fluorescence dyes. Diphenylhexatriene (DPH) intercalates axial between the acyl-chains of phospholipids and labels the hydrocarbon core of the membrane. Its trimethylammonium derivative TMA-DPH localizes in the hydrophilic region of the phospholipids. The lipophilic pyrene molecule labels predominantly the hydrophobic area of the membranes and thus its motion is a measure of membrane bulk fluidity.

2. β-AMYLOID PEPTIDES ALTER MEMBRANE FLUIDITY IN MOUSE AND RAT BRAIN MEMBRANES

Several studies have examined the effects of Aβ peptides on membrane fluidity in model and biological membranes of mice, rats and humans (Table 1). Effects of Aβ on membrane fluidity have been proposed as contributing to disruption in different cell functions. $A\beta_{1-40}$ was reported to increase polarization of DPH in small unilamillar vesicles (SUV) that indicates a more ordered membrane state (Chauhan *et al.*, 1993). This observation was confirmed for $A\beta_{1-40}$ and $A\beta_{1-39}$ peptides using large unilamillar vesicles (Kremer *et al.*, 2000). This study ruled out that self-aggregation of Aβ monomers into aggregates (1-20 μM) exposes hydrophobic sites and induces a decrease in membrane fluidity.

In one of the first studies on Aβ and membrane fluidity using biological tissue, the fragment $A\beta_{25-35}$ which represents Aβ's neurotoxic sequence (Pike *et al.*, 1991; Yankner *et al.*,1990) increased anisotropy of DPH in mouse and rat brain membranes as well as in lymphocyte membranes (Hartmann *et al.*, 1994; Müller *et al.*,1995). $A\beta_{25-35}$ or $A\beta_{1-40}$ in the concentration range of 1-10 μM significantly reduced the acyl-chain flexibility of the tested membranes. A control peptide with the inverse sequence ($A\beta_{35-25}$) was without any effect (Müller *et al.*, 1995). In membranes of young animals, the effect was nearly maximal at 1 μmol/l.

Increasing $A\beta_{25-35}$ concentration tenfold had only little further effect on anisotropy. By contrast, anisotropy of aged mouse brain membranes was less affected by $A\beta_{25-35}$ concentrations up to 1 μmol/l, but showed some changes at a peptide concentration of 10 μmol/l (Müller *et al.*, 1995). Several subsequent studies including brain tissue from humans with and without AD have shown that different Aβ peptides reduce the membrane fluidity (Eckert *et al.*, 2000; Eckert *et al.*, 2001; Kirsch *et al.*, 2002; Müller *et al.*, 1998; Yip *et al.*, 2001). Several neurotoxic Aβ peptides decrease the acyl-chain flexibility of human cortex membranes in a concentration dependent fashion while the non-neurotoxic $A\beta_{1-28}$ was only weakly active (Müller *et al.*, 1998; Pike *et al.*, 1991). The effect of Aβ on membrane fluidity increases with peptide length, is most pronounced for $A\beta_{1-43}$ and can be detected at concentrations as low as 100 nM. Although this concentration is still above the physiological Aβ level in human brain (Pirttilä *et al.*, 1996), it is substantially lower than the *in vitro* concentrations usually employed to demonstrate Aβ's neurotoxic properties (Pike *et al.*, 1991; Yankner *et al.*, 1990).

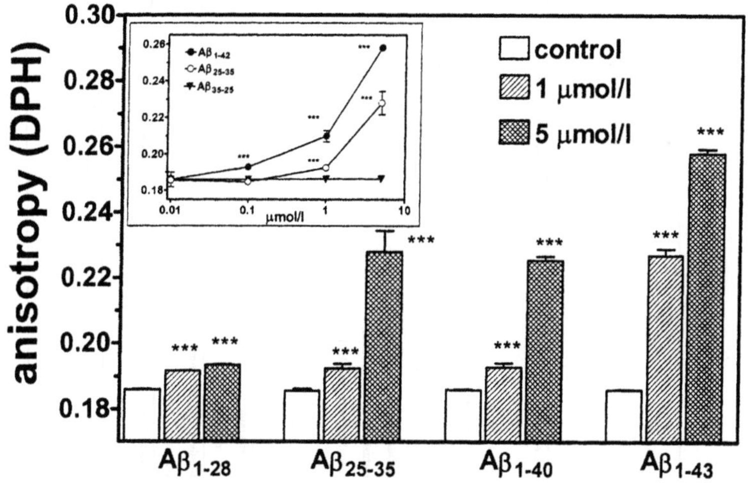

Figure 2. Effect of four Aβ peptides on the anisotropy (DPH) of human frontal cortex membranes. All data are means ± S.D. of 6 experiments each representing an individual brain sample (according to (13). (***p<0,001; n.s. not significant, when compared with the anisotropy without Aβ (paired t-test)). Inset: The effect of increasing concentrations of $A\beta_{25-35}$, $A\beta_{1-42}$ peptides on the anisotropy (DPH) of human frontal cortex membranes. $A\beta_{35-25}$ was used as control. All data are means ± S.D. of 6 experiments each representing an individual brain sample (according to Müller *et al.*, 1995). (***p<0,001, +++p<0,001, ++p<0,01 when compared with the anisotropy without Aβ (paired t-test)).

The effect of Aβ peptides on fluidity of the phospholipid bilayer is more pronounced in the hydrocarbon core than in the region of the hydrophilic heads. For this investigations DPH-anisotropy and TMA-DPH anisotropy were used as markers for the membrane hydrocarbon core and the hydrophilic region of the membrane phospholipid head groups, respectively (Müller *et al.*, 1998).The pathologically relevant fragments $A\beta_{1-42}$ and $A\beta_{1-43}$ affected the membrane in the same manner, but were much more active than the neurotoxic sequence $A\beta_{25-35}$, confirming data with mouse brain membranes (Müller *et al.*, 1995). Diverse disturbing effects of Aβ on different membrane areas were confirmed subsequently using human hippocampal membranes (Eckert *et al.*, 2000). Membranes isolated from brains of AD patients showed reduced baseline fluidity measured as enhanced DPH-anisotropy values compared to controls (Eckert *et al.*, 2000).

Table 1: Effects of Aβ on neuronal membranes

Membrane	Aβ (concentration)	Fluidity	Reference
Mouse brain homogenate	$A\beta_{25-35}$ (1-10μM)	↓ (DPH)	(Hartmann et al.,1994)
Mouse brain homogenate	$A\beta_{25-35}$ (1-10μM)	↓ (DPH)	(Müller et al., 1995)
	$A\beta_{1-40}$ (1-10μM)	↓↓ (DPH)	
	$A\beta_{35-25}$ (10 μM)	↔(DPH)	
mouse brain homogenate 3 month	Aβ 25-35 (5μM)	↓ (DPH); ↓ (TMA-DPH); ↔ (pyrene)	(Kirsch et al., 2002)
Mouse brain homogenate 12 month	$A\beta_{25-35}$ (5μM)	↓ (DPH); ↓ (TMA-DPH); ↔ (pyrene)	
Mouse brain homogenate 22 month	$A\beta_{25-35}$ (5μM)	↓ (DPH); ↓ (TMA-DPH); ↔ (pyrene)	
Mouse SPM[1], 3 month	$A\beta_{35-25}$ (5μM)	↔ (DPH); ↔ (pyrene/annular); ↔ (pyrene/bulk)	(Eckert et al., 2001)
	$A\beta_{25-35}$ (5μM)	↓ (DPH); ↑ (pyrene/annular); ↔ (pyrene/bulk)	
	$A\beta_{1-40}$ (5μM)	↓↓ (DPH); ↑ (pyrene/annular); ↔ (pyrene/bulk)	
	$A\beta_{1-42}$ (5μM)	↓↓↓ (DPH); ↑ (pyrene/annular); ↔ (pyrene/bulk)	
Mouse SPM, 22 month	$A\beta_{35-25}$ (5μM)	↔ (DPH); ↔ (pyrene/annular); ↔ (pyrene/bulk	
	$A\beta_{25-35}$ (5μM)	↓ (DPH); ↑ (pyrene/annular); ↔ (pyrene/bulk)	
	$A\beta_{1-40}$ (5μM)	↓ (DPH); ↑ (pyrene/annular); ↔ (pyrene/bulk)	
	$A\beta_{1-42}$ (5μM)	↓↓ (DPH); ↑ (pyrene/annular); ↔ (pyrene/bulk)	
Rat SPM	$A\beta_{1-40}$ (1μM)	↑ (pyrene/annular); ↑ (pyrene/bulk)	(Mason et al., 1999)
	$A\beta_{1-42}$ (1μM)	↑ (pyrene/annular); ↑ (pyrene/bulk)	
Rat SPM	$A\beta_{1-40}$ (1μM)	↔ (DPH); ↑ (pyrene/annular); ↑ (pyrene/bulk)	(Avdulov et al., 1997a)
	$A\beta_{25-35}$ (1μM)	↔ (DPH); ↑ (pyrene/annular); ↑ (pyrene/bulk)	
Rat SPM cerebrum	$A\beta_{1-40}$ (1 μM)	↑ (pyrene/annular); ↑ (pyrene/bulk)	(Chochina et al., 2001)
SPM cerebellum	$A\beta_{1-40}$ (1 μM)	↔ (pyrene/annular); ↔ (pyrene/bulk)	
Human brain PM	$A\beta_{1-40}$ (5-10)	↓ (DPH)	(Waschuk et al., 2001)
	$A\beta_{1-42}$ (5-10)	↓ (DPH)	
Human brain Golgi	$A\beta_{1-42}$ (5-10)	↑ (DPH)	

Human brain endosomes	$A\beta_{1\text{-}42}$ (5-10)	↑ (DPH)	
	$A\beta_{1\text{-}40}$ (5-10)	↓ (DPH)	
	$A\beta_{1\text{-}42}$ (5-10)	↓ (DPH)	
Human hippocampus, control	$A\beta_{25\text{-}35}$ (1-5μM)	↓ (DPH); ↓ (TMA-DPH)	(Eckert et al., 2000)
Human hippocampus, Alzheimer	$A\beta_{1\text{-}42}$ (1-5μM)	↓↓ (DPH); ↓↓ (TMA-DPH)	
	$A\beta_{25\text{-}35}$ (1-5μM)	↓ (DPH); ↓ (TMA-DPH)	
	$A\beta_{1\text{-}42}$ (1-5μM)	↓↓ (DPH); ↓↓ (TMA-DPH)	
Human lymphocytes	$A\beta_{25\text{-}35}$ (1-10μM)	↓ (DPH)	(Hartmann et al., 1994)
PC 12 membranes	$A\beta_{25\text{-}35}$ (0,01-0,1μM)	↓ (DPH)	(Eckert et al., 2003b)
Brain lipid extract	$A\beta_{1\text{-}40}$ (23 μM)	↓ (DPH)	(Yip et al., 2001)
Artificial liposomes	$A\beta_{1\text{-}39}$ (5-20μM)	↓ (DPH)	(Kremer et al., 2000)
	$A\beta_{1\text{-}40}$ (5-20μM)	↓ (DPH)	

1 SPM = synaptosomal plasma membrane

Many studies demonstrate membrane-perturbing effects of Aβ on neuronal membranes (SPM, synaptosomal plasma membranes). Increased (↑) or decreased (↓) membrane fluidity was assessed using different fluorescent dyes (Pyrene; DPH, diphenylhexatriene; TMA-DPH, Trimethylamino-diphenylhexatriene).

Incubation of hippocampal membranes with $A\beta_{1-42}$ and $A\beta_{25-35}$ lead to a concentration dependent decrease of the acyl-chain flexibility. The disordering effect was much more pronounced for the physiological peptide $A\beta_{1-42}$ compared with the synthetic fragment $A\beta_{25-35}$. The amyloid peptides affected membranes from AD patients and controls in the same fashion and to about the same extent. While the observed differences of the acyl-chain flexibility between control and AD membranes were still apparent in the presence of even high concentrations of $A\beta_{25-35}$, treatment of hippocampal membranes with high concentrations of $A\beta_{1-42}$ abolished the differences of DPH anisotropy between control and AD tissues (Eckert *et al.*, 2000).

The above-mentioned studies used a rather crude brain membrane fraction. Investigations of Aβ effects using purified synaptosomal/synapatic (SPM) as well as mitochondrial membranes isolated from brains of young and aged mice confirmed that Aβ decreases the flexibility of membrane acyl-chains (Eckert *et al.*, 2001). From this data it appears that Aβ has a rigidifying effect on membranes. However, it has been reported that Aβ disorders or increases fluidity of membranes (Avdulov *et al.*, 1997a; Chochina *et al.*, 2001; Eckert *et al.*, 2001; Mason *et al.*, 1999). $A\beta_{1-40}$ increase both annular fluidity and bulk fluidity in SPM using energy transfer from protein tryptophan residues and excitation of pyrene (annular fluidity) and pyrene excitation alone (bulk fluidity) (Avdulov *et al.*, 1997a; Chochina *et al.*, 2001; Mason *et al.*, 1999). Aβ peptide also increased the lateral mobility of pyrene in brain mitochondria membranes isolated from young and aged mice. Infact $A\beta_{1-42}$ caused the strongest effects on the bulk-fluidity in mitochondrial membranes (Eckert *et al.*, 2001). So. there is no agreement on the effects of Aβ proteins on membrane fluidity. Acyl-chain flexibility was increased in human brain Golgi membranes in the presence of $A\beta_{1-40,42}$ but just the opposite effect was noted in endosomal, lysosomal and mitochondrial membranes from the same human samples (Waschuk *et al.*, 2001). Certainly, differences in effects of Aβ on fluidity could result from variations in tissue source and preparation, and whether Aβ is soluble or aggregated. However, it is most likely that the differences in effects of Aβ on fluidity are largely the result of differences in the location of the fluorescence probes in the membrane environment and the lifetime of the fluorescence probes. The majority of studies examining Aβ and fluidity have used steady-state fluorescence of DPH and TMA-DPH as well as pyrene fluorescence. Generally, studies reporting a reduction in fluidity induced by Aβ have measured polarization or anisotropy of DPH or TMA-DPH. In contrast, studies finding that Aβ increases membrane fluidity have measured energy transfer and excitation of pyrene. Structurally, DPH and pyrene differ and this difference can influence their behaviour in membranes (Kaiser *et al.*, 1998; Mulders *et al.*, 1986; Wood *et al.*, 2003). DPH is a rod like structure whose axis is parallel to the acyl groups and pyrene is spherical in structure and is positioned at the terminal end of the acyl groups. As the lifetime of a fluorescent probe establishes the duration of time for which the probe interacts in its environment such as membranes, the longer lifetime of pyrene may increase interaction with different areas of the membrane (Lakowicz *et al.*, 1983). The most parsimonious explanation for the reported differences in Aβ effects using fluorescence of DPH or TMA-DPH and pyrene is that the fluorescent probes are reporting on behavior of Aβ in different membrane environments as a function of the location and of the lifetime of the fluorescence probes.

3. Aβ INTERACTIONS WITH MEMBRANES

Localization of $A\beta_{1-42}$ on the cell surface plasma membranes of neurones from the brains of patients with AD was revealed using electron microscopy (Yamaguchi *et al.*, 2000). The authors concluded that $A\beta_{1-42}$ deposition on the cell surface plasma membrane was an initial event in formation of diffuse plaques that gradually develop into fibril amyloid. Similar findings were recently reported in aged dogs, which are an accepted model of AD, since dogs exhibit age-dependent cognitive decline that is correlated with the accumulation of Aβ (Torp *et al.*, 2000). Neuronal labelling indicated that $A\beta_{1-42}$ was associated with the neuronal plasma membrane and it was suggested that the peptide might be produced at the dendritic plasma membrane. Two pools of insoluble Aβ were identified in human prefrontal cortices (Oshima *et al.*, 2001). One pool was located in a cholesterol-enriched low density membrane domain while the second pool comprised extracellular Aβ deposits. In the same study, it was shown that mice that over express the amyloid-precursor protein (APP) and develop plaques, low-density membrane domains showed accumulation of $A\beta_{1-42}$. Other studies have also reported that APP and Aβ were associated with cholesterol-rich low-density membrane domains (Hayashi *et al.*, 2000; Oshima *et al.*, 2001). Studies *in vitro*, using model membranes and biological membranes, have also shown that Aβ associated with membranes. Electrostatic binding of Aβ to phospholipid polar head groups has been proposed to be a mechanism that may contribute to Aβ neurotoxicity (McLaurin *et al.*, 1997; Terzi *et al.*, 1997). Circular dichroism spectroscopy showed that $A\beta_{1-40}$ interacted with negatively charged unilamellar vesicles but no effect was detected in vesicles containing deuterated phosphatidylcholine mixed with phosphatidylglycerol using deuterium NMR and that $A\beta_{1-40}$ did not penetrate into membranes (Terzi *et al.*, 1997). $A\beta_{1-40}$ had a more pronounced effect on disrupting and aggregating negatively charged lipid vesicles than zwitterionic vesicles (McLaurin and Chakrabartty, 1997). Conversely, studies of Aβ peptides in both model membranes and biological membranes indicated that soluble or fresh Aβ partitioned into the hydrophobic core of membranes (Mason *et al.*, 1996; Mason *et al.*, 1999). Liposomes consisting of 1-palmitoyl-2-oleoylphosphatidylcholine revealed that the peptide fragment $A\beta_{25-35}$ localized in the membrane hydrocarbon core (Mason *et al.*, 1996). Moreover, soluble $A\beta_{1-40}$ and aggregated $A\beta_{1-40}$ were found to differ in their location in rat synaptic plasma membranes (SPM) (Mason *et al.*, 1999). Soluble $A\beta_{1-40}$ intercalated into the hydrophobic region of SPM. Aggregated $A\beta_{1-40}$ was positioned at the polar head group region of the membrane (Figure 3).

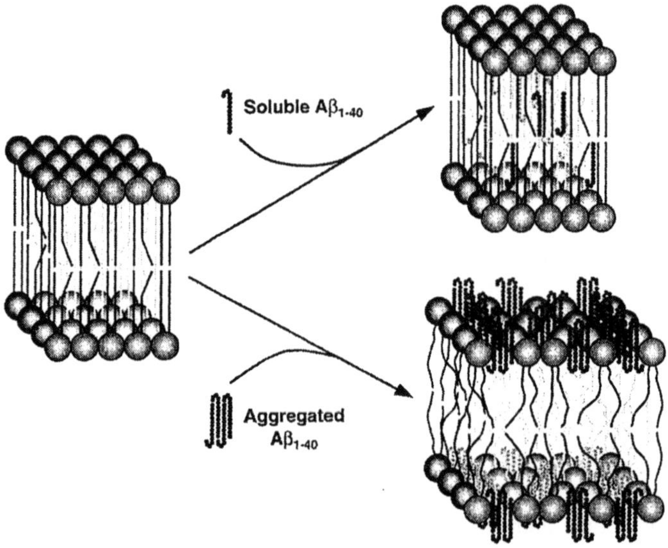

Figure 3. A schematic representation of the proposed molecular membrane interactions of soluble versus aggregated $A\beta_{1-40}$ with synaptic plasma membranes, based on the results of this study. Soluble $A\beta_{1-40}$ intercalates deep into the plasma membrane hydrocarbon core, whereas aggregated $A\beta_{1-40}$ interacts with the membrane bilayer at the head group/water interface, resulting in a marked reorganization of the lipid bilayer, including increased *trans-gauche* isomerizations. Adapted from (Mason *et al.*,1999).

Differences in results of the aforementioned studies may have occurred as a consequence of variations in peptide structure and dissimilarities in structure of liposomes versus biological membranes. Results of studies using the Aβ peptide fragment 25–35 have to be viewed cautiously in view of the fact that this fragment is not present *in vivo*. Nevertheless, it can be seen from Table 1 that Aβ does have a physico-chemical interaction with membranes. It has been previously proposed that Aβ may interact with lipids both by hydrophobic interactions and electrostatic interactions at the membrane surface, particularly with negatively charged phospholipids (McLaurin and Chakrabartty, 1997). The cytofacial lipid monolayer of biological membranes is negatively charged as a result of enrichment of phosphatidylinositol and phosphatidylserine as compared to the zwitterionic exofacial monolayer. Electrostatic interaction of Aβ with phospholipid polar head groups may be greater in the cytofacial lipid monolayer than the exofacial monolayer and this prediction is consistent with the affinity of Aβ for negatively charged lipids (McLaurin and Chakrabartty, 1997; Terzi *et al.*, 1997).

4. THE RELEVANCE OF CHOLESTEROL

Cholesterol is known to be an essential modulator of the physico-chemical state and functional activity of biological membranes and hence, plays an essential role in the regulation of synaptic function and cell plasticity. The majority of cellular cholesterol resides within the plasma membrane (PM). The cell tightly regulates levels and distribution of PM cholesterol (Palacino *et al.*, 2001; Wood *et al.*, 2002). The plasma membrane lipid bilayer consists of two monolayers that are asymmetric in lipid distribution, electrical charge, fluidity, function and possibly the localization of lipid rafts. Within the PM, 70-85% of free cholesterol reside in the cytofacial bilayer monolayer, whereas only 15-30% join the exofacial leaflet (for review, refer to Wood *et al.*, 1999). Even in this outer membrane domain, the intra-membrane distribution of cholesterol follows a strict organization into structural pools and is altered *e.g.* during aging (Igbavboa *et al.*, 1996). Exofacial cholesterol builds up lateral membrane domains like kinetic pools or lipid rafts (Wood *et al.*, 1999). Lipid rafts are specialized plasma membrane micro domains highly enriched in glycosphingolipids and multiple membrane proteins (Eckert *et al.*, 2003a). Strong evidence suggest that cholesterol condenses the packing of sphingolipid molecules and thus cholesterol-sphingolipid microdomains form a separate lipid-ordered phase in the exofacial leaflet (Brown and London, 1998). Functionally, lipid rafts are thought to be involved in intracellular trafficking of proteins and lipids, secretory and endocytic pathways as well as in cell-surface proteolysis and signal-transduction-pathways (Simons and Ikonen, 2000; Simons and Toomre, 2000). Reducing cholesterol levels appear to decreases the production of Aβ (Bodovitz *et al.*, 1996; Kojro *et al.*, 2001). On the other hand, there are data indicating that cholesterol may act to attenuate the effects of Aβ.

It has been shown that membrane cholesterol reduces the disordering effects of Aβ on mouse brain membranes *in vitro* (Mattson *et al.*, 1993) and that endogenous membrane cholesterol differently modulates Aβ effects on brain membranes from AD patients (Eckert *et al.*, 2000). Moreover, cholesterol protects PC12 cells from Aβ toxicity *in vitro* (Arispe *et al.*, 2002a; Eckert *et al.*, 2003b) and inhibits the effects of Aβ on cellular calcium signaling (Mattson *et al.*, 1993). Accordingly, enhanced cholesterol levels and rather reduced disordering effects of Aβ peptides in SPM membranes isolated from aged mice brains were reported (Eckert *et al.*, 2001). To further evaluate the role of cholesterol in regulating the membrane disordering effects of Aβ, cholesterol was depleted from SPM using methyl-β-cyclodextrin (MβCD), a water soluble cyclic oligosaccharide (Kirsch *et al.*, 2002). Cholesterol-cyclodextrin-complexes were used to enrich membrane cholesterol. Depletion of cholesterol using 16 mmol/l MβCD significantly reduced the cholesterol content of about 50% in SPM isolated from brains of 12 middle-aged (12 months) and aged (22 months) mice. Accordingly, using MβCD-inclusion-complexes SPM cholesterol levels could be enhanced up to about 150%. However, SPM from aged mice were not equally accessible to external cholesterol enrichment that might be due to the high physiological cholesterol load of these membranes *per se*. Effects of $A\beta_{1-42}$ on the interior membrane region of cholesterol-modulated SPM from middle age and aged mice determined by DPH anisotropy measurement were negatively correlated with the cholesterol content as shown in Figure 4.

Brain membranes from middle-aged mice were more susceptible to $A\beta_{1-42}$. Cholesterol depletion of brain membranes from middle-aged and aged mice significantly enhances the effects of $A\beta_{1-42}$ on DPH anisotropy values; cholesterol enrichment strongly decreases the effect. Analogously, the disordering effects of $A\beta_{1-42}$ on the hydrophilic region of neuronal membranes determined by TMA-DPH anisotropy measurements were again strongly dependent on the cholesterol content of membranes from middle-aged and aged mice (Kirsch *et al.*, 2002). Hence, weaker interactions of Aβ peptides with cholesterol-loaded neuronal membranes should lead to more moderate disturbances on membrane-related signal transduction processes, *e.g.* calcium signaling. This hypothesis is underlined by earlier findings evaluating functional effects of Aβ on signal cascade pathways. Aβ induced changes in free intracellular calcium levels were significantly lower in dissociated brain cells from mice that were enriched in cholesterol (Hartmann *et al.*, 1993).

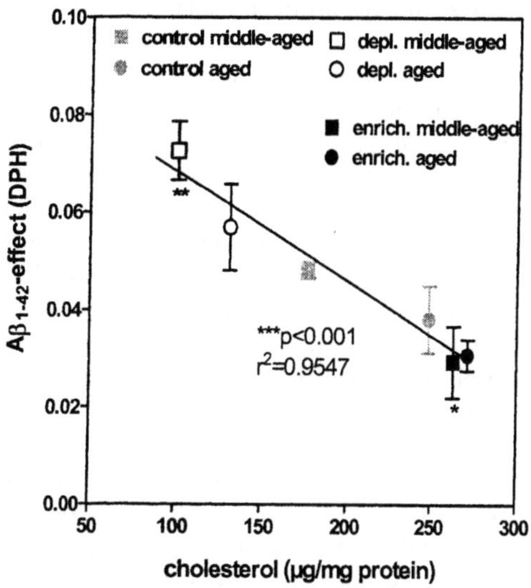

Figure 4. Cholesterol modulation in SPM isolated from brains of middle-aged (12 months) and aged (22 months) mice according to (23). Correlation of cholesterol levels and the effects of $A\beta_{1-42}$ (1 µmol/l) on DPH anisotropy values of cholesterol modulated SPM isolated from brains of middle-aged and aged mice. Means ± SEM; n=4-5, *p<0.05, **p<0.01 significant vs. control (t-test). ***p<0.001 significant correlation (Pearson). depl. = depletion; enrich. = enrichment.

5. FUNCTIONAL SIGNIFICANCE

Aβ-induced changes of fluidity parameters in neuronal membranes are connected with effects on the activity of different enzymes and on signal-transduction processes. It is well recognized that changes in the physico-chemical state of the membrane can markedly alter activity of various membrane proteins (Burger *et al.*, 2000). Effects of Aβ on membrane fluidity have been proposed to contribute to disruption in different cell functions.

Aβ enhances tyrosine kinase activity (McDonald *et al.*, 1997) and effects the activity of PLC (Hartmann *et al.*, 1996; Shimohama *et al.*, 1995). Several studies confirmed Aβ's amplifying effect on calcium signaling in different cell types, including neurons and astrocytes (Hartmann *et al.*, 1994; Haughey *et al.*., 2003; Mattson and Sherman, 2003). At low concentrations, Aβ alone does not affect baseline Ca^{2+}, but enhances Ca^{2+} influx after various stimuli (Hartmann *et al.*, 1993), suggesting the activation of endogenous Ca^{2+} channels (Hartmann *et al.*, 1994). Observations that Aβ produces at high concentrations Ca^{2+} fluxes through artificial planar lipid bilayers suggest that Aβ may also disrupt or disturb membrane structure or integrity leading to channel like alterations of the membrane (Arispe *et al.*, 1993; Arispe *et al.*, 2002b). Recently, it has been shown using patch clamp analysis that low nanomolar concentrations of protofibillar $A\beta_{1-42}$ induce reversible Ca^{2+} dependent increase in spontaneous action potentials and membrane depolarization (Ye *et al.*, 2003). Aβ's effects on intracellular calcium homeostasis could be mediated by G-protein regulated phosphoinoside hydrolysis and adenylate cyclase pathways (Cowburn *et al.*, 2001; Soomets *et al.*, 1999), that have been shown to be altered in AD post-mortem brains, including impaired agonist and G-protein regulation of phospholipase C (PLC), decreased phosphokinase C (PKC) levels and activity, and a reduced number of receptor sites for the second messenger, Ins(1,2,5)P3 (Cowburn *et al.*, 2001; Garrido *et al.*, 2002; Olariu *et al.*, 2002). However, PKC seems to modulate Aβ's neurotoxic effects. Activation of PKC by phorbolesters protected rat hippocampal neurons from Aβ toxicity (Garrido *et al.*, 2002). This effect was accomplished by inhibition of glycogen synthase kinase-beta activity that led to the accumulation of cytoplasmic β-catenin and transcriptional activation of Wnt target genes. Classical activators of the Wnt pathway mimicked PCK activation (Garrido *et al.*, 2002). Accordingly, chronic intracereboroventricular infusion of $A\beta_{1-40}$ decreased the activity of PKC in the hippocampus of rats and impaired memory (Olariu *et al.*, 2002).

6. CONCLUSIONS

Aβ not only disturbs membrane properties of synaptosomal plasma membranes but also of mitochondrial membranes. Brain aging also differently modulates both Aβ effects and itself modifies the fluidity of intracellular membranes. Aβ-induced changes in membrane fluidity can be explained by physico-chemical interactions of the peptide with membrane components (Kanfer *et al.*, 1999). According to X-ray diffraction analysis Aβ localizes in the membrane hydrocarbon core (Mason *et al.*, 1996). Using phospholipid vesicles it was shown that Aβ interacts electrostatically with phospholipids (Fletcher and Keire, 1997) and binds to gangliosides (Matsuzaki and Horikiri, 1999). Moreover, it has been demonstrated that Aβ adsorbs to membrane components such as fatty acids and cholesterol (Avdulov *et al.*, 1996; Avdulov *et al.*, 1997b).

Aggregation of Aβ peptides leads to β-sheet-like protein structures (Soto *et al.*, 1994). Effects of Aβ on neuronal membranes are clearly related to peptide aggregation. The extent of the formation of β-sheet structures defines Aβ's membrane disturbing properties. Accordingly, Aβ toxicity correlates with the aggregation state of the peptide (Hirakura *et al.*, 1998); blocking peptide aggregation abolishes Aβ`s toxic properties (Drouet *et al.*, 1999; Soto *et al.*, 2000). Aggregated Aβ incorporates into membrane bilayers (Mason *et al.*, 1996) and disturbs membrane structure. This leads to restricted membrane fatty-acid mobility and consequently to decreased membrane fluidity, as indicated by enhanced DPH and TMA-DPH anisotropy values. Changes in membrane fluidity affect membrane components (*e.g.* ion channels) and thereby modulate signal transduction processes (Gimpl *et al.*, 1997; Scheuer *et al.*, 1996) that at least partly contribute to Aβ's effects in the brain. Effects of Aβ on membrane properties are also dependent on peptide length. The synthetic $A\beta_{25-35}$ possess only weak effects on membrane properties (Müller *et al.*, 2001). Compared to $A\beta_{1-40}$, the full-length $A\beta_{1-42}$ effects membrane fluidity much stronger and also shows higher neurotoxicity (Müller *et al.*, 2001; Pike *et al.*, 1995). These properties possibly contribute to genetically determined forms of AD that are associate with relatively high $A\beta_{1-42}$ brain levels and early development of AD symptoms (Haass and De Strooper, 1999; Sinha and Lieberburg, 1999).

All findings together are of particular interest in view of the neurotoxic properties of Aβ and its intracellular biology. Moreover, the available data point to membrane alterations in AD brain as possible therapeutic target.

REFERENCES

Arispe, N., and Doh, M., 2002, Plasma membrane cholesterol controls the cytotoxicity of Alzheimer's disease AbetaP (1-40) and (1-42) peptides. *FASEB J.,* **16:** 1526-1536.

Arispe, N., Rojas, E., and Pollard, H.B., 1993, Alzheimer disease amyloid beta protein forms calcium channels in bilayer membranes: blockade by tromethamine and aluminum. *Proc. Natl. Acad. Sci. USA.*, **90:** 567-571.

Avdulov, N.A., Chochina, S.V., Igbavboa, U., O'Hare, E.O., Schroeder, F., Cleary, J.P., and Wood,W.G., 1997a, Amyloid beta-peptides increase annular and bulk fluidity and induce lipid peroxidation in brain synaptic plasma membranes. *J. Neurochem.*, **68:** 2086-2091.

Avdulov, N.A., Chochina, S.V., Igbavboa, U., Vassiliev, A.V., Maletta, G.J., and Wood, W.G., 1996, Selective binding of cholesterol and fatty acids to amyloid beta peptide 1-40 aggregates: A fluorescent study. *Soc. Neurosc. Abst.,* **22:** 2112.

Avdulov, N.A., Chochina, S.V., Igbavboa, U., Warden, C.S., Vassiliev ,A.V., and Wood, W.G., 1997b, Lipid binding to amyloid beta-peptide aggregates: Preferential binding of cholesterol as compared with phosphatidylcholine and fatty acids. *J. Neurochem.,* **69:** 1746-1752.

Bodovitz, S., and Klein, W.L., 1996, Cholesterol modulates alpha-secretase cleavage of amyloid precursor protein. *J. Biol. Chem.* **271:** 4436-4440.

Brown, D.A., and London, E., 1998, Structure and origin of ordered lipid domains in biological membranes. *J. Membr. Biol.,* **164:** 103-114.

Burger, K., Gimpl, G., and Fahrenholz,F., 2000, Regulation of receptor function by cholesterol. *Cell Mol. Life Sci,* **57:** 1577-1592.

Butterfield, D.A., Yatin, S.M., Varadarajan, S., and Koppal, T., 1999, Amyloid beta-peptide-associated free radical oxidative stress, neurotoxicity, and Alzheimer's disease. *Meth..Enzymol.* **309:** 746-768.

Chauhan, A., Chauhan, V.P.S., Brockerhoff, H., and Wisniewski, H.M., 1993, Effect of amyloid beta - protein on membrane properties. In: Alzheimer's Disease: Advances in Clinical and Basic Research. Corain, B, Iqbal, K, Nicolini, M, Winblad, B, Wisniewski,H, and Zatta, P., (Eds.), John Wiley & Sons Ltd, New York, pp. 431-439.

Chochina, S.V., Avdulov, N.A., Igbavboa, U., Cleary, J.P., O'Hare, E.O., and Wood,W.G., 2001, Amyloid beta-peptide(1-40) increases neuronal membrane fluidity. Role of cholesterol and brain region. *J. Lipid Res.* **42:** 1292-1297.

Cowburn, R.F., O'Neill, C., Bonkale, W.L., Ohm, T.G., and Fastbom, J., 2001, Receptor-G-protein signalling in Alzheimer's disease, Biochem. Soc. Symp., XXX: 163-175.

Drouet, B., Pincon-Raymond, M., Chambaz, J., and Pillot, T., 1999, Laminin 1 attenuates beta-amyloid peptide Abeta(1-40) neurotoxicity of cultured fetal rat cortical neurons. *J. Neurochem.* **73:** 742-749.

Eckert, G.P., Cairns, N.J., Maras, A., Gattaz, W.F., and Müller,W.E., 2000, Cholesterol modulates the membrane-disordering effects of beta-amyloid peptides in the hippocampus: specific changes in Alzheimer's disease. *Dement. Geriatr. Cogn. Disord.* **11:** 181-186.

Eckert, G.P., Igbavboa, U., Muller, W.E., and Wood,W.G., 2003a, Lipid rafts of purified mouse brain synaptosomes prepared with or without detergent reveal different lipid and protein domains. *Brain Res.* **962:** 144-150.

Eckert, G.P., Kirsch, C., Leutz, S., Wood, W.G., and Muller, W.E., 2003b, Cholesterol modulates amyloid beta-peptide's membrane interactions. *Pharmacopsychiatry,* **36 Suppl 2:** S136-S143.

Eckert, G.P., Wood, W.G., and Müller, W.E., 2001, Effects of aging and beta-amyloid on the properties of brain synaptic and mitochondrial membranes. *J. Neural Transm.* **108:** 1051-1064.

Fletcher, T.G., and Keire,D.A., 1997, The interaction of beta-amyloid protein fragment (12-28) with lipid environments. *Protein Sci.,* **6:** 666-675.

Garrido, J.L., Godoy, J.A., Alvarez, A., Bronfman, M., and Inestrosa, N.C., 2002, Protein kinase C inhibits amyloid beta peptide neurotoxicity by acting on members of the Wnt pathway. *FASEB J.,* **16:** 1982-1984.

Gimpl, G., Burger, K., and Fahrenholz, F., 1997, Cholesterol as modulator of receptor function. *Biochemistry* **36:** 10959-10974.

Haass, C., and De Strooper, B., 1999, The presenilins in Alzheimer's disease-proteolysis holds the Key. *Science,* **286:** 916-919.

Hartmann, H., Eckert, A., Crews, F.T., Müller, W.E., 1996, b-amyloid amplifies PLC activity and Ca^{2+} signalling in fully differentiated brain cells of adult mice. *Amyloid* **3:** 234-241.

Hartmann, H., Eckert, A., and Müller, W.E., 1993, Beta-Amyloid protein amplifies calcium signalling in central neurons from the adult mouse. *Biochem. Biophys. Res. Commun.* **194:** 1216-1220.

Hartmann, H., Eckert, A., and Müller, W.E., 1994, Apolipoprotein E and cholesterol affect neuronal calcium signalling: The possible relationship to beta-amyloid neurotoxicity. *Biochem. Biophys. Res. Commun.* **200:** 1185-1192.

Haughey, N.J., and Mattson, M.P., 2003, Alzheimer's amyloid beta-peptide enhances ATP/gap junction-mediated calcium-wave propagation in astrocytes. *Neuromolecular. Med.* **3:** 173-180.

Hayashi, H., Mizuno, T., Michikawa, M., Haass, C., and Yanagisawa, K., 2000, Amyloid precursor protein in unique cholesterol-rich microdomains different from caveolae-like domains. *Biochim. Biophys. Acta* **1483:** 81-90.

Hirakura, Y., Satoh, Y., Hirashima, N., Suzuki, T., Kagan, B.L., and Kirino, Y., 1998, Membrane perturbation by the neurotoxic Alzheimer amyloid fragment beta 25-35 requires aggregation and beta-sheet formation. *Biochem. Mol. Biol. Int.* **46:** 787-794.

Igbavboa, U., Avdulov, N.A., Schroeder, F., and Wood, W.G., 1996, Increasing Age Alters Transbilayer Fluidity and Cholesterol Asymetry in Synaptic Plasma Membrane of Mice. *J. Neurochem.* **66:** 1717-1725.

Kaiser, R.D., and London, E., 1998, Location of diphenylhexatriene (DPH) and its derivatives within membranes: comparison of different fluorescence quenching analyses of membrane depth. *Biochemistry* **37:** 8180-8190.

Kanfer, J.N., Sorrentino, G., and Sitar, D.S., 1999, Amyloid beta peptide membrane perturbation is the basis for its biological effects. *Neurochem. Res.* **24:** 1621-1630.

Kawahara, M., and Kuroda, Y., 2000, Molecular mechanism of neurodegeneration induced by Alzheimer's beta-amyloid protein: channel formation and disruption of calcium homeostasis. *Brain Res. Bull.* **53:** 389-397.

Keller, J.N., Pang, Z., Geddes, J.W., Begley, J.G., Germeyer, A., Waeg, G., and Mattson, M.P., 1997, Impairment of glucose and glutamate transport and induction of mitochondrial oxidative stress and dysfunction in synaptosomes by amyloid beta-peptide: Role of the lipid peroxidation product 4- hydroxynonenal. *J. Neurochem.* **69:** 273-284.

Kirsch, C., Eckert, G.P., and Müller, W.E., 2002, Cholesterol attenuates the membrane perturbing properties of b-amyloid peptides. *Amyloid* **9:** 149-159.

Kojro, E., Gimpl, G., Lammich, S., Marz, W., and Fahrenholz,F., 2001, Low cholesterol stimulates the nonamyloidogenic pathway by its effect on the alpha-secretase ADAM 10. *Proc. Natl. Acad. Sci. USA* **98:** 5815-5820.

Kremer, J.J., Pallitto, M.M., Sklansky, D.J., and Murphy, R.M., 2000, Correlation of beta-amyloid aggregate size and hydrophobicity with decreased bilayer fluidity of model membranes. *Biochemistry* **39:** 10309-10318.

Lakowicz, J.R.,1983, Fluorescence Polarisation. In: Lakowicz, JR., (Ed.), Principles of Fluorescence Spectroscopy. Plenum Press, New York, pp. 112-151.

Mason, R.P., Estermyer, J.D., Kelly, J.F., and Mason, P.E., 1996, Alzheimer's disease amyloid beta peptide 25-35 is localized in the membrane hydrocarbon core: x-ray diffraction analysis. *Biochem. Biophys. Res. Commun.* **222:** 78-82.

Mason, R.P., Jacob, R.F., Walter, M.F., Mason, P.E., Avdulov, N.A., Chochina, S.V., Igbavboa, U., and Wood, W.G., 1999, Distribution and Fluidizing Action of Soluble and Aggregated Amyloid Beta - Peptide in Rat Synaptic Plasma Membranes. *J. Biol. Chem.* **274:** 18801-18807.

Matsuzaki, K., and Horikiri, C., 1999, Interactions of amyloid beta-peptide (1-40) with ganglioside-containing membranes. *Biochemistry* **38:** 4137-4142.

Mattson, M.P., and Sherman,M., 2003, Perturbed signal transduction in neurodegenerative disorders involving aberrant protein aggregation. *Neuromolecular. Med.* **4:** 109-132.

Mattson, M.P., Tomaselli, K.J., and Rydel, R.E., 1993, Calcium-destabilizing and neurodegenerative effects of aggregated beta-amyloid peptide are attenuated by basic FGF. *Brain Res.* **621:** 35-49.

McDonald, D.R., Brunden, K.R., and Landreth,G.E., 1997, Amyloid fibrils activate tyrosine kinase-dependent signaling and superoxide production in microglia. *J. Neurosci.* **17:** 2284-2294.

McLaurin, J., and Chakrabartty, A., 1996, Membrane disruption by Alzheimer beta-amyloid peptides mediated through specific finding to either phospholipids or gangliosides - Implications for neurotoxicity. *J. Biol. Chem.* **271:** 26482-26489.

McLaurin, J., and Chakrabartty, A., 1997, Characterization of the interactions of Alzheimer beta-amyloid peptides with phospholipid membranes. *Eur. J. Biochem.* **245:** 355-363.

Mulders, F., van Langen, H., Van Ginkel, G., and Levine, Y.K., 1986, The static and dynamic behaviour of fluorescent probe molecules in lipid bilayers. *Biochim. Biophys. Acta* **859:** 209-218.

Müller, W.E., Eckert, G.P., Scheuer, K., Cairns, N.J., Maras, A., and Gattaz,W.F., 1998, Effects of b-amyloid peptides on the fluidity of membranes from frontal and parietal lobes of human brain. High potencies of A beta 1- 42 and A beta 1-43. *Amyloid* **5:** 10-15.

Müller, W.E., Kirsch, C., and Eckert, G.P., 2001, Membrane-disordering effects of beta-amyloid peptides. *Biochem. Soc.Trans.* **29:** 617-623.

Müller, W.E., Koch, S., Eckert, A., Hartmann, H., and Scheuer, K., 1995, β-Amyloid peptide decreases membrane fluidity. *Brain Res.* **674:** 133-136.

Olariu, A., Yamada, K., Mamiya, T., Hefco, V., and Nabeshima,T., 2002, Memory impairment induced by chronic intracerebroventricular infusion of beta-amyloid (1-40) involves downregulation of protein kinase C. *Brain Res.* **957:** 278-286.

Oshima, N., Morishima-Kawashima, M., Yamaguchi, H., Yoshimura, M., Sugihara, S., Khan, K., Games, D., Schenk, D., and Ihara,Y., 2001, Accumulation of Amyloid beta-Protein in the Low-Density Membrane Domain Accurately Reflects the Extent of beta-Amyloid Deposition in the Brain. *Am. J. Pathol.* **158:** 2209-2218.

Palacino, J.J., Murphy, M.P., Murayama, O., Iwasaki, K., Fujiwara, M., Takashima, A., Golde, T.E., and Wolozin, B., 2001, Presenilin 1 regulates beta-catenin-mediated transcription in a glycogen synthase kinase-3-independent fashion. *J. Biol. Chem.* **276:** 38563-38569.

Pike, C.J., Walencewicz, A.J., Kosmoski ,J., Cribbs, D.H., Glabe, C.G., and Cotman, C.W., 1995, Structure-activity analyses of beta-amyloid peptides: contributions of the beta 25-35 region to aggregation and neurotoxicity. *J. Neurochem.* **64:** 253-265.

Pike, C.J., Walencewicz A.J., Glabe, C.G., and Cotman, C.W., 1991, Aggregation-related toxicity of synthetic beta-amyloid protein in hippocampal cultures. *Eur. J. Pharmacol.* **207:** 367-368.

Pirttilä, T., Soininen, H., Mehta, P.D., Heinonen, O., Lehtimaeki, T., Bogdanovic, N., Paljaervi, L., Kim, K.S., Kosunen, O., Winblad, B., Riekkinen, P.S., and Wisniewski, H.M., 1996, Apolipoprotein E genotype and amyloid load in Alzheimer disease and control brains. *Neurobiol. Aging* **18:** 121-127.

Scheuer, K., Maras, A., Gattaz, W.F., Cairns, N., Förstl, H., and Müller, W.E., 1996, Cortical NMDA Receptor Properties and Membrane Fluidity are Altered in Alzheimer's Disease. *Dementia* **7:** 210-214.

Shimohama, S., and Matsushima, H., 1995, Signal transduction mechanisms in Alzheimer disease. *Alzheimer. Dis. Assoc. Disord.* **9 Suppl 2:** 15-22.

Simons, K., and Ikonen, E., 2000, How cells handle cholesterol. *Science* **290:** 1721-1726.

Simons, K., and Toomre, D., 2000 Lipid Rafts and Signal Transduction. *Nature Rev. Mol. Cell Biol.* **1:** 31-39.

Sinha, S., and Lieberburg, I., 1999, Cellular mechanisms of beta-amyloid production and secretion. *Proc. Natl. Acad. Sci. USA* **96:** 11049-11053.

Small, D.H., and McLean, C.A., 1999, Alzheimer's disease and the amyloid beta protein: What is the role of amyloid? *J. Neurochem.* **73:** 443-449.

Soomets, U., Mahlapuu, R., Tehranian, R., Jarvet, J., Karelson E., Zilmer, M., Iverfeldt, K., Zorko, M., Graslund, A., and Langel, U., 1999, Regulation of GTPase and adenylate cyclase activity by amyloid beta- peptide and its fragments in rat brain tissue. *Brain Res.* **850:** 179-188.

Soto, C., Branes, M.C., Alvarez, J., and Inestrosa, N.C., 1994, Structural determinants of the Alzheimer's amyloid beta-peptide. *J. Neurochem.* **63:** 1191-1198.

Soto, C., Saborio, G.P., and Permanne, B., 2000, Inhibiting the conversion of soluble amyloid-beta peptide into abnormally folded amyloidogenic intermediates: relevance for Alzheimer's disease therapy. *Acta Neurol. Scand.* **Suppl, 176:** 90-95.

Terzi, E., Holzemann, G., and Seelig, J., 1997, Interaction of Alzheimer beta-amyloid peptide(1-40) with lipid membranes. *Biochemistry* **36:** 14845-14852.

Torp, R., Head, E., Milgram, N.W., Hahn, F., Ottersen, O.P., and Cotman,.W., 2000, Ultrastructural evidence of fibrillar beta-amyloid associated with neuronal membranes in behaviorally characterized aged dog brains. *Neuroscience* **96:** 495-506.

Waschuk, S., Elton, E.A., Darabie, A.A., Fraser, P.E., and McLaurin, J., 2001, Cellular membrane composition defines Ab-lipid interactions. *J. Biol. Chem.* **275:** 33561-33568.

Wood, W.G., Eckert, G.P., Igbavboa, U., and Müller, W.E., 2003, Amyloid beta-protein interactions with membranes and cholesterol: causes or casualties of Alzheimer's disease. *Biochim Biophys Acta* **1610:** 281-290.

Wood, W.G., Schroeder, F., Avdulov, N.A., Chochina, S.V., and Igbavboa, U., 1999, Recent advances in brain cholesterol dynamics: transport, domains, and Alzheimer's disease. *Lipids* **34:** 225-234.

Wood, W.G., Schroeder, F., Igbavboa, U., Avdulov, N.A., Chochina, S.V., 2002, Brain membrane cholesterol domains, aging and amyloid beta-peptides. *Neurobiol. Aging* **23:** 685.

Yamaguchi, H., Maat-Schieman, M.L., van Duinen, S.G., Prins, F.A., Neeskens, P., Natte, R., and Roos, R.A., 2000, Amyloid beta protein (Abeta) starts to deposit as plasma membrane-bound form in diffuse plaques of brains from hereditary cerebral hemorrhage with amyloidosis-Dutch type, Alzheimer disease and nondemented aged subjects. *J. Neuropathol. Exp. Neurol.* **59:** 723-732.

Yankner, B.A., Duffy, L.K., Kirschner, D.A., 1990, Neurotrophic and neurotoxic effects of amyloid - beta protein: Reversal by tachykinin neuropeptides. *Science* **250:** 279-282.

Ye, C.P., Selkoe, D.J., and Hartley, D.M., 2003, Protofibrils of amyloid beta-protein inhibit specific K+ currents in neocortical cultures. *Neurobiol. Dis.* **13:** 177-190.

Yeagle,P.L., 1989, Lipid regulation of cell membrane structure and function. *FASEB J.* **3:** 1833-1842.

Yip, C.M., Elton, E.A., Darabie, A.A., Morrison, M.R., and McLaurin, J.,2001, Cholesterol, a modulator of membrane-associated Abeta-fibrillogenesis and neurotoxicity. *J. Mol. Biol.* **311:** 723-734.

Chapter 17

The Role of Alzheimer Aβ Peptides in Ion Transport Across Cell Membranes

Vernon M. Ingram
Department of Biology, M.I.T, 31 Ames Street, Cambridge, MA 02139, USA

Abstract: Accumulation of beta amyloid (Aβ) fibrils in senile plaques and cerebral blood vessel walls is characteristic of Alzheimer's disease (AD). We discuss several models that seek to explain the neurotoxic consequences, in particular the manner in which the neurotoxicity promotes cell dysfunction and cell death by an increase in cytosolic calcium ion concentration. To base correctly a new therapy on *in vitro* experiments, one must choose the right model mechanism.

Key words: Alzheimer's disease, Aβ peptide, ionophore, misfolding, β-sheets, calcium homeostasis, AMPA receptor.

1. INTRODUCTION

The pathology of Alzheimer's disease (AD) was first defined by Alois Alzheimer in 1907. He saw by light microscopy "senile plaques" (see Figure 1) and also "tangles" in stained sections from post-mortem brains of his patients. Much later the chemistry of the plaque components was described (Glenner and Wong, 1984; Sisodia *et al.*, 1990). It turned out that the plaques were composed of fibrils formed by the aggregation of short peptides, mostly 40 and 42 amino acids long, which had a perfectly normal amino acid sequence, common to both. They represented a part of a much larger transmembrane "amyloid precursor protein" (APP), commonly found but of unknown function. For a variety of reasons, some genetic, some unknown, these peptides are overproduced in Alzheimer's Disease. This matters, because the two overexpressed peptides, Aβ1-40 and Aβ1-42, once liberated from their parent APP, refold and aggregate spontaneously to a neurotoxic form. They attack paritcular neurons in the CNS, causing them to become dysfunctional and eventually to die in large numbers. It should be remembered that the plaques are extracellular and exert their toxic effect from the outside. This distinguishes AD from other amyloidoses, such as Huntington's disease, for example, which are intracellular phenomena. A part of the consequent dysfunction is thought to be the hyperphosphorylation of the microtubule-asssociated protein tau (Ferreira *et al.*, 1997) which then aggregates to form the "tangles" that fill then neuronal cell body and produce dysfunctional neurons.

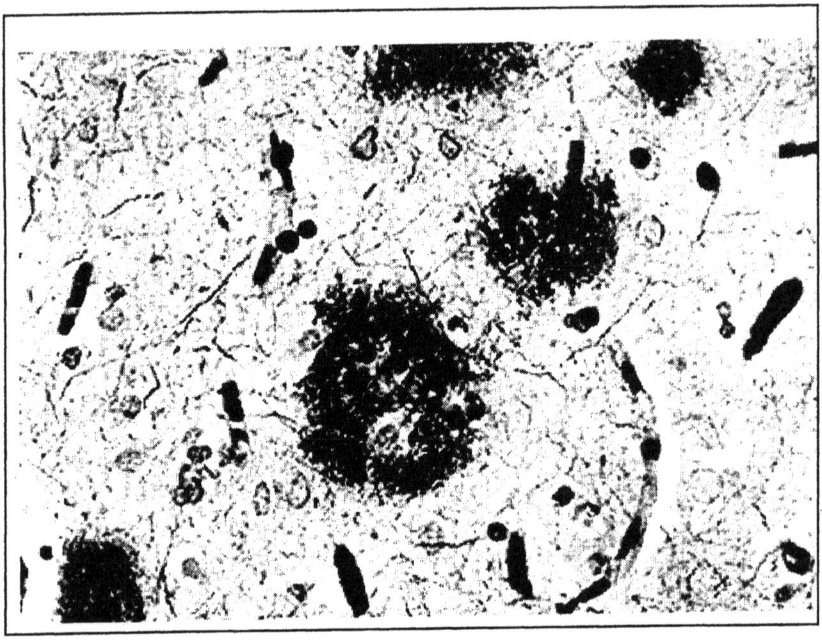

Figure 1. Plaques of Alzheimer's Disease (The Internet Pathology Laboratory). Stained section of an AD brain. The plaque is an accumulation of Aβ fibrils; note that the surrounding neuritis are also stained, indicating an interaction with plaque fibrils and neighboring neuritis.

The pathological over-production of this family of amyloid peptides is due to a variety of causes, some of which are genetically controlled, giving rise to the so-called Familial Alzheimer's disease, which are fortunately rare. Clearly, there are other causes, the mechanisms of which are not yet clearly understood. The accumulation of the Aβ1-42 peptide seems to be the most important phenomenon in causing disease, although this notion is somewhat controversial.

It is characteristic of both Aβ1-40 and Aβ1-42 that once released from the APP they rapidly and spontaneously refold to oligomeric aggregates that quickly form long fibrils with high β-sheet content. A currently popular view is that this "misfolding" produces a new conformation with new properties. Current research is devoted to understanding the molecular mechanism of Aβ neurotoxicity.

While these considerations place the emphasis on understanding the origin and pathological role of the Aβ peptides, it ignores the formation and function of the intracellular tangles. These fill the axons and somata of affected neurons and thereby interfere with axonal transport, synapse formation and signalling. It is unclear whether Aβ plaque formation and tau tangle formation are independent processes, or whether tangle formation is a downstream event that follows Aβ formation and Aβ aggregation, as this author believes. Further complicating this issue is the existence of a number of tau mutations that seem to favor tau hyperphosphorylation and that are associated with tangle formation and with severe dementia. The events leading to tangle formation will not be considered further in this review.

We here review the mechanisms of Aβ peptide toxicity that are currently under discussion. We must first devise a useful definition of "Aβ peptide toxicity". In many literature quotations this is taken to mean the induction of neuronal cell death by apoptosis or necrosis. These events occur in the central nervous system. Generally little notice is paid to the always present glial cells, their role in Alzheimer's and their fate. But there is a general gliosis associated with Alzheimer's, often a "reactive astrocytosis". Is there no glial cell death in AD? For the purposes of our discussions, and ignoring the uncertain fate of glial cells, we will assume that in AD Aβ peptide toxicity is sufficient to lead to the death of neurons, at least in the first instance.

Another important notion not usually mentioned is the important intermediate stage between initial peptide-neuron interaction and eventual cell death. It seems highly unlikely that a normal neuron "dies" instantly when attacked by the toxin. We clearly need a new definition of neuronal cell death. It seems more likely that there will be a series of stages when a neuron becomes gradually more and more dysfunctional. We have no idea how long this process might take or what its various steps will be. For the purposes of this review, we will define "Aβ peptide toxicity" as the mechanistic step that sends the neuron into the death process.

Although one's intuition is to find one single initiating mechanism, it is quite possible that there are two, perhaps three, quite different mechanisms that operate at different stages of Alzheimer's disease or that are found simultaneously in different brain cell types. The discussion that follows explores four different neurotoxic mechanisms below. The proposed mechanisms all point to Aβ1-42 and Aβ1-40 as the best targets for therapeutic intervention, which makes it all the more important to understand these mechanisms.

1.1 Increased cytosolic [Ca^{2+}] causes cell dysfunction and cell death

While the action of aggregated Aβ peptides on neurons is the precipitating factor in causing internal [Ca^{2+}] to rise by the mechanisms discussed below, the real damage to the cell is done by the increased calcium itself. It triggers a number of important downstream effects. For example, Rao *et al.* (2004) write: "The endoplasmic reticulum (ER) regulates protein synthesis, protein folding and trafficking, cellular responses to stress and intracellular calcium (Ca^{2+}) levels. Alterations in Ca^{2+} homeostasis and accumulation of misfolded proteins in the ER cause ER stress that ultimately leads to apoptosis. Prolonged ER stress is linked to the pathogenesis of several different neurodegenerative disorders." Other examples of the role of Ca^{2+} levels is described by Agell *et al.* (2002) and Mattson *et al.* (1992) The literature is extensive enough to reinforce the notion that disturbance of the tightly controlled calcium homeostasis is pivotal in bringing about deleterious downstream effects in Alzheimer's disease.

2. Aβ PEPTIDES RAISE CYTOSOLIC Ca^{2+} CONCENTRATIONS BY AN UNKNOWN MECHANISM

Mattson *et al.* (1992) reported experiments using human cerebral cortical cell cultures to test the hypothesis that "β-amyloid can destabilize neuronal calcium regulation and render neurons more vulnerable to environmental stimuli that elevate intracellular calcium levels." The environmental stimuli used are glutamate and other excitatory amino acids. The β-amyloid peptides involved are Aβ1-38 and Aβ25-35, shorter versions of the main peptides now known to be involved in AD, Aβ1-40 and Aβ1-42. More recent experiments from other laboratories (*eg* Teplow, Ingram) indicate that the state of aggregation of the peptide is crucial in determining whether an aggregate is toxic to the neuron or not. Mattson and Furukawa (2003) in their chapter found that over a period of several days these peptides by themselves did not shorten survival times of the neurons, but they did exacerbate the killing effect of glutamate during the same time. The first time point recorded was 1 day. The relation of these findings to the pathology of AD is not obvious, since Aβ25-35 is not found in the disease and Aβ1-38 is a minor product. Nevertheless these findings stimulated a large number of experiments.

Mattson and Furukawa (2003) summarize more than a decade of work from this group. Their ideas are centered on the notion that the Alzheimer Aβ peptides disturb cellular calcium homeostasis and cause oxidative changes in neuronal cell membranes. They see a large number of deleterious processes in AD as down-stream from this phenomenon. The initiating event is said to be that "during the process of self-aggregation, Aβ generates hydrogen peroxide and hydroxyl radicals and the production of these reactive oxygen species induces the peroxidation of lipids in the plasma membrane which can impair the function of membrane ion-motive ATPases (Na^+/K^+- and Ca^{2+}-ATPases, and glucose and glutamate transporters), resulting in membrane depolarization and a decrease in cellular ATP levels." Disruption of cellular calcium homeostasis follows. Evidence for these downstream cellular events is given in the summary and in the group of papers that are cited. These experiments involve substantial time delays from application of Aβ peptide and reading of the assay. The initiating event remains in doubt. Although the statement is made in Mattson and Furukawa that "During the process of self-aggregation, Aβ generates hydrogen peroxide and hydroxyl radical......" followed by the downstream events mentioned above, no direct evidence is given in the review or in the cited references to substantiate this important notion concerning initiation.

Behl *et al.* (1994) [*see also* Chapter 4] produced evidence that "the cytotoxic action of Aβ on neurons results from free radical damage to susceptible neurons". Using PC12 neuronal cells, B12 cells derived from rat brain tumors and CNS primary cultures they showed that Aβ causes the overproduction of H_2O_2, leading to cell death. The Aβ peptides used were only Aβ1-40 and Aβ25-35. (At the time of these experiments the importance of Aβ1-42 overproduction in the pathogenesis of AD was not generally appreciated.) Also, the question of whether the Alzheimer peptides are aggregated or not is not addressed in this early paper. Since the toxicity assays are done after considerable delay, one can assume that a measure of aggregation has taken place during the Aβ challenge. They show an increase in H_2O_2 production which could "lead directly to cell death, probably via hydroxyl radical induced oxidative damage". These authors demonstrate cytotoxicity in three ways: the MTT assay, thought to reflect early redox changes in the cell, LDH release, a sign of necrotic cell death, and the Trypan Blue assay which measures loss of membrane integrity. Oxidative damage is clearly demonstrated in these experiments, as is the role of various anti-oxidants in preventing cell death, as measured in their assays. There were striking time lapses before measurements were taken, ranging from several hours to 1 day. Therefore, all the biochemical events described or postulated might be downstream from an initiating event or events. They remain important possibilities in the pathogenesis of AD.

The notion that Alzheimer pathogenesis is primarily due to or at least involves disturbance of calcium homeostasis is very appealing. In particular the impairment of long-term potentiation by Aβ peptides is intriguing, since it provides a mechanistic link between Aβ toxicity and the characteristic memory deficits of AD (Ferreira *et al.*, 1997).

Glabe (2004) recently pointed to the ability of Aβ peptide oligomers to effect ion transport across membranes. He writes "... We also examined the effect of soluble (*amyloid*) oligomers on membrane conductance. We found that soluble oligomers and not low MW species or fibrils from all types of amyloids tested specifically increase the conductivity of the membrane. The conductivity is not ion selective and does not appear to be due to pore or channel formation. This increase in conductivity could have a detrimental effect on a broad range of biological processes, ...". The actual mechanism remains to be determined.

3. Aβ PEPTIDES RAISE CYTOSOLIC Ca^{2+} CONCENTRATIONS BY INFLUX THROUGH AN IONOPHORE FORMED BY Aβ PEPTIDES

Starting with the early work of Arispe and colleagues (Arispe *et al.* 1993a,b; Durell *et al.*, 1994) the Alzheimer Aβ1-40 peptide has been reported to form calcium-specific ionophore "holes" in artificial membranes, allowing a Ca^{2+} current to flow down its electrochemical gradient. Some of these experiments have been repeated in membranes from a hypothalamic cell line. Models of possible ionophore configurations have been proposed based on certain assumptions of the α-helical content of Aβ1-40. The putative ionophores are not blocked by the usual antagonists of voltage-gated calcium channels (VGCCs), but are blocked by Al^{3+} and by high concentrations of $Tris^{+}$. At their simplest, such ionophores should be effective Ca^{2+}-channels in all cells, neuronal and non-neuronal alike. But Alzheimer's is highly regionalized and is especially active in killing neurons in particular brain nuclei.

Our early patch clamp experiments (Sanderson *et al.*, 1997) using hNT neuronal cells support such a mechanism for Ca^{2+} influx caused by the synthetic Aβ25-35 peptide. This is a short version of the naturally occurring Aβ1-40, Aβ1-42 peptides, but does not itself occur in AD or in normal brains. Models for the conformation of this peptide have been proposed, but curiously they postulate mostly α-helices in the aggregated Aβ25-35, even though this is a peptide region generally assumed to be in β-sheet conformation. This question remains unresolved.

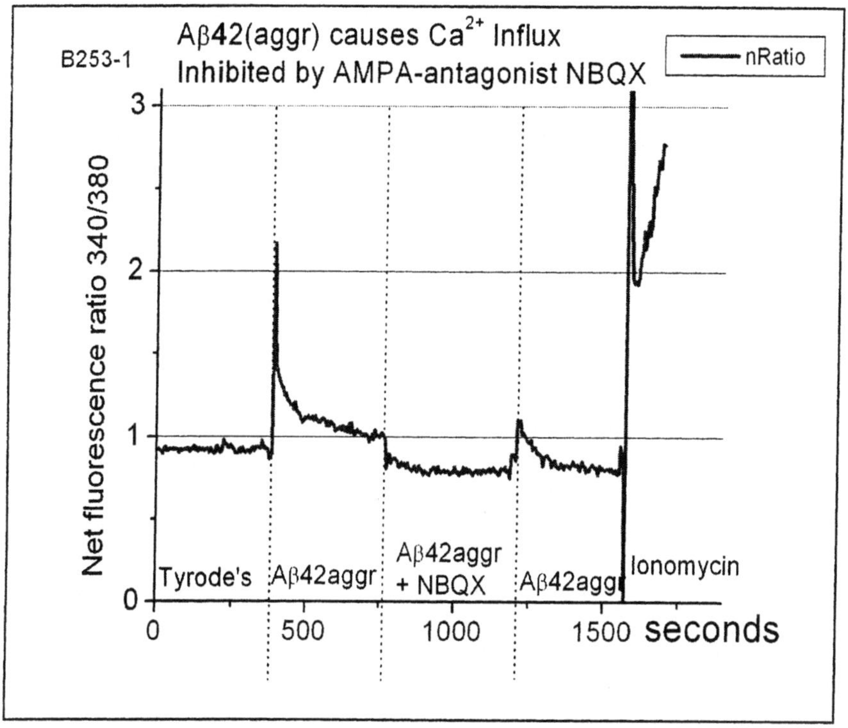

Figure 2. Aggregated 10 μM Aβ1-42 raises cytosolic Ca^{2+} in CATH.a cells. The AMPA antagonist NBQX completely block Aβ1-42 calcium influx into CATH.a cells, an effect that is partially reversible.

4. Aβ PEPTIDES RAISE CYTOSOLIC Ca^{2+} CONCENTRATIONS BY ACTIVATION OF LIGAND-ACTIVATED ION CHANNELS

Recent experiments from our laboratory (Blanchard *et al.*, 1997; 2000) have established a third mechanism of Aβ neurotoxicity that is active on neuronal cells in culture. This is an immediate and very early mechanism, producing a measurable influx of external Ca^{2+} ions within a few seconds.

Our proposal for this third mechanism states that aggregated Aβ1-42, and probably also Aβ1-40, interact specifically with Ca^{2+}-permeant α-amino-3-hydroxy-5-methylisoxazolepropionic acid (AMPA) receptors on the surface of certain neurons, thereby allowing a dramatic immediate influx of external Ca^{2+} ions.

As can be seen in Figure 2, Ca^{2+} ions flow into the cell within seconds of application. This does not occur when the external Ca^{2+} concentration is zero, *i.e.* it is not due to release of Ca^{2+} ions from internal storage. It is completely inhibited by NBQX, a specific AMPA-receptor antagonist. Earlier experiments with CNQX, another AMPA antagonist, confirm this observation. These observations prove the direct and crucial involvement of Ca^{2+}-permeant AMPA receptors in this mechanism.

As was said earlier, both Aβ1-40 and Aβ1-42 self-aggregate and form cytotoxic aggregates that eventually lead to neuronal death (Blanc *et al.*, 1997; Yankner *et al.*, 1990). This process would account for the enormous neuronal loss in AD, affecting specific types of neurone that carry the specific AMPA-receptor involved.

Donut-like oligomers/aggregates of the Aβ1-42 peptide are very common at early incubation times at, pH7.4 (Blanchard *et al.*, 2000, and unpublished data). Certainly *in vitro* these donuts quickly change into protofibrils and then into fibrils. It is conceivable that in accordance with the ionophore mechanism, as described above, these donuts might insert into neuronal cell membranes and conduct Ca^{2+} influx as ionophores. In our hands, this is not the case. Aβ1-42 must be pre-incubated for at least 24 hours before the preparation will promote Ca^{2+}-ion influx. By this time they have turned into protofibrils and some fibrils. This transition into protofibrils and fibrils is likely to occur *in vivo,* as *in vitro*. Furthermore, donuts, the early oligomeric forms, do not contain β-sheet structures, as do protofibrils and fibrils. Such a scheme could explain the neurotoxicity (deleterious Ca^{2+}-ion influx) as well as the singular distribution of cell loss.

As seen in the electron microscope (e.g. by negative staining)

$(A\beta 1\text{-}42)_1$ → $(A\beta 1\text{-}42)_n$ → $(A\beta 1\text{-}42)_{protofibrils}$ → $(A\beta 1\text{-}42)_{fibrils}$

not visible "donuts"[1] ---------neurotoxic----------

Only certain neurons have the specific receptors that are sensitive to Aβ1-42 protofibrils and fibrils. The aggregated peptide $(A\beta 1\text{-}42)_{fibrils}$ apparently opens Ca^{2+}-permeant AMPA receptors, causing Ca^{2+}-influx only

[1] donut-like aggregates are very commonly observed at very early Aβ incubation times, at pH7.4 (*see also* Chapter 1)

in those neurons that have this type of receptor. This view would explain the nucleus-specific nature of Alzheimer's disease, since it is known that the distribution of, for example, a calcium-permeant AMPA-receptor is quite restricted.

It seems that AD pathology correlates with a much higher proportion of the Aβ1-42 species. The accumulated Aβ-peptides and their apparent mechanism of action are obvious targets for the development of therapeutic strategies (see Chapter 19).

5. Aβ PEPTIDES MODULATE SIGNALLING MECHANISMS

There are very interesting reports that the Alzheimer Aβ peptide can inhibit Long Term Potentiation (LTP) in brain slices (Wang *et al.*, 2002). Transgenic mice over-expressing APP gene show greatly diminished LTP *in vivo*, although regular transmission is normal (Chapman *et al.*, 1999). It is not known whether any of the above ionic transmembrane mechanisms are involved in LTP.

6. CONCLUSIONS

Three of the four models described agree that the end result of Aβ1-42 action on neuronal cells is a destabilization of cytosolic calcium homeostasis in the direction of increased [Ca^{2+}]. The mechanisms, however, differ with respect to neuronal specificity and the initiation of calcium dyshomeostasis. Only one, involving the action of fibrils of aggregated Aβ1-42 peptide on certain AMPA-receptors is effective within a few seconds.

Much remains to be done. At present, all the proposed mechanisms are based on *in vitro* experiments, yet we apply them to the *in vivo* situation. Currently that is the only thing that we can do. But there is hope! The direct photography of Alzheimer plaques in the lining mouse brain (Bacskai *et al.*, 2003) gives us hope that a refinement of that technique might make possible the detection of at least early protofibrils. That would substantiate part of the mechanism #4 above, since visualization is through dyes that specifically interact with β-sheet-containing structures. A time course, using such a technique, would be very helpful

It remains to be seen whether therapeutic candidates for treating Alzheimer's disease, that are based on *in vitro* findings, will be successful in animal models and in human trials.

REFERENCES

Agell, N., Bachs, O., Rocamora, N., and Villalonga, P., 2002, Moduclation of the Ras/Raf/MEK/ERK pathway by Ca (2+), and calmodulin. *Cell Signal* **14:** 649-654.

Arispe, N., Rojas, E., Pollard, H.B., 1993a, Alzheimer disease amyloid beta protein forms calcium channels in bilayer membranes: blockade by tromethamine and aluminium. *Proc. Natl. Acad. Sci. USA* **90:** 567-571.

Arispe, N., Rollard, H.B., and Rojas, E., 1993b, Giant multilevel cation channels formed by Alzheimer disease amuyloid beta-protein [A beta P-(1-40)] in bilayer membranes. *Proc. Natl. Acad. Sci. USA* **90:** 10573-10577.

Bacskai, B.J., Hickey, G.A., Skoch, J., Kajdasz, S.T., Wang, Y., Huang, G.F., Mathis, C.A., Klunk, W.E., and Hyman, B.T., 2003, Four-dimensional multiphoton imaging of brain entry, amyloid binding, and clearance of an amyloid-beta ligand in transgenic mice. *Proc. Natl. Acad. Sci. USA* **100:** 12462-12467.

Behl, C., Davis, J.B., Lesley, R., and Schubert, D., 1994, Hydrogen peroxide mediates amyloid β protein toxicity. *Cell* **77:** 817-827.

Blanc, E.M., Toborek, M., Mark, R.J., Hennig, B., and Mattson, M.P., 1997, Amyloid beta-peptide induced cell monolayer albumin permeability, impairs glucose transport, and induces apoptosis in vascular endothelial cells. *J. Neurochem.* **68:** 1870-1881.

Blanchard, B.J., Konopka, G., Russell, M., and Ingram, V.M., 1997, Mechanis and prevention of neurotoxicity caused by beta-amyloid peptides: relation to Alzheimer's disease. *Brain Res.* **776:** 40-50.

Blanchard, B.J., Hiniker, A.E., Lu, C.C., Margolin, Y., Yu, A.S., and Ingram, V.M., 2000, Elimination of amyloid beta neurotoxicity. *J. Alzheimer's Dis.* **2:** 137-149.

Chapman, P.F., White, G.L., Jones, M.W., Cooper-Blacheter, D., Marshall, V.J., Irizarry, M., Younkin, L., Good, M.A., Bliss, T.V., Hyman, B.T., Younkin, S.G., and Hsiao, K.K., 1999, Impaired synaptic plasticity and learning in aged amyloid parecursor protein transgenic mice. *Nature Neurosci.* **2:** 271-276.

Durell, S.R., Guy, H.R., Arispe, N., Rojas, E., Pollard, H.B., 1994, Theoretical models of the ion channel structure of amyloid beta-protein. *Biophys. J.* **67:** 2137-2145.

Ferriera, A., Lu, Q., Orecchio, L., and Kosik, K. S., 1990, Selective phosphorylation of adult tau isoformas in mature hippocampal neurons exposed to fibrillar A beta. *Mol. Cell. Neurosci.* **9:** 220-224.

Glabe, C.G., 2004, Common structure and mechanism of amyloid pathogenesis in degenerative diseases. Abst.#112,. *Biophys. J.,* **86:** *Symp Suppl.,#1.*

Glenner, G.G., and Wong, C.W., 1984, Alzhemer's disease and Down's syndrome; sharing a unique cerebrovascular fibril protein. *Biochim. Biophys. Res. Commun.* **122:** 1131-1135.

Mattson, M.P., and Chan, S.L., 2003, Calcium orchestrates apoptosis. *Nat. Cell Biol.* **5:** 1051-1061, and comment: *ibid* pp. 1041-1043.

Mattson, M.P., and Furukawa, K., 2003, Dysregulation of calcium signaling in Alzheimer's pathogenesis: Impact on neuronal synaptic plasticity and cell death, neurogenesis, and glial functions. In: *Alzheimer's Disease and Related Disorders: Research Advances,* Eds. Iqubal, K., and Winblad, *B.*, Ana Asian Intl. Acad. Aging, Bucharest, Romania.

Mattson, M. P., Cheng, B., Davis, D., Bryant, K., Lieberburg, I., and Rydel, R., 1992, β-Amyloid peptides destabilize calcium homeostasis and render human cortical neurons vulnerable to excitotoxicity. *J. Neurosci.* **12:** 376-389.

Rao, R.V., Ellerby, H.M., and Bredesen, D.E., 2004, Coupling endoplasmic reticulum stress to the cell death program. *Cell Death Differ.* **11:** 372-380.

Sanderson, K.L., Butler, L., Ingram, V.M., 1997, Aggregates of a beta-amyloid peptide are required to induce calcium currents in neuron-like human teratocarcinoma cells: relation to Alzheimer's disease. *Brain Res.* **744:** 7-14.

Sisodia, S.S., Koo, E.H., Beyreuther, K., Unterbeck, A., and Price, D.L., 1990, Evidence that beta-amyloid protein in Alzheimer's disease is not derived by normal processing. *Science* **248:** 492-495.

Wang, H.W., Pasternak, J.F., Kuo, H., Ristic, H., Lambert, M.P., Chromy, B., Viola, K.L.,. Klein, W.L., Stine, W.B., Krafft, G.A., Trommer, B.L., 2002, Soluble oligomers of beta amyloid (1-42) inhibit long-term potentiation but not long-term depression in rat dentate gyrus. *Brain Res.* **924:** 133-140.

Yankner, B.A., Duffy, L.K., Kirschner, D.A., 1990, Neurotrophic and neurotoxic effects of amyloid beta protein: reversal by tachykinin neuropeptides. *Science* **250:** 279-282.

Chapter 18

Amyloid Inhibitors and β-Sheet Breakers

Claudio Soto and Lisbell Estrada
Protein Misfolding Disorders Laboratory, Department of Neurology, University of Texas Medical Branch, Galveston, TX, USA

Abstract: Compelling evidence indicates that a key pathological event in Alzheimer's disease is the misfolding and aggregation of normal soluble amyloid-β peptide into β-sheet-rich oligomeric structures which have a neurotoxic activity and ability to form insoluble amyloid deposits that accumulate in the brain. β-sheet breakers constitute a new class of drugs that are designed to specifically bind amyloid-β peptide blocking and/or reversing the misfolding process. In this article we review this approach and summarize the data supporting the view that β-sheet breakers could be serious candidates to combat this devastating disease.

Key words: Alzheimer's disease; amyloid; β-sheet breakers; therapy

1. INTRODUCTION

Alzheimer's disease (AD) is the most common cause of late-life dementia, and represents one of the leading causes of death in the developed world. AD is a progressive and devastating neurodegenerative disorder characterized by memory loss, and a variety of cognitive disabilities. This disease is a major social and health care problem, because of its increasing prevalence, long duration and high cost of care. Despite important efforts aimed at elucidating its underlying molecular mechanisms, there is still no effective treatment available for this disease.

The major neuropathological changes in the brains of AD patients are neuronal death, particularly in regions related to memory and cognition and the presence of abnormal intra- and extracellular abnormal protein aggregates, known as neurofibrillary tangles and amyloid plaques, respectively (Terry, 1994; Selkoe, 1997). Tangles are paired helical filaments composed largely of abnormally hyperphosphorylated tau protein (Terry, 1994). Amyloid plaques deposit in AD brain parenchyma in the form of senile plaques and around the cerebral vessels walls (Selkoe, 1997). Amyloid deposits have revealed to be essentially composed of aggregated amyloid-β peptide (Aβ). This 39-42 residue peptide is derived from the processing of a large ubiquitous type I transmembrane protein called APP (amyloid precursor protein), whose gene is located on chromosome 21 (Selkoe, 1997). The Aβ domain, which starts on the cell surface and ends within the membrane is sequentially cleaved by β- and γ-secretase at the N- and C-terminus, respectively (Selkoe, 1997). Although the role of tangles and amyloid plaques in the pathogenesis of the disease is not fully understood they represent a typical signature of AD and their presence is needed to make a definitive postmortem diagnosis of the disease.

2. IS AMYLOID THE CAUSE OF AD PATHOGENESIS?

In spite of numerous studies aimed at understanding the aetiology of AD, the triggering events leading to neuropathology remain to be clearly defined. Compelling evidence has accumulated in the last fifteen years indicating that amyloid may be the cause of neurodegeneration in AD or at least a central event in the pathogenesis (Selkoe, 1997; Selkoe, 2000b). These evidences come primarily from studies of individuals affected by the familial form of the disease or by Down's syndrome patients. Almost all patients with Down's syndrome (trisomy 21), which contain three copies of the APP gene, develop AD neuropathology at an early age, and the presence of diffuse amyloid deposits has been shown to precede other alterations (Mann, 1989). The fact that mutations in the amyloid protein precursor (APP) gene are associated to familial AD is a strong indication of the importance of amyloid in the pathogenesis of the disease (Selkoe, 2000a; Hardy *et al.*, 1998). In addition to *APP*, three other genes have been so far linked with AD: apolipoprotein E (*apoE*, chromosome 19), presenilin 1 (*PS1*, chromosome 14) and presenilin 2 (*PS2*, chromosome 1). Mutations or polymorphisms in these genes induce an increase in the production or amyloidogenicity of Aβ and therefore appear to be implicated in the disease through the formation of amyloid (Selkoe, 2000a; Selkoe, 2000b; Younkin, 1995; Fraser *et al.*, 2000; Price *et al.*, 1998). Moreover, transgenic mice that express high levels of human mutant APP progressively develop many of the pathological hallmarks of AD, including cerebral amyloid deposits, neuritic dystrophy, astrogliosis and behavioral alterations (Price *et al.*, 1998; Van Leuven, 2000).

Based on these findings it was thought during several years that mature amyloid plaques deposited in brain parenchyma were the culprit of neurodegeneration and disease. This idea was further supported by results *indicating* that amyloid aggregates are toxic to neuronal cells in culture and that citotoxicity was dependent upon Aβ aggregation (Pike *et al.*, 1993; Lorenzo and Yankner, 1994). However, in the last couple of years this view has been changing progressively in favour of soluble misfolded oligomeric intermediates as the toxic specie (Caughey and Lansbury, 2003; Walsh *et al.*, 2002). Recent in vitro experiments obtained in cell culture experiments suggest that the soluble oligomeric intermediates (also known as Aβ-derived diffusible ligands) and protofibrils could be even more toxic than amyloid fibrils (Walsh *et al.*, 1999; Lambert *et al.*, 1998). Moreover, temporal studies of the appearance of AD-like alterations in APP transgenic animals have shown that significant cerebral damage and cognitive impairment appear before amyloid plaque detection (Moechars *et al.*, 1999). The nature of the toxic Aβ specie is very important to design efficient therapeutic strategies based on arresting the negative influence of the Aβ pathway. Indeed, some scientists have proposed that the deposition of amyloid plaques could even be considered a protective event allowing the deposition and isolation of the toxic abnormally folded Aβ oligomers (Caughey and Lansbury, 2003).

A clear mechanism to explain the role of Aβ in neurodegeneration is still missing (Yankner, 1996; Soto, 2003). Cell culture experiments involving the conversion of soluble Aβ into β-sheet rich Aβ have suggested that amyloid deposits, protofibrils and/or soluble oligomers could be directly toxic to neurons (Selkoe, 1997; Soto, 2003; Caughey and Lansbury, 2003). Various mechanisms have been proposed to explain Aβ neurotoxicity, including mitochondrial redox activity impairment leading to increased free radicals, intracellular Ca^{2+} increase, ion channels formation, or signal transduction by interaction with specific cellular receptors (Soto, 2003; Yankner, 1996). Alternatively, amyloid could be indirectly toxic through a locally induced chronic inflammatory response, with the abnormal Aβ structures acting as an irritant that triggers inflammation (Soto, 2003; McGeer and McGeer, 1998). Indeed, AD brains are characterized by microglia and astrocyte activation, as well as by numerous inflammatory proteins, including complement factors, pro-inflammatory cytokines and acute phase-reactant proteins (McGeer and McGeer, 1998). However, it remains to be clarified whether one central mechanism mediates amyloid toxicity or if it is dependent on an interplay of several direct and indirect processes.

3. AMYLOID AGGREGATION INHIBITORS

Aβ misfolding and aggregation is probably the first pathological processes in AD. This fact and the abundant knowledge accumulated about the molecular mechanism of amyloid formation make the inhibition of Aβ misfolding and oligomerization an attractive therapeutic target for AD. However, uncertainties respect to which Aβ species is mostly involved in the disease pathogenesis complicate this approach, because inhibiting the process in the wrong step may lead to accumulation of toxic intermediates.

Several unrelated small molecules have been shown to prevent and/or reverse Aβ polymerization *in vitro*. Among these compounds is possible to mention the following (Fig. 1): congo red (Lorenzo and Yankner, 1994), hexadecyl-N-methylpiperidinium bromide (Wood *et al.*, 1996), small sulfonated anions (Kisilevsky *et al.*, 1995), benzofuran-based compounds (Allsop *et al.*, 2001), rifampicin (Tomiyama *et al.*, 1994), melatonin (Pappolla *et al.*, 1998), nicotine (Salomon *et al.*, 1996), estrogen (Hosoda *et al.*, 2001), glycosaminoglycans mimetics (Gervais *et al.*, 2001), nitrophenols (De Felice *et al.*, 2001), tetracycline (Forloni *et al.*, 2001), anthracycline 4'-iodo-4'-deoxydoxorubicin (Merlini *et al.*, 1995), clioquinol (Cherny *et al.*, 2001), ibuprofen (Lim *et al.*, 2000) and N,N'-bis(3-hydroxyphenyl)pyridazine-3,6-diamine (Nakagami *et al.*, 2002). The activity of many of these compounds has also been shown in vivo using AD animal models and even some of them are currently under clinical evaluation in AD patients. However, the usefulness of these small molecules as amyloid inhibitors is compromised by their lack of specificity and their (in most of the cases) unknown mechanism of action, which makes it difficult to improve them. In addition, many of these compounds are highly toxic.

A more rational approach to make inhibitors of amyloid formation is to design specific peptide ligands based on the well-known self-recognition ability of Aβ and on the study of the structural requirements for Aβ aggregation (Soto, 1999; Findeis, 2002; Mason *et al.*, 2003). This knowledge was used to develop peptide inhibitors that contain the self-recognition motif, which binds Aβ and partially inhibits amyloid fibril formation *in vitro*. Several different peptides around the sequence 16-22 of Aβ are under development (Soto, 1999; Findeis, 2002; Mason *et al.*, 2003).

Tjernberg and co-workers showed that the Aβ(16–20) peptide is able to bind full-length Aβ and prevent its assembly into fibrils (Tjernberg *et al.*, 1996).

Figure 1

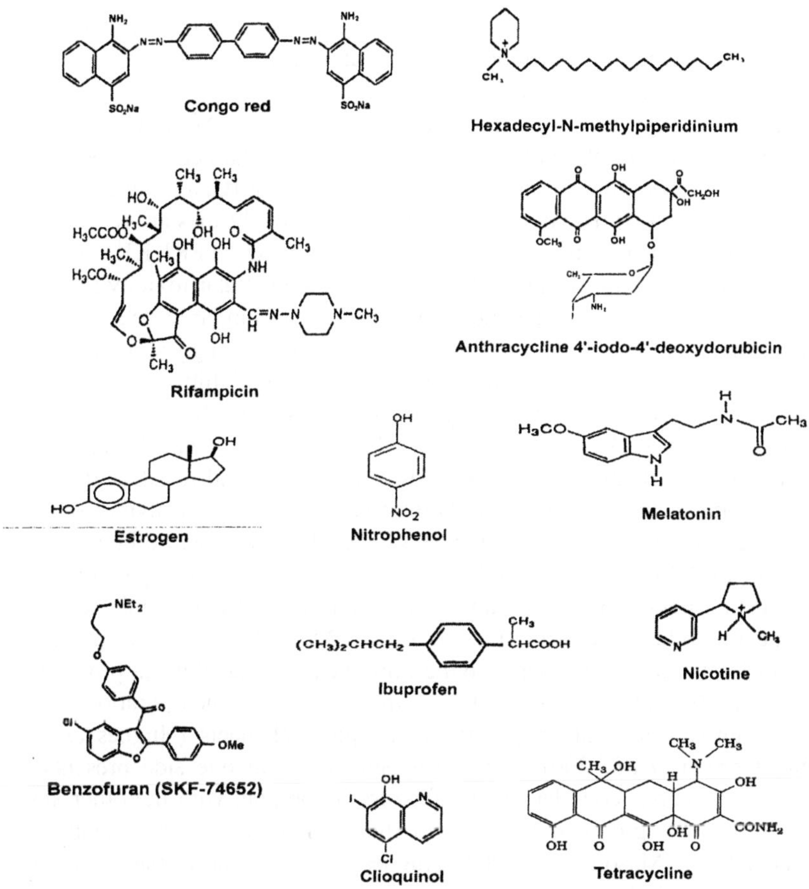

Figure 1. Chemical structure of some small-molecule amyloid inhibitors. Several structurally diverse small chemical molecules have been reported to inhibit either amyloid fibril formation or amyloid neurotoxicity in vitro. Some of them have also been shown active in animal models of AD. Unfortunately, their lack of specificity and, in many cases, toxicity diminishes their therapeutic use.

Using molecular graphics simulations, they hypothesized that it binds stereospecifically and in an antiparallel conformation to Aβ. However, Aβ (16–20) spontaneously aggregates into amyloid-like fibrils and thus its use as an inhibitor might be problematic. Therefore, several groups began to modify this sequence to produce peptide derivatives containing the self-recognition motif, but at the same time a disrupting element enhancing their inhibitory activity. Our approach using β-sheet breaking amino acids was the first to lead to modified peptides with inhibitory activity (Soto *et al.*, 1996). This strategy will be discussed in the next section.

Based on the fact that the major force driving Aβ aggregation is hydrophobicity, Murphy and colleagues have added charged residues to the ends of the recognition motif as a disrupting element (Ghanta *et al.*, 1996; Pallitto *et al.*, 1999). Having shown that at least three lysines are required as an appropriate disrupting element, the compound (KLVFFKKKK) showed activity in altering fibril morphology and reducing cellular toxicity in vitro. The anionic disrupting compound KLVFFEEEE had similar effects, whereas the neutral compound KLVFFSSSS was ineffective, suggesting that the charged nature of the disrupting element is critical (Pallitto *et al.*, 1999). Another approach to produce peptide inhibitors have been reported by Findeis and co-workers (Findeis *et al.*, 1999; Findeis *et al.*, 2001). Their strategy was to retain a peptide sequence that could bind to Aβ and add a bulky group, such as a steroid, at its terminus to hinder Aβ polymerization. The all-D-amino acid peptide cholyl-LVFFA-OH was shown to be a potent inhibitor of Aβ polymerization, but was cleared up almost completely upon hepatic first pass, possibly because the cholyl group was recognized as an endogenous bile component (Findeis *et al.*, 2001). Several other peptides with different chemical groups having improved pharmacological properties are under development. Various teams are studying the incorporation of *N*-methyl amino acids into peptides as disrupting elements (Hughes *et al.*, 2000; Gordon *et al.*, 2001). The idea behind is that one side presents a hydrogen-bonding `complementary' face to the protein, with the other side having *N*-methyl groups in place of backbone NH groups, thus presenting a `blocking' face (Mason *et al.*, 2003). Hughes *et al.* have shown that *N*-methyl derivatives of Aβ(25–35) are able to prevent aggregation and inhibit toxicity in PC-12 cells (Hughes *et al.*, 2000). Meredith and co-workers investigated *N*-methylated peptides corresponding to 16–22 and subsequently 16–20 sequence of Aβ. These peptides can prevent Aβ fibrils from forming and break down preformed fibrils (Gordon *et al.*, 2001). N-methylation contributes additionally to make the peptides more resistant to proteolysis than regular peptides (Adessi and Soto, 2002). More recently, Kapurniotu and colleagues reported an Aβ1-28 analog constrained by an internal cycle between residues Lys17 and Ala21. This modified peptide inhibited Aβ aggregation and cytotoxicity (Kapurniotu *et al.*, 2003).

4. β-SHEET BREAKERS

β-sheet breakers have emerged in the last few years as the prototype class of compounds inhibiting and reversing protein misfolding and aggregation. Indeed, although we coined this name to refer specifically to our compounds rationally-designed to break β-sheets, in recent years the β-sheet breaker concept has been used to refer to any compound with amyloid inhibitory properties, such as some of those described in the previous section.

Our approach to produce β-sheet breakers uses the Aβ self-recognition motif (17-20) to achieve binding and specificity, but replacing a residue

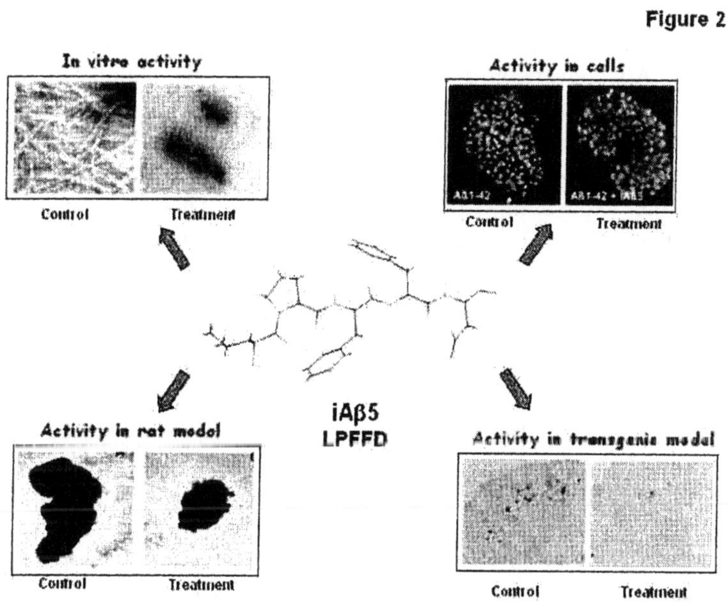

Figure 2. β-sheet breaker peptides as a promising approach for the treatment of AD. A 5 residues peptide (chemical structure in center panel) has been shown to be active in several in vitro, cellular and in vivo models. In vitro studies showed that the compound was able to inhibit and dissolve amyloid plaques in the test tube, as studied by electron microscopy (top, left panel) among other techniques. In cell cultures, the peptide prevented neuronal death induced by amyloid (top, right panel) as studied by staining procedures that allow to distinguish healthy (green) from death (orange) cells. In vivo experiments were done by using two different animal models; a rat model of cerebral amyloidosis (bottom, left panel) and a double transgenic mice model of AD (bottom, right panel). In both models, the 5 residues b-sheet breaker showed a significant activity on preventing and dissolving amyloid plaques.

important for forming β-sheets by an amino acid thermodynamically unable to fit inside this structure (Soto *et al.*, 1996; Soto, 1999). Valine at position 18 of Aβ plays an important role on stabilizing β-sheet folding in Aβ (Soto *et al.*, 1995), but seems not to be absolutely necessary for self-recognition. Therefore, we decided to replace this amino acid by proline, a residue that because of its particular chemical structure is an efficient β-sheet breaker (Wood *et al.*, 1995; Kim and Berg, 1993). A prototype 5-residue β-sheet breaker peptide (iAβ5 Seq: LPFFD) has been tested in different in vitro, cellular and *in vivo* assays (Figure 2). The *in vitro* activity was quantitatively demonstrated using a thioflavine T binding assay, which is a commonly used fluorometric method to quantify amyloid, and qualitatively confirmed by electron microscopic examination of amyloid fibrils (Soto *et al.*, 1996; Soto *et al.*, 1998). This peptide appeared not only to inhibit the abnormal Aβ conversion and subsequent amyloid formation, but also to dissolve preformed fibrils (Soto *et al.*, 1996; Soto *et al.*, 1998). Human neuroblastoma cell culture experiments revealed that iAβ5 was able to prevent neuronal death induced by the formation of β-sheet-rich oligomeric Aβ structures (Soto *et al.*, 1998). Two animal models have been employed to monitor the activity of β-sheet breaker peptides *in vivo*. In the first model, amyloid deposition was induced by injecting non-aggregated Aβ1-42 in rat brain. After some time, a single fibrillar lesion with ultrastructural properties similar to AD amyloid plaques could be observed at the site of injection together with some AD-typical neurodegenerative features such as extensive neuronal shrinkage, astrocytosis and microglial activation (Soto *et al.*, 1998; Sigurdsson *et al.*, 2000). Co-injecting iAβ5 with Aβ1-42 reduced cerebral Aβ accumulation and completely prevented the formation of fibrillar amyloid-like lesions (Soto *et al.*, 1998). In a second experiment in the rat model, iAβ5 even proved to be able to induce a significant reduction in the size of preformed Aβ fibrils when it was injected into the cerebral amygdala 8 days after injecting Aβ at the same place (Sigurdsson *et al.*, 2000). A reversion of the associated cerebral histopathological changes, such as neuronal shrinkage and microglial activation, was also observed in this experiment. No effect was detected by injecting unrelated peptides (used as controls) under the same conditions.

5. FROM β-SHEET BREAKER PEPTIDES TO β-SHEET BREAKER DRUGS

Like most short unmodified peptides, β-sheet breaker peptides are very prone to peptidase degradation and exhibit very short half-lives *in vivo* (Adessi *et al.*, 2003; Adessi and Soto, 2002). For example, incubation of iAβ5 in human plasma revealed a half-life of only 5 minutes (Permanne *et al.*, 2002a). Since no metabolites could be detected, it was concluded that degradation was complete and most probably due to exopeptidases. In order to block this enzymatic activity, iAβ5 was end-protected by N-terminal acetylation and C-terminal amidation, yielding iAβ5p (Ac-LPFFD-NH_2). iAβ5p appeared to be very stable both in human plasma and CSF, with little or no degradation after 24 hours of incubation (Permanne *et al.*, 2002a). This better stability made possible the evaluation of iAβ5p efficacy in a more relevant transgenic animal model by reasonable routes of administration. A double transgenic mouse model overexpressing human APP with the London mutation (V717I) and human PS1 with the A246E mutation develops many AD pathological features, such as extensive deposition of amyloid plaques, neuritic dystrophy, astrogliosis and some degree of tau neuropathology (Dewachter *et al.*, 2000). Such animals were treated with iAβ5p via intracerebroventricular infusion (icv) or intraperitoneal injection (ip, three times a week) for a period of 8 weeks. This treatment resulted in a significant lowering of cerebral amyloid load (67.3% and 46% for icv and ip, respectively) compared to vehicle administration, as well as a decreased brain inflammation and neuronal loss (Permanne *et al.*, 2002b). However, by contrast with *in vitro* stability experiments, pharmacokinetic studies in rats revealed an *in vivo* iAβ5p half-life of only 37 minutes. In order to further increase enzymatic stability, we identified the major cleavage sites in vivo and protected these peptide bonds by introducing various chemical modifications around them (Adessi *et al.*, 2003). From the numerous newly engineered iAβ5p derivatives, one peptide appeared to have the same *in vitro* activity as iAβ5p in inhibiting amyloid fibril formation and amyloid neurotoxicity, while exhibiting a much higher stability in rat brain homogenate and a 10-fold greater half-life after intravenous administration to mice (Adessi *et al.*, 2003).

Minimizing the weakness of a peptide by specific chemical modifications may increase the drug-like properties of the compound, but in order to have an orally available and highly effective drug, the design of a non-peptide mimetic is needed. For this purpose, it is important to know the mechanism of action of the lead compound, the chemical groups responsible for activity and the tridimensional structure of the active compound (Adessi and Soto, 2002; Adessi and Soto, 2002; Moore, 1994).

Although numerous experiments involving various β-sheet breaker peptides have shown that these drug candidates were very efficient both *in vitro* and *in vivo*, the mechanism underlying their activity remains mostly unclear. β-sheet breaker peptides appear to bind to the central self-recognition motif of Aβ through hydrophobic interactions. Hydrophobicity is also the major driving force for Aβ-Aβ interaction through the self-recognition sequence (Soto *et al.*, 1994). Indeed, the replacement of hydrophobic residues by hydrophilic ones in this region of Aβ has been shown to prevent amyloid formation, thereby suggesting that aggregation of Aβ monomers is driven by hydrophobic interactions (Hilbich *et al.*, 1992). Since β-sheet breaker peptides exhibit a certain degree of homology to the self-recognition region of Aβ, they may compete with Aβ monomers to bind this region through similar interactions. Consistent with this view, molecular modelling studies of the interaction between the three-dimensional structure of Aβ1-40 and iAβ5p, both elucidated by nuclear magnetic resonance, suggest that the β-sheet breaker binds to a hydrophobic pocket in the central region of Aβ in an antiparallel way (unpublished observations).

From our structure-activity studies, it seems that Pro and Phe at positions 2 and 3, respectively, play a fundamental role in the activity of β-sheet breaker peptides (Adessi *et al.*, 2003). It has been shown previously by several groups that Phe 19 (equivalent to the Phe 3 of iAβ5p) is critical for Aβ-Aβ interaction and amyloid formation (Hilbich *et al.*, 1992; Tjernberg *et al.*, 1996; Wood *et al.*, 1995). It is likely that the phenylalanines (second and third position) and leucine (first position) residues are responsible for selective binding of the β-sheet breaker peptides to Aβ. Also an ionic interaction between positively and negatively charged residues in Aβ and iAβ5 appears to be important for binding affinity. On the other hand, the Pro residue does not seem to play a role on binding, but is critical for activity. These results suggest the importance of maintaining a key balance between the sequence homology between Aβ and the β-sheet breaker, and the presence of amino acids able to disrupt β-sheet folding. Indeed, a β-sheet breaker peptide with full sequence homology to Aβ would have a strong affinity but would not be able to block the pathological conformational change. On the other hand, a compound with too many proline residues to inhibit β-sheet structures would be inactive because of low affinity due to poor homology.

With all this information in hand, now the aim is to design compounds with similar chemical groups in the correct tridimensional position but tied together by a non-peptidic scaffold (Adessi and Soto, 2002; Moore, 1994). Although this is a challenging task, the small size of the peptide lead and the abundant knowledge of its structure and properties, coupled with the accelerated advance on scaffold chemistry, make it feasible.

REFERENCES

Adessi, C., Frossard, M.J., Boissard ,C., Fraga, S., Bieler, S., Ruckle, T., Vilbois, F., Robinson, S.M., Mutter, M., Banks, W.A., and Soto,C., 2003, Pharmacological profiles of peptide drug candidates for the treatment of Alzheimer's disease. *J. Biol. Chem.* **278:** 13905-13911.

Adessi, C. and Soto, C., 2002, Converting a peptide into a drug: Strategies to improve stability and bioavailability. *Curr. Med. Chem.* **9:** 963-978.

Allsop, D., Gibson G., Martin, I.K., Moore S., Turnbull, S., and Twyman, L.J., 2001, 3-p-Toluoyl-2-[4'-(3-diethylaminopropoxy)-phenyl]-benzofuran and 2-[4'-(3-diethylaminopropoxy)-phenyl]-benzofuran do not act as surfactants or micelles when inhibiting the aggregation of beta-amyloid peptide. *Bioorg. Med. Chem. Lett.* **11:** 255-257.

Caughey, B. and Lansbury, P.T., 2003, Protofibrils, pores, fibrils, and neurodegeneration: separating the responsible protein aggregates from the innocent bystanders. *Annu. Rev. Neurosci.* **26:** 267-298.

Cherny,R.A., Atwood,C.S., Xilinas,M.E., Gray,D.N., Jones,W.D., McLean,C.A., Barnham,K.J., Volitakis,I., Fraser,F.W., Kim,Y., Huang,X., Goldstein,L.E., Moir,R.D., Lim,J.T., Beyreuther,K., Zheng,H., Tanzi,R.E., Masters,C.L., and Bush,A.I., 2001, Treatment with a copper-zinc chelator markedly and rapidly inhibits beta-amyloid accumulation in Alzheimer's disease transgenic mice. *Neuron* 30: 665-676.

De Felice,F.G., Houzel,J.C., Garcia-Abreu,J., Louzada,P.R., Jr., Afonso,R.C., Meirelles,M.N., Lent,R., Neto,V.M., and Ferreira,S.T., 2001, Inhibition of Alzheimer's disease beta-amyloid aggregation, neurotoxicity, and in vivo deposition by nitrophenols: implications for Alzheimer's therapy. *FASEB J.* 15: 1297-1299.

Dewachter,I., Van Dorpe,J., Smeijers,L., Gilis,M., Kuiperi,C., Laenen,I., Caluwaerts,N., Moechars,D., Checler,F., Vanderstichele,H., and Van Leuven,F., 2000, Aging increased amyloid peptide and caused amyloid plaques in brain of old APP/V717I transgenic mice by a different mechanism than mutant presenilin1. *J. Neurosci.* 20: 6452-6458.

Findeis,M.A., 2002, Peptide inhibitors of beta amyloid aggregation. *Curr. Top. Med. Chem.* 2: 417-423.

Findeis,M.A., Lee,J.J., Kelley,M., Wakefield,J.D., Zhang,M.H., Chin,J., Kubasek,W., and Molineaux,S.M., 2001, Characterization of cholyl-leu-val-phe-phe-ala-OH as an inhibitor of amyloid beta-peptide polymerization. *Amyloid.* 8: 231-241.

Findeis,M.A., Musso,G.M., Arico-Muendel,C.C., Benjamin,H.W., Hundal,A.M., Lee,J.J., Chin,J., Kelley,M., Wakefield,J., Hayward,N.J., and Molineaux,S.M., 1999, Modified-peptide inhibitors of amyloid beta peptide polymerization. *Biochemistry* 38: 6791-6800.

Forloni,G., Colombo,L., Girola,L., Tagliavini,F., and Salmona,M., 2001, Anti-amyloidogenic activity of tetracyclines: studies in vitro. *FEBS Lett.* 487: 404-407.

Fraser,P.E., Yang,D.S., Yu,G., Levesque,L., Nishimura,M., Arawaka,S., Serpell,L.C., Rogaeva,E., and George-Hyslop,P., 2000, Presenilin structure, function and role in Alzheimer disease. *Biochim. Biophys. Acta* 1502: 1-15.

Gervais,F., Chalifour,R., Garceau,D., Kong,X., Laurin,J., Mclaughlin,R., Morissette,C., and Paquette,J., 2001, Glycosaminoglycan mimetics: a therapeutic approach to cerebral amyloid angiopathy. *Amyloid.* 8 Suppl 1: 28-35.

Ghanta, J., Shen, C.L., Kiessling, L.L., and Murphy, R.M., 1996, A strategy for designing inhibitors of beta-amyloid toxicity. *J. Biol. Chem.* **271:** 29525-29528.

Gordon, D.J., Sciarretta, K.L., and Meredith, S.C., 2001, Inhibition of beta-amyloid(40) fibrillogenesis and disassembly of beta-amyloid(40) fibrils by short beta-amyloid congeners containing N-methyl amino acids at alternate residues. *Biochemistry* **40:** 8237-8245.

Hardy, J., Duff, K., Hardy, K.G., Perez-Tur, J., and Hutton,M., 1998, Genetic dissection of Alzheimer's disease and related dementias: amyloid and its relationship to tau. *Nat. Neurosci.* **1:** 355-358.

Hilbich, C., Kisters-Woike, B., Reed, J., Masters, C.L., and Beyreuther, K., 1992, Substitutions of hydrophobic amino acids reduce the amyloidogenicity of Alzheimer's disease beta A4 peptides. *J. Mol. Biol.* **228:** 460-473.

Hosoda, T., Nakajima, H., and Honjo, H., 2001, Estrogen protects neuronal cells from amyloid beta-induced apoptotic cell death. *Neuroreport* **12:** 1965-1970.

Hughes, E., Burke R.M., and Doig, A.J., 2000, Inhibition of toxicity in the beta-amyloid peptide fragment beta -(25-35) using N-methylated derivatives: a general strategy to prevent amyloid formation. *J. Biol. Chem.* **275:** 25109-25115.

Kapurniotu, A., Buck, A., Weber, M., Schmauder, A., Hirsch, T., Bernhagen, J., and Tatarek-Nossol, M., 2003, Conformational restriction via cyclization in beta-amyloid peptide Abeta(1-28) leads to an inhibitor of Abeta(1-28) amyloidogenesis and cytotoxicity. *Chem. Biol.* **10:** 149-159.

Kim, C.A. and Berg, J.M., 1993, Thermodynamic beta-sheet propensities measured using a zinc-finger host peptide. *Nature* **362:** 270.

Kisilevsky, R., Lemieux, L.J., Fraser, P.E., Kong, X., Hultin, P.G., and Szarek,W.A., 1995, Arresting amyloidosis in vivo using small-molecule anionic sulphonates or sulphates: implications for Alzheimer's disease. *Nat. Med.* **1:** 143-148.

Lambert, M.P., Barlow, A.K., Chromy, B.A., Edwards, C., Freed, R., Liosatos, M., Morgan, T.E., Rozovsky, I., Trommer, B., Viola, K.L., Wals, P., Zhang, C., Finch, C.E., Krafft, G.A., and Klein,W.L., 1998, Diffusible, nonfibrillar ligands derived from Abeta1-42 are potent central nervous system neurotoxins. *Proc. Natl. Acad. Sci. USA* **95:** 6448-6453.

Lim, G.P., Yang, F., Chu, T., Chen, P., Beech, W., Teter, B., Tran, T., Ubeda, O., Ashe, K.H., Frautschy, S.A., and Cole, G.M., 2000, Ibuprofen suppresses plaque pathology and inflammation in a mouse model for Alzheimer's disease. *J. Neurosci.* **20:** 5709-5714.

Lorenzo, A. and Yankner, B.A., 1994, Beta-amyloid neurotoxicity requires fibril formation and is inhibited by congo red. *Proc. Natl. Acad. Sci. USA* **91:** 12243-12247.

Mann, D.M., 1989, Cerebral amyloidosis, ageing and Alzheimer's disease; a contribution from studies on Down's syndrome. *Neurobiol. Aging* **10:** 397-399.

Mason, J.M., Kokkoni, N., Stott, K., and Doig, A.J., 2003, Design strategies for anti-amyloid agents. *Curr. Opin. Struct. Biol.* **13:** 526-532.

McGeer, E.G. and McGeer, P.L., 1998, The importance of inflammatory mechanisms in Alzheimer disease. *Exp. Gerontol.* **33:** 371-378.

Merlini, G., Ascari, E., Amboldi, N., Bellotti, V., Arbustini, E., Perfetti, V., Ferrari, M., Zorzoli, I., Marinone, M.G., and Garini, P., 1995, Interaction of the anthracycline 4'-iodo-4'-deoxydoxorubicin with amyloid fibrils: inhibition of amyloidogenesis. *Proc. Natl. Acad. Sci. USA* **92:** 2959-2963.

Moechars, D., Dewachter, I., Lorent, K., Reverse, D., Baekelandt, V., Naidu,A., Tesseur, I., Spittaels, K., Haute, C.V., Checler, F., Godaux, E., Cordell, B., and Van Leuven, F., 1999, Early phenotypic changes in transgenic mice that overexpress different mutants of amyloid precursor protein in brain. *J. Biol. Chem.* **274:** 6483-6492.

Moore,G.J., 1994, Designing peptide mimetics, *Trends Pharmacol. Sci.* **15:** 124-129.

Nakagami,Y., Nishimura, S., Murasugi, T., Kaneko, I., Meguro, M., Marumoto, S., Kogen, H., Koyama, K., and Oda, T., 2002, A novel beta-sheet breaker, RS-0406, reverses amyloid beta-induced cytotoxicity and impairment of long-term potentiation in vitro. *Br. J. Pharmacol.* **137:** 676-682.

Pallitto, M.M., Ghanta, J., Heinzelman, P., Kiessling, L.L., and Murphy, R.M., 1999, Recognition sequence design for peptidyl modulators of beta-amyloid aggregation and toxicity. *Biochemistry* **38:** 3570-3578.

Pappolla, M., Bozner, P., Soto, C., Shao, H., Robakis, N.K., Zagorski, M., Frangione, B., and Ghiso, J., 1998, Inhibition of Alzheimer beta-fibrillogenesis by melatonin. *J. Biol. Chem.* **273:** 7185-7188.

Permanne, B., Adessi, C., Fraga, S., Frossard, M.J., Saborio, G.P., and Soto, C., 2002a, Are beta-sheet breaker peptides dissolving the therapeutic problem of Alzheimer's disease? *J. Neural Transm.* **Suppl** 293-301.

Permanne, B., Adessi, C., Saborio, G.P., Fraga, S., Frossard, M.J., Van Dorpe, J., Dewachter, I., Banks, W.A., Van Leuven, F., and Soto, C., 2002b, Reduction of amyloid load and cerebral damage in a transgenic mouse model of Alzheimer's disease by treatment with a beta-sheet breaker peptide. *FASEB J.* **16:** 860-862.

Pike, C.J., Burdick, D., Walencewicz, A.J., Glabe, C.G., and Cotman, C.W., 1993, Neurodegeneration induced by beta-amyloid peptides in vitro: the role of peptide assembly state. *J. Neurosci.* **13:** 1676-1687.

Price, D.L., Tanzi, R.E., Borchelt, D.R., and Sisodia, S.S., 1998, Alzheimer's disease: genetic studies and transgenic models. *Annu. Rev. Genet.* **32:** 461-493.

Salomon, A.R., Marcinowski, K.J., Friedland, R.P., and Zagorski, M.G., 1996, Nicotine inhibits amyloid formation by the beta-peptide. *Biochemistry* **35:** 13568-13578.

Selkoe, D.J., 1997, Alzheimer's disease: genotypes, phenotypes, and treatments. *Science* **275:** 630-631.

Selkoe, D.J., 2000a, The genetics and molecular pathology of Alzheimer's disease: roles of amyloid and the presenilins. *Neurol. Clin.* **18:** 903-922.

Selkoe, D.J., 2000b, The origins of Alzheimer disease - A is for amyloid. *Jama: Journal of the American Medical Association* **283:** 1615-1617.

Sigurdsson, E.M., Permanne, B., Soto, C., Wisniewski, T., and Frangione, B., 2000, In vivo reversal of amyloid-beta lesions in rat brain. *J. Neuropathol. Exp. Neurol.* **59:** 11-17.

Soto, C., 1999, Plaque busters: strategies to inhibit amyloid formation in Alzheimer's disease. *Mol. Med. Today* **5:** 343-350.

Soto, C., 2003, Unfolding the role of Protein Misfolding in Neurodegenerative Diseases. *Nature Rev. Neurosci.* **4:** 49-60.

Soto, C., Branes, M.C., Alvarez, J., and Inestrosa, N.C., 1994, Structural determinants of the Alzheimer's amyloid beta-peptide. *J. Neurochem.* **63:** 1191-1198.

Soto, C., Castano, E.M., Frangione, B., and Inestrosa, N.C., 1995, The alpha-helical to beta-strand transition in the amino-terminal fragment of the amyloid beta-peptide modulates amyloid formation. *J. Biol. Chem.* **270:** 3063-3067.

Soto, C., Kindy, M.S., Baumann, M., and Frangione, B., 1996, Inhibition of Alzheimer's amyloidosis by peptides that prevent beta-sheet conformation. *Biochem. Biophys. Res. Commun.* **226:** 672-680.

Soto, C., Sigurdsson, E.M., Morelli, L., Kumar, R.A., Castano, E.M., and Frangione, B., 1998, Beta-sheet breaker peptides inhibit fibrillogenesis in a rat brain model of amyloidosis: implications for Alzheimer's therapy. *Nature Med.* **4:** 822-826.

Terry, R.D., 1994, Neuropathological changes in Alzheimer disease. *Prog. Brain Res.* **101:** 383-390.

Tjernberg, L.O., Naslund, J., Lindqvist, F., Johansson, J., Karlstrom, A.R., Thyberg, J., Terenius, L., and Nordstedt, C., 1996, Arrest of beta-amyloid fibril formation by a pentapeptide ligand. *J. Biol. Chem.* **271:** 8545-8548.

Tomiyama, T., Asano, S., Suwa,Y., Morita, T., Kataoka, K., Mori, H., and Endo, N., 1994, Rifampicin prevents the aggregation and neurotoxicity of amyloid beta protein in vitro. *Biochem. Biophys. Res. Commun.* **204:** 76-83.

Van Leuven, F., 2000, Single and multiple transgenic mice as models for Alzheimer's disease. *Progress in Neurobiology* **61:** 305-312.

Walsh, D.M., Hartley, D., Kusumoto, Y., Fezoui, Y., Condrom, M.M., Lomakin, A., Benedek, G.B., Selkoe, D.J., and Teplow, D.B., 1999, Amyloid beta-protein fibrillogenesis. Structure and biological activity of protofibrillar intermediates. *J. Biol. Chem.* **274:** 25945-25952.

Walsh, D.M., Klyubin, I., Fadeeva, J.V., Rowan, M.J., and Selkoe, D.J., 2002, Amyloid-beta oligomers: their production, toxicity and therapeutic inhibition. *Biochem. Soc. Trans.* **30:** 552-557.

Wood, S.J., MacKenzie, L., Maleeff, B., Hurle, M.R., and Wetzel, R., 1996, Selective inhibition of Aβ fibril formation. *J. Biol. Chem.* **271:** 4086-4092.

Wood, S.J., Wetzel, R., Martin, J.D., and Hurle, M.R., 1995, Prolines and amyloidogenicity in fragments of the Alzheimer's peptide beta/A4. *Biochemistry* **34:** 724-730.

Yankner, B.A., 1996, Mechanisms of neuronal degeneration in Alzheimer's disease. *Neuron* **16:** 921-932.

Younkin, S.G., 1995, Evidence that Aβ 42 is the real culprit in Alzheimer's disease. *Ann. Neurol.* **37:** 287-288.

Chapter 19

Cholesterol and Alzheimer's Disease: Statins, Cholesterol Depletion in APP Processing and Aβ Generation

Tobias Hartman
Center for Molecular Biology Heidelberg (ZMBH), university of Heidelberg, Im Neuenheimer Feld 282, D-69120 Heidelberg, Germany

Abstract: Molecular and more specifically subcellular analyses of the neurodegenerative mechanisms involved in Alzheimer's disease (AD) had been considered most of the time an interplay of proteins and genes. However, some of the observations could not be explained this way. Recently, a number of research groups found the missing link ... lipids! Among the variety of lipids that had been investigated, most investigations had been focused on cholesterol and some derivatives. A recent statistic found that for every primary research article on AD and cholesterol/statins, approximately two reviews were published. This clearly reflects as much the interest in this topic, as it gives evidence that this field is still in its juvenile phase and most aspects have yet to be covered or clarified. To date there is evidently no final answer to whether this approach will eventually provide a therapeutic solution to treat or prevent AD. At the end of the day such answers can only be obtained from clinical studies and to date only two studies with a suitable design have published their results, one of them with preliminary results only. This review focuses on what we know about the cellular mechanisms involved in the AD-lipid connection and what kinds of problematic issues; theoretical and practical, are at hand.

Key words: Cholesterol, neurodegeneration, amyloid, risk factor, therapy, molecular mechanism, disease prevention, animal model, sub cellular compartments, lipid trafficking, statin.

1. INTRODUCTION

Cholesterol has its own reputation and is best known as a risk factor in cardio-vascular diseases. Accordingly, cholesterol is the best-studied lipid, however very little is known about its role in brain. Recognition as straightforward risk factor in AD came late, the overall risk it contributes is small but significant (Kivipelto *et al.*, 2002; Kivipelto *et al.*, 2001; Pappolla *et al.*, 2003). There is more indirect indication that point towards cholesterol. But this risk factor is beneficial and very strongly reduces the AD risk. These are cholesterol-lowering drugs called statins. In retrospective studies, patients who received statins were strongly underrepresented in the AD, but not the control group(s) (Jick *et al.*, 2000; Rockwood *et al.*, 2002; Wolozin *et al.*, 2000). And there is more to this. Carriers of the ε4 allele of cholesterol (and other lipids) transporting protein ApoE develop AD symptoms years before others(Corder *et al.*, 1993), since ApoE alleles and cholesterol seem to work hand in hand when it comes to AD progression (Evans *et al.*, 2004). ApoE e4 is the major genetic risk factor in AD (St George-Hyslop, 2000), it therefore clearly deserves some attention. Cholesterol turnover in adult human cerebral neurons is comparably low; overall, brain cholesterol half-life is approx. 6 month (Bjorkhem and Meaney, 2004; Dietschy and Turley, 2001). Thus brain cholesterol is very stable as compared to blood cholesterol in which the steady state level is fully adjusted after a maximum of only 6 weeks statin treatment. Cholesterol is also not shuffled between the peripheral blood system and neurons or glia in any reasonable amount. This is even more remarkable, considering that the central nervous system represents approx. 2% of the entire body weight but contains approx. one quarter of the cholesterol. Therefore, regarding cholesterol, the brain might be seen as a largely autonomous organ. The need for cholesterol in neurons may change significantly upon stimulation. For instance, synapses are very rich in cholesterol and some of the traffic along axons and dendrites may critically depend on cholesterol to form the appropriate raft domains (Mauch *et al.*, 2001; Pfrieger, 2003a, 2003b). Under conditions of increased cholesterol need the limited capability of neurons to synthesize cholesterol might be insufficient to respond adequately. Astrocytes, on the other hand are well able to produce and store surplus cholesterol rapidly (Pfrieger, 2003a). At this very moment ApoE comes into play, as it delivers the cholesterol produced in astrocytes to neurons (Goritz *et al.*, 2002). Of course neurons also have to remove excess cholesterol, *e.g.* following synapse formation. This is done in a number of ways; cholesterol export *via* ApoE is one of them. The cholesterol-ApoE complex is then taken up by astrocytes and the exchange cycle is completed. There are a number of ApoE mutations known, three of these, ApoE ε2, ε3 and ε4 occur so frequent that they are regarded as alleles. The point mutation in the ε4 allele results in altered receptor binding and consequences to deliver and remove cholesterol to or from neurons. Why should this be of relevance to AD? ApoE4 might leave neurons transiently with increased cholesterol, which might therefore be one risk factor in AD. Indeed there is now crystal-clear evidence from a number of studies that cholesterol modulates Aβ production (Hartmann, 2001). This gets furthermore emphasized, as some studies indicate that ApoE alters aggregation of Aβ in an allele specific manner (Bales *et al.*, 1999; LaDu *et al.*, 1995; Ma *et al.*, 1994; Näslund *et al.*, 1995; Wisniewski *et al.*, 1994).

2. INTRACELLULAR ASPECTS

One of the earliest indications that cholesterol might be relevant to Aβ generation came from a study on rabbits (Sparks, 1996; Sparks *et al.*, 1994). Animals fed with a high cholesterol diet formed amyloid deposits. Histological analyses revealed that these deposits stained positively with an Aβ directed antibody. No sequencing or Western-blot of this material had ever been done, which raised some doubts if these are true (intracellular) Aβ deposits. Some time later it was found that primary mouse neurons transfected with an amyloid precursor protein (APP) expressing Semliki forest virus, could be prevented from producing Aβ as soon as the cellular cholesterol content was reduced (Simons *et al.*, 1998). In fact, this reduction was achieved by culturing these cells in the presence of lovastatin, the very first statin approved for clinical use. Statins block the activity of 3-hydroxy-3-methylglutaryl-CoA (HMG–CoA) reductase (Alberts *et al.*, 1980). This enzyme is not only essential for cholesterol biosynthesis; it is also the central point of a regulatory cascade used to stabilize cholesterol homeostasis involving the SREBP/SCAP system (Brown and Goldstein, 1999; Brown *et al.*, 2000). In cell-culture based experiments lovastatin or simvastatin (another statin) is usually given at equilibrium concentration (Fassbender *et al.*, 2001). At this concentration the cholesterol biosynthesis pathway is reproducibly blocked (see Figure 1). This however results in cellular toxicity, since some essential nonsteroidal products produced in pathways branching from of the main cholesterol biosynthetic pathway are not produced anymore. Addition of small amounts of mevalonate (the product of HMG–CoA reductase) prevents this (Langan and Volpe, 1987; Rao *et al.*, 1999). Apparently, mevalonate is shuttled mainly to the side pathways and thus prevents toxicity without causing significant cholesterol production. One interesting observation is that one of the branching pathways following HMG–CoA reductase activity leads to geranylgeranyl-pyrophosphate (Meske *et al.*, 2003; Sawamura *et al.*, 2003). If the HMG–CoA-reductase pathway is blocked, irrespective of cholesterol levels, abnormal tau phosphorylation occurs. This is a very unexpected link between lipids and Alzheimer's disease that certainly warrants careful consideration.

Since cholesterol is abundant its in cells and turnover in neurons is limited, statin treatment was originally not effective to reduce cellular cholesterol levels. This treatment was therefore supplemented with a rather harsh but effective and specific cholesterol extraction method. For this 5 mM methyl-β-cyclodextrin (CDX) was added to the cell culture media, CDX rapidly extracts cholesterol from the plasma membrane. However, this has quite a number of consequences: 1) CDX in the originally applied concentration (5mM for 20 min.) is fully sufficient to completely abrogate Aβ production in most cell types, including primary neurons (Bergmann, unpublished observation), it was therefore not possible to establish whether statins do affect Aβ production when applied in combination with CDX. Indeed lovastatin alone did not prevent Aβ formation (Simons *et al.*, 1998), most likely because after 48 hours of treatment internal cholesterol stores remain sufficiently intact. Eventually, statin treatment was lengthened up to 72 hours, which strongly reduced Aβ production, showing that statins do in fact prevent Aβ production *in vitro* (Fassbender *et al.*, 2001). 2) Another consequence is that CDX treatment enabled further insight into the cellular mechanisms involved to be obtained. Both statins and CDX produced cholesterol reduction by entirely separate mechanisms: statins prevent *de novo* cholesterol synthesis, but CDX physically removes cholesterol from cells. The outcome is identical – Aβ production declines. It was therefore concluded that the levels of cellular cholesterol and not its precursors are related to lowering of Aβ production (Fassbender *et al.*, 2001). This was later confirmed by the treatment of transgenic mice (Refolo *et al.*, 2001) and guinea pigs (Fassbender and Hartmann unpublished observation) with BM15.766. BM15.766 inhibits the final step of cholesterol synthesis and results in reduction of cerebral Aβ levels in both animal models.

3. INTRACELLULAR Aβ AND CHOLESTEROL

Cholesterol is not evenly distributed throughout the cells. Cholesterol is synthesized at the endoplasmic reticulum (ER) and trafficked to the plasma membrane. Cholesterol concentration is very low in the ER and increases continuously along the secretory pathway. Its concentration is highest at the plasma membrane, especially in lipid micro domains or raft domains

Figure 1. (On facing page) The cholesterol biosynthetic pathways. Indicated are enzymic steps of interest for Aβ generation and tau phosphorylation.

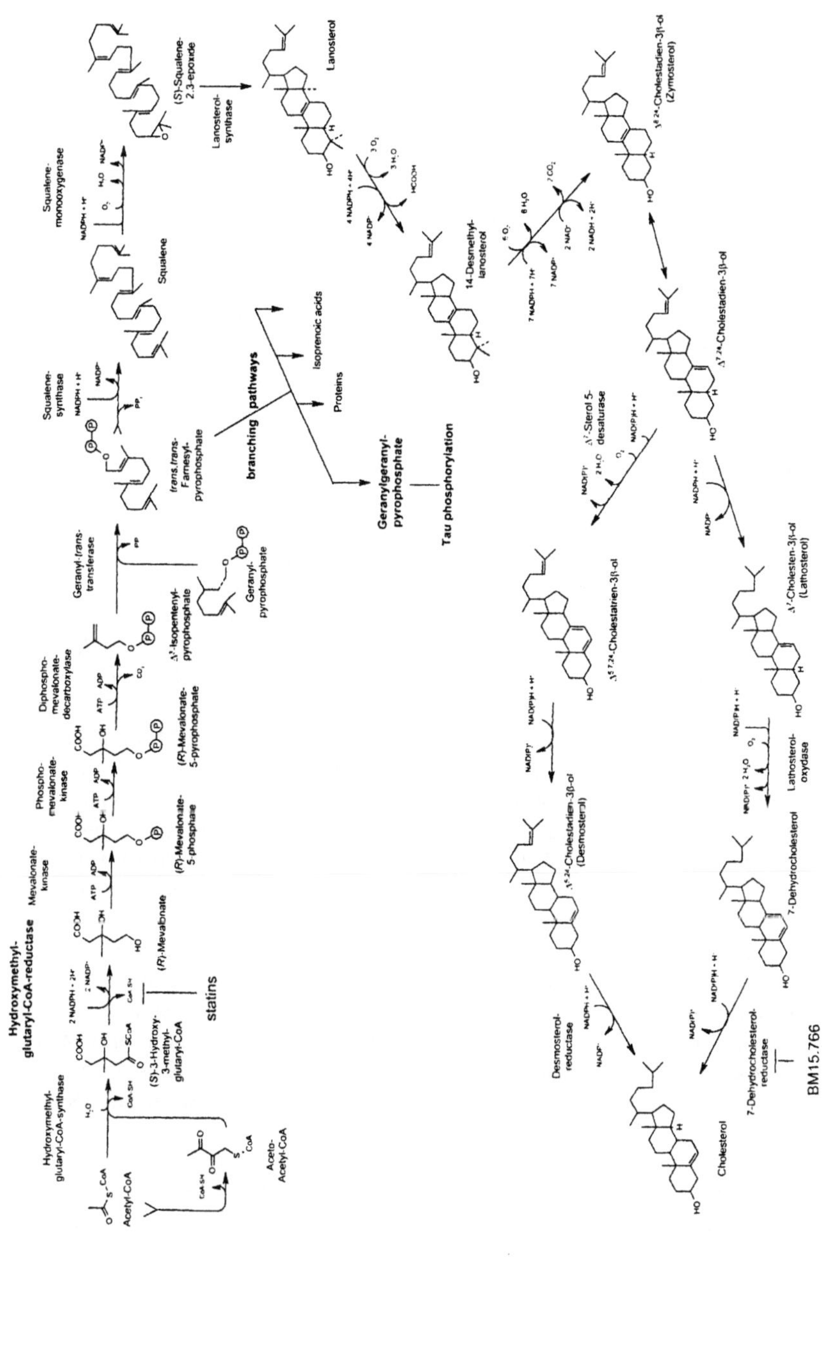
Acetyl-CoA
Aceto-Acetyl-CoA
Hydroxymethyl-glutaryl-CoA-synthase
(S)-3-Hydroxy-3-methyl-glutaryl-CoA
Hydroxymethyl-glutaryl-CoA-reductase
statins
(R)-Mevalonate
Mevalonate-kinase
(R)-Mevalonate-5-phosphate
Phospho-mevalonate-kinase
(R)-Mevalonate-5-pyrophosphate
Diphospho-mevalonate-decarboxylase
Δ^3-Isopentenyl-pyrophosphate
Geranyl-pyrophosphate
Geranyl-trans-transferase
trans,trans-Farnesyl-pyrophosphate
Squalene-synthase
Squalene
Squalene-monooxygenase
(S)-Squalene-2,3-epoxide
Lanosterol-synthase
Lanosterol
branching pathways
Isoprenoic acids
Proteins
Geranylgeranyl-pyrophosphate
Tau phosphorylation
14-Desmethyl-lanosterol
$\Delta^{8,24}$-Cholestadien-3β-ol (Zymosterol)
$\Delta^{7,24}$-Cholestadien-3β-ol
Δ^7-Sterol 5-desaturase
$\Delta^{5,7,24}$-Cholestatrien-3β-ol
Δ^7-Cholesten-3β-ol (Lathosterol)
Lathosterol-oxydase
$\Delta^{5,24}$-Cholestadien-3β-ol (Desmosterol)
7-Dehydrocholesterol
Desmosterol-reductase
7-Dehydrocholesterol-reductase
Cholesterol
BM15.766

(Simons and Ikonen, 1997). There are several routes for cholesterol up-take but eventually all cholesterol is trafficked back to the ER. From here cholesterol is either redistributed or esterified and stored in lipid-droplets. Puglieli *et al.* found that esterified cholesterol too, affects Aβ production (Puglielli *et al.*, 2001). However, it appears that this is due to a new cleavage site in APP that is activated by intracellular esterified cholesterol levels and presumably results in retargeting of APP towards a non-amyloidogenic degradation pathway and might be independent of the effects, cholesterol has on APP-secretase activity (Kovacs, personal communication).

The Niemann-Pick C1 (NPC1) protein mediates one of the critical pathways of cholesterol up-take. NPC1 is essential for transfer of cholesterol and a number of other lipids from late endosomes into a Rab7 positive compartment; involved in cholesterol sorting (Cruz and Chang, 2000; Cruz *et al.*, 2000; Liscum *et al.*, 1989). From here cholesterol is transferred to the ER. NPC1 activity can be suppressed by U1866A or Imipramine. This results in an accumulation of cholesterol in endosomes and a parallel increase in γ-secretase activity (Runz *et al.*, 2002). However, as detailed analyses showed, the intracellular Aβ accumulation occurs in a compartment separate from that of the cholesterol accumulation. Aβ accumulates in the actual Rab7 positive compartment that is normally involved in cholesterol sorting, while cholesterol remains in endosomes. The most likely cause of increased Aβ production is a shift in presenilin (PS) localization from the ER into the Rab7 positive compartment. This is also confirmed by increased γ-secretase activity. However, at the same time β-secretase activity is reduced (Runz *et al.*, 2002; Yamazaki *et al.*, 2001). The finding that PS localization depends on NPC1 activity and cholesterol trafficking *in vitro* and *in vivo* (Burns *et al.*, 2003) is especially remarkable, in the context of a connection between AD and lipids. The function attributed to the Rab7 positive compartment is that of lipid trafficking. Purification of these vesicles revealed a relative small variety of proteins; indicating that one of the biological functions of one of the enzymes central to AD pathogenesis might be associated with lipid trafficking or lipid homeostasis. It is important to highlight that this compartment links endosomes with ER and at least upon inhibition of NPC1 function some ER proteins are present in this compartment as well. This shift in localization is therefore not as provoking as it appears on first sight. This is also confounded by small amounts of PS already colocalizing with Rab7 during normal cholesterol trafficking (Runz *et al.*, 2002).

The reason for reduced β-secretase activity is not entirely clear, but might be related to very strong cholesterol increase in endosomes; the site of major β-secretase activity. There are a number of other findings that link β-secretase activity to cholesterol rich rafts, which may contribute to reduced β-secretase activity (Cordy *et al.*, 2003; Ehehalt *et al.*, 2003; Marlow *et al.*, 2003; Riddell *et al.*, 2001).

Since γ–secretase scission has to be preceded by β-secretase cleavage (Hartmann, 1999), γ–secretase activity cannot be judged independently of β-secretase activity when APP is used as substrate. Therefore γ–secretase activity had to be judged by using a truncated APP construct; C99 (Tienari *et al.*, 1996), which is identical to the β-secretase cleavage product. In this respect it is important to note that due to the strong inhibition of β-secretase activity no increase in Aβ can be observed when cholesterol retrograde trafficking is blocked and APP is used as substrate (Runz *et al.*, 2002).

One way of analyzing β- and γ–secretase activity together is to use a construct that can be cleaved independently by both enzymes. The C111 construct (Tienari *et al.*, 1996) contains both cleavage sites but is sufficiently short at the amino-terminal side to be processed by γ–secretase independently. Here depressed β-secretase and increased γ–secretase activity was observed upon NPC1 inhibition. Moreover, this substrate can also be used to address, which amyloidogenic secretase is mainly affected by cholesterol and what pools of intracellular Aβ are reduced? Neurons produce intracellular Aβ42 early in the secretory pathway, mainly in the ER (Hartmann *et al.*, 1997). The TGN produces mainly Aβ40, which is not present in the ER, and low amounts of Aβ42 (Grimm *et al.*, 2003). Cholesterol depletion with CDX affects primarily the plasma membrane, followed by some cholesterol depletion in the late secretory pathway due to membrane trafficking, but cholesterol levels in the Golgi and earlier compartments remain unchanged. With C111 it is now possible to specifically address these Aβ pools. As can be expected from the cholesterol distribution, Aβ42 levels are only slightly altered indicating that the intracellular γ–secretase 42 activity remains resistant to such treatment. The β- and γ–secretase 40 activities, however, are strongly reduced, confirming that most β- and γ–secretase 40 activities are present in the late secretory pathway (Bergmann and Hartmann unpublished observation).

Statins seem to act very differently here. Since they are slow acting and internal cholesterol stores have to be reduced before the Aβ metabolism is affected statins reduce secretase activity more uniformly and do not discriminate between Aβ40 and Aβ42. Nevertheless, it cannot be excluded that some statins might shift the Aβ40/42 ratio as well. The reason might be that γ–secretase cleaves the APP trans-membrane domain (APP-TMD) in a site, but non-sequence specific manner. The exact cleavage site is determined by the integration of APP-TMD into the membrane. γ-cleavage always occurs at the membrane center, most likely at the lipid leaflet interface (Grziwa *et al.*, 2003). In thin membranes, *e.g.* the ER membrane, the APP-TMD shifts towards the C-terminus, resulting in a shift of the cleavage site as well. In case of the ER this is by exactly two amino acids, resulting in Aβ42 production rather than Aβ40 that would be typical for Aβ produced in the thick plasma membrane. Alteration exclusively in one membrane leaflet should have similar effects. Kirsch *et al.* found that some statins may do exactly this (Kirsch *et al.*, 2003). Lovastatin and pravastatin, but not simvastatin, reduced cholesterol specifically in the exofacial membrane leaflet. Nevertheless, it remains to be seen whether this is of *in vivo* relevance, currently no such changes in Aβ ratio upon statin treatment were observed in clinical trial.

Whereas cholesterol reduction decreases β- and γ–secretase activity, it has just the opposite effect on α-secretase activity, which should result in further decreased Aβ production (Kojro, 2001). Aspects of the non-amyloidogenic pathway are described in detail in Chapter 5.

4. *IN VIVO* ASPECTS

Several attempts have been made to address this question *in vivo*. One of the most difficult aspects is standard animal models for AD, the transgenic mouse model handles exchange to peripheral organs and brain cholesterol differently from humans (Fassbender *et al.*, 2001). Therefore statin based animal experiments appear to be very difficult to perform and may or may not reflect the human situation. The only published amyloid mouse model treated thus far with statins, resulted in an increased Aβ production in female, but not male, mice (Park *et al.*, 2003). While statins effects on mouse cerebral Aβ metabolism remain a critical issue an interesting note has to be added in respect to ApoE. Refolo and Pappolla found that statins decrease glia derived ApoE levels (Petanceska *et al.*, 2003). Moreover, atorvastatin, which does not cross the blood brain barrier effectively, also affects ApoE brain levels. It thus appears possible that statins not need to cross the blood brain barrier to affect AD pathology.

This situation is somewhat more straightforward in animals with a cholesterol homeostasis system more closely related to humans (Fassbender *et al.*, 2001; Horsmans *et al.*, 1990). In guinea pigs, high dosage simvastatin treatment resulted in very effective, fast and reversible decrease of Aβ levels in cerebrospinal fluid and brain tissue. After three weeks of treatment Aβ levels were reduced by 50%. Although animals showed the typical adverse effects of statin treatment, but no significant signs of toxicity were observed (Fassbender *et al.*, 2001). When treatment was continued for another 3 weeks, toxicity parameters did not increase further, however the further prolongation of this animal trial was interrupted when clear signs of toxicity appeared following further extension of the treatment duration (Fassbender and Hartmann unpublished observation). Unlike blood cholesterol levels, statins are not expected to decrease total brain cholesterol levels in any measurable manner, as confirmed by analysis of brain material from experimental animals. Significant alterations in cholesterol precursors, as expected from HMG–CoA reductase inhibition, and cholesterol degradation; indicative for slower cholesterol re-supply, were observed. Total cholesterol levels, however, remained within experimental error margins. It is important in this context to realize that brain cholesterol is not evenly distributed. Large amounts of cholesterol are present in specialized membranes that are unlikely to participate in Aβ production; other cholesterol fractions are present as esterified cholesterol. Thus alterations might occur in membranes engaged in Aβ production, but this would be revealed only by more specialized assays. Indeed it is known that statins reduce the cholesterol content in synaptosomal membrane fractions. Moreover, investigations on lipids other than cholesterol indicate that secretases are actively involved in lipid homeostasis.

Feeding transgenic mice with cholesterol enriched diet or treatment with drugs other than statins seems to be less problematic. In a number of studies cholesterol feeding was found to increase amyloid load (Levin-Allerhand *et al.*, 2002; Refolo *et al.*, 2000; Shie *et al.*, 2002), but one study (Howland *et al.*, 1998) found an inverse effect. Interestingly, the more directly cholesterol-lowering drug, BM15.766 too resulted in strongly decreased Aβ accumulation. Strengthening the interpretation that, at least in these animal models, Aβ reductions are due to cholesterol reduction, not due to other statin effects.

5. RELEVANCE TO AD?

There are a number of retrospective epidemiological studies that focused on statins and dementia. Uniformly these studies reported an approx. 70% reduction in disease prevalence or incidence. It is unclear from these studies to which extend AD is affected, as dementia is a rather broad group of diseases. Since AD represents the vast majority of all dementia cases it would be very difficult to perceive any situation in which AD would have been excluded from the protective statin effects. There are however, a number of important considerations that need to be kept in mind. Obviously, epidemiological studies are not designed to reveal cause – effect relationships. Moreover, these were retrospective studies and it is reasonable to assume that the statin group received this treatment for a specific cause, most likely due to high blood LDL-C levels. Statins are now the most widely used pharmaceuticals and risk benefit considerations are particularly good.

Nevertheless, when statins were introduced into the market, a bias towards medication of the well-educated population might have occurred. There a number of additional issues some have already been resolved by the more recent studies, but some still need to be addressed. One of the important points relates again to disease specificity, vascular aspects often coincide with AD and this is the point where statin properties including cholesterol lowering, vascular protection and immunomodulation may provide protection beyond Aβ lowering (Hofman *et al.*, 1997; Kalaria and Ballard, 1999). It should also be noted that large prospective studies, like the PROSPER study (Shepherd *et al.*, 2002), which attempted to address potential dementia aspects. In these studies no effect on cognition was observed, but doubts had been raised whether cognitive performance had been tested adequately to detect early stages of AD.

If one considers Aβ generation as the molecular target to treat AD, prevention rather than treatment of clinically manifest AD is certainly preferable. During the pre-clinical phase Aβ triggered neurodegeneration should be rather mild and the self-healing capacity of the brain should further assist as soon as the toxic impact of Aβ is reduced together with lowered Aβ production. However, unlike epidemiological studies dedicated AD-statin trials are performed with patients well inside the clinical phase of AD. Given the duration of AD, the long half-life time of cerebral cholesterol and comparatively the mild impact statins have on brain cholesterol, treatment duration and statin dosage become critical. We therefore selected for our small prospective clinical trial a dosage of 80mg simvastatin and 6-month duration (Simons *et al.*, 2002). Statin treated patients showed decreased cerebrospinal fluid Aβ levels, which correlated with declined cholesterol degradation and cognitive decline was absent as assessed by MMSE. Importantly, patients in the mild AD group (MMSE >20) responded best to the treatment and no difference between treated and placebo group was observed by ADAS-Cog. The correlation between Aβ decline and cholesterol degradation is of especial interest and may help in future to separate responders from non-responders early on. Cholesterol cannot be degraded in the human brain, instead it is modified to a slightly more hydrophilic derivative that passes the blood brain barrier and is ultimately transformed to bile acids and then secreted. The two major brain cholesterol degradation products are 27S-OH-cholesterol and 24S-OH-cholesterol (cerebrosterol). AD-patients usually have increased cerebrosterol levels (Lutjohann and von Bergmann, 2003; Papassotiropoulos *et al.*, 2002; Schonknecht *et al.*, 2002) and two intronic polymorphic forms of the cholesterol modifying enzyme, CYP46, are presumably associated with AD (Kolsch *et al.*, 2002; Papassotiropoulos *et al.*, 2003). Cerebrosterol is produced almost exclusively in the brain and thus allows monitoring brain cholesterol drainage. Under conditions of limited cholesterol re-supply, e.g. during statin treatment, cerebrosterol secretion is down regulated. Evidence for statin dependent down regulation of Aβ production also comes from a second study. Here blood Aβ levels were measured from volunteers treated with controlled released lovastatin. Decreased Aβ was observed in the 40 mg and 60 mg treatment group, but not in the 10 mg or 20 mg groups (Buxbaum *et al.*, 2002). Cognitive parameters were not studied. In another small trial presented at the Springfield symposium 2004, patients were treated for 12 months with 80 mg atorvastatin. Atorvastatin is less efficient in brain penetration than simvastatin but it is approx. twice as efficient in lowering blood cholesterol levels. In this study two thirds of the subjects responded positively and cognition either remained stable or improved in half of the patients as determined by ADAS-Cog. The complete details of this study are not available so far. There is also a number of *post-hoc* and prospective studies that used low dosages of various statins and sometimes patients with hypercholesterolemia were employed. In agreement with the aforementioned studies no effect was observed on Aβ or cognitive performance (Hoglund *et al.*, 2004; Ishii *et al.*, 2003; Sjogren *et al.*, 2003).

6. OUTLOOK

All APP-secretase activities are modulated by cellular lipid composition. At least some undergo altered intracellular trafficking and γ–secretase localizes to a cholesterol/lipid sorting compartment. One additional proteases apparently cleaves APP specifically depending on cholesterol ester levels. Abnormal HMG–CoA reductase transcripts, CYP46 and ApoE are candidates for genetic AD risk factors and the cholesterol biosynthetic pathway is involved in tau phosphorylation. There are several additional factors concerning non-cholesterol lipids, thus outside of the scope of this review, but all of this points towards an inherent link between the molecular biology of AD and lipid biology. This clearly highlights the importance of lipids in AD and their role as important risk factors. It might also indicate that additional disturbances in lipid homeostasis are present in AD and that the discovery of the role of cholesterol in AD represents only the tip of the iceberg. In respect to treatment this should also raise the investigators attention to potential side effects. However, if applied rationally, then this will allow to design even more intricate treatment strategies to down regulate the amyloidogenic APP processing pathway even more effectively.

REFERENCES

Alberts, A. W., Chen, J., Kuron, G., Hunt, V., Huff, J., Hoffman, C., Rothrock, J., Lopez, M., Joshua, H., Harris, E., Patchett, A., Monaghan, R., Currie, S., Stapley, E., Albers-Schonberg, G., Hensens, O., Hirshfield, J., Hoogsteen, K., Liesch, J., and Springer, J., 1980, Mevinolin: a highly potent competitive inhibitor of hydroxymethylglutaryl-coenzyme A reductase and a cholesterol-lowering agent. *Proc Natl Acad Sci U S A,* **77:** 3957-3961.

Bales, K. R., Verina, T., Cummins, D. J., Du, Y., Dodel, R. C., Saura, J., Fishman, C. E., DeLong, C. A., Piccardo, P., Petegnief, V., Ghetti, B., and Paul, S. M., 1999, Apolipoprotein E is essential for amyloid deposition in the APP(V717F) transgenic mouse model of Alzheimer's disease. *Proc Natl Acad Sci U S A,* **96:** 15233-15238.

Bjorkhem, I., and Meaney, S., 2004, Brain cholesterol: long secret life behind a barrier. *Arterioscler Thromb Vasc Biol,* **24:** 806-815.

Brown, M. S., and Goldstein, J. L., 1999, A proteolytic pathway that controls the cholesterol content of membranes, cells, and blood. *Proc Natl Acad Sci U S A,* **96:** 11041-11048.

Brown, M. S., Ye, J., Rawson, R. B., and Goldstein, J. L., 2000, Regulated intramembrane proteolysis: a control mechanism conserved from bacteria to humans. *Cell,* **100:** 391-398.

Burns, M., Gaynor, K., Olm, V., Mercken, M., LaFrancois, J., Wang, L., Mathews, P. M., Noble, W., Matsuoka, Y., and Duff, K., 2003, Presenilin redistribution associated with aberrant cholesterol transport enhances beta-amyloid production in vivo. *J Neurosci,* **23:** 5645-5649.

Buxbaum, J. D., Cullen, E. I., and Friedhoff, L. T., 2002, Pharmacological concentrations of the HMG-CoA reductase inhibitor lovastatin decrease the formation of the Alzheimer beta-amyloid peptide in vitro and in patients. *Front Biosci,* **7:** a50-59.

Corder, E. H., Saunders, A. M., Strittmatter, W. J., Schmechel, D. E., Gaskell, P. C., Small, G. W., Roses, A. D., Haines, J. L., and Pericak Vance, M. A., 1993, Gene dose of apolipoprotein E type 4 allele and the risk of Alzheimer's disease in late onset families. *Science,* **261:** 921-923.

Cordy, J. M., Hussain, I., Dingwall, C., Hooper, N. M., and Turner, A. J., 2003, Exclusively targeting beta-secretase to lipid rafts by GPI-anchor addition up-regulates beta-site processing of the amyloid precursor protein. *Proc Natl Acad Sci U S A,* **100:** 11735-11740.

Cruz, J. C., and Chang, T. Y., 2000, Fate of endogenously synthesized cholesterol in Niemann-Pick type C1 cells. *J Biol Chem,* **275:** 41309-41316.

Cruz, J. C., Sugii, S., Yu, C., and Chang, T. Y., 2000, Role of Niemann-Pick type C1 protein in intracellular trafficking of low density lipoprotein-derived cholesterol. *J Biol Chem,* **275:** 4013-4021.

Dietschy, J. M., and Turley, S. D., 2001, Cholesterol metabolism in the brain. *Curr Opin Lipidol,* **12:** 105-112.

Ehehalt, R., Keller, P., Haass, C., Thiele, C., and Simons, K., 2003, Amyloidogenic processing of the Alzheimer beta-amyloid precursor protein depends on lipid rafts. *J Cell Biol,* **160:** 113-123.

Evans, R. M., Hui, S., Perkins, A., Lahiri, D. K., Poirier, J., and Farlow, M. R., 2004, Cholesterol and APOE genotype interact to influence Alzheimer disease progression. *Neurology,* **62:** 1869-1871.

Fassbender, F., Simons, M., Bergmann, C., Stroick, M., Lütjohann, D., Keller, P., Runz, H., Kühl, S., Bertsch, T., von Bergmann, K., Hennerici, M., Beyreuther, K., and Hartmann, T., 2001, Simvastatin strongly reduces Alzheimer's disease Aβ42 and Aβ40 levels in vitro and in vivo. *Proc Natl Acad Sci U S A,* **98:** 5856-5861.

Goritz, C., Mauch, D. H., Nagler, K., and Pfrieger, F. W., 2002, Role of glia-derived cholesterol in synaptogenesis: new revelations in the synapse-glia affair. *J Physiol Paris,* **96:** 257-263.

Grimm, H. S., Beher, D., Lichtenthaler, S. F., Shearman, M. S., Beyreuther, K., and Hartmann, T., 2003, gamma-Secretase cleavage site specificity differs for intracellular and secretory amyloid beta. *J Biol Chem,* **278:** 13077-13085.

Grziwa, B., Grimm, M. O., Masters, C. L., Beyreuther, K., Hartmann, T., and Lichtenthaler, S. F., 2003, The Transmembrane Domain of the Amyloid Precursor Protein in Microsomal Membranes Is on Both Sides Shorter than Predicted. *J Biol Chem,* **278:** 6803-6808.

Hartmann, T., 1999, Intracellular biology of Alzheimer's disease amyloid beta peptide. *Eur Arch Psychiatry Clin Neurosci,* **249:** 291-298.

Hartmann, T., 2001, Cholesterol, Abeta and Alzheimer's disease. *TINS,* **24:** 45-48.

Hartmann, T., Bieger, S. C., Bruhl, B., Tienari, P. J., Ida, N., Allsop, D., Roberts, G. W., Masters, C. L., Dotti, C. G., Unsicker, K., and Beyreuther, K., 1997, Distinct sites of intracellular production for Alzheimer's disease A beta40/42 amyloid peptides. *Nat-Med,* **3:** 1016-1020.

Hofman, A., Ott, A., Breteler, M. M., Bots, M. L., Slooter, A. J., van Harskamp, F., van Duijn, C. N., Van Broeckhoven, C., and Grobbee, D. E., 1997, Atherosclerosis, apolipoprotein E, and prevalence of dementia and Alzheimer's disease in the Rotterdam Study. *Lancet,* **349:** 151-154.

Hoglund, K., Wiklund, O., Vanderstichele, H., Eikenberg, O., Vanmechelen, E., and Blennow, K., 2004, Plasma levels of beta-amyloid(1-40), beta-amyloid(1-42), and total beta-amyloid remain unaffected in adult patients with hypercholesterolemia after treatment with statins. *Arch Neurol,* **61:** 333-337.

Horsmans, Y., Desager, J. P., and Harvengt, C., 1990, Biochemical changes and morphological alterations of the liver in guinea-pigs after administration of simvastatin (HMG CoA reductase-inhibitor). *Pharmacol Toxicol,* **67:** 336-339.

Howland, D. S., Trusko, S. P., Savage, M. J., Reaume, A. G., Lang, D. M., Hirsch, J. D., Maeda, N., Siman, R., Greenberg, B. D., Scott, R. W., and Flood, D. G., 1998, Modulation of secreted beta-amyloid precursor protein and amyloid beta- peptide in brain by cholesterol. *J Biol Chem,* **273:** 16576-16582.

Ishii, K., Tokuda, T., Matsushima, T., Miya, F., Shoji, S., Ikeda, S., and Tamaoka, A., 2003, Pravastatin at 10 mg/day does not decrease plasma levels of either amyloid-beta (Abeta) 40 or Abeta 42 in humans. *Neurosci Lett,* **350:** 161-164.

Jick, H., Zornberg, G. L., Jick, S. S., Seshadri, S., and Drachman, D. A., 2000, Statins and the risk of dementia. *Lancet,* **356:** 1627-1631.

Kalaria, R. N., and Ballard, C., 1999, Overlap between pathology of Alzheimer disease and vascular dementia. *Alzheimer Dis Assoc Disord,* **13 Suppl 3:** S115-123.

Kirsch, C., Eckert, G. P., and Mueller, W. E., 2003, Statin effects on cholesterol micro-domains in brain plasma membranes. *Biochem Pharmacol,* **65:** 843-856.

Kivipelto, M., Helkala, E. L., Laakso, M. P., Hanninen, T., Hallikainen, M., Alhainen, K., Iivonen, S., Mannermaa, A., Tuomilehto, J., Nissinen, A., and Soininen, H., 2002, Apolipoprotein E epsilon4 allele, elevated midlife total cholesterol level, and high midlife systolic blood pressure are independent risk factors for late-life Alzheimer disease. *Ann Intern Med,* **137:** 149-155.

Kivipelto, M., Helkala, E. L., Laakso, M. P., Hanninen, T., Hallikainen, M., Alhainen, K., Soininen, H., Tuomilehto, J., and Nissinen, A., 2001, Midlife vascular risk factors and Alzheimer's disease in later life: longitudinal, population based study. *Bmj,* **322:** 1447-1451.

Kojro, E., Gimpl, G., Lammich, S., März, W., Fahrnholz, F., 2001, Low cholesterol stimulates the nonamyloidogenic pathway by its effect on the alpha-secretase ADAM 10. *Proc Natl Acad Sci USA*, **98:** 5815-5820.

Kolsch, H., Lutjohann, D., Ludwig, M., Schulte, A., Ptok, U., Jessen, F., von Bergmann, K., Rao, M. L., Maier, W., and Heun, R., 2002, Polymorphism in the cholesterol 24S-hydroxylase gene is associated with Alzheimer's disease. *Mol Psychiatry*, **7:** 899-902.

LaDu, M. J., Pederson, T. M., Frail, D., Reardon, C. A., Getz, G. S., and Falduto, M. T., 1995, Purification of apolipoprotein E attenuates isoform-specific binding to β-amyloid. *J. Biol. Chem.*, **270:** 9039-9042.

Langan, T. J., and Volpe, J. J., 1987, Cell cycle-specific requirement for mevalonate, but not for cholesterol, for DNA synthesis in glial primary cultures. *J Neurochem*, **49:** 513-521.

Levin-Allerhand, J. A., Lominska, C. E., and Smith, J. D., 2002, Increased amyloid- levels in APPSWE transgenic mice treated chronically with a physiological high-fat high-cholesterol diet. *J Nutr Health Aging*, **6:** 315-319.

Liscum, L., Ruggiero, R. M., and Faust, J. R., 1989, The intracellular transport of low density lipoprotein-derived cholesterol is defective in Niemann-Pick type C fibroblasts. *J Cell Biol*, **108:** 1625-1636.

Lutjohann, D., and von Bergmann, K., 2003, 24S-hydroxycholesterol: a marker of brain cholesterol metabolism. *Pharmacopsychiatry*, **36 Suppl 2:** S102-106.

Ma, J., Yee, A., Brewer, H. B., Das, S., and Potter, H., 1994, Amyloid-associated proteins α_1-antichymotrypsin and apolipoprotein E promote assembly of Alzheimer β-protein into filaments. *Nature*, **372:** 92-94.

Marlow, L., Cain, M., Pappolla, M. A., and Sambamurti, K., 2003, Beta-secretase processing of the Alzheimer's amyloid protein precursor (APP). *J Mol Neurosci*, **20:** 233-239.

Mauch, D. H., Nagler, K., Schumacher, S., Goritz, C., Muller, E. C., Otto, A., and Pfrieger, F. W., 2001, CNS synaptogenesis promoted by glia-derived cholesterol. *Science*, **294:** 1354-1357.

Meske, V., Albert, F., Richter, D., Schwarze, J., and Ohm, T. G., 2003, Blockade of HMG-CoA reductase activity causes changes in microtubule-stabilizing protein tau via suppression of geranylgeranylpyrophosphate formation: implications for Alzheimer's disease. *Eur J Neurosci*, **17:** 93-102.

Näslund, J., Thyberg, J., Tjernberg, L. O., Wernstedt, C., Karlström, A. R., Bogdanovic, N., Gandy, S. E., Lannfelt, L., Terenius, L., and Nordstedt, C., 1995, Characterization of stable complexes involving apolipoprotein E and the amyloid ß peptide in Alzheimer's disease brain. *Neuron*, **15:** 219-228.

Papassotiropoulos, A., Lutjohann, D., Bagli, M., Locatelli, S., Jessen, F., Buschfort, R., Ptok, U., Bjorkhem, I., von Bergmann, K., and Heun, R., 2002, 24S-hydroxycholesterol in cerebrospinal fluid is elevated in early stages of dementia. *J Psychiatr Res*, **36:** 27-32.

Papassotiropoulos, A., Streffer, J. R., Tsolaki, M., Schmid, S., Thal, D., Nicosia, F., Iakovidou, V., Maddalena, A., Lutjohann, D., Ghebremedhin, E., Hegi, T., Pasch, T., Traxler, M., Bruhl, A., Benussi, L., Binetti, G., Braak, H., Nitsch, R. M., and Hock, C., 2003, Increased brain beta-amyloid load, phosphorylated tau, and risk of Alzheimer disease associated with an intronic CYP46 polymorphism. *Arch Neurol*, **60:** 29-35.

Pappolla, M. A., Bryant-Thomas, T. K., Herbert, D., Pacheco, J., Fabra Garcia, M., Manjon, M., Girones, X., Henry, T. L., Matsubara, E., Zambon, D., Wolozin, B., Sano, M., Cruz-Sanchez, F. F., Thal, L. J., Petanceska, S. S., and Refolo, L. M., 2003, Mild hypercholesterolemia is an early risk factor for the development of Alzheimer amyloid pathology. *Neurology*, **61:** 199-205.

Park, I. H., Hwang, E. M., Hong, H. S., Boo, J. H., Oh, S. S., Lee, J., Jung, M. W., Bang, O. Y., Kim, S. U., and Mook-Jung, I., 2003, Lovastatin enhances Abeta production and senile plaque deposition in female Tg2576 mice. *Neurobiol Aging,* **24:** 637-643.

Petanceska, S. S., DeRosa, S., Sharma, A., Diaz, N., Duff, K., Tint, S. G., Refolo, L. M., and Pappolla, M., 2003, Changes in apolipoprotein E expression in response to dietary and pharmacological modulation of cholesterol. *J Mol Neurosci,* **20:** 395-406.

Pfrieger, F. W., 2003a, Cholesterol homeostasis and function in neurons of the central nervous system. *Cell Mol Life Sci,* **60:** 1158-1171.

Pfrieger, F. W., 2003b, Role of cholesterol in synapse formation and function. *Biochim Biophys Acta,* **1610:** 271-280.

Puglielli, L., Konopka, G., Pack-Chung, E., Ingano, L. A., Berezovska, O., Hyman, B. T., Chang, T. Y., Tanzi, R. E., and Kovacs, D. M., 2001, Acyl-coenzyme A: cholesterol acyltransferase modulates the generation of the amyloid beta-peptide. *Nat Cell Biol,* **3:** 905-912.

Rao, S., Porter, D. C., Chen, X., Herliczek, T., Lowe, M., and Keyomarsi, K., 1999, Lovastatin-mediated G1 arrest is through inhibition of the proteasome, independent of hydroxymethyl glutaryl-CoA reductase. *Proc Natl Acad Sci U S A,* **96:** 7797-7802.

Refolo, L. M., Pappolla, M. A., LaFrancois, J., Malester, B., Schmidt, S. D., Thomas-Bryant, T., Tint, G. S., Wang, R., Mercken, M., Petanceska, S. S., and Duff, K. E., 2001, A cholesterol-lowering drug reduces beta-amyloid pathology in a transgenic mouse model of Alzheimer's disease. *Neurobiol Dis,* **8:** 890-899.

Refolo, L. M., Pappolla, M. A., Malester, B., LaFrancois, J., Bryant-Thomas, T., Wang, R., Tint, G. S., Sambamurti, K., and Duff, K., 2000, Hypercholesterolemia accelerates the Alzheimer's amyloid pathology in a transgenic mouse model. *Neurobiol Dis,* **7:** 321-331.

Riddell, D. R., Christie, G., Hussain, I., and Dingwall, C., 2001, Compartmentalization of beta-secretase (Asp2) into low-buoyant density, noncaveolar lipid rafts. *Curr Biol,* **11:** 1288-1293.

Rockwood, K., Kirkland, S., Hogan, D. B., MacKnight, C., Merry, H., Verreault, R., Wolfson, C., and McDowell, I., 2002, Use of lipid-lowering agents, indication bias, and the risk of dementia in community-dwelling elderly people. *Arch Neurol,* **59:** 223-227.

Runz, H., Rietdorf, J., Tomic, I., de Bernard, M., Beyreuther, K., Pepperkok, R., and Hartmann, T., 2002, Inhibition of intracellular cholesterol transport alters presenilin localization and amyloid precursor protein processing in neuronal cells. *J Neurosci,* **22:** 1679-1689.

Sawamura, N., Gong, J. S., Chang, T. Y., Yanagisawa, K., and Michikawa, M., 2003, Promotion of tau phosphorylation by MAP kinase Erk1/2 is accompanied by reduced cholesterol level in detergent-insoluble membrane fraction in Niemann-Pick C1-deficient cells. *J Neurochem,* **84:** 1086-1096.

Schonknecht, P., Lutjohann, D., Pantel, J., Bardenheuer, H., Hartmann, T., von Bergmann, K., Beyreuther, K., and Schroder, J., 2002, Cerebrospinal fluid 24S-hydroxycholesterol is increased in patients with Alzheimer's disease compared to healthy controls. *Neurosci Lett,* **324:** 83-85.

Shepherd, J., Blauw, G. J., Murphy, M. B., Bollen, E. L., Buckley, B. M., Cobbe, S. M., Ford, I., Gaw, A., Hyland, M., Jukema, J. W., Kamper, A. M., Macfarlane, P. W., Meinders, A. E., Norrie, J., Packard, C. J., Perry, I. J., Stott, D. J., Sweeney, B. J., Twomey, C., and Westendorp, R. G., 2002, Pravastatin in elderly individuals at risk of vascular disease (PROSPER): a randomised controlled trial. *Lancet,* **360:** 1623-1630.

Shie, F. S., Jin, L. W., Cook, D. G., Leverenz, J. B., and LeBoeuf, R. C., 2002, Diet-induced hypercholesterolemia enhances brain A beta accumulation in transgenic mice. *Neuroreport,* **13:** 455-459.

Simons, K., and Ikonen, E., 1997, Functional rafts in cell membranes. *Nature,* **387:** 569-572.

Simons, M., Keller, P., De Strooper, B., Beyreuther, K., Dotti, C. G., and Simons, K., 1998, Cholesterol depletion inhibits the generation of beta-amyloid in hippocampal neurons. *Proc Natl Acad Sci USA,* **95:** 6460-6464.

Simons, M., Schwarzler, F., Lutjohann, D., von Bergmann, K., Beyreuther, K., Dichgans, J., Wormstall, H., Hartmann, T., and Schulz, J. B., 2002, Treatment with simvastatin in normocholesterolemic patients with Alzheimer's disease: A 26-week randomized, placebo-controlled, double-blind trial. *Ann Neurol,* **52:** 346-350.

Sjogren, M., Gustafsson, K., Syversen, S., Olsson, A., Edman, A., Davidsson, P., Wallin, A., and Blennow, K., 2003, Treatment with simvastatin in patients with Alzheimer's disease lowers both alpha- and beta-cleaved amyloid precursor protein. *Dement Geriatr Cogn Disord,* **16:** 25-30.

Sparks, D. L., 1996, Intraneuronal beta-amyloid immunoreactivity in the CNS. *Neurobiol Aging,* **17:** 291-299.

Sparks, D. L., Scheff, S. W., Hunsaker, J. C., 3rd, Liu, H., Landers, T., and Gross, D. R., 1994, Induction of Alzheimer-like beta-amyloid immunoreactivity in the brains of rabbits with dietary cholesterol. *Exp Neurol,* **126:** 88-94.

St George-Hyslop, P. H., 2000, Molecular genetics of Alzheimer's disease. *Biol Psychiatry,* **47:** 183-199.

Tienari, P. J., De Strooper, B., Ikonen, E., Simons, M., Weidemann, A., Czech, C., Hartmann, T., Ida, N., Multhaup, G., Masters, C. L., Van Leuven, F., Beyreuther, K., and Dotti, C. G., 1996, The beta-amyloid domain is essential for axonal sorting of amyloid precursor protein. *Embo J,* **15:** 5218-5229.

Wisniewski, T., Castaño, E. M., Golabek, A., Vogel, T., and Frangione, B., 1994, Acceleration of Alzheimer's fibril formation by apolipoprotein E *in vivo. Am. J. Pathol.,* **145:** 1030-1035.

Wolozin, B., Kellman, W., Ruosseau, P., Celesia, G. G., and Siegel, G., 2000, Decreased prevalence of Alzheimer disease associated with 3-hydroxy-3- methyglutaryl coenzyme A reductase inhibitors. *Arch Neurol,* **57:** 1439-1443.

Yamazaki, T., Chang, T. Y., Haass, C., and Ihara, Y., 2001, Accumulation and aggregation of amyloid beta-protein in late endosomes of Niemann-pick type C cells. *J Biol Chem,* **276:** 4454-4460.

Chapter 20

Phosphorylated Amyloid-β: the Toxic Intermediate in Alzheimer's Disease Neurodegeneration

Nathaniel G. N. Milton
Department of Clinical Neurosciences, Royal Free & University College Medical School, University College London, Rowland Hill Street, London, UK

Abstract: Phosphorylated Amyloid-β (Aβ) was identified in Alzheimer's disease (AD) brain. Using an anti-sense peptide approach the human cyclin-dependent kinase-1 (CDK-1) was identified as being responsible for Aβ phosphorylation. The phosphorylated Aβ peptide showed increased neurotoxicity and reduced ability to form Congo red-positive fibrils. Mutation of the serine 26 residue and inhibition of Aβ phosphorylation by the CDK-1 inhibitor olomoucine prevented Aβ toxicity, suggesting that the phosphorylated Aβ peptide represents a toxic intermediate. Cannabinoids prevented phosphorylated Aβ toxicity. The results from this study suggest that Aβ phosphorylation could play a role in AD pathology and represent a novel therapeutic target.

Keywords: Phosphorylation, cyclin-dependent kinase, mitogen-activated protein kinase cannabinoid, olomoucine, anti-sense peptide

1. INTRODUCTION

The Amyloid-cascade hypothesis for Alzheimer's Disease (AD) has been central to AD based research and is well supported by the observations of mutations in the amyloid-β precursor protein (APP) gene causing early onset AD (Selkoe, 1999; Milton, 2004a). Within this cascade the role of phosphorylation is placed downstream of amyloid-β (Aβ) deposition, in part due to the observations that deposition of hyperphosphorylated tau protein in the form of neurofibrillary tangles (NFT's) appears after amyloid plaques in AD. However, recent observations point to a role for phosphorylation in the cleavage of the APP to produce Aβ (Lee *et al.*, 2003) and to a direct role of phosphorylation in the cytotoxicity of Aβ (Milton, 2001a; 2002a; 2004b). Results have suggested that the actions of the β-APP cleavage enzymes (BACE), which cleave APP at the N-terminus of Aβ, are influenced by APP phosphorylation (Lee *et al.*, 2003). Protein kinases including cyclin-dependent kinases (CDK: Alvarez *et al.*, 2001; Copani *et al.*, 1999; Giovanni *et al.*, 1999), mitogen-activated protein kinase (MAPK: Abe and Saito, 2000) and protein kinase C (PKC: Nakai *et al.*, 2001) have also been implicated in the mechanisms underlying Aβ cytotoxicity.

Phosphorylation defects are a major pathological feature of AD and the form of tau responsible for NFT deposition is hyperphosphorylated (Buee *et al.*, 2000). Many of the kinases mentioned above, which influence Aβ cytotoxicity, show increased expression and activity within the AD brain. Since the profound neuronal loss seen in AD is associated with NFT's and Aβ containing plaques these two proteins have been central to the search for the cause of and cure for AD. The diagnosis of AD is dependent on detection of neurofibrillary pathology in the form of NFT and/or Aβ containing neuritic plaques (Braak *et al.*, 1993; Jellinger, 1998). The observations of phosphorylation defects and their role in the actions of both Aβ and tau suggest that phosphorylation may be an important process in the pathology of AD and a target for therapeutic intervention.

Two questions raised from this are:

(a) Is Aβ itself phosphorylated?

(b) Does Aβ increase the activity of protein kinases?

This chapter will address the recent research which attempts to answer these questions and will review the recent evidence that points to the importance of links between Aβ and protein kinases.

2. PHOSPHORYLATED AMYLOID-β PEPTIDE

The main amino acid residues of proteins that can be phosphorylated are serine, threonine and tyrosine. The kinase enzymes can be sub-grouped into the serine/threonine kinases and the tyrosine kinases. The Aβ 1-43 sequence, the longest of the major forms associated with AD, contains serines at residues 8 and 26, a tyrosine at residue 10 and a threonine at residue 43. The NetPhos 2.0 computer program, which predicts phosphorylation sites in proteins (Blom *et al.*, 1999), was used to determine whether any of these Aβ residues were potential targets for phosphorylation. The NetPhos 2.0 scores were obtained from the output score of the ensemble of neural networks trained on that acceptor residue type and a value > 0.5 was considered significant. The scores for Aβ serine 8, tyrosine 10 and serine 26 were 0.963, 0.870 and 0.787 respectively, suggesting that these Aß residues have a significant chance of being phosphorylated (Milton, 2001a).

Neuronal extracts were prepared to determine whether a phosphorylated form of Aβ was present. The extraction techniques were optimised to preserve phosphoproteins and Aβ (Milton 2001a; 2004b). Immunoprecipitated and chromatographically purified Aβ-like material from AD brains and human NT-2 (NTera2/D1) neurons was found to contain a serine phosphorylated form of Aβ-like immunoreactivity. SDS-PAGE analysis of this material demonstrated that the serine-phosphorylated Aβ-like material was a similar size (4 kDa) to synthetic Aβ (Figure 1). The levels of immunoreactive phosphorylated Aβ, detected with a two-site immunoassay, represented approximately 20% of the total immunoreactive Aβ in extracts of AD brains (Milton, 2001a).

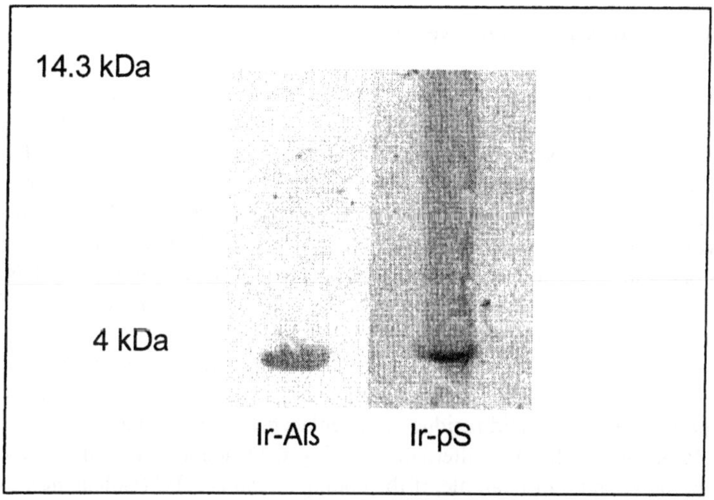

Figure 1. Western blot analysis of immunoprecipitated Aβ from AD brain. Samples were run on a 15% SDS-PAGE gel prior to transfer to nitrocellulose and staining with anti-Aβ monoclonal antibody ALI-01 (Ir-Aβ) or with anti-phosphoserine polyclonal antibody (Ir-pS).

These results indicate that Aβ can be phosphorylated by neurons and is present in AD brains. The purified Aβ from these tissues did not contain larger fragments of APP, as indicated by SDS-PAGE analysis.

2.1 Which kinase phosphorylates the amyloid-β peptide?

The presence of a serine phosphorylated Aβ form in AD brains and human NT-2 neurons indicates that Aβ phosphorylation involves a serine/threonine protein kinase. The two serine residues are located at positions 8 and 26. These sites correspond to a potential casein kinase II (CKII) substrate sequence (Pinna, 1990) and a PKC site (Kishimoto *et al.*, 1985; Woodgett *et al.*, 1986). However, neither of these kinases has been shown to phosphorylate Aβ *in vitro* (Chauhan *et al.*, 1993; Lee *et al.*, 2004). The activities of both CKII and PKC are modified by Aβ suggesting that interactions occur, however, the failure of these kinases to phosphorylate Aβ suggests that a different kinase phosphorylates Aβ.

The human genome contains over 500 protein kinases (Manning *et al.*, 2002) and clearly a screening assay to identify potential kinases that could phosphorylate Aβ is a prerequisite.

2.1.1 Anti-sense peptide screening

The principles of the anti-sense approach are summarised as follows: Anti-sense peptide sequences are derived from the complementary strand of DNA encoding a given protein, read in the same open reading frame (ORF). They can also be derived directly from the amino acid sequence of a protein, *via* reverse translation to produce a complementary DNA sequence. However, due to the degeneracy of the genetic code, there is typically more than one anti-sense sequence for any one protein. The complementary DNA strand for each individual amino acid can be read in either the forward 3'-5' or reverse 5'-3' direction, adding further degeneracy to the potential anti-sense peptide sequences. The sequence of the complementary DNA strand may encode a stop codon. In these circumstances, it is necessary to introduce an appropriate amino acid residue. The replacement amino acid residue will usually be derived from an alternative coding sequence for the amino acid of the coding strand. For example, if the coding strand is ATC (isoleucine), the complementary strand is a stop codon TAG. Isoleucine is also encoded by ATA, the complement of which encodes tyrosine (TAT). Therefore, tyrosine is used at the position corresponding to the stop codon. For database screening an unknown residue (X) can be used in the anti-sense peptide sequence. Anti-sense peptides have been shown to bind with high affinity to the given protein due to hydropathic interactions. Anti-sense peptides have also been shown to have sequence similarity to receptor binding sites and compounds, such as antibodies, that specifically bind such anti-sense peptides, have been used to isolate receptors (Bost and Blalock, 1989).

The binding of Aβ to itself can occur in both parallel and anti-parallel orientations (Serpell, 2000) with consequent interactions between for example two N-terminals in parallel binding or an N and a C terminus in anti-parallel binding. If binding of a peptide to an anti-sense peptide sequence were to occur in an anti-parallel orientation then the anti-sense peptide would have to be synthesized in the anti-parallel direction with the C terminus occupying the N-terminus of the resultant peptide. Similarly an anti-parallel binding interaction between a binding protein and a peptide may be identified by comparison of the anti-sense peptide sequence in the C-terminus to N-terminus orientation with the binding protein sequence in the normal N-terminus to C-terminus orientation. Since protein interactions can occur in using both parallel and anti-parallel binding it is necessary to use the two anti-sense peptides derived from the DNA sequence in both N-terminus to C-terminus orientation and the C-terminus to N-terminus orientation for screening purposes.

Previous studies have demonstrated the potential of Anti-Sense peptide screening to identify Aβ interactions with enzymes (Milton *et al.*, 2001). The interaction of Aβ with the ABAD or ERAB protein was characterised using this technique and identified ERAB residues 99-108 as the site of interaction with Aβ (Milton *et al.*, 2001; Milton, 2004b). This region of ERAB was confirmed as the Aβ binding site using mutational analysis and X-ray crystallography (Lustbader *et al.*, 2004). The anti-sense approach was applied to identification of the kinase involved in Aβ phosphorylation (Milton, 2001a; 2004b).

2.1.2 Identification of a potential interaction between Cyclin-Dependent Kinase-1 and Amyloid-β

Figure 2 illustrates the four potential Aβ 1-43 anti-sense peptides (AßAS) which could be used in searches to identify proteins having sequence similarity to AβAS and hence possibly interacting via hydropathic binding with the Aβ peptide. An initial BLAST comparison (Altschul *et al.*, 1997) of the protein kinase domain consensus sequence (Pfam 00069) identified the forward AβAS 3-30 sequence, derived by reading the complementary (non-coding) strand of DNA from the region encoding the Aβ 1-43 peptide in the 3'-5' direction, as having 23% identity and 46% similarity with 100-127 region.

The forward AβAS 1-43 sequence was used in a subsequent BLAST search of the human genome encoded proteins. The results showed that the forward AβAS 3-30 sequence had 46% identity and 68% similarity with the human CDK-1 enzyme 105-132 region. This region of CDK-1 contains the Aspartic acid residue (128) that has been suggested to transfer a Phosphate group from ATP to the CDK-1 substrates and constitutes an active site of the enzyme (DeBondt *et al.,* 1993). The alignment of the AβAS 26 residue, which corresponds to the Aβ serine 26 residue, with this active site residue in CDK-1 indicated that Aβ 1-43 could be phosphorylated by CDK-1 and suggested that the serine 26 residue would be the target.

Human Aß 1-43	DAEFRHDSGYEVHHQKLVFFAEDVGSNKGAIIGLMVGGVVIAT
Forward AßAS N-C	LRLKAVLSPILQVVVFNHKKRLLHPSLFPRXXPEYHPPQQYRC
Forward AßAS C-N	CRYQQPPHYEPXXRPFLSPHLLRKKHNFVVVQLIPSLVAKLRL
Reverse AßAS N-C	ICFESMVXSIFNMMLFQHEKCFIHTXVFTCDNSEHHATNDYRC
Reverse AßAS C-N	CRYDNTAHHESNDCTFVXTHIFCKEHQFLMMNFISXVMSEFCI

Figure 2. Amyloid-β anti-sense sequences derived from the complementary DNA strand read in the 3'-5' direction (Forward) or 5'-3' direction (Reverse). N-C = N terminus to C terminus sequence, C-N = C terminus to N terminus sequence

The BLAST comparison between the forward AβAS and human CDK-1 (Accession No. GI 87058) also identified three other regions of sequence similarity. CDK-1 residues 56 to 63 showing 50% identity and 75% similarity with AβAS 20-27; CDK-1 residues 95 to 99 showing 80% identity with AβAS 37-41; and CDK-1 residues 229 to 238 showing 40% identity and 50% similarity with AβAS 33-42.

The CDK family shows considerable similarity around the active site and these enzymes are thought to be proline directed kinases. Figure 3 illustrates the sequence similarities between the CDK family members and the forward AβAS 3-30 sequence.

The principle of anti-sense peptides is that the hydropathic character of a peptide derived from the coding strand will be opposite to that derived from the complementary strand. Therefore, even though the actual anti-sense amino acid sequence will be very different from that derived from the coding strand, there will be a relationship in respect of the hydropathic character (Bost and Blalock, 1989). Because an anti-sense peptide will, in general, have a hydropathy profile opposite to that of the corresponding sense peptide, it is expected that both will undergo protein-protein interactions. Indeed binding peptides can be derived based on opposite hydropathy scores (Villain *et al.,* 2000).

Aß (3 - 30)	E F R H D S G Y E V H H Q K L V F F A E D V G S N K G A
AßAS(F) (3 - 30)	L K A V L S P I L Q V V V F N H K K R L L H P S L F P R
CDK-1 (105 - 132)	V K S Y L Y Q I L Q G I V F C H S R R V L H R D L K P Q
CDK2 (104 - 131)	I K S Y L F Q L L Q G L A F C H S H R V L H R D L K P Q
CDK3 (104 - 131)	I K S Y L F Q L L Q G V S F C H S H R V I H R D L K P Q
CDK4 (117 - 144)	I K D L M R Q F L R G L D F L H A N C I V H R D L K P E
CDK5 (103 - 130)	V K S F L F Q L L K G L G F C H S R N V L H R D L K P Q
CDK6 (122 - 149)	I K D M M F Q L L R G L D F L H S H R V V H R D L K P Q
CDK7 (114 - 141)	I K A Y M L M T L Q G L E Y L H Q H W I L H R D L K P N
CDK8 (128 - 155)	V K S L L Y Q I L D G I H Y L H A N W V L H R D L K P A
CDK9 (126 - 153)	I K R V M Q M L L N G L Y Y I H R N K I L H R D M K A A
CDK10 (140 - 167)	V K C I V L Q V L R G L Q Y L H R N F I I H R D L K V S
PCTK1 (263 - 290)	V K L F L F Q L L R G L A Y C H R Q K V L H R D L K P Q
PCTK2 (290 - 317)	V K L F L Y Q I L R G L A Y C H R R K V L H R D L K P Q
PCTK3 (148 - 175)	V K I F M F Q L L R G L A Y C H T R K I L H R D L K P Q
KKIALRE (104 - 131)	V K S I T W Q T L Q A V N F C H K H N C I H R D V K P E
CDC2L (200 - 227)	I K S F M R Q L M E G L D Y C H K K N F L H R D I K C S
MAPK-1 (126 – 153)	I C Y F L Y Q I L R G L K Y I H S A N V L H R D L K P S

Figure 3. Comparison of Cyclin Dependent Kinase family member sequences with the forward Amyloid-β Anti-sense Sequence 3-30 sequence and Amyloid-β 3-30 sequence. The boxed in region contains the active site Aspartic acid (D) residue of the CDK family members. Underlined amino acids are identical to the βAS sequence.

The hydropathic profiles of Aβ 3-30 and CDK-1 105-132 were opposite (Figure 4) suggesting that a protein-protein interaction between these compounds could occur. The alignment of the active site of CDK-1 with Aβ serine 26 provides a potential mechanism an Aβ phosphorylation reaction mediated *via* hydropathic interactions between Aβ and the kinase. Since the CDK family of kinases share similar structural features around the ATP binding and phosphate group transfer residues, it is possible that Aβ could be phosphorylated by other CDK kinases, and this may explain why different groups have shown roles for different CDK enzymes in Aβ cytotoxicity.

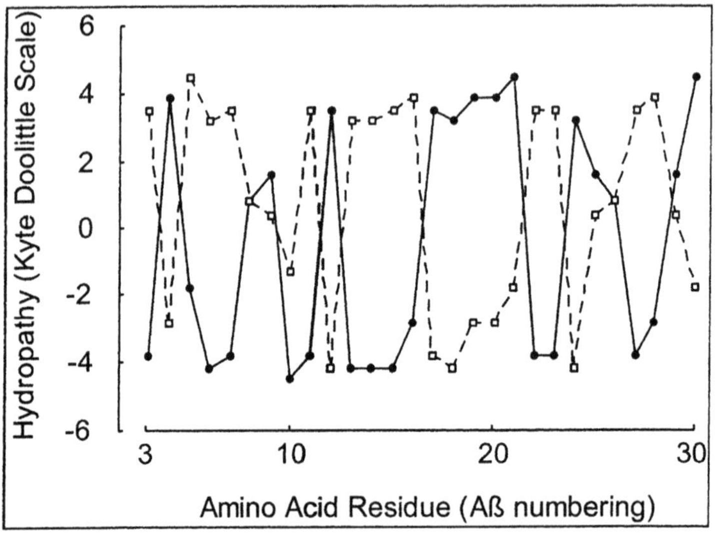

Figure 4. Comparison of the hydropathic profiles of Cyclin Dependent Kinase-1 and Amyloid-β. The Kyte Doolitle scores Hydropathy scores for CDK-1 105-132 region (closed circles - solid line) and the Aβ 3-30 region (open squares – dashed line) were determined.

Cyclins are co-factors for CDK-1, which are required for enzyme activity. These proteins also contain substrate recognition sequences which may play a role in the recruitment of substrate molecules to the active CDK-1/cyclin complex. The CDK-1/cyclin B1 enzyme complex plays a role in cell division and is a major active form purified from cells. To test if this protein contained an Aβ binding site the AβAS reverse peptide sequence, read in the C to N direction was used in a BLAST comparison with the cyclin B1 (Accession No GI 116176) protein sequence. Results showed a region with 30% sequence identity and 43% sequence similarity between this AβAS peptide and the cyclin B1 257-285 region.

2.1.3 Biochemical characterisation of the interaction between Cyclin-Dependent Kinase-1 and Amyloid-β

Interactions between Aβ peptides and CDK-1 were characterised using binding assays. Since the MAPK-1 protein kinase shows some sequence similarity and is also inhibited by the CDK-1 inhibitors such as olomoucine (Senderowicz & Sausville 2000) this kinase was also tested. Binding studies indicated that Aβ 1-42, Aβ 1-40 and Aβ 25-35 bound to human CDK-1 and also to human MAPK-1 (Figure 5).

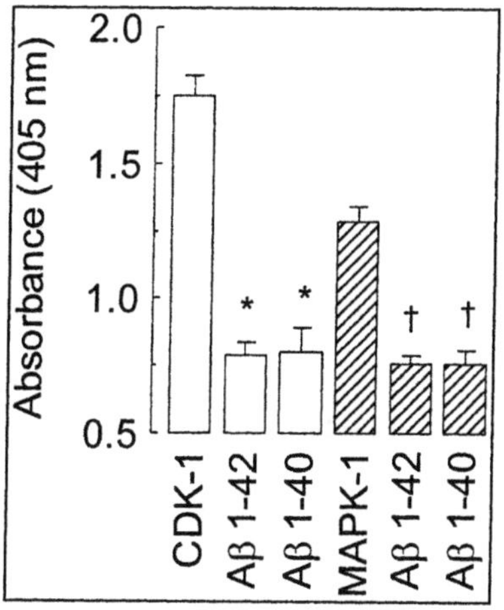

Figure 5. Binding of Amyloid-β 1-40 peptide to Human Cyclin Dependent Kinase-1 and Mitogen-Activated Protein Kinase-1. Biotinylated Aβ 1-40 was incubated on CDK-1 (open columns) or MAPK-1 (hatched columns) coated plates for 24 h and binding determined by ELISA. * = $P < 0.05$ vs Biotinylated Aβ binding to CDK-1; † = $P < 0.05$ vs Biotinylated Aβ binding to MAPK-1

Peptides containing Aβ residues 17-28 could inhibit the binding of Aβ to CDK-1. The AβAS 14-23 but not the AβAS 27-36 peptide also inhibited binding of Aβ to CDK-1. The binding of Aβ 1-40 to CDK-1 was concentration dependent and showed an affinity constant (K_D) of 12.7 ± 4.3 μM. The inhibition of Aβ binding to CDK-1 by the Aβ 17-28 but not the Aβ 31-35 fragments indicated that the CDK-1 56-63 region identified in the BLAST search above could also contribute to Aβ binding. The alignment within this CDK-1 region of Aβ residue 23 (a negatively charged Aspartic acid residue) with the positively charged Arginine 59 of CDK-1 suggested that a charge-based interaction may occur at this location. The tertiary structure of CDK-1 suggests that the 56-63 region plays a role in interactions with the substrate bound to the active site region surrounding CDK-1 residue 128. To confirm binding to the active site region a synthetic peptide (CDKP1) corresponding to CDK-1 residues 119-133 was tested for ability to bind Aβ peptides. Results showed that Aβ 1-40 and 25-35 both bound to CDKP1. Binding of Aβ 25-35 to the CDKP1 peptide could be inhibited by peptides containing the Aβ 17-28 sequence and by either anti-Aβ 17-28 or anti-CDKP1 antibodies.

Binding of Aβ to MAPK-1 was also concentration dependent and showed an affinity constant of (K_D) of 273 ± 41 μM. Binding of Aβ to this kinase was inhibited by the AβAS 14-23 and Aβ 17-28 containing peptides, suggesting that a similar binding interaction was involved.

Binding assays using cyclin B1 coated plates showed that cyclin B1 bound to biotinylated Aβ 1-40 and 25-35. The binding was inhibited by Aβ 31-35 containing peptides. The affinity constant for Aβ 1-40 binding to cyclin B1 was 2.3 ± 0.5 μM. The binding could be inhibited by the forward AβAS 27-36 but not the AβAS 14-23 peptide (Milton 2002a; 2004b). This suggested that the CDK-1/Cyclin B1 enzyme may be a prime candidate for phosphorylation of Aβ.

2.1.4 Amyloid-β phosphorylation by Cyclin-Dependent Kinase-1

Having identified two kinases which bind the Aβ peptide the next step was to determine whether either could phosphorylate the Aβ peptide. Using biotinylated Aβ 1-40 as a substrate it was determined that both active CDK-1/Cyclin B1 and MAPK-1 could phosphorylate the Aβ peptide. Results showed that Aβ 1-42, Aβ 1-40 and Aβ 25-35 incorporated ^{32}P from $\gamma^{32}P$-ATP in the presence of CDK-1/Cyclin-B1. Both CDK-1/Cyclin-B1 and MAPK-1 caused the appearance of phosphorylated serine residues in Aβ 1-42, Aβ 1-40 and Aβ 25-35. Phosphorylation of Aβ by both CDK-1 and MAPK-1 was inhibited by olomoucine, a purinergic CDK-1 inhibitor, the CDKP1 peptide, Aβ 12-28 and Aβ 17-28. The levels of phosphorylated serine residues in Aβ were significantly greater when incubated with CDK-1 than with MAPK-1, (Figure 6). Kinetic analysis of the reaction between Aβ and CDK-1 showed that the phosphorylation was concentration dependent and the Michaelis constant (K_M) for the phosphorylation of Aβ 1-40 was 5.2 μM, which compared with a K_M of 2.7 μM for the H1 peptide substrate (Milton, 2001a; 2004b).

The lack of a proline residue in the Aβ peptide sequence indicates that Aβ does not contain a normal CDK-1 or MAPK-1 recognition sequence. The structural features of many CDK substrates have been characterised (Brown *et al.,* 1999; Knockaert *et al.,* 2002). These features include the presence of β-turn regions containing the target serine or threonine residue (Mager, 1998). The serine 26 residue in Aβ is located within a β-turn region and this structural feature may be important for the Aβ phosphorylation reaction. The lack of a substrate consensus sequence in the Aβ peptide that the enzyme-substrate complex formation between Aβ and the kinase is mediated via a novel mechanism. The proposed mechanism for Aβ phosphorylation is a direct hydropathic interaction between Aβ and the active site of the kinase.

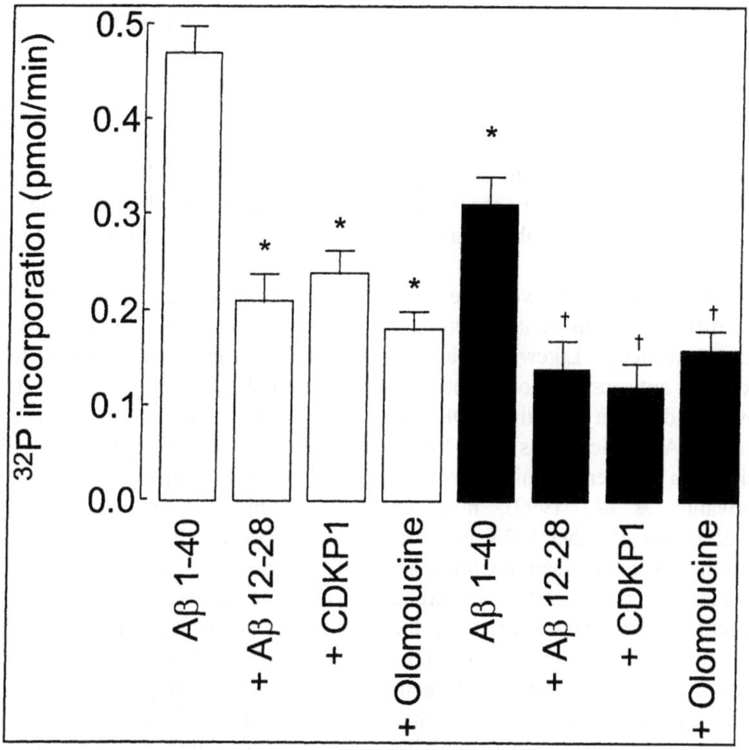

Figure 6. Phosphorylation of Amyloid-β by Cyclin-Dependent Kinase-1 and Mitogen-Activated Protein Kinase 1. Biotinylated Aβ was incubated with CDK-1 (open columns) or MAPK-1 (closed columns) in the presence of γ^{32}P-ATP for 30 min and ^{32}P incorporation determined. * = $P < 0.05$ vs Aβ incubated with CDK-1; † = $P < 0.05$ vs Aβ incubated with MAPK-1

2.2 Is Cyclin-Dependent Kinase-1 relevant to Alzheimer's disease pathology?

The levels of the CDK-1 protein plus associated cyclins are elevated in Alzheimer's disease brains (Pei *et al.*, 2002; Smith *et al.*, 1999; Vincent *et al.*, 1997), suggesting a role for this kinase in the disease process. The CDK kinases also play an important role in the cell cycle and are proline directed serine/threonine kinases (Hutchinson and Glover, 1995). The CDK-1 protein kinase plays a role in the G2 to M phase of the cell cycle and is homologous to the yeast cell division control protein 2. The CDK enzymes require a phosphorylation step and binding to a cyclin for activity. CDK-1 is also known to phosphorylate the tau protein (Buee *et al.*, 2000), which forms a key part of the neurofibrillary tangles associated with the disease pathology.

The CDK-1 protein kinase phosphorylates the APP and causes changes in the relative levels of the Aβ 1-40 and Aβ 1-42 isoforms *in vitro* (Ando *et al.*, 1999). These observations all point to a role for CDK enzymes in Alzheimer's disease and pose the question as to whether the disease could be caused by cell cycle abnormalities associated with the disease pathology (Nagy, 2000).

The similarities between the CDK structures and the interactions between the different components of the cell cycle mean that CDK inhibitors may lack specificity. Likewise manipulation of CDK expression, using for example antisense oligonucleotides, may have indirect effects on both the levels and activity of other CDK enzymes. The ability of CDK inhibitors to prevent Aβ toxicity has been linked to CDKs 4, 5 and 6 using chemical inhibitors and gene manipulation (Alvarez *et al.*, 2001; Copani *et al.*, 1999; Giovanni *et al.*, 1999), however, such studies cannot exclude the involvement of other CDK enzymes in the response. The human CDK-1 gene has been localized to Chromosome 10 and the recent sequencing of the human genome identified its precise location (Nazarenko *et al.*, 1991). Recent studies have suggested that Chromosome 10 also contains an Alzheimer's disease linked locus (Bertram *et al.*, 2000; Ertekin-Taner *et al.*, 2000; Myers *et al.*, 2000). The gene for CDK-1 is 1.3 Mbps distal to the polymorphism marker D10S589 that has been linked to elevated Aβ levels (Ertekin-Taner *et al.*, 2000) and lies within a region linked to late onset Alzheimer's disease (Myers *et al.*, 2000). The proximity suggests that CDK-1 could be a candidate gene responsible for this observed linkage (Milton 2001a) and indeed a polymorphism in the splice site after exon 6 of the CDK-1 gene shows an increased frequency in AD (Johansson *et al.*, 2003).

2.3 Does amyloid-β directly influence the activity of Cyclin-Dependent Kinase-1?

Since Aβ binds to both the CDK-1 and cyclin B1 components of the active enzyme it is possible that Aβ modulates the activity of the kinase. This was tested by performing kinase activity measurements using a biotinylated Histone H1 substrate peptide (PKTPKKAKKL) and measurement of incorporation of ^{32}P from ^{32}P-ATP. Results showed that Aß 1-40, 17-35, 25-35 and 31-35 all increased the phosphorylation of the H1 peptide by CDK-1/cyclin B1 (Milton 2002a; 2004b), suggesting that Aβ could activate the kinase. The fragments capable of activation were the same as those which inhibited Aβ 1-40 binding to cyclin B1 and these results suggest that the binding to cyclin B1 may be a mechanism for the enzyme activation. The binding of Aβ to cyclin-B1 is suggestive of the recruitment of substrate peptides to CDK-2 by cyclin-A (Schulman *et al.*, 1998) and may also play a role in the activation of CDK-1 by Aβ peptides.

The Aβ 31-35 sequence which interacts with Cyclin B1 and activates the CDK-1 enzyme is the minimal fragment of Aβ that causes neurotoxicity (Milton 2001b) and also interacts with other enzymes including catalase (Milton 1999). The ability of Aβ to directly activate the CDK-1 enzyme could also contribute to the other major pathological feature of AD pathology – tau phosphorylation, which is in part mediated by CDK-1 (Buee *et al.*, 2000).

2.4 How might amyloid-β phosphorylation play a role in Alzheimer's disease pathology?

The two major features of Aβ which are thought to contribute to its role in AD pathology are its ability to aggregate into the insoluble fibrils that are found in amyloid plaques and its ability to kill neurons.

2.4.1 Effects of phosphorylation on the aggregation of amyloid-β

Following the isolation of the Aβ peptide from the AD plaques by Glenner and Wong (1984) this peptide has been implicated as a causal agent in AD. The ability of synthetic Aβ to aggregate and form fibrils *in vitro* suggests that this 40-43 amino acid peptide may have a causal role in plaque formation (Kheterpal *et al.*, 2000). The levels of phosphorylated Aβ in Alzheimer's Brain extracts suggests that it may not be present in the Aβ aggregates, which comprise the majority of the total Aβ in such extracts (Mattson, 1997; Selkoe, 1999). The Aβ 17-35 region contains the CDK-1 and cyclin-B1 binding sites, the CDK-1 target residue (serine 26), plus the cytotoxic domain and the hydrophobic core region that plays a role in peptide aggregation (Mager, 1998). The aggregation characteristics of Aβ 17-35, a serine 26 to alanine mutated Aβ 17-35 derivative (S26A) and a phosphoserine 26 (pS26) containing Aβ 17-35 derivative were therefore tested using a spectral assay for determination of Aβ-fibril concentrations within Aβ solutions (Klunk *et al.*, 1999). Results showed that a 50 μM solution of the Aβ 17-35 contained 18.9 ± 1.5 μM Aβ fibrils after a 24 hr incubation, the Aβ 17-35 S26A peptide contained 19.1 ± 1.4 μM fibrils and the Aβ 17-35 pS26 peptide contained 10.8 ± 0.3 μM fibrils ($P < 0.05$ vs Aβ 17-35; students t-test; n=6 for each group). This suggests that phosphorylation alters the ability of the peptide to aggregate, possibly by causing an increase in the levels of soluble (oligomeric) forms.

2.4.2 Neurotoxicity of phosphorylated Aβ

The effects of Aβ 1-40, Aβ 17-35, an S26A mutated Aβ 17-35 derivative, a pS26 containing Aβ 1-40 derivative and a pS26 containing Aβ 17-35 derivative on MTT activity (Shearman, 1999) in human NT-2 neurons were tested to see if phosphorylation affected cytotoxicity. The pS26 containing Aβ 1-40 derivative caused a dose dependent neurotoxicity which was significantly more potent than that caused by the Aβ 1-40 peptide (Figure 7). The concentrations of pSAβ 1-40 and Aβ 1-40 required to cause a drop to 50% MTT reduction were calculated as 330 nM and 6.1 μM respectively from the dose response curves. The Aβ 17-35 S26A peptide was significantly less toxic than the Aβ 17-35 peptide whilst the pS26 containing Aβ 17-35 peptide was significantly 10-100 fold more toxic to human NT-2 neurons (Milton, 2001a). Similar results were obtained when cytotoxicity was determined by measuring cell viability by trypan blue dye exclusion. Olomoucine dose dependently inhibited the toxicity of Aβ 17-35 but not the phosphorylated Aβ 17-35 derivative, suggesting that phosphorylation of the peptide may play a direct role in the cytotoxicity.

Neurons exposed to Aβ 17-35 peptides showed increased levels of ir-Aβ. Measurement of ir-pSAβ in the same cell extracts showed that cells exposed to Aβ 17-35 and the pS26 containing Aβ 17-35 derivative both contained increased amounts of ir-pSAß whilst cells exposed to the Aβ 17-35 S26A peptide showed no difference to control cells. These results demonstrate that neurons can phosphorylate exogenously added Aβ. The increase in ir-pSAβ levels when cells were treated with Aβ was prevented by the CDK-1 inhibitor olomoucine, confirming the involvement of CDK-1 or a related kinase in the response.

A previous study has shown that substitution of the serine 26 residue with an alanine residue in Aβ 25-35 has no effect on the toxicity of the peptide (Sato *et al.*,1995) suggesting that phosphorylation of the serine 26 residue is not essential for Aβ cytotoxicity. However, the serine 26 residue is within a turn region of the Aβ molecule (Mager, 1998) and phosphorylation may alter the structure sufficiently to cause the observed changes in cytotoxicity and aggregation. The ability of the CDK inhibitor olomoucine to prevent the both cytotoxicity and phosphorylation of Aβ peptides indicates that phosphorylation of Aβ by CDK like enzymes could be crucial to the cytotoxicity of the Aβ peptide. Such a mechanism for toxicity involving phosphorylation of Aβ is supported by the failure of Aβ 17-35 S26A to elicit a full cytotoxic response and the increased cytotoxicity of Aß 17-35 pS26 derivative (Milton, 2001a). Olomoucine can inhibit many members of the CDK family plus MAPK enzymes (Senderowicz and Sausville, 2000) and as such these *in vitro* observations cannot directly confirm a role for CDK-1 in

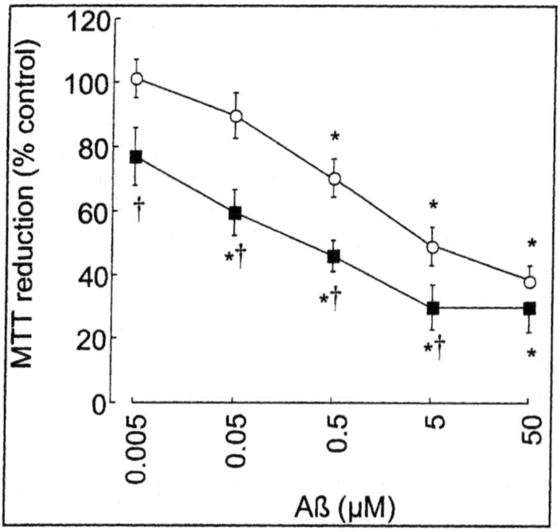

Figure 7. Neurotoxicity of Amyloid-β 1-40 and Phosphorylated Amyloid-β 1-40. Aβ 1-40 (open circles) and Phosphorylated Aβ 1-40 (closed squares) were added to NT-2 neurons in the presence or absence of test compounds for 24 h and MTT reduction determined. * = $P < 0.05$ vs Control; † = $P < 0.05$ vs Aß 1-40, students t-test n=6.

NT-2 neurons. Since the Aβ 17-35 pS26 form has a reduced ability to aggregate it is possible that phosphorylated Aβ, in its non-fibrillar form, mediates the cytotoxic response. Changes in the structure of Aβ induced by phosphorylation may also allow the cytotoxic Aβ 31-35 region within the peptide to act on a downstream target protein, possibly cyclin-B1 or catalase. The concentration of phosphorylated Aβ in NT-2 neuronal cell extracts are similar to those found in Alzheimer's brain extracts (Milton, 2001a) and probably represent Aβ derived from the intracellular pool, which has been suggested to play an important role in neurodegeneration and aging (Skovronsky *et al.*, 1998).

A second mechanism for cytotoxicity mediated by Aβ activation of CDK-like enzymes is supported by the ability of Aβ 31-35, which lacks a residue that could be phosphorylated, to increase CDK-1 activity plus the ability of olomoucine to prevent the cytotoxicity of Aβ 31-35. The failure of Aβ to activate MAPK and PKC directly suggests that the roles of these other kinases in Aβ cytotoxicity (Abe & Saito, 2000; Nakai *et al.*, 2001) are not mediated by the same mechanism or occur downstream of CDK activation, since both kinases are themselves activated by phosphorylation. The ability of Aβ to increase CDK-1 phosphorylation of substrate molecules suggests that Aβ could contribute to the tau hyperphosphorylation observed in Alzheimer's disease. The cytotoxicity of an Aβ 25-35 derivative with an S26A mutation (Sato *et al.,* 1995) further supports a role for unphosphorylated Aβ in the cytotoxicity. The failure of olomoucine to inhibit the cytotoxicity of the phosphorylated Aβ 17-35 pS26 derivative, however, suggests that the Aβ phosphorylation reaction itself may play a role in cytotoxicity. The ability of Aβ to increase H_2O_2 levels and antioxidants to prevent Aβ cytotoxicity (Milton 2004a) suggests that the Aβ phosphorylation plus CDK activation mechanisms are not necessarily the only routes *via* which Aβ kills cells. However, the toxicity of both Aβ and H_2O_2 can be blocked by the kinase inhibitor staurosporine (Tan *et al.,* 1997), which also effects CDK activity family (Senderowicz & Sausville, 2000), and it is therefore possible that these mechanisms are in fact linked.

3. PHOSPHORYLATED AMYLOID-β TOXICITY PREVENTION?

The Aβ peptide is cytotoxic to both neuronal and non-neuronal cells (Mattson, 1997), with the 31-35 region thought to be a functional cytotoxic domain (Milton, 1999; Milton, 2002b). The cytotoxicity of Aβ is thought to be dependent on aggregation of the peptide (Ueda *et al.,* 1994). Intracellular Aβ binding to the endoplasmic reticulum Aβ-binding protein (ERAB; Yan *et al.,* 1997) has also been suggested to play a role in toxicity. The mechanisms underlying the cytotoxicity are however unclear since a large number of compounds including anti-oxidants (Miranda *et al.,* 2000), caspase inhibitors (Nakagawa *et al.,* 2000), protein phosphatase inhibitors and protein kinase inhibitors (Tan *et al.,* 1997) can all inhibit cytotoxicity *in vitro.* These observations all point to a role for Aβ in the neurodegeneration associated with AD. Antioxidants, Aβ-antisense peptides, anti-Aβ antibodies, Aβ binding peptides, a caspase inhibitor, an HMG-CoA reductase inhibitor, a G-protein inhibitor, kinase inhibitors and a synthetic cannabinoid were tested for their ability to prevent both pS26 Aβ 1-40 and Aβ 1-40 induced toxicity. The Aβ-antisense peptides, anti-Aβ antibodies, Aβ binding peptides protected against both pSAβ 1-40 and Aβ 1-40, see Figure 8.

The kinase inhibitor, olomoucine, which inhibits both CDK and MAPK enzymes, was effective against Aβ but not the pS26 derivative suggesting that phosphorylation of Aβ is a key step in the toxicity of the peptide. Cannabinoids have been suggested as neuroprotective agents and potential therapeutic agents for the treatment of neurodegenerative disorders. Receptor mediated actions of cannabinoids can be antagonised by MAPK inhibitors and activation of MAPK via CB_1 receptors has been shown to prevent Aβ neurotoxicity (Milton 2002b). The synthetic cannabinoid (WIN 55,212-2 mesylate), was effective at preventing the neurotoxicity of the pS26 derivative of Aβ 1-40.

The protective actions of anandamide, noladin ether and WIN 55,212-2 mesylate against pS26 derivative of Aβ 1-40 could be prevented by cannabinoid receptor antagonists. In neuronal cells the CB_1 cannabinoid receptor antagonist AM251 was able to totally block cannabinoid protection whilst the CB_2 cannabinoid receptor antagonist AM630 had no affect. In myeloma cells neither antagonist prevented the protection by anandamide but a combination of both AM251 and AM630 was effective. The anandamide uptake inhibitor AM404 and the vanilloid receptor antagonist capsazepine had no affect on the protective actions of either anandamide or noladin ether. The MAPK kinase inhibitor PD98059 had no affect on pS26 Aβ 1-40 toxicity and prevented the protective action of both cannabinoids. Neuronal cells showed similar increases in the levels of total immunoreactive (ir)-Aβ after incubation with either Aβ 1-40 or pS26 Aβ

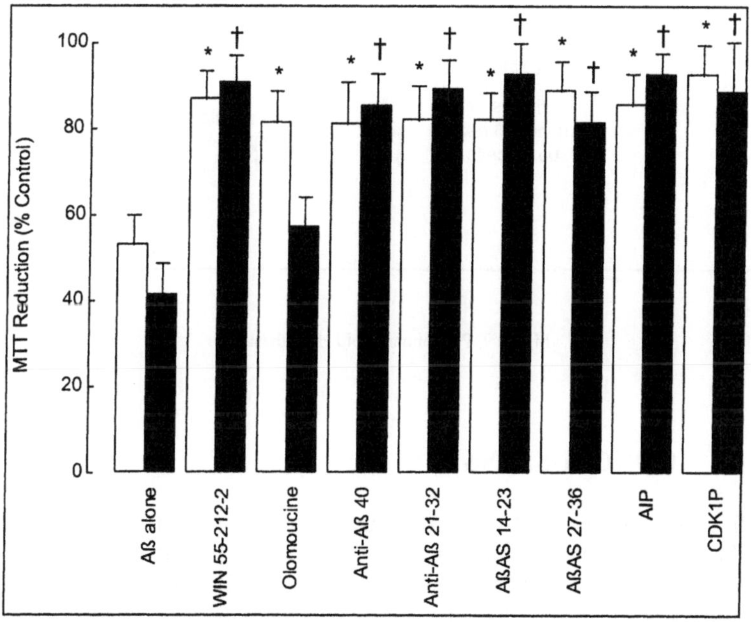

Figure 8. Prevention of Phosphorylated Amyloid-β neurotoxicity. Aβ 1-40 (open columns) and Phosphorylated Aβ 1-40 (closed columns) were added to NT-2 neurons in the presence or absence of test compounds for 24 h and MTT reduction determined. * = $P < 0.05$ vs Aß alone; † = $P < 0.05$ vs Phosphorylated Aß alone, students *t*-test, n=6.

1-40. The levels of ir-pSAβ were also elevated after both treatments. The increase in ir-pSAß caused by Aβ 1-40 was abolished by pretreatment of the cells with olomoucine but was unaffected by anandamide, noladin ether or WIN 55,212-2 mesylate, confirming that the cannabinoids were acting downstream of Aβ phosphorylation.

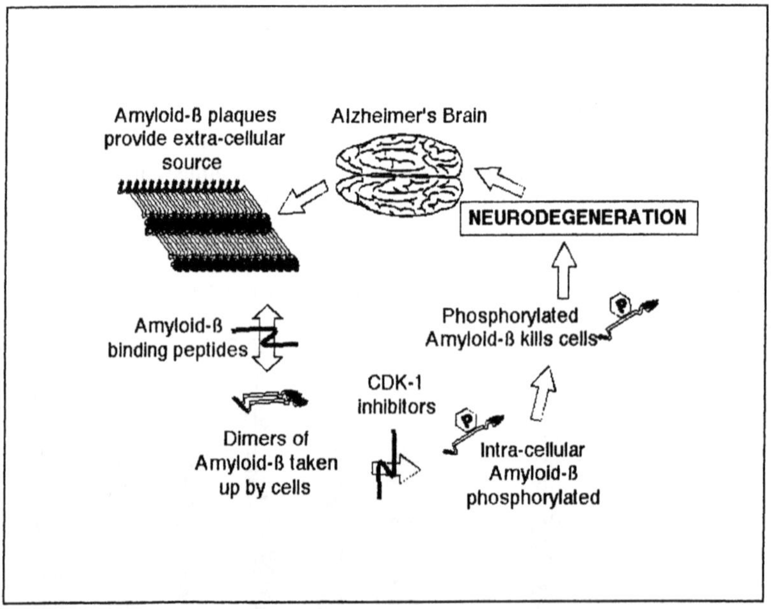

Figure 9. Phosphoamyloid Hypothesis

4. CONCLUSIONS

In conclusion, the results from this study demonstrate that the Aβ peptide can be phosphorylated by CDK-1 and MAPK-1. The increased toxicity of phosphorylated Aβ and its apparent intracellular localization suggest that it may be a key component involved in Aβ cytotoxicity and hence potentially in the neurodegenerative changes characteristic of Alzheimer's disease. The ability of Aβ to directly increase the activity of CDK-1 may also contribute to the phosphorylation abnormalities characteristic of Alzheimer's disease, particularly the tau protein which is hyperphosphorylated in the disease. The actions of phosphorylated Aβ and its role in AD pathology are illustrated in

Figure 9. The ability of compounds to prevent Aβ phosphorylation or its cytotoxic actions may represent the future of AD therapeutic advances.

ACKNOWLEDGMENTS

The author thanks G.K. Wilcock, S.H. MacGowan, A.S. Milton, N.M. Bascombe-Milton, C.N. Bascombe-Milton, L.R. Francis and J. Rawlinson for discussion. The research described was supported by Insight Biotechnology Ltd and NeuroDelta Ltd.

REFERENCES

Abe, K., and Saito, H., 2000, Amyloid ß neurotoxicity not mediated by the mitogen-activated protein kinase cascade in cultured rat hippocampal and cortical neurons. *Neurosci. Lett.* **292**: 1-4.

Altschul, S. F., Madden T.L., Alejandro A. Schaeffer, A.A., Zhang, J., Zhang, Z., Miller, W., and Lipman D.J., 1997, Gapped BLAST and PSI-BLAST: a new generation of protein database search programs. *Nucleic Acids Res.* **25**: 3389-3402.

Alvarez, A., Munoz, J.P., and Maccioni, R.B., 2001. A Cdk5-p35 stable complex is involved in the ß-amyloid-induced deregulation of Cdk5 activity in hippocampal neurons. *Exp. Cell Res.* **264**: 266-274.

Ando, K., Oishi, M., Takeda, S., Iijima, K., Isohara, T., Nairn, A.C., Kirino, Y., Greengard, P., and Suzuki, T., 1999, Role of Phosphorylation of Alzheimer's Amyloid Precursor Protein during Neuronal Differentiation. *J. Neurosci.* **19**: 4421–4427.

Bertram, L., Blacker, D., Mullin, K., Keeney, D., Jones, J., Basu, S., Yhu, S., McInnis, M.G., Go, R.C.P., Vekrellis, K., Selkoe, D.J., Saunders, A.J., and Tanzi, R.E., 2000, Evidence for genetic linkage of Alzheimer's Disease to chromosome 10q. *Science* **290**: 2302-2303.

Blom, N., Gammeltoft, S., and Brunak, S., 1999, Sequence- and structure-based prediction of eukaryotic protein phosphorylation sites. *J. Mol. Biol.* **294**: 1351-1362.

Bost, K.L., and Blalock, J.E., 1989, Preparation and use of complementary peptides. *Methods Enzymol.* **168**: 16-28.

Braak, H., Braak, E., and Bohl, J., 1993, Staging of Alzheimer-related cortical destruction. *Eur. Neurol.* **33**: 403-8.

Brown, N.R., Noble, M.E.M., Endicott, J.A., and Johnson, L.N., 1999, The structural basis for specificity of substrate and recruitment peptides for cyclin-dependent kinases. *Nat. Cell Biol.* **1**: 438-443.

Buee, L., Bussiere, T., Buee-Scherrer, V., Delacourte, A., and Hof, P.R., 2000, Tau protein isoforms, phosphorylation and role in neurodegenerative disorders. *Brain Res. Rev.* **33**: 95–130.

Chauhan, A., Chauhan, V.P., Murakami, N., Brockerhoff, H., and Wisniewski, H.M., 1993, Amyloid beta-protein stimulates casein kinase I and casein kinase II activities. *Brain Res.* 629: 47-52.

Copani, A., Condorelli, F., Caruso, A., Vancheri, C., Sala, A., Giuffrida-Stella, A.M., Canonico, P.L., Nicoletti, F, and Sortino, M.A., 1999, Mitotic signaling by beta-amyloid causes neuronal death. *FASEB J.* **13**: 2225-2234.

De Bondt, H.L., Rosenblatt, J., Jamcarik, J., Jones, H.D., Morgan, D.O., and Kim, S-H., 1993, Crystal structure of cyclin-dependent kinase-2. *Nature* **363**: 595-597.

Ertekin-Taner, N., Graff-Radford, N., Younkin, L.H., Eckman, C., Baker, M., Adamson, J., Ronald, J., Blangero, J., Hutton, M., and Younkin, S.G., 2000, Linkage of plasma Aß42 to a quantitative locus on chromosome 10 in late-ons*et al.,*zheimer's disease pedigrees. *Science* **290**: 2303-2304.

Giovanni, A., Wirtz-Brugger, F., Keramaris, E., Slack, R., and Park, D.S., 1999, Involvement of cell cycle elements, cyclin dependent kinases, pRb, and E2F DP, in B-amyloid-induced neuronal death. *J. Biol. Chem.* **274**: 19011-19016.

Glenner, G.G., and Wong C.W., 1984, Alzheimer's disease: initial report of the purification and characterization of a novel cerebrovascular amyloid protein. *Biochem. Biophys. Res. Commun.* **120**: 885–890.

Hutchison, C., and Glover D.M. (eds), 1995, *Cell Cycle Control*, IRL Press, Oxford, UK.

Jellinger, K.A., 1998, The neuropathological diagnosis of Alzheimer disease. *J. Neural. Transm. Suppl.* **53**: 97-118.

Johansson, A., Hampel, H., Faltraco, F., Buerger, K., Minthon, L., Bogdanovic, N., Sjogren, M., Zetterberg, H., Forsell, L., Lilius, L., Wahlund, L.O., Rymo, L., Prince, J.A., and Blennow, K., 2003, Increased frequency of a new polymorphism in the cell division cycle 2 (cdc2) gene in patients with Alzheimer's disease and frontotemporal dementia. *Neurosci. Lett.* **340**: 69-73.

Kishimoto, A., Nishiyama, K., Nakanishi, H., Uratsuji, Y., Nomura, H., Takeyama, Y., and Nishizuka, Y., 1985, Studies on the phosphorylation of myelin basic protein by protein kinase C and adenosine 3':5'-monophosphate-dependent protein kinase. *J. Biol. Chem.* **260**: 12492-12499.

Kheterpal, I., Zhou, S., Cook, K.D., and Wetzel, R., 2000, Aß amyloid fibrils possess a core structure highly resistant to hydrogen exchange. *Proc. Natl. Acad. Sci. USA*. **97**: 13597-13601.

Klunk, W.E., Jacob, R. F., and Manson, R.P., 1999, Quantifying amyloid by congo red spectral assay. *Methods Enzymol.* **309**: 285-304.

Knockaert, M., Greengard, P., and Meijer, L., 2002, Pharmacological inhibitors of cyclin-dependent kinases. *Trends Pharmacol. Sci.* **23**: 417-25.

Lee, M.S., Kao, S.C., Lemere, C.A., Xia, W., Tseng, H.C., Zhou, Y., Neve, R., Ahlijanian, M.K., and Tsai, L.H., 2003, APP processing is regulated by cytoplasmic phosphorylation. *J. Cell Biol.* **163**: 83-95.

Lee, W., Hyun Boo, J., Whan Jung, M., Dai Park, S., Ho Kim, Y., Kim, S.U., and Mook-Jung, I., 2004, Amyloid beta peptide directly inhibits PKC activation, *Mol. Cell Neurosci.* **26**: 222-31.

Lustbader, J.W., Cirilli, M., Lin, C., Xu, H.W., Takuma, K., Wang, N., Caspersen, C., Chen, X., Pollak, S., Chaney, M., Trinchese, F., Liu, S., Gunn-Moore, F., Lue, L.F., Walker, D.G., Kuppusamy, P., Zewier, Z.L., Arancio, O., Stern, D., Yan, S.S.,and Wu, H., 2004, ABAD directly links Abeta to mitochondrial toxicity in Alzheimer's disease. *Science* **304**: 448-52.

Mager, P. P., 1998, Molecular simulation of the primary and secondary structures of the Aß(1-42)-peptide of Alzheimer's disease. *Med. Res. Rev.* **18**: 403–430.

Manning, G., Whyte, D.B., Martinez, R., Hunter, T., and Sudarsanam, S., 2002, The protein kinase complement of the human genome. *Science* **298**: 1912-34.

Mattson, M.P., 1997, Cellular actions of ß-amyloid precursor protein and its soluble and fibrillogenic derivatives. *Physiol. Rev.* **77**: 1081-1132.

Milton, N.G.N., 1999, Amyloid-ß binds catalase with high affinity and inhibits hydrogen peroxide breakdown. *Biochem. J.*, **344**: 293-6.

Milton, N.G.N., 2001a, Phosphorylation of amyloid-ß at the serine 26 residue by human cdc2 kinase. *NeuroReport* **12**: 3839-44.

Milton, N.G.N., 2001b, Inhibition of catalase activity with 3-amino-triazole enhances the cytotoxicity of the Alzheimer's amyloid-ß peptide. *NeuroToxicology* **22**: 767-74.

Milton, N.G.N., 2002a, The amyloid-ß peptide binds to cyclin B1 and increases human cyclin-dependent kinase-1 activity. *Neurosci. Lett.*, **322**: 131-3.

Milton, N.G.N., 2002b, Anandamide and noladin ether prevent neurotoxicity of the human amyloid-ß peptide. *Neurosci. Lett.*, **332**: 127-30.

Milton, N.G.N., 2004a, Role of hydrogen peroxide in the aetiology of Alzheimer's Disease: implications for treatment. *Drugs Aging*, **21**: 81-100.

Milton, N.G.N., 2004b, Peptides for use in the treatment of Alzheimer's disease. *United States Patent Application Publication Number* US2004/0072753 A1.

Milton, N.G.N., Mayor, N.P., and Rawlinson, J., 2001, Identification of amyloid-ß binding sites using an antisense peptide approach. *NeuroReport*, **12**: 2561-6.

Miranda, S., Opazo, C., Larrondo, L.F., Munoz, F.J., Ruiz, F., Leighton, F., and Inestrosa N.C., 2000, The role of oxidative stress in the toxicity induced by amyloid ß-peptide in Alzheimer's disease. *Prog. Neurobiol.* **62**: 633-648.

Myers, A., Holmans, P., Marshall, H., Kwon, J., Meyer, D., Ramic, D., Shears, S., Booth, J., DeVrieze, F.W., Crook R., Hamshere, M., Abraham, R., Tunstall, N., Rice, F., Carty, S., Lillystone, S., Kehoe, P., Rudrasingham, V., Jones, L., Lovestone, S., Perez-Tur, J., Williams, J., Owen, M.J., Hardy, J.,and Goate, A.M., 2000, Susceptibility locus for Alzheimer's disease on chromosome 10. *Science* **290**: 2304-2305.

Nagy, Z., 2000, Cell cycle regulatory failure in neurones: causes and consequences. *Neurobiol. Aging* **21**: 761–769.

Nakagawa, T., Zhu, H., Morishima, N., Li, E., Xu, J., Yankner, B.A., and Yuan, J., 2000, Caspase-12 mediates endoplasmic-reticulum-specific apoptosis and cytotoxicity by amyloid-ß. *Nature* **403**: 98-103.

Nakai, M., Tanimukai, S., Yagi, K., Saito,N., Taniguchi, T., Terashima, A., Kawamata, T., Yamamoto, N., Fukunaga, K., Miyamoto, E., and Tanaka, C., 2001, Amyloid ß protein activates PKC-δ and induces translocation of myristoylated alanine-rich C kinase substrate (MARCKS) in microglia. *Neurochem. Int.* **38**: 593-600.

Nazarenko, S.A., Ostroverhova, N.V., and Spurr, N.K., 1991, Regional assignment of the human cell cycle control gene CDC2 to chromosome 10q21 by in situ hybridization. *Hum. Genet.* **87**: 621-622.

Pei, J.J., Braak, H., Gong, C.X., Grundke-Iqbal, I., Iqbal, K., Winblad, B., and Cowburn, R.F., 2002, Up-regulation of cell division cycle (cdc) 2 kinase in neurons with early stage Alzheimer's disease neurofibrillary degeneration. *Acta Neuropathol. (Berl).* **104**: 369-76.

Pinna, L.A., 1990, Casein kinase 2: an 'eminence grise' in cellular regulation? *Biochim. Biophys. Acta* **1054**: 267-284.

Sato, K., Wakamiya, A., Maeda, T., Noguchi, K., Takashima, A., and Imahori, K., 1995, Correlation among secondary structure, amyloid precursor protein accumulation, and neurotoxicity of amyloid beta(25-35) peptide as analyzed by single alanine substitution. *J. Biochem (Tokyo)* **118**: 1108-1111.

Schulman, B., Lindstrom, D.L., and Harlow, E., 1998, Substrate recruitment to cyclin-dependent kinase 2 by a multipurpose docking site on cyclin A. *Proc. Natl. Acad. Sci. USA.* **95**: 10453-10458.

Selkoe, D. J., 1999, Translating cell biology into therapeutic advances in Alzheimer's disease. *Nature* **399** Suppl: A23-A31.

Senderowicz, A.M., and Sausville, E.A., 2000, Preclinical and clinical development of cyclin-dependent kinase modulators. *J. Natl. Cancer Inst.* **92**: 376-387

Serpell, L.C., 2000, Alzheimer's amyloid fibrils: structure and assembly. *Biochimica et Biophysica Acta* **1502**: 16-30.

Shearman, M., 1999, Toxicity of protein aggregates in PC12 cells: 3-(4,5-dimethylthiazol-2-yl)-2,5,-diphenyltetrazolium bromide assay. *Methods Enzymol.* **309**: 716-723.

Skovronsky, D.M., Doms, R.W., and Lee, V.M., 1998, Detection of a novel intraneuronal pool of insoluble amyloid beta protein that accumulates with time in culture. *J. Cell Biol.* **141**: 1031-1039.

Smith, M.Z., Nagy, Z., and Esiri, M.M., 1999, Cell cycle-related protein expression in vascular dementia and Alzheimer's disease. *Neurosci. Lett.* **271**: 45-48.

Tan, S., Maher, P., and Schubert, D., 1997, The role of protein phosphorylation in ß-amyloid toxicity. *Brain Res.* **765**: 159-163.

Ueda, K., Fukui, Y., and Kageyama, H., 1994, Amyloid ß protein-induced neuronal cell death: neurotoxic properties of aggregated amyloid ß protein. *Brain Res.* **639**: 240-244.

Villain, M., Jackson, P.L., Manion, M.K., Dong, W.J., Su, Z., Fassina, G., Johnson, T.M,. Sakai, T.T., Krishna, N.R., and Blalock, J.E., 2000, De novo design of peptides targeted to the EF hands of calmodulin. *J. Biol. Chem.* **275**: 2676-2685.

Vincent, I., Jicha, G., Rosado, M., and Dickson, D.W., 1997, Aberrant expression of mitotic Cdc2/Cyclin B1 kinase in degenerating neurons of Alzheimer's disease brain. *J. Neurosci.* **17**: 3588-3598.

Woodgett, J.R., Gould, K.L.,and Hunter, T., 1986, Substrate specificity of protein kinase C. Use of synthetic peptides corresponding to physiological sites as probes for substrate recognition requirements. *Eur. J. Biochem.* **161**: 177-184.

Yan, S.D., Fu, J., Soto, C., Chen, X., Zhu, H., Al-Mohanna, F., Collison, K., Zhu, A., Stern, E., Saido, T., Tohyama, M., Ogawa, S., Roher, A., and Stern D., 1997, An intracellular protein that binds amyloid-beta peptide and mediates neurotoxicity in Alzheimer's disease. *Nature* **389**: 689-695.

Index

MIX
Papier aus verantwortungsvollen Quellen
Paper from responsible sources
FSC® C105338

If you have any concerns about our products,
you can contact us on
ProductSafety@springernature.com

In case Publisher is established outside the EU,
the EU authorized representative is:
Springer Nature Customer Service Center GmbH
Europaplatz 3, 69115 Heidelberg, Germany

Printed by Libri Plureos GmbH
in Hamburg, Germany